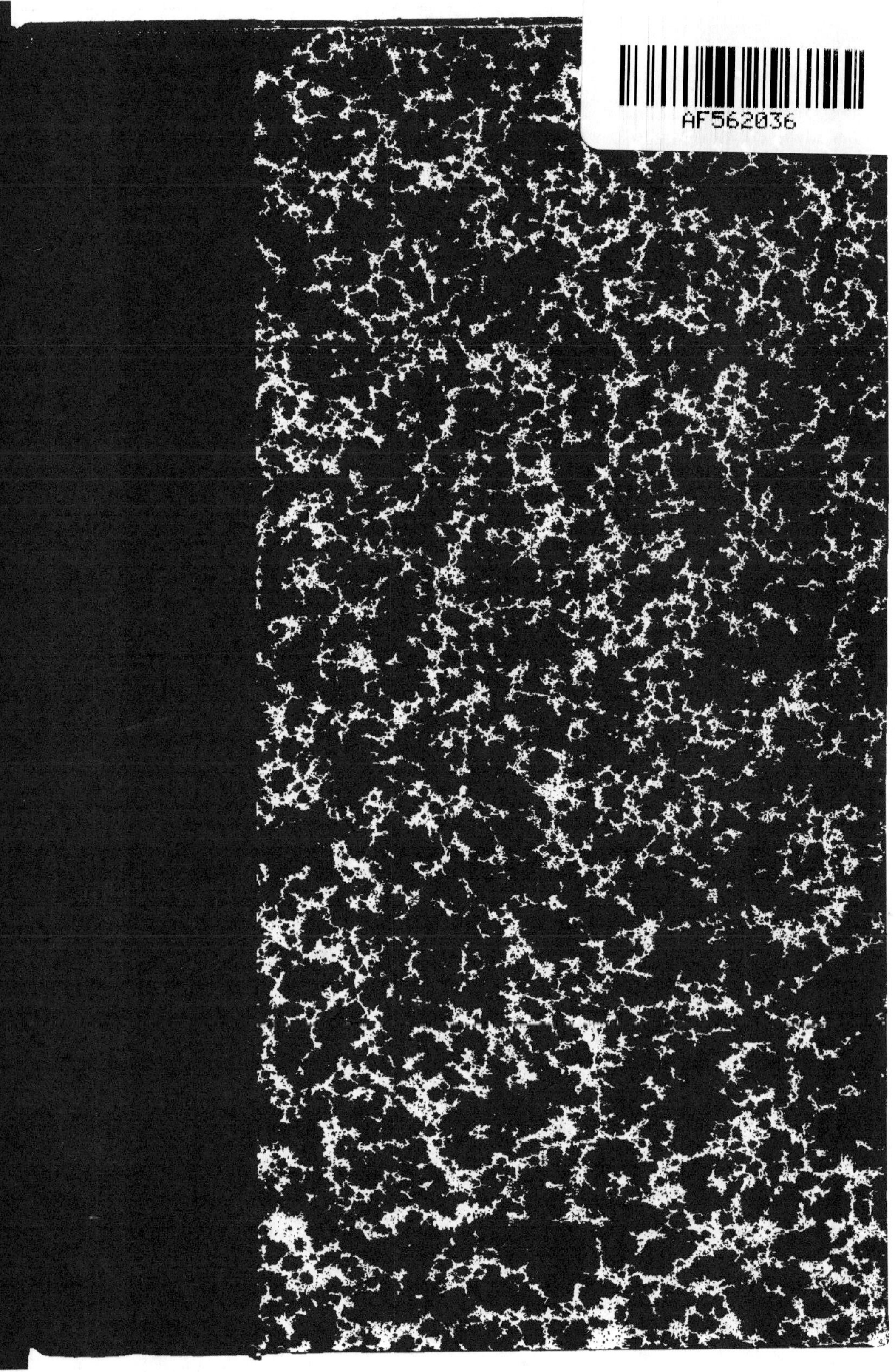

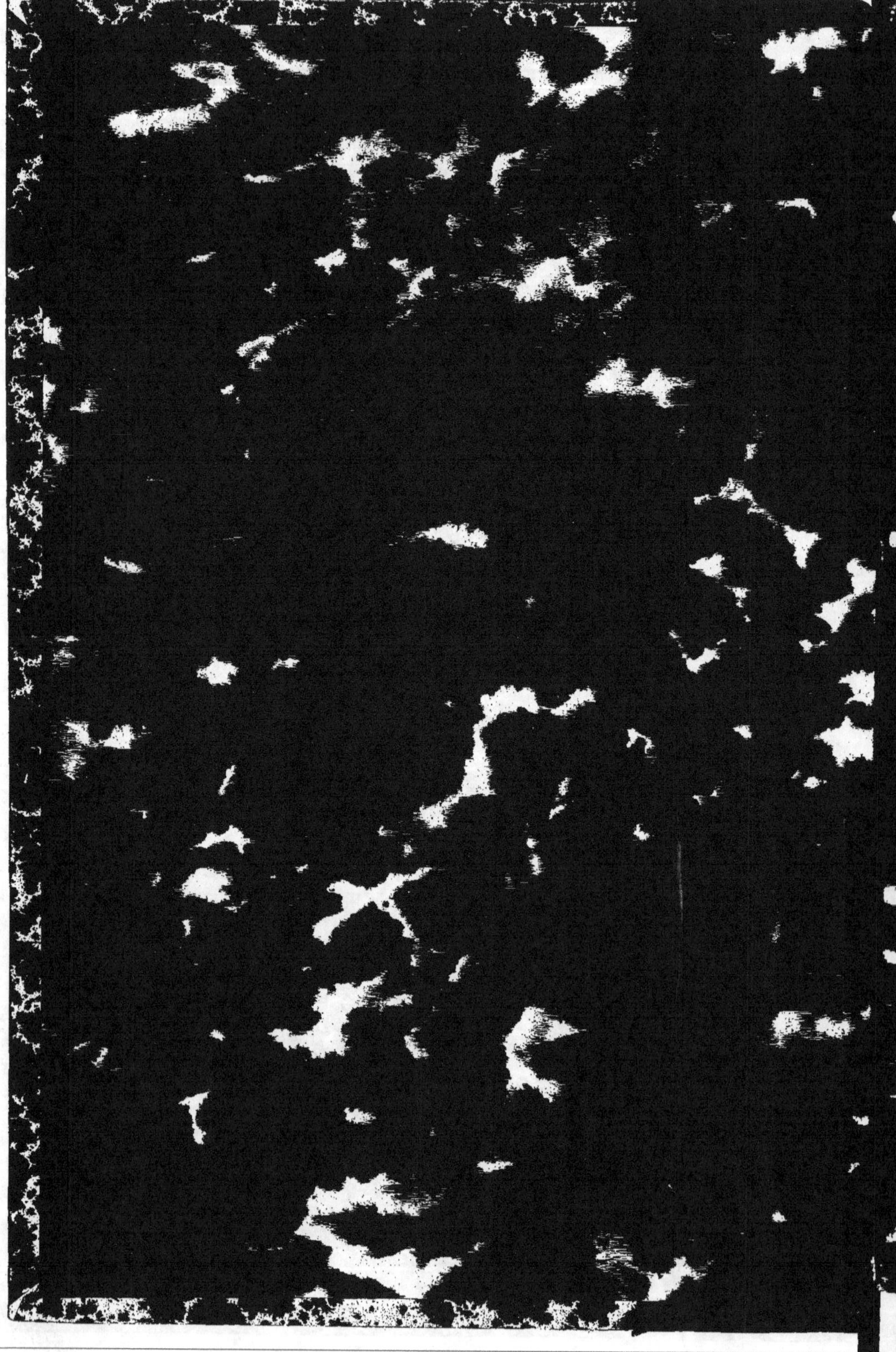

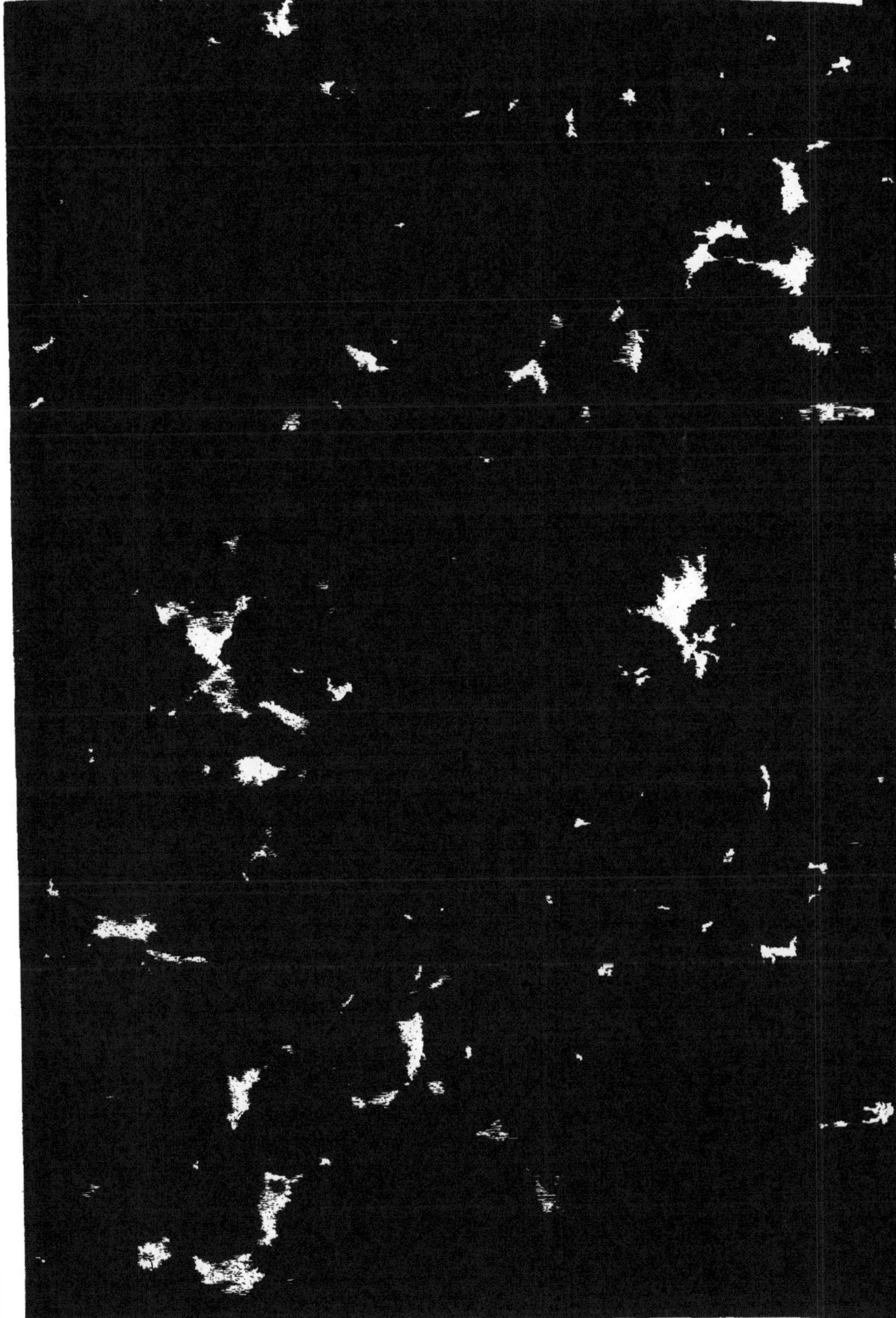

Ier CONGRÈS INTERNATIONAL D'ENTOMOLOGIE

IER CONGRÈS INTERNATIONAL

D'ENTOMOLOGIE

BRUXELLES, AOÛT 1910

VOLUME II

MÉMOIRES

BRUXELLES

HAYEZ, IMPRIMEUR DES ACADÉMIES ROYALES

Rue de Louvain, 112

30 OCTOBRE 1911

« Springtails » (Collembola). Their economic importance, with notes on some unrecorded instances of damage,

by FRED. V. THEOBALD, M. A.; F. E. S.; Hon. F. R. H. S., etc.

The importance of the Collembola or « Springtails », one of the two divisions of the order Aptera, as destructive insects has been referred to by Sir JOHN LUBBOCK (now Lord AVEBURY) and JOHN CURTIS, and since then in several parts of the world references have been made to them in connection with the damage they do to plants. The order, however, has been more or less ignored by economic entomologists.

Recently some fresh facts have come to light through the researches of Professor CARPENTER of Dublin in connection with Apteran injury, and several new forms of damage are recorded here.

Their insignificant size is made up for by their enormous numbers at certain times and in certain places.

That very many are non-injurious we know, but of these, how many may under altered circumstances become of economic importance we do not know. Certain species, especially in the genera *Sminthurus* and *Lipura*, are undoubtedly quite harmful to plant life, attacking both leaves and stems; others damage roots (*Templetonia*) and others seem to so disturb the soil (*Achorutes*) that seedlings suffer.

Although their mouth parts are weak, the mandibles are sufficiently powerful to do much harm. Moreover in a second way these minute insects may do harm, namely, by so opening plant tissue that it is easily invaded by bacteria and fungoid pests, and as many are found feeding on diseased and fungus growths, it is quite possible that they may act as distributors of bacterial and vegetal disease germs.

On several occasions countless numbers of Collembola have been sent me from sewage works. What economic importance they have, there has not in any way been worked out. The enormous masses of them that sometimes occur in sewage farms is remarkable.

Direct injury seems to have been traced to the following species :

List of injurious species.

* *Lipura armata* TULLBERG.
Lipura ambulans LINNÆUS.
Lipura burmeisteri LUBBOCK.
* *Lipura fimetaria* LINNÆUS.
Achorutes armatus NICOLET.
Achorutes manubrialis TULLBERG.
Achorutes longispinus TULLBERG.
Achorutes rufescens NICOLET.
Achorutes purpurescens LUBBOCK.
Anurida granaria NICOLET.
* *Isotoma palustris* MULLER.
Isotoma tenella REUTER.
* *Orchesella cincta* LINNÆUS.
* *Entomobrya nivalis* TULLBERG.
Entomobrya multifasciata TULLBERG.
Lepidocyrtus sp.?
* *Sminthurus fuscus* LINNÆUS.
* *Sminthurus luteus* LUBBOCK.
* *Sminthurus niger* LUBBOCK.
* *Sminthurus pruinosus* TULLBERG.
Sminthurus hortensis.
* *Templetonia nitida* TEMPLETON.
* *Degeeria annulata* FABRICIUS.

The species personally observed are marked with an asterisk.

LUBBOCK in 1872 said : « Some species of Sminthurus live on the leaves of plants » (1).

---

(1) LUBBOCK, Mon. Collembola and Thysanura, Ray Soc. 1872.

CURTIS (1) in 1883 calls them « Ground Fleas », a term I have often heard used by gardeners in the South of England.

He refers to *Sminthurus solani* (2) as follows : « It is not bigger than a small grain of sand, and either of a deep ochreous hue with black eyes, or is as black as soot with ochreous horns. The head is large, like a great mask and attached by a slender neck; the eyes are placed on each side of the crown; the horns more than half the length of the body, slender, elbowed and four jointed (the terminal joint appears to be divided into six lobes), the trunk and body are united, forming a large globose mass with a forked tail doubled under the latter for leaping; the six legs are rather short and apparently triarticulate.

» These minute animals are nourished by eating the parenchyma of the green leaves, but some species feed on fungi. » This refers to attack on the potato crop.

Again (p. 433) CURTIS tells us that « in Nova Scotia the crop of turnips and cabbages are principally destroyed whilst in the second leaf, by some *Sminthurus*, the size of a pin's head and nearly globular. It hops with great agility by means of its forked tail, and may be found on every square inch of all old cultivated ground, but it is not plentiful on new land ».

MURRAY (3) says : « Our young gherkins will be found shrivelling, and on examination it will be seen that they have been stripped of great portions of their skin that has been browsed or rasped away by these little creatures. Nor are their ravages confined to frames, although the greater heat there seems to suit their delicate semi-transparent bodies. But in the open borders they carry on the same work on succulent roots and plants, specially where anything has happened to diminish the vitality of the plant.

» On carrots, for instance, that are suffering from rust, they will be found browsing on the sound parts. »

COLLINGE (4) records damage to bulbs by Collembola, especially to hyacinth, narcissus and tulip bulbs.

« The nature of the injury is the same, he says, in all cases and consists in scraping away the epidermis and then the softer tissue

---

(1) *Farm Insects*, p. 432, 1883.

(2) This is evidently *Sminthurus luteus* of LUBBOCK.

(3) *Economic Entomology*, Aptera. p. 404.

(4) *Journ. Econ. Biol.*, IV, N° 3, p. 86, 1909.

until a distinct hole or depression is formed. After, rapid decomposition of the plant tissues takes place, due to inroads of fungi. »

Connection with potato disease.

In the « Gardeners' Chronicle », volume VIII, page 702, one of the *Podura* was accused of helping in the spread of potato disease. Needless to say it plays no important part in it.

The statement concerning Podura is quoted as a matter of interest : « First in an early stage of its existence, it lives on decayed vegetable matter, which it collects by burrowing into the earth; secondly it occurs in numbers sufficient to cover nearly the whole surface of the earth; thirdly it collects, as a means of existence, a substance which is poisonous to vegetables. It has a power to infuse this into living plants by burrowing into the parenchyma. The poison is circulated into the system, vital action becomes suspended, mildew immediately follows, and in less than three days some of the plants attacked are a dead vegetable matter, food for the offspring of the newly discovered *Podura*. »

From recent observations I have made it seems that it is quite likely that *Sminthurus luteus* plays some part in distributing potato disease. It is small, however, and more a matter of interest than of any economic importance.

Sminthurus luteus LUBBOCK damaging currants, and potatoes.

Since 1904 specimens of damaged currant shoots (fig. 1) and young potato tops have constantly been sent me from several localities in the South and Midlands of England. During the year 1910 similar damage was done to apples in Kent. This was due mainly to Hemiptera and not to Aptera. It was not until last year, however, that I traced the cause of the damage. It proved to be the small yellow springtail known as *Sminthurus luteus* of LUBBOCK which swarmed on the currants.

In 1909 a large grower near Horsmonden, Kent, reported the great damage done by this insect. This year a correspondent writes from Hernhill near Faversham that « a great many of our Fay's early red currants have the younger leaves damaged, and I find the same in nursery stock ». Another grower near

Maidstone reported the damaged as being very severe, and found the yellow Springtails on the foliage in great numbers.

These small yellow Aptera are most active in their mature stages, skipping from the foliage at the least movement, and in consequence, unless one proceeds with very great care, nothing will be found on the plants. The adults are bright orange yellow with black eyes, the immature stages pale yellow and the dark eyes not so prominent. The young forms, some the size of a pin point, evidently just hatched, are almost white. The immature insects are not active like the adults, but gradually become more and more agile. Whilst the adults are difficult to capture except by sweeping, the immature ones can easily be taken in a glass tube.

They appear to feed exclusively on the under sides of the foliage, where they eat away the lower epidermis and the soft mesophyll tissue beneath, leaving the upper epidermis intact.

The eaten areas (fig. 2) are at first mainly confined to small patches between the veins of the leaves, sometimes the veins are eaten through, and thus larger patches are formed. These areas vary from 1 to 3 mm. in length and show as smaller pale spots on the upper surface of the leaves of a yellow or rusty brown hue. Later, these patches or spots burst, and ragged holes are formed, the edges heal and a ragged, stunted appearance of the foliage is produced, as shown in the figure (fig. 1).

When young leaves are attacked, the resultant damage is very serious, and growth is quite checked.

Although these Springtails work most rapidly in June and early July, they may be found on the young leaves in May, and if in any numbers it seems to me that at this time they do the most harm.

I have found them on the foliage until the second week in September.

The amount of damage done by each of these minute insects, like a grain of sand in size, is considerable; one on a leaf will cause the damage shown in figure 2 (2), in a couple of days, and if left this leaf soon becomes a riddled ragged structure seen in figure 2(3).

The surface of the mid vein and larger lateral veins is also gnawed and presents a rusty appearance, and in some cases had clearly caused the leaves to curl as seen in figure 3.

Experiments were made by placing a dozen of this *Sminthurus luteus* on healthy shoots on bushes under muslin tents, and in all cases in a couple of weeks the result seen was the same as that shown in figure 1. Not only is young growth checked, but in bad cases the year's fruit shanks off entirely.

Damage to apples.

The damage on apples (fig. 4) is very similar to that caused by the yellow Springtail to currant foliage. The young tender leaves are gnawed in small areas, either the upper or lower epidermis being left intact, and later splitting. This attack has only been reported to me once from Kent during the last year (fig. 5).

Most of this damage is caused by Hemiptera.

The yellow Sminthurus on potatoes.

For some years a similar disease has been sent me attacking potatoes. This also could not be explained, as there were no signs of insect or fungus on the foliage or haulm.

Owing to the creatures appearing in abundance in my garden, I was able to trace this common form of damage to the same Sminthurus as attacks the currant (fig. 5 and 6). Normal potato foliage from the same bed is shown in the photograph.

That other causes (1) may produce similar results there is no doubt, but that the yellow Springtail (fig. 7) is the chief cause of this disease in currants and potatoes there is also no doubt. I have seen a similar attack in gooseberries, but on the bush being examined, a casual examination revealed no Springtails.

Damage to turnip leaves by *Sminthurus luteus* and *S. niger* LUBBOCK, etc.

Miss ORMEROD (2) records *Sminthuri* as damaging the turnip crop in Aberdeenshire in 1894. She said that in July a correspondent wrote as follows : « My turnips, and others in the neighbourhood (Lumphanan, Aberdeenshire) are being a good deal damaged not only by turnip flea, but by something else, which

(1) I have traced very similar damage to a green *Lygus*.

(2) *Rept. and Obs. Inj. Ins. for 1894*, p. 110, 1905.

I never observed before ». The correspondent later wrote : « Not only the first two leaves were eaten, but the rough leaf was also attacked », and he described the pests as being a dirty yellow colour, and also black. « They jump on being approached. So numerous were they in one field that they rose before one like a little cloud. »

Miss ORMEROD identified one as *S. luteus* and assumed the dark species to be *S. niger*.

In Cheshire, one of my old students, Mr. SAUNDERS, biologist at Harper Adams Agricultural College, noticed damage done to young turnips in 1909 and 1910 by Springtails. They eat out round holes in the seed leaves similar to the attack of turnip flea beetle.

He found the same insect common on *Polygonum* and other weeds. Although they did little actual damage, they were noticed to feed in a most persistent manner.

He informed me they had been identified as *Sminthurus hortensis*.

Sminthurus pruinosus TULLBERG **attacking beans and sweet peas**.

The damage done by this dusky Sminthurus was first noticed by Mr. BLAKEY of Redditch, and some of his notes are included here.

The damage to bean leaves (fig. 8) seems somewhat different to that done to sweet pea foliage (fig. 9). In the former, the leaf is characteristically eaten away in small areas which may later split into larger spaces, as seen in the photograph of a french bean leaf taken by Mr. BLAKEY.

In sweet peas the leaves are eaten in larger patches of longer form, as seen in figure 9.

I have not seen the attack on french beans (fig. 8), but the attack on sweet peas is very common, and usually starts on the lowest leaves, and gradually the attack works upwards.

This dusky Springtail (fig. 10) also attacks turnips in a similar way. It is just as shy as *S. luteus*, but if one proceeds carefully, it can be watched feeding on the under side of the foliage.

The young are not very active, and are paler in colour than the adults and may easily be seen on the lower surface of the leaves, whilst the adults skip rapidly away.

In one case observed on peas, considerable damage was done to the foliage of the white everlasting pea, the plants being quite ruined, and a new set of leaves and shoots were sent up and but poor blossom resulted in consequence. I have seen the same attack in culinary peas, especially in the William Hurst variety, and it certainly lessened the yield, although the damage was confined to the young leaves.

June and July seem to be the months when this Springtail is most abundant. In one case the damage was very noticeable in May. Mr. BLAKEY, writing me on the second of June this year, says the same species « is now on the turnips ». Probably this is the same as the attack recorded by Miss ORMEROD from Scotland.

Mr. SAUNDERS reported to me a bad attack on dwarf beans in 1910, in Cheshire, some of the attacked plants being quite killed.

Mr. BLAKEY sends the following notes from near Redditch :

« I found the Sminthurus this year in the same garden, but they did not do so much damage to the beans. In the garden under my charge I have found them on sweet peas, turnips, radishes and lettuce. We grow sweet peas here rather extensively and I do a good deal of cross breeding. One batch was planted out early in April after being grown in pots. After they had been planted about a fortnight, I noticed some of the leaves damaged as the one sent. It was the leaves nearest the ground that were damaged the most. I thought at first it was small slugs. I had to be away for a time, when I came back I made a careful examination of the leaves and ground beneath and there found the Sminthurus by the hundreds. One of my garden hands was brushing the paths in the kitchen garden when he called my attention to a number of small insects, as he called them, that came out of the box edging when moved by his brush. I found thousands of these Sminthurus there. This was near the sweet peas. There were some young turnip plants on the opposite side, and I noticed the leaves were somewhat perforated, but thought it was the flea beetle which we generally get on our turnips, but I found in addition to the flea numbers of these Sminthurus, and as the turnip plant was a fairly good material to make observations upon I arranged myself to watch them feed. I noticed they attacked the upper as well as the lower surface of the leaf; one would start at a certain spot, pierce the epidermis, work round and round; if they were browsing on the upper surface, they left the lower epidermis, if on the lower, vice versa. When one had got a sufficient area destroyed, another would come to its aid, until

I have seen half a dozen or more on one spot. From what I can see this pest is increasing very much and ought to be looked after, and if they shelter in such places as box edging, it ought to be sprayed or dusted with some material that will destroy or make their habitation untenable. I have been in the habit of using vaporite for soil pests. In this case, instead of using it in the usual way, i. e. dry, I dissolved some in hot water, then diluted 1/4 lb. to 4 gallons, watered the plants and surrounding ground with a fine rose can. This seems to be effectual. No doubt other things would answer the purpose. »

An attack of *Sminthurus* sp.? on mangolds was noticed by Mr. SAUNDERS in 1909 in Cheshire.

In 1908 the same Springtail attacked maize. The young plants 4 to 18 inches high were attacked in July. They ate out longitudinal holes between the veins, much as in the damage done in peas.

The attacked maize sent me showed markedly twisted and deformed leaves, and the whole growth was said to be stunted.

Little harm was done after August, and the crop was only slightly below the normal. Another, but slighter attack, followed in 1909. Mr. SAUNDERS informed me that the species was identified for him by Mr. COLLINGE as *Sminthurus hortensis*.

A Sminthurus attacking tobacco in America.

The following note occurs in « Insect Life » (1) :

« *Sminthurus hortensis*. These active little insects were extremely abundant in Indiana the present season.

» About La Fayette I observed them feeding upon young cucumber plants, the injured parts of plant not being affected by other insects or fungus.

» They were also reported by Mr. C. G. BOEMER as injuring young tobacco in Switzerland County during the month of May. »

Sminthurus damage in Australia.

FROGGATT (2) says : « Another species belonging to the genus

---

(1) RILEY and HOWARD, *Insect Life*, vol. 3, p. 151, 1891.

(2) *Australian Insects*, p. 10.

Sminthurus allied to *S. viridis*, an European form, but probably an undescribed native species, appeared in great numbers in lucerne paddocks in S. Australia in 1896, where they did a great deal of damage by eating the surface of the leaves, swarming over the fields in countless numbers. »

Collembola attacking orchids.

Species of *Orchesella* have been known to attack orchids (1). The first record (2) of damage was sent to the British Museum by a correspondent who stated that they not uncommonly cleared off the whole of the contents of a pot of seed as soon as it is was sown and germinated. The chief species has recently been found to be *Orchesella cincta* LINNÆUS.

They were noticed to thrive most where there was damp moss to dwell in; unfortunately an essential feature in orchid cultivation. They proved to be equally troublesome in cool houses (50° to 60° F.) as in hot ones (65° to 85° F.). The seed was devoured directly it germinated, and later they attacked the small bulblets as soon as they were formed, leaving only the shell.

After the early stages they are harmless. Careful removal of all moss and fumigation with hydrocyanic acid gas of the pots which may be dried for a *short* time will clean this pest out.

White Springtails (*Lipura ambulans* LINNÆUS).

In 1904 and 1905 CARPENTER (3) stated he examined in County Dublin roots of kidney-beans badly gnawed by *Lipuræ* and a few *Achorutes*. In 1906 CARPENTER (4) recorded damage by this white Springtail in April from County Armagh to roots of garden flowers and succulent vegetables.

In County Dublin CARPENTER (5) records this species in 1907 as

---

(1) F. V. THEOBALD, *Second Report on Economic Zoology* (Brit. Mus.), p. 76, 1904.

(2) F. V. THEOBALD *First Report on Econ. Zool.* (B. M.), pp. 108-112, 1903.

(3) G. H. CARPENTER, *Economic Proceedings of the Royal Dublin Society*. I, pt. 6, p. 293, 1905. (Plate XXVI A.)

(4) *Ibidem*, I, pt. 8, p. 340, 1906.

(5) *Ibidem*, I, pt. 2, p. 442, 1907.

damaging peas, beans and other vegetables in a garden, eating across the young green shoots just below the ground.

There is no doubt that this species frequently does much harm, and other cases of Lipurid damage are here recorded.

Lipura damaging roots of celery and cauliflower.

A bad attack of *Lipura* sp. (1) was reported from Pembrokeshire in 1903. The specimens of insects sent were much damaged, and so not then identified. They attacked the roots and stems below ground of the cauliflowers, and the leaf stalks of the celery; in some cases so badly that the outer tissues were eaten away all around, and the plants killed. In this case soot and lime mixed in to the soil had the desired effect of clearing out the Aptera (2).

Mr. SAUNDERS tells me he found *Lipura ambulans* doing much harm to celery in 1909 in Cheshire.

The same species was also noticed attacking seakale and asparagus in 1910 in Cheshire. Two whole beds of asparagus were eaten up by them, the creatures swarming in the beds.

Lipura fimetaria LINNÆUS.

This is found throughout the year in damp earth, often engaged in feeding upon carrots, potatoes and other roots. It is white and of a velvety texture, and often occurs in swarms in the soil.

In all cases I have seen it has been associated with plants already diseased.

COLLINGE (3) records *Lipura ambulans* LINN. on narcissus bulbs injured by the stem eelworm, and also examples of *Achorutes armatus* NICOLET, from nurserey gardens near Birmingham.

Damage to strawberry roots by *Templetonia* (*Heteromurus*) *nitida* TEMPLETON.

In June of this year (1910) a correspondent writing from East

(1) This is now known to have been *Lipura ambulans* LINNÆUS.

(2) F. V. THEOBALD, *Second Report on Economic Zoology* (Brit. Mus.), p. 158, 1904.

(3) *Report on the injurious Insects and other animals observed in the Midland Counties during 1905*, p. 10, 1906.

Peckham, Kent, said some of his plantation strawberries were not doing well, and on digging them up found great numbers of springtails around them. Professor CARPENTER identified these for me, and found them to be *Templetonia nitida*, which is recorded later as found on and breeding in corpses in Europe (fig. 11).

Similar damage was traced in my own garden, the roots and bases of the leaves being eaten away in patches, some to such an expent that the plants died. This pale yellowish to pinkish species literally swarmed around the plants. They disappeared as soon as the soil became dry without any special treatment, but recurred again after heavy rains, and the plants had a good dressing of fine lime hoed in around them, which again drove the Springtails away. They were found in numbers sheltering under some seed boxes near the strawberry beds.

Degeeria annulata FABRICIUS **feeding on currant leaves.**

This species (fig. 12), which is a most active runner, is also possibly connected in some way with the disease in currants, etc., caused by *S. luteus*.

In one lot of currant bushes I had constantly under observation these Aptera swarmed on the leaves in the second week of July, and were watched browsing off the upper epidermis, whilst the yellow Sminthurus were feeding below. Considerable variation in the dark markings were noticed, but the general pattern answered to LUBBOCK's figure (plate 31, Mon. Collembola and Thysanura 1872). In August of the same year they still further increased on the currants, and seemed to take the place of *Sminthurus luteus*. I could not find them on neighbouring plants, but many were to be found on the soil.

Mushrooms attacked by *Achorutes rufescens* NICOLET.

In October 1908, information was sought regarding the great damage done to mushrooms in the beds of the North downs mushroom quarries. Professor CARPENTER examined the specimens, and said they were *Achorutes rufescens* of NICOLET. The damage was again reported in August 1909. Mr. W. F. TAYLOR then wrote that they were « still causing great trouble, and they seem difficult to kill without injuring mushrooms ». This species swarmed in the caves and was found in the soil long before the mushrooms had started growing. The young fungi were gnawed at once, and those that did not rot under the attack were mere stunted things unfit for market. The Aptera watched fed mainly on the tender upper skin, but later they were found in and on the gills.

Cabbages attacked by *Achorutes purpurescens* LUBBOCK.

An *Achorutes* identified by Professor CARPENTER as *A. purpurescens* of Lubbock has been found on one of the farms belonging to the S. E. Agricultural College in great numbers on the roots of cabbages (fig. 12). They were kept under observation, and it was noticed that the gnawing of the roots and the stem at the ground level was due entirely to these Aptera. In some places the outer skin of the stems at ground level was quite gnawed away. Backward plants were destroyed by them, but their attack on well set healthy plants was very slight.

Achorutes armatus NICOLET **and** *Achorutes longispinus* TULLBERG **damaging plants.**

CARPENTER (1) refers to the two Achorutes mentioned above as doing damage to roots and seeds of healthy plants.

The former in 1902 was recorded from County Longford, Ireland, as swarming over and partly devouring fruit left lying on the ground.

In June 1903 he also received bean seeds eaten from the outside by two species (*A. armatus* and *L. ambulans*) which swarmed in the surrounding soil.

In the summer of 1904 he records from County Dublin bean seedlings damaged by *Achorutes longispinus* and *Lipura ambulans*.

---

(1) *Proceedings of the Association of Economic Biologists*, Vol. 1, part. 1, p. 14, 1905.

Achorutes preventing germination of seeds.

GUTHRIE (1) mentions that a Collembolan, probably belonging to the genus *Achorutes*, was so abundant in soil containing seeds in America that the plants had little or no chance to root, and many died.

Achorutes armatus and *Templetonia nitida* found on and breeding in corpses in Europe.

Whilst dealing with this genus, the following note may be added of its breeding in corpses with *Templetonia*.

« La faune des tombeaux » (P. MEGNIN, *Compt. rendus de l'Acad. des Sciences,* V, 105, N° 20, Nov. 14, 1887, pp. 350 to 351).

Entomobrya nivalis LINNÆUS attacking hops.

This is said to be the same as *Degeeria annulata* LUBBOCK (non FABRICIUS), but as far as I can see is quite distinct.

There was no connection as far as I could see between the specimens damaging hops which were identified by Professor CARPENTER as *E. nivalis* and true *Degeeria annulata.*

It is a common insect found on all kinds of plants, and under logs of wood. It occurs under the bark of trees, where it probably lays its eggs, also in damp moss and grass. It also occurs on low bushes, furze, heather, etc. Very large numbers appeared in a hop-garden near Chilham, Kent in 1907.

These Aptera attacked the «burr» of the hops, and usually started their work at the tips of the bine, the damage extending down to the lower parts in many cases. The tender bine was also damaged, as well as the « burr » and young foliage.

Those watched were seen in the act of feeding, and at the least movement they fell or sprang to the soil. The damage was done mainly at night.

During daytime great numbers were found sheltering under

---

(1) *The Collembola of Minnesota,* p. 4, 1903.

clods of earth. The damage done was very severe, and immediate steps were taken to clear the Aptera out (1).

A Podurid sp. ? attacking mushrooms.

One of the *Poduridæ* was reported to me in 1908 as being very harmful to mushrooms in the caves of the North Downs Mushroom Company, near Reigate. The mushrooms were covered with them, and they gnawed the skin, and so not only spoiled great numbers for market, but actually ruined large numbers. Many of the attacked fungi were deformed in very marked manner.

Watering the beds with nicotine before the mushrooms appeared above ground seemed to be successful.

French growers use lysol as a means of destroying insects in mushroom beds.

Isotoma tenella REUTER destroying tobacco seedlings in Ireland.

This small dark greyish Collembola has been recorded by CARPENTER (2) as occurring in multitudes on seedling tobacco leaves at Kilkenny in Ireland.

The surface of the leaves showed distinct abrasions, and CARPENTER found that the digestive tracts of the Aptera contained chloroplasts.

This species was first described from Finland, later it was found in Germany, and then by CARPENTER as a destructive insect in Ireland.

They were easily killed with a nicotine fumigator in spite of their feeding on tobacco.

A Collembola as a poultry pest.

A single case of a species of *Isotoma* attacking poultry has been

---

(1) F. V. THEOBALD, *Report on Economic Zoology for year ending April 1st, 1908*, pp. 100-101.

(2) G. H CARPENTER, Injurious Insects and other animals observed in Ireland during the year 1907. (*Economic Proceedings of the Royal Dublin Society*, Vol. 1, pt. 15, p. 574, 1908.)

recorded. A photograph of a species of this genus was sent me with a note that it had occurred in great numbers in fowl's nests, especially old and dirty ones, and that it had been a great torment to the hens (1). No further observations have been made.

Collembola infesting houses and destroying wheat rust.

MARLATT (2) records a *Lepidocyrtus* infesting houses in the United States.

A Sminthurus (? species) was found by WEBSTER in breeding cages of wheat, and feeding on uredo spores of the common wheat rust *Puccinia rubigovera*. It was shown also to distribute spores so any good done is neutralised (3).

Prevention and treatment of Collembola on fruit and hops, etc.

The fact that Collembola have no spiracles and breath by means of the whole body surface makes them intolerant to dry conditions as a rule.

In consequence dressings of lime or fine lime and soot are often successful.

CURTIS (4) says: « As these 'ground fleas' will not remain on damp ground, they may be expelled by sprinkling salt over the land after the seed is sown, and well rolled down, or a thin layer of seaweed is a perfect security against them (Halifax Times). » This is contrary to general experience.

Other observers besides myself have noticed that the Springtails object to *dryness*. This is certainly my experience in the species I note here, and in all cases of bad attack it has been in damp seasons, and where plants are in damp positions, or grown under damp conditions that damage has been done. Any drying agents seem at once to drive them away. Salt or seaweed would have the reverse effect, but nevertheless may act in other ways

(1) F. V. T. *Parasitic Diseases of Poultry*, p. 37, 1896.
(2) *Canad. Entomol.*, XXVIII, 1896.
(3) *Insect Life*, 2, p. 259, 1889-1890.
(4) *Farm Insects*, p. 433.

In the attack on currants, which I had fortunately in my own garden and so could keep under constant observation, it soon became evident that the little yellow Springtail worked most in shaded moist spots, and that once the sun dried up the foliage and soil a check came in the damage. This was equally noticeable in sweet peas being damaged by Sminthuri.

Seeing and being informed of the damage done by these Aptera to currants, and that it was becoming of no mean importance, I at once tried to clear them out without tracing the egg stage, which I had hoped to.

Six lots of bushes were treated as follows, and one left as a check :

1. Two bushes were heavily washed with the Woburn nicotine wash and soap, and the ground beneath was also sprayed.

2. Two bushes were washed in the same way, but not the ground. The bushes were then netted over.

3. Two bushes were sprayed with arsenate of lead, and also the soil.

4. Two bushes with quassia and soft soap, and also the soil drenched with the same.

5. Two bushes were sprayed with arsenate of lead, and the soil beneath heavily dusted with fine lime and soot.

6. Two bushes were sprayed with nicotine wash and heavily dusted on the soil beneath with lime and soot. Others were left as checks.

In 1, 2, 3 and 4 considerable good was done for a time, but in 2 there were a number of Springtails on the bushes again in two days, and in 3 very much the same occurred a few days later.

In 1 and 4 the results seemed somewhat better than in 2 and 3. But in none of these were the insects checked as one would wish to see done.

In 5 and 6 the results were practically the same, viz. that no trace of the Aptera could be found, and no further damage was noticeable.

From this it appears that (A) the Springtails may be poisoned with arsenate of lead (Swift's paste) or nicotine at the strength used (viz. 1 1/5 oz. of nicotine to 10 gallons of water), that (B) the ground treatment is necessary also, and (C) as they do not feed on the soil, that the drying caustic effects of lime and soot are better than the mere spraying of the ground by the several insecticides tried, which all have a more or less baneful effect upon the Aptera on the foliage.

It has been found that where they attack asparagus, seakale, etc., they may be kept down by being trapped with half scooped out orange rinds.

The smearing of hop bines near the ground with soft soap or better molasses stops the hop pest ascending. Also spraying with nicotine and soft soap, but the former method is apparently best.

The photographs illustrating this article were taken by M. Edenden of Wye and M. Blakey of Redditch, F. V. T.

Fig. 1. — Currant shoot damaged by *Sminthurus luteus* Lubbock.

Fig. 2. — (1) Normal red currant leaf Lubbock
(2 3) Two leaves damaged by *Sminthurus luteus*.

Fig. 3. — (1) Currant leaf damaged by *Sminthurus luteus*.
(2) Normal currant leaf.

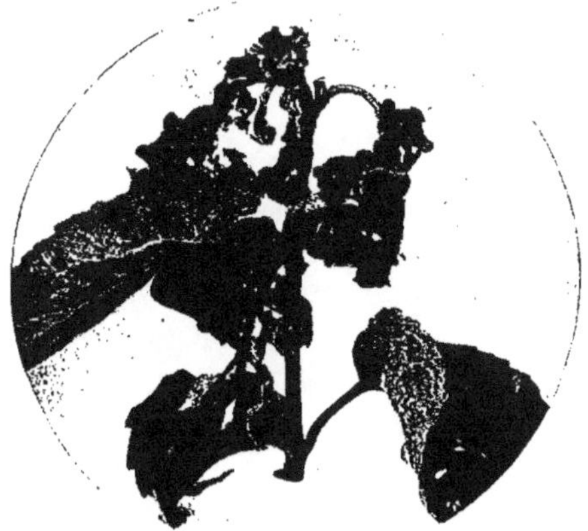

Fig. 4. — Apple shoot damaged by *Sminthurus luteus*.

Fig. 5. — Potato foliage damaged by *Sminthurus luteus*

Fig. 7 — *Sminthurus luteus* Lubbock.

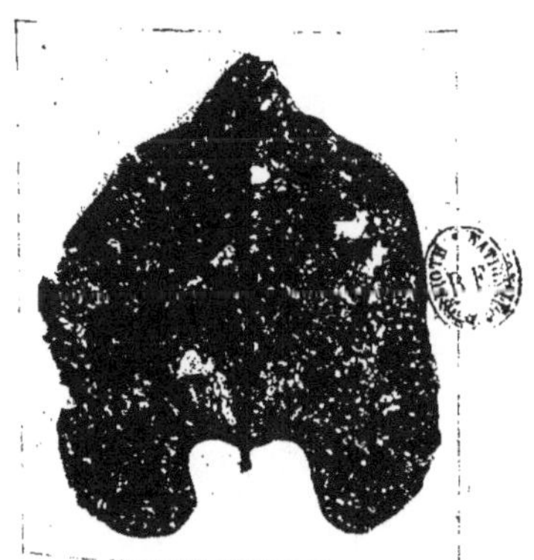

Fig. 8 — Damage to bean leaves

Fig. 6. — Normal potato foliage and blossom.

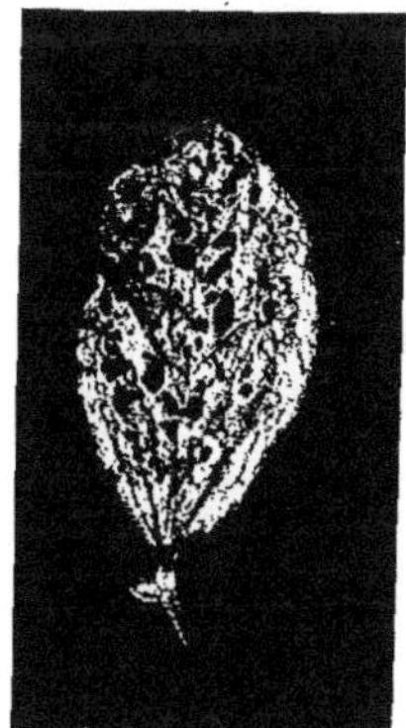

Fig. 9 — Damage to pea leaves.

Fig. 10. — *Sminthurus pruinosus* Tullb.

Fig 11. — *Templetonia nitida* Templ.

Fig. 12. — *Degeeria annulata* Fabr.

Fig 12 — *Achorutes purpurescens* Lubb.

Les Calliphorines à larves cuticoles des animaux domestiques,

par L. GEDOELST (Bruxelles),
Professeur à l'École de médecine vétérinaire de Cureghem.

Les myiases cutanées que l'on observe chez l'homme et les animaux sont dues à des insectes appartenant à trois familles de *Tachinidæ* : les *Sarcophaginæ*, les *Calliphorinæ* et les *Hypoderminæ.* AUSTEN (1) a publié récemment une revue très complète des Diptères qui causent les myiases chez l'homme. Nous nous occuperons ici des espèces appartenant à la famille des *Calliphorinæ* dont les larves se développent au niveau du tégument chez nos animaux domestiques.

Ces cas de parasitisme se rencontrent dans les diverses régions du globe : en Europe, en Asie, en Afrique, en Amérique et en Australie.

En Europe, ils semblent dus presque exclusivement à des représentants du genre *Calliphora* : *C. vomitoria* L., et du genre *Lucilia* : L. *Cæsar* L. et *L. sericata* MEIGEN. Les accidents que cette dernière espèce cause chez les moutons, en Hollande et en Angleterre notamment, sont trop connus pour que nous ayons à insister à leur sujet. Les larves de Calliphores et de Lucilies ne paraissent constituer que des parasites accidentels; ces mouches pondent normalement leurs œufs sur les matières organiques en voie de décomposition, occasionnellement elles le font au niveau de plaies infectées. C'est dans cette dernière condition qu'on a signalé le parasitisme des larves de *Calliphora vomitoria* et de

L. Cæsar. Il semble que l'infestation par les larves de *L. sericata* a un déterminisme analogue. Les femelles de cette espèce pondent en effet leurs œufs, au pourtour de l'anus, dans les excréments qui restent adhérer à la laine, et les larves qui en éclosent percent ensuite la peau fine de cette région.

Calliphora vomitoria et *Lucilia Cæsar* se rencontrent de même en Amérique, mais à cette partie du globe appartient en propre une mouche redoutable, la *Chrysomyia macellaria* FABR., qui est répandue depuis le nord des États-Unis jusqu'à la République Argentine et possède des mœurs analogues à celles de nos Lucilies; elle pond également ses œufs dans les plaies ou les introduit dans les cavités naturelles de l'homme et des animaux. Il en éclot des larves que les Américains désignent communément sous le nom de Screw-worm, à raison de la forme annelée du corps. Nous en devons une excellente description à R. BLANCHARD (2).

D'Asie nous ne possédons que peu de documents concernant les *Calliphorinæ* à larves cuticoles chez les animaux domestiques. Nous pouvons citer l'observation du Dr DEPIED (3), faite au Tonkin. Après avoir signalé deux cas de myiase chez des soldats annamites, il ajoute : « J'observai sous le ventre d'un cheval, en avant et à gauche des parties génitales, un véritable trou qui me parut ressembler extérieurement aux nids de larves de mes deux tirailleurs annamites. Une injection faite dans la plaie amena en effet la sortie de larves... Les larves que je fis éclore comme précédemment donnèrent, bien qu'elles m'eussent paru plus volumineuses, des mouches *absolument semblables* à celles que j'avais obtenues déjà, absolument pareilles à la *Lucilia hominivorax* ». Cette détermination est certainement inexacte, car *Lucilia hominivorax* est synonyme de *Chrysomyia macellaria*, et l'on sait que les représentants du genre *Chrysomyia* appartiennent exclusivement au nouveau monde. Il est probable que DEPIED a eu affaire à une espèce de *Pycnosoma*, genre tout à fait voisin du genre *Chrysomyia* et que certains auteurs identifient même avec ce dernier.

Cette idée gagne en vraisemblance si on la rapproche de la déclaration du Dr PATTERSON (4), qui dit qu'en Assam les larves des « Green-bottles », c'est-à-dire des mouches vertes, se rencontrent fréquemment dans les plaies des bovidés, plus rarement des chiens et des chats. D'autre part, GAIGER (5) signale la présence de larves de Diptères, qu'il rapporte avec doute au genre *Lucilia*, dans les plaies des animaux domestiques dans l'Inde : chevaux, bœufs, chameaux, chiens, etc. Ces larves sont très fréquentes chez le chien.

En Australie, le rôle de la *Lucilia sericata* est rempli par la *Calliphora oceanicæ* R. D. Cette espèce, qui est répandue dans toute l'Australie, dépose ses œufs dans la toison souillée des moutons, et ses larves pénètrent dans la peau, où elles causent des lésions fort étendues. FROGGATT (6) pense que la *Calliphora villosa* R. D. peut déterminer des accidents analogues chez les moutons australiens. Aux îles Hawaï, VAN DINE et NÖRGAARD (7) signalent une autre espèce de Calliphore, *C. dux*, dont les mœurs sont semblables. Elle dépose ses œufs sur la laine des moutons; les larves éclosent au bout de vingt-quatre heures et rampent vers la peau, à la surface de laquelle elles se meuvent en amas serrés d'une manière constante; elles déterminent ainsi de l'inflammation avec suppuration et ulcération. Une fois la peau détruite, elles s'enfoncent dans tous les sens, formant sous le tégument des poches, des fistules, et le mouton succombe à la septicémie ou à l'épuisement. Quand les larves sont mûres, soit en moins de dix jours, elles se laissent tomber sur le sol et s'y transforment en pupes. Tout le développement ne demande pas plus de trois semaines.

En Afrique, les Calliphorines à larves cuticoles paraissent nombreuses. Une des plus anciennement connues est celle dont la larve était communément désignée sous le nom de ver de Cayor. EM. BLANCHARD l'avait dénommée *Ochromyia anthropophaga*. Dans ces dernières années, GRÜNBERG a reconnu que ces larves africaines appartiennent à un nouveau genre, le genre *Cordylobia*, dont quatre espèces sont actuellement connues. Les larves cuticoles de trois d'entre elles ont été observées : ce sont celles de *C. anthropophaga* EM. BLANCHARD, *C. murium* DÖNITZ et *C. Rodhaini* GED. Il est vraisemblable que toutes les espèces de ce genre déterminent des myiases cutanées. *C. anthropophaga* est la seule espèce qui ait été rencontrée jusqu'ici chez les animaux domestiques (chien, chat, chèvre, chameau, mulet, cheval). On peut prévoir que d'autres espèces de *Cordylobia* seront observées dans des conditions analogues. C'est probablement à une espèce de ce genre qu'il faut rapporter les cas de myiase observés par VILLELA (8), à Praia (île du Cap-Vert), chez l'homme et les animaux, et la même espèce que WELLMAN (9) avait déjà observée, en 1906, chez le chien dans cette même localité et qu'il déclare identique à des larves cuticoles recueillies sous la peau d'un enfant à Benguella. A raison de l'intérêt qui s'attache à l'observation de VILLELA, nous reproduisons ici la description qu'il donne de la mouche et de sa larve :

L'imago mesure de 15 à 20 millimètres de longueur. Sa couleur générale est jaune pâle, légèrement brune sur le thorax et brun foncé au niveau des troisième, quatrième et cinquième segments abdominaux. La tête est forte et presque de la même largeur que le thorax, de coloration jaunâtre, sauf les yeux qui sont rouge-brun. La trompe est jaune, courte, placée sous la tête et impropre à la piqûre. Les palpes sont également jaunes et se terminent en massue qui porte quelques poils noirs. Les antennes sont jaunes, et leur troisième article porte une soie (arista) chargée de poils noirs qui vont en diminuant de longueur à mesure qu'ils s'éloignent de la base, qui est assez épaisse. La surface dorsale du thorax montre, à droite et à gauche de la ligne médiane et sur chacun des segments, une tache brunâtre et porte de nombreux poils noirs et quelques soies de la même couleur, particulièrement longues sur les parties latérales. Les sutures transversales du thorax sont nettes.

L'abdomen compte cinq segments couverts de poils noirs; les deux premiers segments sont de couleur jaune pâle et les trois suivants de couleur brune; les deux derniers sont ornés, sur leur bord postérieur, de quelques soies identiques à celles du thorax. Le second segment est le plus long de tous et ne présente rien de remarquable.

Les ailes sont aussi de couleur jaune pâle avec des nervures très nettes et ne se superposent pas quand l'insecte est au repos. La quatrième nervure longitudinale se recourbe vers la troisième.

Les pattes sont de la même couleur que le thorax; elles sont couvertes de poils et présentent quelques soies, surtout à l'extrémité distale de chaque article.

Les caractères de la larve sont les suivants : elle est blanche, compte onze segments, dont le dernier est un peu plus grand que les autres et présente deux pores terminaux très nets. Le premier segment est divisé transversalement en deux parties, la seconde logeant la partie antérieure qui est de forme conique; les pièces buccales sont réduites à deux crochets durs et noirs, placés de part et d'autre de la ligne médiane. Cette extrémité peut être projetée ou rétractée et sert de point de fixation quand l'animal progresse. La partie la plus large du corps correspond aux deux derniers segments. La face dorsale est ornée de nombreuses et petites épines disposées sans régularité et dirigées en rétroversion; la face ventrale est dépourvue de fausses pattes et est très semblable à la face dorsale.

D'après VILLELA, cette mouche dépose sur la peau un œuf d'où éclot une larve qui pénètre dans le tissu sous-cutané. La lésion ainsi produite est bénigne; c'est une tumeur inflammatoire ayant l'apparence d'un furoncle, d'où sort la larve après six à sept jours.

Nous pouvons citer ensuite les larves d'*Auchmeromyia luteola* FABR., dont le parasitisme spécial s'exerce probablement aussi bien sur certains animaux domestiques que sur l'homme. Mais les genres *Cordylobia* et *Auchmeromyia* ne paraissent pas être les seuls en Afrique qui renferment des espèces à larves parasites. La preuve en est fournie par une observation du Dr ROVERE (10), qui a recueilli à maintes reprises, sur des bovidés, des larves dont il a eu le grand mérite de suivre le développement de manière à en obtenir l'imago. Celui-ci, soumis pour identification au Prof[r] BEZZI, a été reconnu comme étant *Chrysomyia* (*Pycnosoma*) *megacephala* FABR. ROVERE a fait de ces larves une étude très complète, qui apporte une contribution importante à nos connaissances sur la biologie des mouches du genre *Pycnosoma* et relate le premier cas bien observé de parasitisme des larves de ces mouches, qui étaient généralement considérées comme pondant leurs œufs sur les matières organiques en voie de décomposition. Nous n'avons trouvé que de rares mentions du parasitisme des larves de *Pycnosoma*.

SANT'ANNA (11) signale qu'à Lourenço-Marquès abondent de petites mouches mesurant de 7 à 9 millimètres, de couleur verte, à reflets bronzés sur le thorax et les deux derniers anneaux abdominaux, à ailes hyalines et à pattes noires, qu'il rapporte avec quelque doute à *Pycnosoma putorium* WIED., qui est une petite espèce fort commune sur la côte occidentale d'Afrique, et il ajoute : « Elles déposent leurs larves sur les substances animales en voie de putréfaction et parfois dans les plaies et ulcères souillés et infectés, comme il est fréquent de le voir chez les indigènes traités par leurs médicastres ».

D'autre part, BRUMPT (12) rapporte que les larves de *Pycnosoma putorium* parasitent souvent l'homme et les animaux en Abyssinie.

Les cas de myiase cutanée dus à *Pycnosoma megacephala* paraissent devoir être assez fréquents au Congo. Depuis que le Dr ROVERE a étudié le premier cas en notre laboratoire, nous avons reçu du Dr BRODEN, de Léopoldville, de semblables larves provenant de différents cas observés par lui sur des bovidés. Mais *P. megacephala* n'est certainement pas la seule espèce qui puisse se développer dans de semblables conditions, car dans l'envoi du Dr BRODEN nous avons trouvé trois autres espèces de larves recueillies en grand nombre, chez des bœufs, au niveau de plaies superficielles de la peau et couvertes d'une croûte plus ou moins épaisse. Nous les distinguerons par les désignations larve 1, larve 2 et larve 3, en attendant leur détermination définitive dont notre confrère SURCOUF a bien voulu se charger (1).

La larve 1 reproduit très exactement les caractères de conformation attribués par ROVERE à sa larve de *Pycnosoma megacephala;* elle n'en diffère que par des dimensions plus faibles; elle mesure en moyenne 12 millimètres de long sur $2^{mm}5$ de large; sa pupe atteint $7^{mm}5$ de long sur 3 millimètres de large.

La larve 2 rappelle aussi la larve de ROVERE; elle en diffère cependant en ce qu'elle est moins trapue :

Le maximum de largeur du corps s'observe au niveau du huitième ou neuvième anneau, à partir duquel le corps s'amincit régulièrement en avant, tandis qu'en arrière les anneaux 10 et 11 conservent sensiblement la même largeur, et le douzième est à peine réduit. Le corps aurait ainsi une forme cylindro-conique élancée, s'il ne présentait une double incurvation en forme de S allongé, qui amène l'anneau céphalique vers la face ventrale et reporte l'anneau anal vers la face dorsale. L'incurvation antérieure n'affecte que les deux premiers anneaux, tandis que l'incurvation postérieure intéresse tout le restant du corps, de sorte que, vue par le côté, la larve montre une ligne dorsale concave et une ligne ventrale convexe. Il en résulte que l'aire stigmatique postérieure est ainsi dirigée qu'elle s'observe presque de champ quand on examine la larve par la face dorsale.

L'armature épineuse est moins puissante que dans la larve de ROVERE : les épines sont plus petites et les ceintures épineuses sont moins proéminentes, du moins les deuxième, troisième et quatrième. Ces ceintures, au nombre de 11, occupent le bord antérieur des segments 2 à 12. Sur la face ventrale, les ceintures 5 à 11 sont traversées par un sillon transversal qui est dépourvu d'épines et qui semble les dédoubler; à la face dorsale, les ceintures sont simples et sensiblement aussi développées qu'à la face ventrale, sauf la ceinture 9, qui est réduite à une couple de rangées d'épines, et les ceintures 10 et 11, qui font défaut.

Les épines sont disposées assez régulièrement en quinconce, surtout sur les ceintures antérieures; elles sont dirigées en rétroversion.

Sur les faces latérales, les ceintures 5 à 10 détachent des champs

---

(1) En réalité, les matériaux recueillis par le Dr BRODEN se rapportent à quatre espèces différentes de *Pycnosoma*, comme il résulte de l'étude des imagos faite par SURCOUF. Notre larve 1 appartient à *Pycnosoma putorium*, notre larve 2 à *Pycnosoma marginale* et notre larve 3 à une espèce nouvelle; la quatrième espèce, qui n'était représentée qu'à l'état de puparium et d'imago, est également nouvelle. La description de l'imago des deux espèces nouvelles sera faite ultérieurement par notre excellent ami SURCOUF, que nous remercions vivement pour la détermination qu'il a bien voulu faire de ces matériaux diptérologiques.

latéraux spinulés, qui augmentent de taille du cinquième au septième pour diminuer du huitième au dixième, qui est fort réduit.

La face ventrale des anneaux 5 à 11 est traversée par un double sillon.

Le segment céphalique porte deux renflements antennaires séparés à leur base par un sillon et ornés d'une double papille ocellaire proéminente; deux crochets buccaux noirs flanquent à droite et à gauche l'orifice buccal, qui se présente sous la forme d'une fente longitudinale bordée par deux lèvres latérales, dont la surface présente une série de stries rayonnantes. En arrière, l'anneau céphalique est limité par un bourrelet spinulé qui constitue la première ceinture épineuse. Ce bourrelet, à raison de l'incurvation de l'extrémité antérieure, est plus développé vers la face dorsale que vers la face ventrale; il est armé de minuscules épines serrées, disposées régulièrement en séries obliques. Sur les faces latérales du second segment viennent s'ouvrir les stigmates antérieurs, qui se présentent sous forme de deux organes flabelliformes fort proéminents, sur le bord libre desquels sont disposés onze à treize petits pertuis.

Le segment anal comporte deux parties : une partie ventrale, qui continue la ligne du corps et sur laquelle s'observe l'anus sous forme d'une fente longitudinale bordée de deux lèvres proéminentes entourées par un rebord de contour triangulaire, armé de petites épines, et flanquée à droite et à gauche d'une forte papille conique dirigée latéralement en dehors. La partie dorsale est coupée obliquement du haut en bas et d'avant en arrière, de manière à constituer une aire ovalaire délimitée par un rebord épais portant douze papilles, six dorsales et six ventrales, et armé de minuscules épines disposées en séries rectilignes nombreuses. Vers la partie supérieure de cette aire, s'observent les stigmates postérieurs sous forme de deux plaques assez régulièrement arrondies : le péritrème dessine une ligne noirâtre entourant trois fentes stigmatiques à peu près rectilignes, convergeant légèrement en bas et en dedans.

La coloration du corps varie du blanc sale au brun plus ou moins foncé. Les dimensions peuvent atteindre 18 millimètres en longueur et 3 millimètres en largeur.

La pupe, de coloration brun foncé, a la forme assez régulière d'un tonnelet avec un rétrécissement au niveau du cinquième anneau; la spinulation s'y reconnaît avec les caractères décrits chez la larve. Le tégument du puparium est épais et opaque.

La pupe mesure 10 millimètres de long sur $3^{mm}5$ à 4 millimètres de large.

La larve 3 se fait remarquer par le faible développement de son armature chitineuse qui ne s'observe qu'à l'aide d'un grossissement assez fort. Elle est de coloration blanc crême et présente douze segments. Le corps est plus ou moins claviforme avec l'extrémité incurvée vers la face ventrale, cette incurvation intéressant les trois premiers segments; la ligne dorsale du corps est rectiligne, tandis que la ligne ventrale est légèrement convexe. Le diamètre maximum s'observe au niveau des anneaux 9 et 10, il se maintient sans réduction apparente en arrière et en avant jusqu'au sixième anneau; l'amincissement du corps ne s'accuse qu'à partir du cinquième anneau et augmente sur les trois segments antérieurs dont l'ensemble simule un cone incurvé. Les anneaux 6 à 11 présentent à la face ventrale un sillon transversal.

Onze ceintures épineuses formées de toutes petites épines en rétroversion, associées en séries plus ou moins nombreuses disposées en lignes transversales ondulées, séparent les anneaux du corps sans qu'il soit possible de décider si elles occupent leur bord antérieur ou leur bord postérieur. Elles entourent complètement le corps, sauf les neuvième et dixième qui sont interrompues au niveau de la face dorsale; déjà la huitième n'y est plus représentée que par quelques rares séries de minuscules épines. A la face ventrale, les ceintures 5 à 11 sont traversées par une ligne dépourvue d'épines.

Les champs latéraux épineux font défaut.

Sur les faces latérales du deuxième segment se voient les stigmates antérieurs sous forme d'organes flabelliformes légèrement proéminents, au bord libre desquels viennent s'ouvrir huit pertuis.

Le segment céphalique porte deux renflements antennaires coniques surmontés de deux papilles munies d'une tache ocellaire. Deux crochets noirs flanquent, à droite et à gauche, la bouche qui apparaît sous la forme d'une fente longitudinale. Les parties latérales montrent des stries rayonnantes.

L'extrémité postérieure du corps est coupée obliquement de bas en haut et d'avant en arrière pour former l'aire stigmatique. Cette aire est délimitée par un rebord orné de multiples épines minuscules et hyalines et surmonté de douze petites papilles. A la face ventrale du douzième segment se trouve situé l'anus sous la forme d'une fente longitudinale délimitée par un double bourrelet et entourée d'une aire chargée d'épines. A droite et à gauche s'observe une forte papille conique à direction latérale.

Les stigmates postérieurs sont situés dans la partie supérieure de l'aire stigmatique et se présentent sous la forme de deux plaques délimitées par un péritrème brun, de contour régulièrement circulaire en dehors, légèrement étiré en dedans et en bas, où s'observent les faux stigmates. A l'intérieur du péritrème se voient trois fentes longitudinales presque rectilignes, disposées obliquement de haut en bas et de dehors en dedans.

Les dimensions de la larve peuvent atteindre 14^{mm}5 en longueur et 2 millimètres en largeur.

La pupe a la forme régulière d'un tonnelet avec un léger étranglement au niveau du cinquième segment; le tégument en est brun clair, translucide. Il présente l'ornement épineux de la larve.

La pupe mesure 8 millimètres de long sur 2^{mm}5 de large.

Les conditions dans lesquelles se réalise l'infestation par les larves de *Pycnosoma* ne semblent pas encore précisées. Le Dr Rovere est disposé à admettre que la femelle pond ses œufs en un point quelconque du corps de l'hôte et que les larves pénètrent secondairement dans et sous la peau; le Dr Broden (13), au contraire, se déclare convaincu que la ponte est toujours faite au niveau d'une plaie préexistante, et il appuie sa manière de voir sur un cas observé chez un bœuf, où il recueillit des larves de *Pycnosoma megacephala* au niveau d'une plaie qui avait été produite quelques jours avant, par déchirure de la peau à la base de la queue, en manipulant violemment l'animal pour examen.

Cette manière de voir se trouve corroborée, d'autre part, par la déclaration du Dr Sant'Anna, que nous avons reproduite plus haut. Quoi qu'il en soit, les deux modes d'infestation se trouvent réalisés par les Calliphorines : le premier, par *Lucilia sericata, Calliphora dux, C. oceanicæ* et par *Cordylobia;* le second, par *Lucilia Cæsar, Calliphora vomitoria* et *Chrysomyia macellaria.* De nouvelles observations détermineront le mode exact réalisé par les *Pycnosoma.*

De cette rapide revue que nous venons de faire il résulte que six genres de Calliphorines sont connus dont les larves peuvent se rencontrer à l'état parasitaire chez les animaux domestiques : *Calliphora* en Europe, Amérique et Australie, *Lucilia* en Europe, *Chrysomyia* en Amérique, *Cordylobia* et *Auchmeromyia* en Afrique et *Pycnosoma* en Asie et en Afrique. Il est vraisemblable que d'autres genres de Calliphorines possèdent des mœurs analogues. A raison de l'intérêt qui s'attache à cette question, il est éminemment désirable de voir recueillir toutes les larves cuticoles de Diptères et d'en faire l'éducation de manière à en obtenir les imagos et en permettre ainsi la détermination exacte.

BIBLIOGRAPHIE

1. Austen, E. E., Some dipterous insects which cause myiasis in man. (*Transactions of the Society of tropical medicine and hygiene,* III, 5, 1910, p. 215.)

2. Blanchard, R., Contributions à l'étude des Diptères parasites, 3e série. (*Annales de la Société entomologique de France,* LXV, 1896, p. 641.)

3. Depied, La *Lucilia hominivorax* au Tonkin. (*Archives de médecine navale,* LXVII, 1897, p. 127.)

4. PATTERSON, R. L., An Indian Screw-worm. (*The Indian Medical Gazette*, XLIV, 10, 1909, p. 374.)

5. GAIGER, S. H., A preliminary check list of the parasites of Indian domesticated animals. (*The Journal of tropical veterinary Science*, V, 1, 1910, p. 65.)

6. FROGGATT, W. W., The Sheep Maggot Fly, with notes on other common Flies. (*Agricultural Gazette of N. S. Wales*, January, 1905.)

7. VAN DINE, D. L., et NÖRGAARD, V. A., Abstract of a preliminary report on insects affecting live-stock in Hawaii. *Proceedings Hawaii Live-stock Breeders' Association*. V, 1907, p. 19.)

8. VILLELA, A., Descrição da larva de uma muscidea que apparece na cidade da Praia de Cabo Verde, na epoca das chuvas. (*Archivos de Hygiene e Pathologia exoticas*. I, 3. 1907, p. 402.)

9. WELLMAN, F.-C. Lettre particulière en date du 1er mai 1906.

10. ROVERE, J., Étude de larves cuticoles appartenant au genre « *Chrysomyia* », observées au Congo belge. (*Bulletin agricole du Congo belge*, I, 1, 1910, p. 26.)

11. SANT' ANNA, J. F., Trabalhos do Laboratorio do Hospital de Lourenço Marques; Provincia de Moçambique, Relatorio do Serviço de Saude, Anno de 1908, Lourenço Marquès, 1909.

12. BRUMPT, E., Précis de parasitologie Paris, 1910.

13. BRODEN, A., Lettre particulière en date du 27 janvier 1910.

Ueber die generische Hinzugehörigkeit der bis jetzt beschriebenen Pachygaster-Arten,

von Dr K. KERTÉSZ (Budapest).

Der allgemein bekannte Habitus von *Pachygaster atra* PANZ. spielte eine grosse Rolle dabei, dass sämtliche Arten von auch nur annähernd ähnlicher Form in diese Gattung einverleibt wurden. Dieser Umstand hatte auch zur Folge, dass die Arten von den Autoren nicht so eingehend untersucht wurden, wie es erforderlich gewesen wäre. Abgesehen von LOEW, JAENNICKE, BRAUER und anderen, weise ich nur auf die vor kurzem erschienene Monographie der englischen Arten von VERRALL hin, in welcher sehr wichtige Merkmale, wohl nur aus Unkenntniss der exotischen Formen, als Artcharaktere betrachtet werden.

Bei den *Stratiomyiiden* sind wir fast ausschliesslich auf plastische Merkmale angewiesen, insbesondere spielt der Bau des Kopfes, die Stellung und Form der Fühler, die Beschaffenheit des Schildchens, sowie das Flügelgeäder, eine wichtige Rolle.

Wenn wir nun die als *Pachygaster* beschriebenen Arten näher ins Auge fassen, finden wir unter denselben die verschiedensten Formen vermengt.

Im Jahre 1901 wurden zwei neue Gattungen aufgestellt, die uns am nächsten interessiren. Die eine ist *Neopachygaster* AUSTEN, die zweite *Zabrachia* COQUILLETT. Als Typus für *Neopachygaster* betrachtet AUSTEN die Art *meromelas* L. DUF., dagegen wurde *Zabrachia* von COQUILLETT auf eine neue Art : *polita* COQUILL. gegründet.

Die beiden Arten betreffend, will ich hierorts nicht näher auf die Synonymie eingehen, denn ohne Kenntniss der Typen ist die Frage nicht sicher zu lösen. Ich nehme — nach VERRALL — vorläufig an, dass die von AUSTEN als *Neopachygaster meromelæna* beschriebene Art tatsächlich *orbitalis* WAHLBG. entspricht, und will vorläufig glauben, obzwar es sehr zweifelhaft ist, dass *polita* COQUILL. eine selbstständige Art sei, bemerken will ich nur, dass nach der Gattungsdiagnose unsere *Pachygaster minutissima* ZETT. in die Gattung *Zabrachia* zu stellen wäre.

Betreffs der übrigen Arten, soweit mir dieselben vorliegen, oder wenigstens aus den Beschreibungen bekannt sind, kann ich folgendes feststellen :

P. limbipennis v. D. WULP gehört auf Grund des fast ganz flachen bandartigen Hinterleibes in eine ganz andere Gruppe der Pachygastrinen.

P. lativentris v. D. WULP ist wegen der Abwesenheit des Hinterastes der Radialis und der ganz ungewöhnlichen Kopfform aus der Gattung auszuscheiden.

P. infurcata MEIJ. muss wegen Abwesenheit des Hinterastes der Radialis, des kurzen Kopfes, ausserordentlich hoch stehenden Fühlern und sehr stark entwickelten Punktaugenhöckers ebenfalls ausgeschlossen werden.

P. albipes BRUN. gehört in dieselbe Gattung wie die vorige Art, wie ich das nach einem typischen Exemplar feststellen konnte.

P. pulchra LW. hat — wenigstens im weiblichen Geschlechte, das Männchen kenne ich nicht — einen kurzen Kopf mit hochstehenden Fühlern.

P. maculicornis HINE wird kaum in der Gattung verbleiben können, in welche sie der Autor auch selbst nur provisorisch einreiht.

P. rufitarsis MACQ. gehört, falls die vom Autor gegebene Abbildung des Flügels richtig ist, vielleicht zu den Sarginen.

Es bleiben also zur näheren Besprechung nur vier paläarktische

Arten übrig, nähmlich : *atra* PANZ., *Leachi* CURT., *orbitalis* WAHLBG. und *tarsalis* ZETT.

Als Hauptmerkmal für die Trennung der paläarktischen Arten wird allgemein die Färbung der Flügel betrachtet. Die erste Gruppe bilden die Arten mit ganz klaren Flügeln, die zweite diejenigen mit an der Basalhälfte geschwärzten Flügeln. Letztere Flügelfärbung kenne ich bisher nur bei den Pachygastrinen; dieselbe scheint für diese Subfamilie charakteristisch zu sein und wird in den Gattungen *Blastocera* GERST., *Spyripoda* GERST. und *Craspedometopon* KERT. angetroffen.

Meiner Auffassung nach besitzen wir jedoch viel wichtigere Charaktere, die darauf hinweisen, dass hier generische Trennungen vorzunehmen sind, die ich im Folgendem kurz zusammenfasse :

1 (2) Fühler hoch über der Mitte des Profiles eingefügt. Augen in beiden Geschlechtern breit getrennt; die Stirn nimmt beim ♂ ca. $^1/_5$ der Kopfbreite ein (6 : 31), beim ♀ ca. $^1/_3$ (8 : 29).

Neopachygaster AUST.
Spec. typ. *orbitalis* WAHLBG.

2 (1) Fühler unter der Mitte des Profiles eingefügt. Augen des Männchens zusammenstossend, die des Weibchens breit getrennt.

3 (4) Schildchen so lang wie breit, parabolisch, deutlich gerandet, am Rande mit vielen kleinen Dörnchen bewaffnet, horizontal liegend. Cubitalquerader fehlend. Die oberen Augenfacetten des Männchens viel gröber als die unteren, mit deutlicher Teilungslinie, die des Weibchens gleichgross.

Eupachygaster, nov. gen.
Spec. typ. *tarsalis* ZETT.

4 (3) Schildchen stumpf dreieckig bis halbkreisförmig, nicht gerandet, am Rande ohne Dörnchen, mehr oder weniger aufgerichtet. Cubitalquerader vorhanden. Augenfacetten des Männchens nicht geteilt, in beiden Geschlechtern ziemlich gleichgross.

Pachygaster MEIG.
Spec. typ. *atra* PANZ.

Zu Pachygaster s. str. rechne ich vorläufig auch die Art *Leachi*

CURT., die ich nicht ausscheiden möchte, bevor es mir möglich gewesen einige exotische Formen heranzuziehen. Ich weise aber schon jetzt auf die grossen Unterschiede hin, die in der Flügel- und Beinfärbung und in der sehr verschiedenen Kopfbildung beider Arten bestehen.

The control and disinfection of imported seeds and plants,

by Sir DANIEL MORRIS, K. C. M. G., D. Sc., F. L. S.,

Late Imperial Commissioner of Agriculture in the West Indies.

The object of this paper is to afford information respecting the successive steps that have been taken in the British West Indies to establish a system of control of imported seeds and plants, and prevent the introduction of insect and other pests (1). As these were possibly the first measures of the kind carried out in the tropics a brief account may prove of interest at this first Congress of International Entomology. The earlier efforts in the West Indies were made during the years 1884 to 1898, when all the colonies, with the exception of British Guiana and Barbados, adopted laws for prohibiting the importation of seeds and plants from certain countries. The main object in view was to prevent the introduction of the coffee-leaf disease (Hemileia) from the East Indies. None of the laws, however, provided for the inspection, fumigation or disinfection of seeds or plants on arrival at their several destinations.

In the period from 1898 to 1908 under the advice and with the assistance of the Imperial Department of Agriculture systematic efforts were made to secure as far as possible effective control of all

---

(1) This summary is based on two articles by Mr. H. A. BALLOU, D. Sc., published in the *West Indian Bulletin*, vol. X, pp. 197-234 and pp. 349-372.

insect and fungoid pests. To assist in this direction an entomologist and a mycologist were added to the Department Staff.

It was pointed out that some of the most destructive pests and diseases known in present-day agriculture were those that had been transported from their native localities to other regions. Many of these have in all probability been carried on, or with their food or host plants, and have become established before their presence was known. Insect pests are likely to be much more destructive when transported to new localities, because they are not controlled, as in their native environment, by natural enemies. The subject of preventing the introduction of pests and diseases in connection with the importation of plants has been carefully considered in most countries where agricultural industries are of great importance. It is further pointed out that, as the West Indian Islands and the British Colonies generally depend almost entirely on agriculture for their welfare, it was essential that everything should be done to ensure protection from infestation from surrounding areas.

As an example of the regulations now in force, the following extract taken from the « Instructions for the guidance of officers » at Jamaica indicates the manner in which imported plants, cuttings and their coverings are dealt with :

Immediately on the landing of any plants, cuttings or other articles specified in the Governor's Proclamation of September 7th, 1901, published in the Government Notice No. 278, of the 10th of that month in « The Jamaica Gazette », they shall be taken charge of by the Customs Officer who will give the Wharfinger or other party concerned a receipt therefor, showing the time and date of delivery;

The Customs Officer shall at once notify the Government Chemist in writing of the articles to be fumigated, stating the approximate dimensions thereof and obtain his instructions as to the time at, and place to, which they are to be forwarded for fumigation;

The Customs Officer will then forward the articles accordingly in charge of a Customs Escort, who will remain in attendance during the process of fumigation and afford, or provide, such assistance and labour as the Government Chemist or his officer in charge may require;

Immediately on receipt of the articles, the Government Chemist (or his assistant) shall cause them to be fumigated in the manner and under the conditions prescribed by the Governor in Privy Council;

So soon as this shall have been done and a memorandum showing the time of receipt and delivery furnished to the Customs Escort, the articles shall be taken charge of by the Escort and conveyed to the King's Warehouse or other place, as arranged by the Landing Waiter;

The greatest care must be taken by the Officer in charge of the King's Warehouse to keep plants, cuttings, etc., alive and in good condition;

All expenses of removing the articles to the Government Laboratory, and thence to the King's Warehouse, with any expenses necessarily incurred in keeping the articles in good condition, shall be met by the Importer, all such amounts beings brought to account as King's Warehouse fees as provided by the Customs Regulations on the subject;

Plants, cuttings, etc., should not be forwarded to the King's Warehouse in cases where importers defray expenses of removal, labour, etc., (if any) at once, and at the same time arrange with the Customs Officer to take delivery of the articles immediately after fumigation. This provision will refer more particularly to the plants, etc., brought by passengers and imported through the parcel post, etc.

At British Guiana the following regulations are in force :

Sugar-canes or cuttings thereof from Java, Australia, Fiji, Brazil, and the West Indian Islands shall not be permitted to be imported in any description of earth or soil;

All sugar-canes or cuttings thereof from the above mentioned places are to be inspected by the Government Botanist before being removed from the wharf or stelling at which they are landed, and are not to be removed from that place unless permitted by the Government Botanist in writing;

If, on such inspection, the sugar-cane or cuttings thereof be found to be not free from pests or diseases of any sort already known to occur in the Colony, the sugar-canes or cuttings are to be treated as the Government Botanist may direct before removal from the wharf. If they are found to be infected with any pest or disease not commonly known in this Colony, the sugar-canes or cuttings are to be destroyed under the supervision of the Government Botanist or an officer of the Department of Science and Agriculture delegated for that purpose;

If their removal is authorised by the Government Botanist, the sugar-canes or cuttings thereof are to be planted in a nursery apart from the general cultivation and separate from other varieties of canes, and to be subject from time to time during twelve months from the date of importation to inspection by the Government Botanist, or by an Officer of the Department of Science and Agriculture deputed by the Director for that purpose. If the canes are found at any inspection to be suffering from any pests or diseases already known in the Colony, they shall be treated as may be directed by the Government Botanist, and if suffering from any pests or diseases not commonly known in the Colony, they shall be rooted out and destroyed under the immediate supervision of the inspecting Officer.

At Barbados an Order issued in 1909 provides that « seed cotton » should not be imported or brought into the Colony from any country. In regard to cotton seed the Order provides that :

Cotton seed shall not be imported or brought into this Colony from the U. S. America, Cuba, Porto-Rico, Mexico, or Guatemala, except under a licence from the Governor-in-Executive Committee permitting any quantity, not exceeding 500 lb., to be imported at any one time, on the condition that such quantity is properly fumigated with carbon bisulphide before shipment, and that it will be fumigated or disinfected, or both, at the discretion of the Superintendent of Agriculture, or his Assistant, on its arrival in this Colony.

Cotton seed, from which oil is to be extracted, shall not be imported or brought into this Colony, except under the following conditions :

Such seed shall be fumigated either at the Government fumigatorium or in a building licensed for that purpose by the Governor-in-Executive Committee. Such seed shall be conveyed by the importer, immediately after it has been landed, to the Government fumigatorium or to the licensed building, under the supervision of an officer of the Local Department of Agriculture, who shall be immediately notified thereof by the Comptroller of Customs, and the fumigation shall be done at the expense of the importer in the presence and to the satisfaction of the Superintendent of Agriculture, or his assistant.

The provision for other seeds and plants is as follows :

All plants, cuttings, buds, grafts, bulbs, roots, seeds, and also fruit and vegetables intended for propagation and not for consumption as food, imported into the Colony from any country, together with the packages, boxes, wrappers, soil, or anything whatsoever in which they are imported, shall be delivered by the importer to the Comptroller or other Officer of Customs, or the Colonial Postmaster or other Officer of the Post-Office in order that they may be fumigated or disinfected or both fumigated and disinfected.

Where, however, in the opinion of the Superintendent of Agriculture, or his assistant, fumigation or disinfection or both fumigation and disinfection are not sufficient to destroy any insect pest or fungoid disease on any plant, cutting, bud, graft, bulb, root, seed, fruit and vegetable, or other article sent him for fumigation, he shall cause same article to be destroyed.

In the Windward and Leeward Islands the regulations in force provide for the absolute prohibition of seeds and plants from certain countries or places where injurious diseases are known to exist, but in the case of other countries seeds and plants are allowed to be imported after they have been fumigated or disinfected or both, according to the discretion of the Agricultural Authority.

At Bermuda a law passed in 1889 prohibits the importation of bulbs, except under regulations approved by the Board of Agriculture. The regulations provide for the « cleaning, disinfection, purification, and treatment of imported bulbs ». The term bulbs includes lily, hyacinth, and narcissus bulbs.

The provision of the several regulations above referred to may be summarised as follows : 1° total prohibition; 2° destruction of badly infested plants on arrival; 3° treatment of plants on arrival, to kill pests or diseases which may be known or suspected to be on them, and 4° inspection of suspected plants from time to time for a certain period, with power to destroy such plants if they are found within that time to be affected with any serious pest or disease. The term plant is intended to cover all plant-material, together with all packages, packing, etc. Thus all seeds, bulbs, roots, etc., if intended for propagation, are included, but fruits, vegetables, etc., intended to be used *at once* as food are not included at present.

Most of the laws quoted give the agricultural authority discretion in the matter of prescribing the treatment to be given to imported plants. This might also apply to the exemption from treatment of certain plants considered liable to be the means of introducing pests or diseases, such as tomato, cabbage and other vegetable seeds, and ordinary flower-garden seeds, sent out by reputable seedsmen in Europe or America. These are put up in sealed packets and are not likely to introduce any pest or disease which would seriously affect any of the crops in the West Indies. A problem still to be solved is that of treating plant material intended to be eaten, such as sweet-potatos. The Scarabee of the sweet-potato (*Cryptorynchus batatae*) is found in most, if not all, of the islands, and there is little danger of serious damage resulting from its transportation from one to another, but it affords an example of the way in which an insect pest might be introduced in plant material intended to be eaten. Another sweet-potato Weevil (*Cylas formicarius*) is common in certain of the Southern States and in Jamaica. It has recently been reported as occurring in British Guiana. This insect sometimes occurs as a serious pest, and might easily be transported with the potatos to new localities. *Cylas formicarius* is stated to have been found in Barbados, but there is no evidence of its occurrence at the present time.

The power to provide for careful inspection, and then to destroy plants found to be infested with pests and diseases which are not likely to be controlled by any available means of desinfection, is very useful, and might prove a means of preventing the introduction of new or little known diseases, as well as those which are well known. Various Borers would come under this head. Sweet-potatos infested with the Borer mentioned, and fruit infected by the Fruit-Fly might also be included here.

The Fruit-Fly (*Ceratitis capitata*) has been a serious pest in Bermuda, but legislation has been enacted, requiring the destruction of all fruit in the island for two seasons, and the pest has been brought under control. This was accomplished under the provisions of the Fruit-Fly Destruction Act, 1907, which came into force on March 1st, of that year.

Desinfection of imported plants.

The following is a brief summary of the recommendations that are being carried out in the case of seeds and plants imported into the several West Indian Colonies :

TREATMENT OF PLANT MATERIAL.

Plants. — Growing plants, in pots or packages, bud-wood, grafts, cuttings, suckers, seedlings, ferns, all packing material.

Treatment. — Hydrocyanic acid gas, normal charge (1 oz. potassium cyanide to each 300 cubic feet) for one hour. In the case of ferns and other tender plants exposure should be for a shorter time, say half an hour.

Plants. — Seeds, in packets, or in bulk; plants growing in soil or with roots packed in soil showing evidences of Ants, Mealybugs or Scale-insects, or other insects on roots; plants showing evidences of Borers in stems, or of Maggots under bark. Fruit, pods, etc., from which seeds are to be planted.

Treatment. — Carbon bisulphide (1 lb. to 1,000 cubic feet), if plants can be exposed to the action of the fumes for from twelve to twenty-four hours. A dose of 1 lb. to 200 cubic feet may be used for a short period, say one to two hours. For fumigating flower and garden seeds in packets and small parcels, make pin pricks in the paper, and use a teaspoonfull per cubic foot for two hours.

Plants. — Rose-cuttings, etc., from reliable dealers in temperate climates.

Treatment. — If Scales are seen, use hydrocyanic acid gas treatment. If no pests are seen, dip in tepid water to which a small amount of Bordeaux mixture has been added.

Plants. — Cane tops and cuttings, pine-apple suckers, any cuttings liable to bring fungoid disease as an external parasite.

Treatment. — Bordeaux mixture.

Plants. — Cotton seed for planting.

Treatment. — Corrosive sublimate (one part to 1,000 of water). From boll weevil countries, when entry is permitted, insist on certificate of fumigation before shipment, and on arrival, fumigate with carbon bisulphide, 1 lb. to 200 cubic feet for four hours.

Plants. — Cotton seed in bulk for oil extraction.

Treatment. — Sulphur dioxide applied by means of a Clayton apparatus, using a strength of 5 per cent, under pressure for twenty-four hours.

Die vergleichende Morphologie und Systematik der Coleopteren,

von Profr. HERMANN KOLBE, Berlin.

In älterer Zeit war die morphologische Betrachtung der lebenden Natur der damals üblichen Methode naturwissenschaftlicher Forschung sehr untergeordnet ; es wurden hauptsächlich die spezielle Anatomie und Bionomie der Lebewesen gepflegt. Die Entomologie hatte aber das Glück, dass bei dem Erwachen der Naturwissenschaften im Verlaufe des 17. Jahrhunderts sogleich hervorragende Männer auftraten, welche auf breiter Basis in ausgezeichneten Werken für die Entomotomie und Bionomie die Wege der Forschung vorzeichneten. Wir bewundern die wichtigen und völlig selbstständigen Arbeiten eines GOEDART, Metamorphosis et historia naturalis, Medioburgi 1662-68. 3 vol., editio tertia : Amstelodami 1700 ; REDI, Esperienze intorno alla generazione degl' Insetti, Firenze 1668 ; LISTER, Johannes Goedartus de Insectis (eine neue Ausgabe von GOEDART's Buch, mit Fortsetzungen und Ergänzungen), Londini 1685 ; SWAMMERDAMM, Historia Insectorum generalis, Utrecht 1669 ; idem, Bijbel der Natuure, of Historie der Insecten Accedit praefatio, in qua vitam auctoris descripsit HERMANNUS BOERHAAVE, Med. Prof., Leydae 1737-38. Deutsche Ausgabe unter dem Titel : Bibel der Natur, worinnen die Insecten in gewisse Classen vertheijt, sorgfältig beschrieben, zergliedert, in saubern Kupferstichen vorgestellt, mit vielen Anmerkungen über die Seltenheiten der Natur erläutert etc., Leipzig 1758 ; MALPIGHI, Dissertatio epistolica de Bombyce, Londini 1669 ; und VAN LEEUWENHOEK, De Bombycum erucis, earumque ovis, Leyden 1688 ; id. Arcana Naturæ detecta ope microscopiorum, Delphis Batavorum 1695 ; id. De ovario et cornea oculi Libellulæ, Delphis 1696.

Wir bewundern dieses Eindringen in die Natur und diese Kraft des Schaffens auf ganz ungepflegtem Boden um so mehr, als während des ganzen Mittelalters und bis in das 17. Jahrhundert hinein keine oder kaum eine selbstständige Regung auf dem Gebiete der Naturforschung zu bemerken ist. Nur Wiederholungen der Ueberlieferungen des ARISTOTELES galten als Regel, selten Ergänzungen durch eigene Production (ALBERTUS MAGNUS im 13. Jahrhundert; WOTTON und CONRAD GESNER im 16. Jahrhundert ; ALDROVANDUS : De animalibus insectis libri septem, Bononiæ 1602 ; MOUFET : Insectorum sive minimorum animalium theatrum, Londini 1634 ; JOH. JOHNSTON : Historia naturalis de Insectis libri III, Francofurti ad Moen 1653).

Die Morphologie der Insecten war durch jene Entomotomen des 17. Jahrhunderts indess noch kaum angebahnt. Aber bereits im Anfange des 18. Jahrhunderts gab JOHN RAY (von LINNÉ oft unter dem Namen *Rajus* citiert) seinen « Methodus Insectorum » (Londini 1705) heraus, in welchem das von vielen Seiten zuströmende Material bereits systematisch gegliedert und nomenclatorisch behandelt wurde. Doch weder dieser Vorläufer noch der bald auftretende Reformator der descriptiven Naturwissenschaft, CARL V. LINNÉ, verstanden sich genügend auf die eigentliche Morphologie, die eben noch in den Kinderschuhen steckte. Dies ist die Ursache der Fehler, die im Bereiche der damaligen Insectenkunde noch in auffälliger Weise in der systematischen Stellung vieler Gattungen vorkamen. So stellte LINNÉ die Forficuliden noch zu den Coleopteren, und in der Ordnung der Coleopteren-Arten von *Opatrum* und *Bolitophagus* (Gattungen der Heteromeren) zu *Silpha* (einer Gattung der Staphylinoideen) ; eine Art von *Apate* (einer Gattung der Bostrychoideen) zu *Dermestes* (einer Gattung der Dascylloideen) ; *Diaperis* (eine Gattung der Heteromeren) zu *Chrysomela* (einer Gattung der Phytophagen) ; *Trichodes* (eine Gattung der Trichodermaten) zu *Attelabus* (einer Gattung der Rhynchophoren) ; *Corynetes*, eine nahe Verwandte von *Trichodes*, zu *Dermestes*.

FABRICIUS baute mit grossem Fleisse die Systematik der Insecten weiter aus und zugleich auch die Formenkenntnis der zahlreichen von ihm aufgestellten Genera (in einer Reihe von Werken von 1775 bis 1805).

Aber erst der geniale LATREILLE, dieser scharfsinnige Morphologe und Systematiker, schuf in seinen Werken (« Précis des caractères génériques des Insectes, Bordeaux 1796 »; « Histoire naturelle générale et particulière des Crustacés et des Insectes, Paris 1802-1805 », u. s. w.) grundlegende Anschauungen über die Formenkunde des Insectenkörpers. Aus der Mitte des 18. Jahrhunderts stammen RÉAUMUR's « Mémoires pour servir à l'histoire naturelle et à l'anatomie des Insectes », 6 Bände, Paris 1734-1742; und LYONET's « Traité anatomique de la Chenille qui ronge le bois de saule, IIe éd. Haag, 1762 », sowie andere entomotomische Arbeiten aus den Jahren 1829-1832. Es würde zu weit führen, hier noch andere Morphologen und Systematiker aus dem ersten Viertel oder Drittel des 19. Jahrhunderts aufzuzählen, welche die Formenkunde der Entoma förderten, z. B. LAMARCK, LEACH u. a. Die Morphologie und Systematik machten von jetzt an grosse Fortschritte.

KIRBY und SPENCE lieferten in ihrem bekannten Werke « Introduction to Entomology », neben zahlreichen Schilderungen aus dem Leben der Insecten im III. Bande (1826), eine generelle Beschreibung des Insectenkörpers und seiner Teile, die aber noch an manchen Unvollkommenheiten leidet. Dagegen hatte BURMEISTER in seinem vortrefflichen « Handbuch der Entomologie », I. Bd. (1832) die vergleichende Methode, die eine höhere Entwicklung der Morphologie bedeutet, auf breiter Grundlage eingeführt. Schon vorher hatte SAVIGNY sein ausgezeichnetes Werk über die Mundteile der Gliederfüsser in die Welt gesetzt (« Mémoires sur les animaux sans vertèbres, 1re Partie, 1^r^ Fasc. Théorie des organes de la bouche des Crustacés et des Insectes », Paris 1816). Auch AUDOUIN, ESCHSCHOLTZ, V. BAER, besonders auch STRAUS-DURCKHEIM, ERICHSON und BRULLÉ haben die Morphologie der Insecten gefördert.

WESTWOOD's Werk « Introduction to the modern Classification of Insects » (London 1839-1840) bedeutet für den Ausbau der Morphologie der einzelnen Insectenordnungen einen grossen Fortschritt. Es lässt indess eine allgemeine vergleichende morphologische Uebersicht der Insecten vermissen, legt aber die Morphologie der einzelnen Familien in vorzüglicher Weise dar.

LACORDAIRE schildert in seinem berühmten von CHAPUIS fortgesetzten Werke « Genera des Coléoptères » (Paris 1854-1876) zwar die Familien und Gattungen in morphologischer Beziehung ganz ausgezeichnet, aber man sucht bei ihm vergebens die allgemeine Morphologie der Coleopteren und in vergleichender Weise die Morphologie der Familien.

In neuerer Zeit, mit dem Strome der Darwinistischen Weltanschauung, kam die descendenztheoretische Methode und dadurch ein ganz neuer Aufschwung in der Morphologie auf, der besonders in FRIEDRICH BRAUER seinen Vertreter fand (« Systematisch-zoologische Studien » Wien 1885).

Mir selbst war es vergönnt, auf der Basis der Transmutation und Descendenz die Morphologie der Insecten in meinem Buche « Einführung in die Kenntnis der Insecten » (Berlin 1889-93) zu bearbeiten.

Die *vergleichende Morphologie der Coleopteren* ist es, die uns im folgenden beschäftigen soll.

Die Coleopteren scheiden sich in zwei grosse Gruppen, von denen die eine (die Adephagen) als die phylogenetisch tiefere zu betrachten, während die andere (die Heterophagen) aus der Wurzel der ersteren abzuleiten ist. Es giebt eine Reihe von morphologischen Verhältnissen, welche durch ihre vergleichenden Beziehungen die phylogenetische Deduction in vorstehendem Sinne ergeben. Schon das Larvenstadium liefert hierzu das Beweismaterial. Die Beine der Adelphagenlarve sind 5-gliedrig (Coxa, Trochanter, Femur, Tibia, Tarsus) und am letzten Gliede (dem Tarsus) meistens mit zwei Krallen versehen; die Beine der Heterophagen sind jedoch 4-gliedrig (oder beinlos), und ihr letztes Glied ist einkrallig. Die Viergliedrigkeit der Beine der Heterophagenlarve ist dadurch entstanden, dass das letzte Glied (der Tarsus) mit der oft ziemlich langen und kräftigen (einfachen) Kralle verwachsen ist und mit dieser zusammen krallenförmig erscheint. Spuren der ursprünglichen Trennung finden sich bei den Larven gewisser auf tiefer Stufe stehender Heterophagen, z. B. *Hylecœtus dermestoides*. VERHOEFF fand dass die Krallen der Larve dieser Spezies aus einem besonderen basalen Gliede und der eigentlichen Kralle bestehen. Bei einer jungen *Hylecœtus*-Larve ist der basale Teil der Kralle sogar wirklich abgegliedert. Dieser basale Teil entspricht dem 5. Gliede an den Beinen der Adephagenlarven. Eine Spur des 5. Gliedes findet sich auch an den Beinen der Larven anderer Heterophagen, besonders einiger Spezies von Staphyliniden, Hydrophiliden, Scarabæiden, Lagriiden, Œdemeriden, Tenebrioniden, etc. Bei näherer Untersuchung zeigen sich nämlich an dem basalen Teile der « Kralle » dieser Larven (etwa am Ende des proximalen Drittels oder gegen die Mitte hin) zwei Borsten (setæ), die sich nicht an den Krallen der Adephagenlarven finden. Dieses Borstenpaar muss dem terminalen Borstenpaare des Tarsus der Adephagenlarven entsprechen. SCHIÖDTE, dessen ausgezeichnetes Werk über Verwandlungsstadien der Coleopteren « De Metamorphosi Eleutheratorum Observationes » sehr genaue Figuren und Angaben über die Morphologie der Coleopterenlarven und Nymphen enthält, macht bereits auf ein abgeschnürtes Glied an einem Larvenbeine einer Histeridenspecies, *Hister unicolor*, der zu den Heterophagen gehört, aufmerksam, mit den Worten « pars basalis ungulae, mollior, rudimentum tarsi exhibens ». Das ist auch bei manchen Staphylinidenlarven der Fall (*Bledius*), an deren Beinen eine abgeschnürte « pars basalis » (also der Tarsus) erkennbar ist. Die Verwachsung der Ungula mit dem Tarsus bei den Heterophagen ist also recht deutlich.

Demnach ist die Ableitung der Heterophagen von den Adephagen hiermit erwiesen. Ueberhaupt muss das Larvenstadium in einem richtigen, also auf natürlicher Grundlage beruhenden Systeme wichtige Momente für das System bieten. Denn trotz mancher, auf Convergenz und physiologisch-biologische Ursachen zurückzuführender Aehnlichkeiten oder Unterschiede in der morphologischen Ausbildung, ist die fast elementare Körperbildung der Adephagenlarven, die *Campodea*-Form derselben ein Beweis für die tiefe Stellung der Adephagen gegenüber den Heterophagen, mit Ausnahme der untersten, an die Adephagen sich anschliessenden Stufe derselben (Staphylinoideen). Die Adephagen aber schliessen sich an neuropteroide Insecten an. Wie auffallend ähnlich sind nicht die Adephagenlarven den *Neuropterenlarven*, sowohl in der Körperform als auch teilweise in der Bildung der appendiculären Organe, z. B. der Beine (5-gliedrige Beine mit 2 Krallen)! Auch die Adulti der *Thysanuren* haben eine gleiche Bildung der Beinen.

Die *Imagines der Coleopteren* bieten eine reichliche Anzahl von Characteren für das vergleichend-morphologische Studium, für die Ableitung der Heterophagen von den Adephagen, für die Feststellung der Stufenfolge der grossen Gruppen innerhalb der Heterophagen. Wie die *Segmentierung des Körpers* die Grundlage für die Morphologie desselben ist, dessen tausendfältig verschiedenartige Bildung stets nur von einer Um- und Ausbildung dieser Segmente ausgeht, so ist es vorauszusehen, dass auch die verschiedenen Gruppen der Coleopteren uns einen Einblick in die Transmutation der Segmente des Rumpfes tun lassen. Aus der mehr elementaren Bildung des Körpers in manchen Gruppen der Coleopteren, sowie aus der mehr derivaten Bildung des gedrungenen Körperbaues in anderen Gruppen ist der Schluss zu ziehen, dass die Segmentierung des Körpers in diesen Fällen erhebliche Unterschiede aufweist. Tatsächlich ist die ventrale Basalgegend des Abdomens bei der Mehrzahl der Coleopteren eingebogen und durch das Postpectus und die Coxen des dritten Beinpaares verdeckt. Das Sternit des 1. Abdominalsegments ist überhaupt bei den Coleopteren fast immer so sehr rudimentiert, dass es meist ganz geschwunden ist. Nun kommt aber die wichtige Tatsache, dass bei den *Adephagen das basale* (zum 2. Segment des Abdomens gehörige) *Sternit frei liegt*, d. h. frei sichtbar ist, während dieses Sternit bei den *Heterophagen vollkommen versteckt* liegt, weil es mit dem Sternit des 3. Segments mehr oder weniger verschmolzen, ganz eingebogen und von den Coxen bedeckt ist. Es gehört also das 1. frei sichtbare Sternit des Abdomens bei den Adephagen zum 2., bei den Heterophagen zum 3. Segment. Aber auch in den inferioren Gruppen der Heterophagen sieht man wenigstens Teile des Sternits des 2. Segments frei liegen, oder dieses Sternit ist mit dem Sternit des 3. Segments nicht verschmolzen. Dies ist besonders bei manchen Staphyliniden, ferner bei den Synteliiden, Passaliden, Scarabæiden, Lyciden, Lampyriden, einigen Lymexyloniden und manchen Meloiden der Fall. Bei den larvenförmigen Weibchen einiger Lampyridengenera (*Lampyris, Lamprohiza*) und bei den ebenfalls larvenförmigen Weibchen von Driliden ist sogar das Sternit des 1. Abdominalsegments vorhanden und an den Seiten deutlich sichtbar. In allen übrigen Gruppen, besonders auf den mittleren und superioren Stufen des Systems, also bei den Sternoxien, Bostrychoideen, Heteromeren (einige inferiore Genera ausgenommen), Clavicorniern, Phytophagen und Rhynchophoren, ist die Verschmelzung der Sternite des 2. und 3. Abdominalsegments Gesetz geworden. Hier zählt man die Sternite des Abdomens erst vom 3. ventralen Halbsegment an, weil erst das 3. frei sichtbar ist. Das ist auch bei den Passaliden und Scarabæiden der Fall, obgleich das Sternit des 2. Segments an den Seiten sichtbar ist, wenn man die Elytren etwas hebt. Dagegen zählt man bei den Adephagen, bei denen das Sternit des 2. Segments beiderseits der hinteren Coxen breit und frei sichtbar ist, die Sternite des Abdomens schon vom 2. ventralen Halbsegment an. Diese Gewohnheit ist vom morphologischen Standpuncte aus ganz unhaltbar. Es ist daher hinsichtlich der praktischen Behandlung der Objecte notwendig, für die frei sichtbaren Sternite des Abdomens den Terminus « Ventralplatte » zu gebrauchen und dann nur die frei sichtbaren Sternite zu zählen. In der wissenschaftlichen Morphologie wird man die erste freie Ventralplatte des Abdomens bei den Adephagen als Sternit des 2. Abdominalsegments, bei den meisten übrigen Coleopteren als Sternit des 3. Abdominalsegments bezeichnen.

Die Verschmelzung eines Teiles der Segmente der Coleopteren auf den superioren Stufen des Systems lässt den Schluss zu, dass die Evolution des Insectenkörpers teilweise in der Verschmelzung von Segmenten und in der Verkürzung von Segmentcomplexen besteht. Die verkürzte Körperform findet sich hauptsächlich auf den superioren Stufen des Systems, sowohl auf den oberen Stufen der Adephagen, wie auf den höheren Stufen der meisten Familiengruppen der Heterophagen, am meisten aber auf den obersten Stufen des Systems (Chrysomeliden, Curculioniden).

Bemerkenswert ist auch das Sternit des Hinterkopfes, die *Gula,* welche gewöhnlich als « Kehle » bezeichnet wird. Diese ventrale Platte auf dem Hinterteile des Kopfes unterliegt gleichfalls einer descendenztheoretisch wichtigen Wandelung. Das Sternit des Hinterkopfes ist bei den meisten Coleopteren eine deutliche, beiderseits durch eine Naht von den Seiten des Kopfes geschiedene Platte. Bei den Adephagen ist die Gula sehr deutlich. Gross ist dieses Sternit des Kopfes bei den Scarabaeiden (Lamellicornier) und Cupediden, von denen besonders die letztere Familie sehr inferior erscheint. Dieser Tatsache steht als Extrem das Verhalten der Rhynchophoren gegenüber, bei denen, als höchster Stufe der Coleopteren, die Gula verschwunden ist. In dieser terminalsten Familiengruppe sind die beiden lateralen Suturen der Gula zu einer einzigen Sutur verschmolzen. Und selbst diese unpaare Naht konnte schwinden (bei den Brenthiden, einer lateralen Stufe der Rhynchophoren). Bei den meisten übrigen Coleopteren ist die Gula schmal, selten teilweise reduziert, weil die Nähte alsdann meistens obliteriren und grossenteils miteinander verschmolzen sind.

Mit der derivaten Ausbildung der verschiedenen Körperteile hält auch die Ausbildung des *Prothorax* gleichen Schritt. Bei der

Betrachtung des Prothorax ist es zunächst wichtig, die *marginale Bildung der Seiten* in das richtige Licht zu setzen. Bisher hat man den Wert dieses Teiles der Morphologie unbeachtet gelassen; ich glaube in meinen « vergleichend - morphologischen Untersuchungen an Coleopteren » (1901) zuerst darauf hingewiesen zu haben.

Es ist bemerkenswert, dass der Prothorax auf allen unteren Stufen des Systems an den Seiten deutlich und scharf gerandet ist: so bei den Adephagen, Staphylinoideen, Lamellicorniern, Malacodermaten, Palpicorniern, Dascylloideen, Sternoxien etc. Mit der inferioren Stellung nimmt die primitive Beschaffenheit des Prothorax zu. Das Pronotum ist oft derartig primär *schildförmig*, dass die Seitenränder scharfkantig vorspringen. Diese Bildung ist genau so bei zahlreichen Insecten tiefer und tiefster Rangstufen, besonders bei den Thysanuren, Forficuliden, Blattiden, Perliden etc. Im Gegensatz zu der primitiven Beschaffenheit des Prothorax in jenen Coleopterengruppen ist dieser Körperteil auf den obersten Stufen des Systems, besonders bei den *Rhynchophoren*, an den Seiten ganz ungerandet; *das Notum ist hier mit den Pleuren völlig verschmolzen*, ohne dass eine Naht (Sutur) übrig geblieben ist. Gewöhnlich ist bei den Curculioniden auch nicht einmal eine Spur von einer trennenden Linie zwischen dem Notum und den Pleuren zu sehen. Aber auch bei den Chrysomeliden ist die laterale Kante oft undeutlich; und bei den Cerambyciden ist nur auf den unteren Stufe, bei den Prioniden, eine deutliche marginale (oft gezahnte oder gehöckerte) Kante vorhanden, welche das Notum von den Pleuren trennt. Die übrigen Cerambyciden, die sämmtlich superiore Stufen einnehmen, besitzen jedoch keine marginale Kanten am Prothorax, weisen aber an deren Stelle häufig einen Höcker oder Zahn auf; oder (besser gesagt) die Mitte der Seiten ist in einen Höcker oder Zahn ausgezogen. Es giebt auch in tiefer stehenden Familien einzelne Gattungen, deren Notum und Pleuren fast verschmolzen sind : bei den Carabiden z. B. *Disphæricus* und *Dyschirius,* bei den Staphyliniden z. B. *Pæderus, Stenus, Stilicus*. Alle diese Gattungen sind als derivat anzusehen und nehmen innerhalb ihrer Gruppe eine superiore Stellung ein.

Es gilt hier das Gesetz der Morphologie, dass *die Verschmelzung von Körperteilen eine superiore Stellung* oder mindestens eine speziale derivate Entwickelung bedeutet.

Auch der Grad der *Verschmelzung der übrigen Bestandteile des Prothorax,* nämlich des Sternums mit den Pleuren, und an diesen das Verhältnis des Episternums zu dem Epimeron, sind sehr wichtige morphologische Tatsachen. Am deutlichsten und vollständigsten sind die getrennten Teile des Prothorax bei den Carabiden conservirt, wo die Nähte zwischen den Bestandteilen des Prothorax (mit wenigen Ausnahmen) sehr deutlich erkennbar sind. Bei den Heterophagen tritt eine teilweise oder völlige Verschmelzung dieser Bestandteile des Prothorax ein. Die Cucujiden zeigen in einigen Gattungen noch eine primäre Trennung dieser Teile wie bei den Adephagen. Das bedeutet keine nähere Verwandtschaft, sondern nur eine gleiche morphologische Entwickelungstufe. Auch bei den Lamellicorniern ist am Prothorax das Sternum von den Pleuren noch deutlich getrennt, aber das Epimeron ist mit dem Episternum bereits innig verschmolzen. Das ist auch bei fast allen übrigen Angehörigen der Heterophagen der Fall. In der am höchsten stehenden Abteilung der Rhynchophoren sind diese Teile des Prothorax, also die Pleuren mit dem Sternum und mit dem Notum und die Teile der Pleuren alle miteinander verschmolzen. Ausgenommen davon sind innerhalb der Abteilung der Rhynchophoren nur die auf tiefer Stufe stehenden Rhinomaceriden, bei denen das Sternum und die Pleuren am Prothorax noch durch eine deutliche Sutur getrennt sind. Diese Familie hat auch wegen ihrer schlanken Palpen, des deutlich vortretenden Labrums und der freien Ventralplatten des Abdomens eine tiefe Stellung in der Abteilung der Rhynchophoren.

Zunächst beschäftigen uns nunmehr die *appendiculären* oder *Anhangsorgane des Rumpfes*, von denen die *Antennen* wegen der verschiedenartigen localen Insertion und der mannigfaltigen morphologischen Ausbildung Beachtung von Seiten des Forschers verlangen. Sie tragen in diesen Beziehungen einen deutlichen evolutionistischen Character zur Schau und haben auch in vielen Fällen für die Systematik Wert. Es ist ferner nicht unwichtig, durch die aus dem Studium der Embryonen gewonnenen Resultate den richtigen Weg zu suchen, um die Morphologie der Imagines zu verstehen. Ich suchte und fand diesen Weg bereits im Jahre 1901 (1).

---

(1) Kolbe, H. J., Vergleichend-morphologische Untersuchungen an Coleopteren nebst Grundlagen zu einem System und zur Systematik derselben (*Archiv für Naturgeschichte*, Jahrg. 1901, Beiheft : *Festschrift für* Eduard von Martens, pp. 89—150, mit 2 Taf.), p. 91.

Beim Embryo sind die Antennen hinsichtlich der Lagerung und morphologischen Beschaffenheit den Palpen und Beinen homolog; sie erscheinen noch nicht differenziert und liegen postoral; sie sind auch wie die übrigen Extremitäten nach unten gerichtet. Bei fortschreitender Entwickelung des Embryo rücken die Antennen neben den Mund, um schliesslich vor oder über demselben zu inserieren. (Vergl. KORSCHELT und HEIDER, « Lehrbuch der vergleichenden Entwicklungsgeschichte der wirbellosen Tiere ».)

Wenn wir die vergleichende Morphologie der Antennen der Imagines hinsichtlich ihrer Insertion auf das Embryonalstadium ausdehnen, so finden wir, dass in zahlreichen Coleopterengruppen die Antennen wie beim Embryo unter dem vorspringenden Seitenrande des Kopfes hinter der Basis der Mandibeln inserieren. Diese inframarginale Insertion findet sich als charakteristisches Merkmal bei den Carabiden, Staphylinoideen (abgesehen von Ausnahmen), Scarabæiden, Rhipiceriden, Elateriden, Buprestiden u. s. w. In anderen Familien sind die Mandibeln hinter der Wurzel der Mandibeln eingefügt und bei noch anderen die Stirn hinaufgerückt (also subfrontal oder frontal). Die Cicindelinen mit ihren frontal inserierten Antennen bilden einen superioren Seitenast am Stamme der Carabiden, deren Antennen sonst gesetzmässig inframarginal inseriert sind. Bei den allermeisten Clavicorniern sind die Antennen inframarginal, besonders bei den Cucujiden, Sphaeritiden, Nitiduliden, Ostomiden, Mycetophagiden und Phalacriden, teilweise auch bei den Colydiiden und Lathridiiden. Aber bei den Erotyliden, Endomychiden, einem Teile der Colydiiden, Lathridiiden und Coccinelliden sind die Antennen frontal oder subfrontal inseriert. Jene Clavicornierfamilien und Gruppen haben also eine inferiore, diese eine superiore Stellung im Systeme. Unter den Cerambyciden besitzen die Prioninen mit ihren inframarginalen Antennen eine inferiore Stellung, die Cerambycinen und Lamiinen mit ihren frontalwärts stehenden Antennen eine superiore Stellung im Systeme; ebenso die Chrysomeliden. Die Antennen der Rhynchophoren sind supramarginal (frontal oder subfrontal) inseriert.

Hinsichtlich ihrer Formen zeigen die Antennen eine einfache (elementare) Bildung auf inferioren Stufen (borsten-, faden- und perlschnurförmige Antennen) und differenzierte (langschäftige und am apicalen Ende durch Hypertrophie verstärkte) Antennen auf höheren Stufen, auch auf höheren Stufen inferiorer Familiengruppen (Histeriden, Scarabæiden). Weil die terminale Differenzierung der Antennen auf physiologische Ursachen zurückzuführen ist, so hat diese Bildung für das System nur secundären Wert. Auch als appendiculäre Organe stehen sie den Bildungen des Rumpfes an primärem Werte nach.

Die Theorie von der Verschmelzung getrennter Teile in aufsteigender Evolution bestätigt sich auch in der *Nervatur der Flügel*. Die Nervatur der Flügel der Adephagen enthält gleichfalls Merkmale, welche die inferiore Stellung dieser Coleopteren im Systeme bestätigen. Die Subbrachialis verläuft noch regulär und ist bis in die Basis des Flügels vollständig und deutlich conserviert. Sowohl zwischen den vorderen Adern (Brachialis und ihrem Ramus und der Subbrachialis) wie zwischen der Subbrachialis und der Mediana befinden zich 1 oder 2 Transversaladern. Die Nervatur der Flügel ist also bei den Adephagen noch mehr oder weniger so primär, dass die sogenannten rücklaufenden Adern (venae recurrentes) nicht zur Ausbildung gelangt sind. Die rücklaufenden Adern sind nur in der grossen Subordo der Heterophagen ausgebildet; sie erscheinen wie eine derivate Verbildung, welche bei den Heterophagen herrschend geworden ist. Nur die Staphylinoideen und besonders die Cupediden haben ein primitives Flügelgeäder. Bei den Heterophagen ist die derivate Ausbildung der Flügeladern (Alarvenen) folgendermassen entstanden. Es sind statt der teilweise primär verlaufenden Adern und verbindenden Transversaläderchen einige der Längsadern teils teilweise verschmolzen, teils am Ende obliteriert und secundär durch Transversaladern miteinander verbunden oder am Grunde ausgelöscht. Die Subbrachialis ist am Grunde hier stets obliteriert und ausgelöscht. Die Transversaladern zwischen der Brachialis, dem Brachialramus, der Subbrachialis und der Mediana sind verbildet; die ursprünglichen Transversaladern in der Gegend des Gelenks sind teils obliteriert teils mit den benachbarten Längsadern (Brachialis und Mediana) hakenförmig verbunden, indem sie kräftig ausgebildet wurden und mit diesen Längsadern solide verwuchsen. Die basalwärts verlaufende Fortsetzung des vorderen Hakens ist der ursprüngliche Brachialramus, die Fortsetzung des hinteren Hakens ist die verkürzte Subbrachialis; es sind die beiden sogenannten rucklaufenden Adern (venae recurrentes), welche die Basis des Flügels nicht erreichen. Ihre Herkunft und Ableitung aus dem primitiveren Flügelgeäder der Adephagen in vorstehendem Sinne glaube ich richtig erkannt zu haben. Auf den höheren Stufen des Systems nimmt die Obliteration und Verbildung der Nervatur noch zu, besonders bei den Rhynchophoren.

Die *Morphologie der Tarsen* läuft mit der derivaten Verbildung der Flügelnervatur parallel. Die tiefstehenden inferioren Familien weisen elementare Tarsenbildung auf: 5 Glieder an jedem Tarsus, dessen 4 proximale Glieder einander ähnlich sind. Das ist der Fall bei den Adephagen, Staphylinoideen und Actinorhabden (Scarabæiden, etc.) abgesehen von manchen Ausnahmen, weniger bei den Adephagen, mehr bei den Staphylinoideen. Ferner sind 5 deutliche Glieder auf den unteren Stufen der Symphyogastren Regel, nämlich bei den Malacodermaten, Trichodermaten, Palpicorniern, Dascylloideen, Sternoxien und Bostrychoideen. Ausnahmen kommen auch hier vor, besonders Vorläufer in der Tarsenbildung der terminalen Familien. Eigentümlich ist die zwischenstehende Familiengruppe der Heteromeren mit 5 Gliedern am Tarsus der Vorder- und Mittelbeine und 4 Gliedern am Tarsus der Hinterbeine. Diese Familiengruppe ist augenscheinlich aus der Basis der Symphyogastren abzuleiten. Die obersten Stufen werden von den grossen Familiengruppen der Phytophagen und Rhynchophoren eingenommen. Sie bilden nebst den Clavicorniern die Abteilung der *Anchistopoden*. Bei den Phytophagen und Rhynchophoren ist fast allgemein das 4. (vorletzte) Tarsenglied verkürzt und verkleinert und oft so obliteriert, dass es wenig sichtbar ist und der Tarsus 4-gliedrig erscheint. Aber auf den inferioren Stufen dieser Familiengruppen giebt es noch deutlich 5-gliedrige Tarsen, z. B. bei den Parandrinen und Spondylinen (Phytophagen) und bei den Platypiden (Rhynchophoren). Als eine Vorstufe der Phytophagen erscheinen die Clavicornier, welche erst auf den höheren Stufen die charakteristische Tarsenbildung aufweisen. Auf den inferioren Stufen der Clavicornier befindet sich die Bildung der Tarsen noch auf einem mehr oder weniger primären Standpunkte, da das vorletzte Glied hier weder verkürzt noch verborgen ist. Diesen Zustand finden wir bei den Cucujiden, Monotomiden, Ostomiden, Thorictiden, Mycetophagiden, Cryptophagiden, etc., auch bei einem Teile der Erotyliden. Die Tarsen sind oft deutlich 5-gliedrig (viele Cucujiden, die Ostomiden, ein Teil der Nitiduliden, die Erotyliden). Beachtenswert ist die Tetramerie mancher Clavicornier. Man darf aus der Vergleichung den Schluss ziehen, dass ein Glied der Tarsen verkümmert und ausgefallen sei; denn der weniger gegliederte Tarsus ist vom 5-gliedrigen Tarsus abzuleiten. Das ist z. B. bei den Colydiiden, Mycetophagiden und Lathridiiden der Fall; ferner bei den Endomychiden und Coccinelliden. Den Zustand der Verborgenheit des vorletzten Gliedes der Tarsen nenne ich *Cryptarthrosie*. Mit der Verkümmerung des vorletzten Tarsengliedes ist gewöhnlich eine für den Gebrauch der Füsse vorteilhafte lappenförmige Vergrösserung des drittletzten Gliedes verbunden. Diese Bildung ist in den betreffenden Familien fast allgemein herrschend geworden. Ich sehe in dieser weithin herrschenden Tarsenbildung ein Evolutionsprinzip. Das Evolutionsprinzip der Cryptarthrosie in der Abteilung der Anchistopoden beginnt auf den oberen Stufen der Clavicornier und ist in den Familiengruppen der Phytophagen und Rhynchophoren weit und breit zum Durchbruch gekommen.

Die eben mitgeteilte Tarsenbildung ist wiederum ein deutlicher Ausdruck für die Theorie von der retrograden Transmutation in aufsteigender Linie. Das ist aber nichts anderes als eine Differenzierung einer früheren elementaren Bildung. Vorläufer der Cryptarthrosie treten sporadisch schon auf unteren Stufen des Systems auf, nämlich in der Familiengruppe der Adephagen bei den Hydroporinen (4. Glied der Tarsen des ersten Beinpaares sehr versteckt, äusserst klein, an den Tarsen des zweiten und dritten Beinpaares meistens deutlich), bei den Cleriden in einigen Gattungen (4. Glied der Tarsen des dritten Beinpaares sehr klein), bei den Ptilodactylinen, einer Gruppe der Dascylloideen (4. Glied aller Tarsen sehr klein).

Von *anatomischen Organen*, soweit diese einigermassen systematisch untersucht sind, mögen für unsere Betrachtungen besonders die *Malpighi'schen Gefässe* (vasa Malpighii) genannt werden. Die geringste Zahl von 4 Malpighi'schen Gefässen ist für die Carabiden, Dytisciden, Gyriniden, Staphyliniden, Silphiden, Pselaphiden, Scaphidiiden, Histeriden, Scarabæiden, Telephoriden, Lampyriden, Lyciden, Malachiiden, Hydrophiliden, Elateriden, Buprestiden und Anobiiden characteristisch, also für Familien tieferer systematischer Stellung. Wegen der Vierzahl der Malpighi'schen Gefässe werden die zu diesen Familien gehörigen Coleopteren als *Coleoptera tetranephria* bezeichnet. Es ist gewiss kein Zufall, dass auch gewisse sehr tief stehende Gruppen anderer Insectenordnungen, die z. T. selbständige Ordnungen bilden, nur 4 Malpighi'sche Gefässe aufweisen, nämlich die Thysanopteren,

Mallophagen und Psociden. — Die Coleopteren mit 6 Malpighi'schen Gefässen *(Coleoptera hexanephria)* sind die Melyriden, Cleriden, die Familien der Clavicornier und Heteromeren, sowie der Phytophagen und Rhynchophoren. — Die Zahl der daraufhin untersuchten Gattungen ist verhältnismässig sehr gering. Bei den vorstehenden Angaben gilt nur das Wort : *pars pro toto*.

Für die Schlussfolgerung, dass die niedrigere Zahl der Malpighi'schen Gefässe die inferiore Stellung im Systeme anzeige, ist das postembryonale Verhalten gewisser Orthopteren massgebend. Bekanntlich erfreuen sich die Orthopteren im ausgebildeten Zustande zahlreicher, oft grosse Büschel bildender Malpighi'scher Gefässe. Das ist auch bei den Blattiden der Fall ; aber die jüngsten Larven derselben besitzen nur wenige dieser schlauchförmigen Gefässe, die sich bei den Häutungen des Insects bis zur Ausbildung der Imago beständig vermehren. Auch die jüngsten Larven der Grylliden fallen durch die sehr geringe Zahl von 4 Gefässen auf, die sich nach und nach sehr vermehren. Auch die Larven der Hymenopteren besitzen nur 4 Malpighi'sche Gefässe, die Imagines dagegen bis 150.

Also liefert auch die Zahl der Malpighi'schen Gefässe Beweise für die tiefere Stellung der Caraboideen, Staphylinoideen, Lamellicornier, Malacodermaten, Palpicornier, etc. gegenüber den superioren Familiengruppen, besonders den Phytophagen und Rhynchophoren.

Es ist wiederholt behauptet worden, besonders von FRIEDRICH BRAUER, dass die sogenannte *concentrierte Ganglienkette* mancher Insecten als eine höhere Ausbildung des *Centralnervensystems* anzusehen sei, als die einfach gegliederte Ganglienkette anderer Insecten. Die gegliederte Ganglienkette zeigt in ihrem elementarsten Baue 2 Kopfganglien, 3 Thoracalganglien und 6 bis 7 Abdominalganglien (oder Knoten) ; das ist unter den Coleopteren besonders bei den Carabiden, z. T. auch bei den Lucaniden, dann bei den Malacodermaten, Elateriden, etc. der Fall. Dagegen sind bei den Lamellicorniern (Scarabæiden) und Rhynchophoren die meisten Ganglien miteinander verschmolzen; die Ganglienkette enthält hier 1 oder 2 Kopfganglien, 1 oder 2 Thoracalganglien, aber keine Abdominalganglien, weil diese unter sich und mit dem letzten Thoracalganglion verschmolzen und zu einem einzigen Complex vereinigt sind und im Thorax liegen. Bei inferioren Scarabæiden (Lucaninen, Glaphyrinen) ist jedoch der abdominale

Teil der Ganglienkette gegliedert; die einzelnen Ganglien sind getrennt. Die Lucaniden, deren Körper gestreckt und schlank ist, besitzen 2 Kopfganglien, 3 Thoracalganglien und je nach dem Genus 5 oder 6 Abdominalganglien, von denen die letzteren grösstenteils im Abdomen, zum kleinsten Teile im Thorax liegen. *Glaphyrus* besitzt 2 Kopfganglien, 2 bis 3 Thoracalganglien und 6 Abdominalganglien, von denen die letzteren grösstenteils im Thorax liegen. Auch bei gewissen Curculioniden fand E. BRANDT 2 Abdominalganglien. BLANCHARD hingegen stellte bei anderen Gattungen der Curculioniden 2 Kopfganglien und 2 oder 3 Thoracalganglien fest, mit deren letztem die verschmolzenen Abdominalganglien zu einem kleinen Complex verbunden sind und im Thorax liegen. Dies ist die concentrierte Ganglienkette, die also besonders bei den Scarabæiden und Rhynchophoren den geschilderten Grad der Ausbildung erreicht hat. Indess hat die Concentration dieses Organs für die Systematik nicht den hohen Wert, wie ihn GANGLBAUER annimmt. Denn ich glaube es aussprechen zu dürfen, dass der Bau der Ganglienkette dem Körperbaue conform ist, indem die Ganglienkette in erster Linie in einem gestreckten, lose gegliederten Körper einer auf tiefer Stufe stehenden Coleopterengattung (Carabiden, Staphyliniden, Malacodermaten etc.) elementar bleibt, d. h. dass die Ganglien getrennt bleiben, während in einem concentrierten Körper, dessen Segmente dicht zusammenschliessen (Curculioniden), auch die Knoten (Ganglien) der Ganglienkette alle oder grossenteils miteinander verschmelzen. Bei den Scarabæiden sind zwar die Segmente durchaus nicht so dicht und concret zusammengeschlossen, wie bei den Curculioniden, aber ihr Körper ist in den meisten Gattungen gedrungen, verkürzt und plump. Wohl zum Teil aus diesem Grunde sind die Ganglien der centralen Kette aneinander gedrängt und grossenteils miteinander verschmolzen. Aber wir dürfen nicht die præimaginalen Stadien vergessen. Die Imago übernimmt in morphologischer und physiologischer Beziehung mancherlei aus dem Larvenstadium. So gehen z. B. die centralen Teile des Nervensystems während der Metamorphose direkt aus der Larve in das Imagostadium über (KORSCHELT-HEIDER). Da nun die Ganglienkette der Scarabæiden bereits im Larvenstadium sehr concentriert ist, indem die Verkürzung nach meiner Ansicht aus der dauernden Krümmung des plumpen Larvenkörpers zu erklären ist, so übernimmt die Imago die bereits grossenteils vollzogene

Concentration der Ganglienkette von der Larve. Daraus geht hervor, dass die Concentration der Ganglienkette nur ein secundärer Wert beizumessen ist. Tatsächlich finden wir, dass die Ganglienkette innerhalb einer Familiengruppe oder sogar einer Familie alle oder manche Wandlungen von einer elementaren Bildung (bei inferioren Gattungen) bis zur Vereinigung oder sogar hochgradigen Verschmelzung der Ganglien durchläuft, z. B. bei den Scarabæiden, wie ich schon eben erwähnte. Ich hatte Gelegenheit, diese Verhältnisse in einer neueren Abhandlung (« Mein System der Coleopteren » 1908) eingehender zu besprechen. Ich gebe aus dieser Abhandlung folgende Beweisprobe wieder : « Dass die Concentration der Ganglienkette, also das Zusammenrücken und die Verschmelzung von Ganglienknoten und deren Vereinigung zu einem Complex eine consequente Wirkung des dichten Zusammenschlusses der Körperteile und der Verschmelzung von Segmenten sei, dafür spricht die Tatsache, dass das erste Ganglion des abdominalen Teiles der Ganglienkette meistens mit dem Metathoracalganglion verwachsen ist, und zwar (nach meiner Meinung) deswegen, weil das Sternit des ersten Abdominalsegments geschwunden und das Sternit des zweiten Segments mit demjenigen des dritten Segments verschmolzen ist. Um den Beweis vollgültig zu machen, constatieren wir auch die Gegenprobe, dass bei manchen derjenigen Coleopteren, bei denen das Sternit des zweiten Abdominalsegments frei (nicht mit dem Sternit des dritten Segments verschmolzen) ist, in entsprechender Weise auch das erste abdominale Ganglion vom Metathoracalganglion getrennt ist, z. B. bei Carabiden (*Carabus*) und Malacodermaten (*Lampyris*, *Dictyopterus*). Die Concentration der Ganglienkette geht also mit der Concentration der Rumpfteile Hand in Hand. » *Die Concentration der Ganglienkette ist nur eine Begleiterscheinung des engen Zusammenschlusses der Rumpfsegmente* und der daraus meistens resultierenden Verkürzung des Rumpfes. Ein typisches beweisendes Beispiel ist hierfür der gedrungene Körper der allermeisten Lamellicornier mit der concentrierten Ganglienkette im Gegensatze zu dem langgestreckten Körper der auf einer inferioren Stufe stehenden Lucaniden mit der gestreckten elementaren Ganglienkette. Es liegt schliesslich auch gar keine Veranlassung vor, in der concentrierten Ganglienkette ein Argument für die Annahme finden zu wollen, dass deren Träger infolgedessen besser organisiert oder sogar intelligenter seien als die Besitzer einer elementaren

Ganglienkette; dass also die Intelligenz einer *Cetonia, Melolontha* oder eines *Geotrupes* grösser sei als diejenige eines *Lucanus*. Im Gegenteil, die Scarabæiden stehen ebenso wie die Lucaniden auf einer psychisch niedrigen Stufe. Sie stehen auf einer tieferen Stufe der Intelligenz als z. B. Carabiden und Staphyliniden, welche eine recht elementar gebaute Ganglienkette besitzen. Auch die Curculioniden befinden sich trotz ihrer mehr oder weniger concentrierten Ganglienkette auf einem tiefen psychischen Niveau. Der Ausspruch, dass « die Lamellicornier bei der hohen Differenzierung ihres Nervensystems die höchste Stufe unter den Coleopteren einnehmen », hängt vollständig in der Luft, ist inhaltlos und deswegen zurückzuweisen. Höhere Differenzierung gewisser Teile des Nervensystems finden wir hingegen bei gewissen Hymenopteren, besonders bei *Apis, Bombus, Vespa* und *Formica*, bei denen die Gehirnwindungen am oberen Kopfganglion (ganglion supraœsophageum) die höchste Ausbildung unter den Insecten erreicht haben. Erst bei diesen intelligenten Insecten kann von einer höheren Organisation von Nervenapparaten die Rede sein, nicht aber bei den stumpfsinnigen Lamellicorniern.

Es bleibt demnach nur die Schlussfolgerung übrig, dass *der Concentration der Ganglienkette nur ein secundärer Wert für die Systematik beizumessen ist.*

Besonders erwähnenswert ist schliesslich noch das morphologische Verhalten der *Eiröhren (Ovarialröhren)*. Die beiden in den Eileiter (Oviduct) mündenden Eierstöcke (Ovarien) bestehen je aus einigen oder mehreren Ovarialröhren, in denen die Eier schnurförmig hintereinander gelagert sind. Die für das Wachstum der Eier erforderlichen Nährzellen sind in den Eiröhren lokal in einer oder mehreren Nährkammern (Dotterkammern) angehäuft. Entweder wechseln (alternierend) in jeder Eiröhre die Eikammern und Nährkammern, so dass sich zwischen je zwei Ovarialröhren eine Nährkammer befindet (Adephagen), oder es folgen die Eikammern ohne Nährkammern aufeinander, der Nährstoff hingegen befindet sich am Ende der Ovarialröhre in einer grossen Kammer (Heterophagen). Die Ovarien der *Adephagen* werden als *Ovaria meroistica*, diejenigen der *Heterophagen* als *Ovaria holoistica* bezeichnet (Emery). Die meroistischen Ovarien, wie bei den *Adephagen*, fand ich ebenso bei den Psociden (in mehreren von mir untersuchten Gattungen) und Neuropteren (was auch Emery angiebt); sie finden sich nach letzterem auch bei den Rhynchoten, Dipteren,

Lepidopteren und Hymenopteren. Dagegen haben die *heterophagen Coleopteren* ihre holoistischen Ovarien mit den Dermapteren, Amphibiotica, Orthopteren und Puliciden gemein. Nun sind allerdings die Coleopteren auf die Ovarien keineswegs in grösserem Umfange systematisch untersucht, aber die vorhandenen Resultate werfen ein eigentümliches Licht auf die tiefe Kluft innerhalb der Coleopteren, als ob diese diphyletischen Ursprunges seien. Jedenfalls weisen die Adephagen auf Grund ihrer Ovarien auf die Psociden und Neuropteren hin, die Heterophagen auf die Dermapteren (Dermatopteren) und Orthopteren. Die innerhalb der Subordo der Heterophagen sehr tief stehende Abteilung der Staphylinodeen würde sich dann an die Dermatopteren (Forficuliden) anschliessen, wobei allerdings von manchen Incongruenzen abgesehen wird. Einen weiteren Ausdruck möchte ich diesem Gedanken hier nicht geben, da ich ihn nicht weiter stützen kann. Es ist abzuwarten, wie weitere anatomische (auch embryologische) Befunde unter den noch nicht untersuchten Familien und Gruppen sich hierzu verhalten.

Schliesslich liegen noch vergleichend-morphologische Untersuchungen über die *Hoden (testiculi)* der Coleopteren vor. BORDAS schrieb darüber vor mehreren Jahren in den « Annales des sciences naturelles (Paris 1900) ».

Es kommt vor, dass manche Familientypen inferiorer Stufen in gewissen morphologischen Richtungen, die erst auf höheren Stufen allgemeiner zur Ausbildung gelangen, einseitig vorauseilen. So besitzen gewisse Malacodermaten bereits einen rüsselförmig verlängerten Kopf (*Lycus, Dictyopterus*, etc.), ebenso gewisse Heteromeren aus der Familie der Pythiden (*Salpingus, Rhinosimus, Mycterus*) und einige Cerambyciden (*Rhinophthalmus, Rhinotragus*, etc.). Diese Genera zeigen sonst alle Merkmale eines inferioren Grades, welche ihre nächsten Verwandten und sie selbst charakterisieren. So stimmen auch die Lammellicornier mit den Phytophagen und Rhynchophoren in der Bildung der Hoden überein. Diese bestehen in diesen Familiengruppen aus je einem Büschel rundlicher Follikel, welche alle durch je einen Ductus (Ausführungsgang) in das Vas deferens einmünden (gestielte Follikel). Bei den übrigen Coleopteren (die Adephagen ausgenommen) ist jeder der beiden Hoden aus einen Büschel länglicher Follikel zusammengesetzt, welche dem Vas deferens direct aufsitzen (sitzende Follikel). Ganz abgesondert ist der Bau der Hoden bei den Adephagen; denn hier besteht jeder Hoden aus einem einzigen sehr langen, knäuelförmig aufgewickelten Blindschlauch. Uebrigens ist der Umfang der Kenntnisse von diesen morphologischen Verhältnissen nicht gross; denn es sind nur je eine geringe Anzahl Gattungen von Angehörigen einer grösseren Zahl von Familien untersucht. Aber so viel geht aus den Resultaten dieser Forschungen hervor, dass, wenn die *Hoden mit gestielten Follikeln* eine höhere Transmutationsstufe darstellen, als es die Hoden mit sitzenden Follikeln sind, die Lamellicornier den Familien der Heterorhabden vorausgeeilt sind, obgleich sie auf Grund anderer primärer Charaktere des Rumpfes auf einer inferioren Stufe beharren.

*
* *

Die Ergebnisse der vergleichenden Morphologie der Coleopteren, wie ich sie hier in einigen Zügen darlegen durfte, laufen darauf hinaus, dass wir diese formenreiche und umfangreiche Insectenordnung in zwei grosse Stämme (subordines) teilen müssen: die *Adephagen* und *Heterophagen*. Darüber sind wir uns auch seit Jahren klar und einig. Auch GANGLBAUER und LAMEERE sind derselben Ansicht.

Es ergiebt sich ferner aus der vergleichend-morphologischen Uebersicht, dass die Evolutionsrichtung aller Organe des Coleopterenkörpers den *Adephagen die unterste Stufe* im morphologischen Aufbaue zuweist. Die Adephagen, deren hauptsächliche Vertreter die Carabiden sind, zeigen auf der ganzen Linie ihrer morphologischen Charactere (sowohl der Imagines wie der Larven) eine *inferiore Organisation*. Bei der Imago aller Adephagen ist das erste Sternit des Abdomens (Sternit des zweiten Abdominalsegments) vollkommen frei, den folgenden Sterniten ganz gleich und völlig sichtbar. Am Prothorax sind die Suturen aller componenten Teile noch deutlich (in einigen Gattungen verschmolzen). Die Nervatur der Flügel ist fast elementar. Der Bau der Ovarien ist bei den Adephagen einzig unter den Coleopteren (meroistisch) und ebenso wie bei den Psociden, Neuropteren, Hemipteren, Dipteren, Lepidopteren und Hymenopteren. Auch die Hoden sind einzig in ihrer Art unter den Coleopteren. Die Larven sind campodeaförmig und haben eine primäre Fussbildung; denn die vorhandenen fünf Glieder, aus denen die Beine bestehen, sind deutlich voneinander geschieden und am Ende mit zwei Krallen versehen.

Diese Bildung haben die Adephagenlarven mit den Larven der Neuropteren gemein.

Zu den **Adephagen** gehören die Familien *Carabidæ* (mit den Cicindelinæ), *Paussidæ*, *Rhysodidæ*, *Amphizoidæ*, *Hygrobiidæ*, *Haliplidæ*, *Dytiscidæ*, *Gyrinidæ*.

Die *Heterophagen* zeigen in den meisten Teilen ihrer Organisation, sowohl des Rumpfes wie der appendiculären Organe, *derivate Bildungen*, die von der Organisation der Adephagen abzuleiten sind. Bei der Imago ist das erste Sternit (Sternit des zweiten Segments) des Abdomens meist ganz unsichtbar, weil es eingesenkt, verbildet, verkürzt und mit dem folgenden Segment verschmolzen ist. Erst dieses zweite Sternit ist sichtbar (bei den Adephagen bereits das erste Sternit). Ausgenommen sind untere Stufen der Heterophagen, nämlich die Staphylinoidea (teilweise), die Actinorhabda, inferiore Gattungen der Malacodermaten etc. (s. oben), bei denen das erste Sternit des Abdomens nicht mit dem zweiten verschmolzen, aber doch meist nicht so frei liegt wie bei den Adephagen. Die Suturen an der Unterseite des Prothorax (die Episternen und Epimeren) sind teilweise verschmolzen (vorhanden nur bei den Passandrinen in der Familiengruppe der Clavicornia). Die Nervatur der Flügel ist ebenso deutlich derivat, aber von derjenigen der Adephagen leicht abzuleiten. Bei den Staphilinoideen erscheint die Flügelnervatur recht primitiv ; denn sie ist von derjenigen der Adephagen durch den Ausfall der Transversaladern und die basale Rudimentierung der Subbrachialis direct abzuleiten. Noch primitiver ist die Flügelnervatur der Cupediden. In den übrigen Familiengruppen der Heterophagen haben die Transversaladern durch Verschmelzung mit Längsadern (Principaladern) eine Verbildung und Umbildung der Flügelnervatur hervorgerufen. Auch ist die Subbrachialis nach der Basis zu rudimentiert. Die Ovarien sind ganz abweichend von denjenigen der Adephagen gebaut (holoistisch), aber conform den Ovarien der Dermapteren, Pseudo-Neuropteren, Orthopteren und Puliciden. Bei den Larven ist das einzige Krallenglied mit den Tarsus verschmolzen, so dass das Bein 4-gliedrig ist.

Die **Heterophagen** bestehen aus 2 Hauptabteilungen. Die I. Hauptabteilung möge als HAPLOGASTRA bezeichnet werden wegen ihres hervorragenden Characters des 1. meist freien abdominalen Sternits, der sie den Adephagen nahebringt. Der unterste

Ast der Haplogastren sind die *Staphylinoideen*, deren *campodeaförmige Larven* allein schon eine sehr tiefe Stellung unter den Coleopteren nahelegt. Aber die Fussbildung der Larven ist bereits derivat (durch die Verschmelzung der einzigen Kralle mit dem Tarsus), wodurch sie sich von den ebenfalls campodeaförmigen und ihnen sehr ähnlichen Adephagenlarven unterscheiden. Ebenso ist die Flügelnervatur ziemlich primitiv. Das 1. Sternit des Abdomens ist von dem 2. meist getrennt, frei, allerdings meist verkürzt oder teilweise häutig. Die Pleuren des 1. Sternits sind von den Pleuren des 2. Sternits deutlich getrennt. Die zweite Familiengruppe der Haplogastren sind die *Actinorhabden*, zu denen hauptsächlich die Scarabæiden (auch Lamellicornier genannt) gehören, und die sich besonders durch das freie 1. Sternit bei derivater Flügelnervatur auszeichnen. Dieses Sternit ist selbständig, stark chitinisiert und den folgenden Sterniten ähnlich, aber durch die Coxen grossenteils bedeckt. Auch die Pleuren des 1. Sternits sind von denjenigen des 2. Sternits stets getrennt.

Die Familiengruppe der STAPHYLINOIDEA enthält die Familien *Staphylinidæ*, *Pselaphidæ*, *Scydmænidæ*, ? *Ectrephidæ*, *Silphidæ*, *Catopidæ*, *Anisotomidæ*, *Clambidæ*, ? *Aphænocephalidæ*, *Corylophidæ*, *Trichopterygidæ*, *Hydroscaphidæ*, *Sphæriidæ*, *Scaphidiidæ*, *Leptinidæ*, *Platypsyllidæ*, *Histeridæ*.

Die Familiengruppe der ACTINORHABDA besteht aus den Familien *Synteliidæ*, *Passalidæ*, *Scarabæidæ* (mit Einschluss der Lucaninæ).

Der ganze grosse Rest der Coleopteren gehört zu der II. Hauptabteilung der *Heterophagen*, den SYMPHYOGASTRA. Bei diesen ist das 1. Sternit des Abdomens mit dem 2. Sternit verwachsen, verbildet und eingesenkt. Auch die zugehörigen Pleuren (also die Pleuren des 2. und 3. Segments) sind miteinander verschmolzen; diese verlängerte Pleure liegt dem 1. freien Sternit an. Nur bei vielen inferioren Malacodermaten, besonders bei den Lampyriden und Lyciden, ferner bei einigen Lymexyloniden (*Atractocerus*) und einigen Meloiden *(Meloë)* ist der sternale Basalteil des Abdomens wie bei den Adephagen und den Haplogastren gebildet. Auch bei den auf äusserst inferiorer Stufe stehenden Cupediden verhält sich das 1. Sternit des Abdomens beinahe primär, da es mit den folgenden Sterniten in einer Ebene liegt und von dem 2. Sternit durch eine deutliche Sutur geschieden ist, aber die Coxen des dritten Beinpaares bedecken das 1. Sternit. Die Flügelnervatur der Symphyogastren ist (abgesehen von den Ausnahmen) sehr derivat, wie schon bei den Actinorhabden; nur bei den Cupediden ist sie sehr primitiv, elementar, da sie hauptsächlich aus einfachen Longitudinaladern (Principaladern) und mehreren diese elementar verbindenden Transversaladern besteht. Auffallend einfach ist auch die Flügelnervatur von *Atractocerus* (Lymexylonidæ). Für die Cupediden, deren Flügelnervatur noch elementarer ist als diejenige der Adephagen, habe ich eine besondere Abteilung *Archostemata* angenommen (von στῆμα = Flügelrippe und ἀρχή = Anfang oder ἀρχαῖος = uranfänglich), im Gegensatze zu den Synactostemata, deren Flügelnervatur derivat, d. h. teilweise durch Verbindung oder Verschmelzung und Verkürzung verbildet und von derjenigen der Cupediden abzuleiten ist.

Die ARCHOSTEMATA enthalten nur die eine Familie *Cupedidæ*.

Die SYNACTOSTEMATA lassen sich noch in die *Heterorhabda* (Tarsen mit gleichartigen Gliedern) und die *Anchistopoda* (Tarsen grösstenteils mit ungleichartigen Gliedern, da das vorletzte meistens klein und verborgen und mit dem letzten eng verbunden ist) teilen.

Unter den HETERORHABDEN haben wir die erste Abteilung *Pelmatophila* nebst der zweiten Abteilung *Bostrychoidea*, deren Tarsen gewöhnlich aus der gleichen Anzahl von Gliedern bestehen, und die dritte Abteilung *Heteromera*, bei denen das erste und zweite Beinpaar 5, das dritte Beinpaar 4 Tarsenglieder besitzt.

Die *Heterorhabden* setzen sich aus folgenden Familiengruppen zusammen :

Familiengruppe MALACODERMATA mit den Familien *Drilidæ*, *Lampyridæ*, *Lycidæ*, *Cantharidæ* (*Telephoridæ*).

Familiengruppe TRICHODERMATA mit den Familien *Malachiidæ Melyridæ*, *Corynetidæ*, *Derodontidæ* (*Laricobiidæ*), *Cleridæ*.

Familiengruppe PALPICORNIA mit der einzigen Familie *Hydrophilidæ*.

Familiengruppe DASCYLLOIDEA mit den Familien *Psephenidæ*, *Helodidæ*, *Ptilodactylidæ*, *Eubriidæ*, *Eucinetidæ*, *Dascyllidæ*, *Artematopidæ*, *Lichadidæ*, *Rhipidoceridæ*, *Chelonariidæ*, *Byrrhidæ*, *Nosodendridæ*, *Dermestidæ*, *Heteroceridæ*, *Dryopidæ* (*Parnidæ*), *Helmidæ*, *Georyssidæ*, *Cyathoceridæ*.

Familiengruppe STERNOXIA mit den Familien *Cerophytidæ*, *Cebrionidæ*, *Plastoceridæ*, *Dicronychidæ*, *Elateridæ*, *Eucnemidæ*, *Throscidæ*, *Buprestidæ*.

Familiengruppe BOSTRYCHOIDEA mit den Familien *Lymexylo-*

nidæ, — *Sphindidæ*, *Aspidiphoridæ*, *Cioidæ*, — *Ptinidæ*, — *Lyctidæ*, *Psoidæ*, *Anobiidæ*, *Bostrychidæ*.

Familiengruppe HETEROMERA mit den Familien *Melandryidæ*, *Mordellidæ*, *Rhipidophoridæ*, *Cephaloidæ*, *Œdemeridæ*, *Pythidæ*, *Anthicidæ*, *Pedilidæ*, *Xylophilidæ*, *Pyrochroidæ*, *Meloidæ*, *Salpingidæ*, *Petriidæ*, *Monommidæ*, *Nilionidæ*, *Trictenotomidæ*, *Othniidæ*, *Lagriidæ*, *Cistelidæ (Alleculidæ)*, *Tenebrionidæ*, *Ægialitidæ*, *Tentyriidæ*.

Die ANCHISTOPODEN umfassen die grossen, allermeist auf Pflanzennahrung angewiesenen Familien mit sehr kleinem versteckten vorletzten Tarsengliede, sowie die Vorstufen zu diesen so characterisierten Familien mit elementar gebildeten Tarsen. Diese Abteilung steht unter dem *Evolutionsprinzip der Cryptarthrosie;* sie umfasst die Familiengruppen *Clavicornia*, *Phytophaga* und *Rhynchophora*.

In der *Familiengruppe* CLAVICORNIA mit den Familien *Cucujidæ*, *Monotomidæ*, — *Sphæritidæ*, *Nitidulidæ*, *Byturidæ*, *Ostomidæ (Trogositidæ)*, — *Cryptophagidæ*, *Atomariidæ*, *Mycetophagidæ*, *Phalacridæ*, *Catopochrotidæ*, *Erotylidæ* (incl. *Languriidæ* und *Helotidæ*), — *Lathridiidæ*, *Thorictidæ*, *Gnostidæ*, *Adimeridæ*, *Colydiidæ*, *Endomychidæ*, ? *Pseudocorylophidæ*, *Coccinellidæ* – sind die ersteren Familien meistens mit elementar gebildeten, die letzteren Familien grösstenteils mit derivat gebildeten Tarsen (vorletztes Glied sehr verkleinert, cryptarthrotisch) versehen. Auffallend ist in dieser Vorstufe der genuinen Anchistopoden die Regellosigkeit in der Zahl der Tarsenglieder. Die Tarsen sind 5-gliedrig bei manchen Cucujiden, den Monotomiden, manchen Nitiduliden, den Sphæritiden, Ostomiden und manchen Erotyliden; heteromerisch bei den Männchen anderer Gattungen der Cucujiden, mancher Nitiduliden und vieler Cryptophagiden; 4-gliedrig bei noch einigen anderen Cucujiden, gewissen Nitiduliden, sowie bei den Colydiiden, Endomychiden und Coccinelliden; 3-gliedrig bei einigen Mycetophagiden und Colydiiden und bei den Lathridiiden. Die sehr tief stehenden Cucujiden sind nicht nur teilweise durch einen sehr elementar (wie bei den Adephagen) gebauten Prothorax, sondern auch durch elementar inserierte (inframarginale) und elementar gebildete (faden- oder schnurförmige) Antennen characterisiert. Auch bei den Nitiduliden, Sphæritiden, Ostomiden etc. stehen die Antennen inframarginal, dagegen frontal oder subfrontal bei den Erotyliden, einem Teile der Lathridiiden und Colydiiden, bei den Endomychiden und Coccinelliden. Jene stehen auf einer inferioren, diese auf einer superioren Stufe. Die höhere Stufe dieser Familien ist auch angezeigt durch die cryptarthrotisch geformten Tarsen : eine Bildung, die in den beiden folgenden und letzten Familiengruppen Gesetz wird. Auch hinsichtlich der Malpighi'schen Gefässe gleichen die Clavicornier den Phytophagen und Rhynchophoren; denn sie gehören zu den Hexanephrien, sie besitzen 6 Malpighi'sche Gefässe. Es ist auch bezeichnend, dass die Clavicornier sich von niedrig stehenden Pflanzen ernähren, mycetophag sind.

Die *Familiengruppe* der PHYTOPHAGEN gehört einem merklich höheren Aste des phylogenetischen Stammbaumes an. Die Cryptarthrosie des vorletzten Tarsengliedes ist hier Gesetz geworden; nur die am tiefsten stehenden Gattungen der Prioniden (*Parandra*, etc.) haben elementar gebildete Tarsen.

Durch den Bau der Hoden stehen die Phytophagen bereits auf einer höheren Stufe als die Clavicornier. Die Prioniden tragen primitive Charactere zur Schau, nämlich die inframarginale Stellung der Antennen, die Form der Coxen am Prothorax, welche quer gestellt und direct von einer lang zapfenformigen Bildung abzuleiten sind, ferner die primitive laterale Marginierung des Prothorax. Die Bruchiden haben etwas zapfenförmig vortretende Prothoracalcoxen und gehören wohl einer Vorstufe der Chrysomeliden an. Die ausserordentlich mannigfaltig verzweigten Chrysomeliden zeigen durch ihren Reichtum an Gattungen, Arten und Individuen, dass sie die äussersten und höchsten Zweige des Phytophagenstammes bilden. Sie erscheinen dadurch mehr superieur, die Cerambycinen und Lamiinen inferieur. Und vollends die Prioniden muten mit ihren grossenteils isolirten Gattungstypen von teilweise immensen Dimensionen des körperlichen Volumens recht altertümlich an. Die wenigen zu den Phytophagen gehörigen Familien sind die *Prionidæ, Cerambycidæ, Bruchidæ* und *Chrysomelidæ*.

Die Familiengruppe RHYNCHOPHORA ist der höchste und am weitesten verzweigte Ast am Stamme der Coleopteren. Sie ist die grösste und formenreichste aller Familiengruppen. Alle ihre Angehörigen leben von vegetabilischer Nahrung. Manche Arten sind so individuenreich, dass sie ganze Pflanzengesellschaften vernichten können, also in der Natur eine Herrschaft ausüben. Die Organisation des Körpers und der Anhänge ist teils am generellsten unter allen Coleopteren differenziert, teils durch Verschmelzung von Teilen äusserst extrem ausgebildet. *Ein ganz genereller und exclusiver Character ist das Fehlen der Gula.* Diese ist das Sternit des Hinterkopfes, welches bei den Rhynchophoren durch derivate Rückbildung geschwunden ist. Die beiden lateralen Suturen der Gula rücken nämlich so nahe zusammen, dass sie sich vereinigen und nur die (zuweilen sogar fehlende) mediane Sutur übrig bleibt. *Generelle, aber nur mehr oder weniger exclusive, durch Differenzierung, Reduction oder Coalescenz von Teilen entstandene Charactere der Rhynchophoren* sind : 1. die Ausbildung des Kopfes zu einem *Rostrum*, welches aber bei den Platypiden und Scolytiden nicht oder kaum ausgebildet ist, und andererseits auch in einigen anderen Familiengruppen inferiorer Organisationsstufen in einzelnen Gattungen auftritt; 2. die derivate *Rückbildung des Labrums*, welches nur bei den auf inferiorer Stufe stehenden Rhinomaceriden und Anthotribiden gut entwickelt ist; 3. die *Reduction der Maxillen und Palpen*, welche meistens klein resp. kurz und starr sind, ausser bei den Rhinomaceriden und Anthotribiden ; 4. die *Verschmelzung aller Teile des Prothorax*, ausser bei den Rhinomaceriden, bei denen das Sternum des Prothorax von den Pleuren durch eine Naht getrennt ist, — und den Anthotribiden, deren Pronotum jederseits durch eine laterale Kante begrenzt ist.

Generelle, aber auch in anderen Familiengruppen mehr oder weniger herrschende Charactere, welche durch hohe Differenzierung, Reduction, Coalescenz oder Concentrierung sich herausgebildet haben, sind bei den Rhynchophoren : 1. die postcoxale Verbindung der *Pleuren des Prothorax* mit dem intercoxalen Fortsatze des Sternums; 2. die appendiculäre Verlängerung des Mesosternums an der Aussenseite der mittleren Coxen bis an das Metasternum; 3. die intime Verschmelzung der Sternite der beiden basalen Abdominalsegmente; 4. die cryptarthrotische Rudimentierung des 4. Tarsengliedes an allen Beinen, ausser bei den Platypiden und einem Teile der Tomiciden; 5. die mehr oder weniger fortgeschrittene Concentrierung der Ganglienkette; 6. die Sechszahl der Malpighi'schen Gefässe; 7. die Organisation der Testikeln (mit gestielten Follikeln); 8. die Apodie der Larven (ausser bei Brenthiden).

Zu dieser Familiengruppe gehören die Familien *Rhinomaceridæ (Nemonychidæ), Anthotribidæ; Platypodidæ, Tomicidæ (Scolytidæ, Ipidæ); ? Protherinidæ, ? Aglycyderidæ; Brenthidæ, Oxycorynidæ, Rhynchitidæ, Apionidæ, Brachyceridæ, Curculionidæ.*

In dem vorstehend dargelegten *natürlichen Systeme der Coleo-*

pteren ist der *Aufbau des Coleopterenstammes* nach seiner Ablösung aus der Wurzel neuropteroider Vorfahren dargestellt. Die zahlreichen Familien (über 130) sind auf vergleichend-morphologischer, descendenztheoretischer und phylogenetischer Grundlage in 2 Unterordnungen und 14 Familiengruppen gegliedert und gruppiert. Der Entwicklungsgang der Organe und Organteile, welche die Gliederung in Familiengruppen und Familien begleitet, wurde teilweise erörtert; es ergiebt sich so ein Bild von der Transmutation und Evolution, welche die einzelnen Organe durchlaufen. In jeder Familie krystallisiert sich eine morphologische Stufe unter Begleitung accessorischer Charactere. Es ist hier nicht der Ort, die Charactere der Familien festzustellen. Es soll hier nur die Frage nach der Natürlichkeit der einzelnen Familien berührt werden. Wir finden dabei dass, wenn wir auch alle Charactere der einzelnen Familien auf gewohnter morphologischer und biologischer Grundlage aufzählen, hiermit die Characteristik der Familien noch nicht vollständig gegeben ist. Der morphologische Familiencharacter ist nun zwar jedenfalls das Resultat aus der Summe der einzelnen morphologischen Merkmale. Aber selbst eine gründliche Erforschung der Charactere einer Familie würde wohl nicht völlig den Familiencharacter der Angehörigen einer Familie erschöpfen. Wahrscheinlich müssen wir mit morphologischen Imponderabilien rechnen, welche die messbaren Charactere ergänzen. Es sind die aus der Convergenz resultierenden Bildungen, die aus dem Milieu der Lebensverhältnisse auf die Formen sich geltend machenden Factoren, unter deren Einfluss alle Angehörigen einer Familie stehen. Diese Convergenzcharactere sind die Imponderabilien, welche die Summe der morphologischen Charactere ergänzen. Mit Hülfe dieser morphologischen Imponderabilien erkennt man auch leichter die Zugehörigkeit eines bis dahin noch unbekannten Insects zu seiner Familie.

Aber nicht nur aus den messbaren Bildungen und den morphologischen Imponderabilien erkennen wir die Verwandtschaft des Coleopterons. Es muss auch möglich sein, an dem Körperbau eines noch unbekannten Käfers bald erkennen zu können auf welche Stufe des Systems dieser gehört. Dem schlanken, gracilen, nicht sehr fest gefügten und substantiell weniger condensierten Körper eines Carabiden mit den meist dünnen und langen Beinen, dem vorstehenden Kopfe, den frei sichtbaren Mundteilen und langen faden- oder borstenförmigen Antennen steht der meist fest gefügte, gedrungene, oft schwerfällige Körper eines Curculioniden mit der harten, dicken äusseren Hautbedeckung, den meist kurzen, dicken Beinen, nach unten gerichteten, absonderlich rüsselförmig verlängerten Kopfe, verkürzten und wenig sichtbaren kleinen Mundteilen an der Spitze des Rüssels und derivat gebildeten, am Ende verdickten Antennen gegenüber. Der geschilderte Körperbau der Carabiden einerseits und der Curculioniden anderseits ist nur der Ausdruck für die morphologische Stufe und die Stufe des Systems, auf welcher jeder der beiden Familien steht. Der deutlich umschriebene messbare Character der systematischen Stellung liegt in mehr oder weniger *verborgenen* Bildungen gewisser Teile des Körpers : besonders in dem Vorhandensein oder Fehlen der Gula (hinteres Sternit des Kopfes), in der Bildung der beiden basalen Sternite des Abdomens (frei und sichtbar oder das erste Sternit versteckt und mit dem zweiten verschmolzen), in der teilweise elementaren oder derivaten Beschaffenheit der Flügelnervatur, in dem deutlichen Vorhandensein der Suturen zwischen den elementaren Stücken des Prothorax oder in der nahtlosen Verschmelzung dieser Teile, in der elementaren oder derivaten Beschaffenheit des vorletzten Tarsengliedes (frei und deutlich oder sehr verkleinert und versteckt), etc. Die äussere Characteristik congruirt hier mit der verborgenen Characteristik. Die äusseren Charactere sind so abhängig von den versteckten Characteren, dass die Kenntnis der ersteren zur Erkennung der Familienzugehörigkeit oft genügt.

Was ist eine *systematische Familie*? Ueber diesen Begriff herrscht keine völlige Einigkeit; bisher als verschieden angesehene Familien werden von einigen Entomologen vereinigt, von andern wieder getrennt. Jedenfalls gehören zu einer systematischen und natürlichen Familie alle morphologisch zunächst miteinander verwandten Gattungen, welche zusammen ein natürliches Ganzes bilden und von einem anderen Collectivum zusammengehöriger Gattungen abgesondert sind. Wenn dieses Collectivum von Gattungen *ex radice* von jenem Collectivum von Gattungen abzuleiten ist, so sollten diese beiden Collectiva von Gattungen verschiedene Familien sein; denn sie sind schon von der Wurzel an getrennt. Ein von einer superioren Stufe jenes Collectivums von Gattungen ausgehendes zweites Collectivum von Gattungen würde als eine Unterfamilie oder Gruppe jenes als Familie zu bezeichnenden Collectivums anzusehen sein. Die Cicindeliden z. B. stimmen in allen wesentlichen Characteren mit den Carabiden überein; manche Entomologen halten sie für eine besondere Familie, andere für eine Unterfamilie der Carabiden. Weil nun die *radix* der Carabiden in der Unterfamilie der Carabinæ (Amphizoini, Carabini, Nebriini, etc.) steckt, die Cicindeliden aber sich an die auf der superioren Stufe stehenden Unterfamilien der Scaritinæ, Elaphrinæ und Verwandten anschliesst, so gehören die Cicindeliden als Unterfamilie Cicindelinæ zu den Carabiden. Die Dytisciden aber schliessen sich morphologisch direct an die Unterfamilie der Carabinæ (Carabini, Nebriini, Amphizoini, etc.) an; sie sind also *ex radice* von den Carabiden abzuleiten und bilden deswegen mit Recht eine eigene Familie. Ich habe die Cicindeliden schon vor vielen Jahren mit den Carabiden vereinigt (1).

Die Berechtigung mancher Familien würde also nach dem Gesetze der Descendenz zu beurteilen sein. Es ist notwendig, manche der aufgeführten Familien der Coleopteren nach dieser Norm auf ihre Berechtigung zu prüfen. Den Schlüssel zu diesen Feststellungen finden wir in dem descendenztheoretischen Gesetze, welches der vergleichenden Morphologie der Familien zu Grunde liegt. Wie die Familien aus den Unterfamilien und Gruppen, so baut das ganze System der Coleopteren sich aus den Familien auf der realen Basis der fortschreitenden Transmutation und Descendenz auf. Die vergleichende Morphologie und Systematik stehen auf der Basis der Descendenztheorie. Das natürliche System ist das Resultat aus der vergleichenden Morphologie und den aus der Vergleichung der Organe und Organteile gewonnenen descendenztheoretischen Verhältnissen.

---

(1) Kolbe, H., Natürliches System der carnivoren Coleopteren. (*Deutsche Ent. Zeitschr.*, 1880, pp. 258-280.)

Algunos órganos de las alas de los Insectos,

por el R. P. Longinos Navás, S. J. (Zaragoza).

Introducción.

No tengo la pretensión de dar á conocer cosas enteramente nuevas para mis doctos colegas en Entomología. Mucho de lo que voy á decir lo conocerán ya directamente por experiencia propia; de otros órganos que voy á mencionar tendrán acaso noticias por referencia; si algo nuevo apunto será porque no le habrán prestado atención ó les habrá faltado la ocasión de estudiar ejemplares que presentasen aquellas particularidades que indico.

Mucho menos pretendo hacer un estudio monográfico, que pudiera ser muy extenso y abarcase no sólo la denominación ó descripción breve y somera, que es lo que ahora pretendo, sino también el estudio microscópico, la anatomía y sobre todo la averiguación de las funciones de los órganos que voy á mencionar. Estudio fuera éste de grandísimo interés y que dejo para quien se vea con más ánimos y tiempo de los que á mí me asisten.

A mí me basta mostrar este nuevo campo en que puedan segar copiosa miés aventajados entomólogos, histólogos y fisiólogos. Especialmente la parte fisiológica, que nos es casi del todo desconocida, sin duda daría lugar á investigaciones y preciosos descubrimientos.

Entre los órganos alares dignos de toda nuestra atención escogeré para mi trabajo algunos que apenas han sido tocados más que ligeramente, ú ofrecen alguna particularidad en que podemos fijarnos, omitiendo aquellos que por muy notorios y generales á todos los Insectos se ven con frecuencia tratados por los autores.

La principal utilidad que en estos órganos consideraré será su valor taxonómico.

Finalmente advertiré que mis observaciones han versado sobre especies de Neurópteros (sensu lato); pero me persuado que algunas de ellas podrán extenderse á otros órdenes de Insectos.

1. **Pupila** (latín *pupilla*).

En mi trabajo « Diláridos de España » (Memorias de la Real Academia de Ciencias y Artes de Barcelona, vol. IV, n. 28, Barcelona, 1903) llamé « punctum discoidale » y « punctum pupillatum » la mancha que ya RAMBUR había observado, aunque única, en ambas alas de su especie típica *Dilar nevadensis* (RAMBUR, Névroptères, p. 446, pl. 10, fig. 3, 4). En dicho trabajo hice notar que tal mancha era doble en el ala anterior, única en la posterior, que la había visto en todas las especies de Diláridos y se hallaba también en los Osmílidos.

En mi monografía « Neurópteros de España y Portugal » (Broteria, San Fiel, 1906-1908) llamé *pupila* á esta manchita, por la figura oceliforme que presenta, y este nombre, ó el de pupila discal, empleé en mi monografía de la familia de los Diláridos, p. 7 (Memorias de la Real Academia de Ciencias y Artes de Barcelona, vol. VII, n. 17, Barcelona, 1905) y en otros trabajos.

Existe en todos los géneros que he visto de la familia de los Diláridos (*Dilar, Lidar, Fuentenus, Nepal, Rexavius*), de la de los Osmílidos (*Osmylus, Stenosmylus, Dictyosmylus*), en el *Polystœchotes* de los Hemeróbidos, en los *Neuromus, Chauliodes* y *Corydalis* entre los Siálidos. En las especies que he visto del género *Sialis (fuliginosa, lutaria, sibirica, infumata)* no la he podido distinguir.

Su *número* es vario. Lo ordinario es que existan dos en el ala anterior y una en la posterior. En los Siálidos que las poseen su número aumenta, pues se ven constantemente tres en el ala anterior y dos al menos en la posterior, aunque pueden ser poco visibles; lo son mucho en el *Chauliodes japonicus* MAC LACHL.; y aun en el género *Corydalis*, por lo menos en las especies *Batesi* MAC LACHL. y *cornuta* L. hallo tres bien manifiestas en esta ala. Pero con frecuencia las pupilas típicas se multiplican, viéndose cerca de la principal una ó dos secundarias, ordinariamente más pequeñas.

Su *posición* es bastante precisa. Su campo propio está detrás del primer sector del radio, en el espacio que media entre éste y el procúbito (fig. 1) y que podemos llamar campo intermedio ó *área*

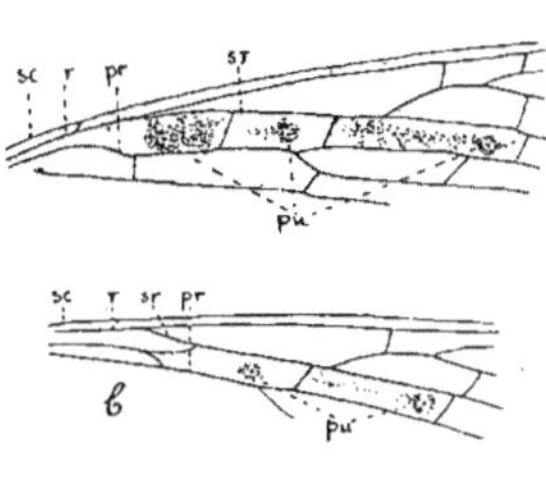

Fig. 1.

Chauliodes japonicus Mac Lachl. Región de las pupilas.
a) Ala anterior. — *b*) Ala posterior; *sc.* subcostal; *r.* radio; *sr.* sector del radio; *pr.* procúbito; *pu.* pupilas.

intermedia del ala. Ella separa dos regiones del ala muy distantas en su estructura: la anterior, con las venas costal, subcostal y radial, y la posterior con los cúbitos y venas posteriores á ellos.

Si hay dos pupilas, la interna suele colocarse más atrás de la raiz del sector mismo y la externa avanza hacia la mitad del ala. Cuando alguna de ellas se divide en dos ó se multiplica en varias, lo ordinario es que la principal ocupe su lugar propio y las secundarias ó alguna de ellas avancen ó retrocedan, introduciéndose á veces entre los sectores del radio ó entre las ramas del procúbito. En los Osmílidos (fig. 2) es constante que la externa se halle entre los dos primeros ramos del sector del radio en ambas alas, pues hay dos pupilas también en el ala posterior. En los Diláridos la externa del ala anterior suele situarse entre los dos primeros sectores.

El *color* de las mismas es muy vario. En los Diláridos se hacen muy visibles ordinariamente por caer dentro de una mancha parda y hallarse ésta más ó menos aislada, pero en otros grupos se desvanecen, hácense incoloras, y sólo se perciben con lente de fuerte aumento. El *Chauliodes japonicus* (fig. 1) forma excepción entre los de su familia, pues las ofrece muy marcadas y á simple vista sensibles, especialmente en el ala posterior, donde campean por la palidez de la membrana.

Que tales pupilas no sean meras manchas, sino *órgano especial*, lo persuade la vista armada de fuerte lente y mejor aún el microscopio. Aun cuando son incoloras se nos presentan como un callo, la membrana en ellas se halla modificada, como estirada y endurecida; las venas se encorvan junto á ella para darle espacio. Confirman este parecer su posición tan fija, su forma tan definida y propia en las diferentes especies y géneros.

Qué órgano sea y cuál *su función* es lo que desconozco completamente, ni entra en mis planes averiguarlo. Que sea glandular ó sensorial, acaso olfativo ó auditivo, lo sospecho, mas dejo á otros la solución de esta duda.

Pero no me cabe duda, por lo que antecede, que semejante órgano debe poseer gran *valor taxonómico* y que hay que tenerlo en cuenta en las clasificaciones, sobre todo si va junto con otros caracteres propios de determinadas especies, géneros y familias, como acontece.

En atención á este órgano propongo que el género *Polystœchotes* BURM. se separe de la familia de los Hemeróbidos, en cuya tribu de los Sisirinos venía colocándose, y se incluya en la familia de los

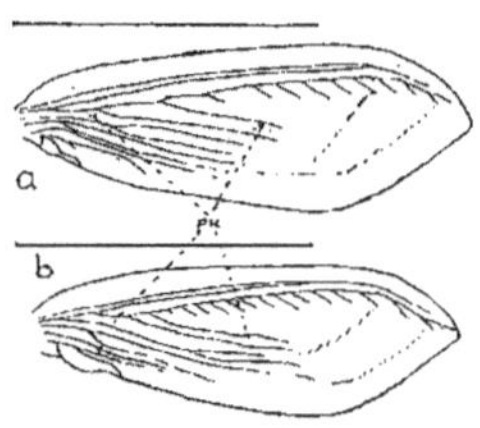

FIG. 2.

Ala anterior derecha de a. *Osmylus hyalinatus* MAC LACHL.
b. *Polystœchotes punctatus* F. (Ala inferior casi esquemático.)

Osmílidos. Efectivamente, la estructura de las alas (fig. 2), con muchos sectores del radio, la malla, la figura entera, es totalmente semejante. Convienen asimismo en otros caracteres tomados de la cabeza, patas, etc.

Atendiendo á varios géneros de la familia de los Siálidos que he podido estudiar en mi colección y poseen dichas pupilas y difieren notablemente en lo demás de otros que no las poseen, v. gr. el género *Sialis*, como es en la estructura de las patas y de las alas, propongo separar estos géneros de la familia de los Siálidos para constituir familia aparte con el nombre de Neurómidos (Neuromidæ).

El nombre se lo da el género *Neuromus* Ramb., 1842. Además de él comprende la nueva familia los géneros *Corydalis*, *Chauliodes* y otros.

Su caractéristica podrá ser la siguiente :

Caput latum, ovale, deplanatum, mandibulis validis, dentatis, antennis longis filiformibus vel pectinatis.

Alæ pupillis instructæ, 3 in ala anteriore, 2 in posteriore.

Pedes cylindrici, similes, tarsis 5-articulatis, articulis cylindricis, haud dilatatis.

Cetera ut in Sialidis.

La diversidad completa de los tarsos me ha confirmado en la separación de este grupo de la familia de los Siálidos.

2. **Estriola** (en latín *striola*).

Llamo estriola una línea pálida situada en el campo intermedio del ala anterior á lo largo del procúbito y del primer sector del radio.

El nombre lo he formado del latín, por la apariencia de pequeña estría que presenta.

No la he visto citada en ningún autor, á pesar de su frecuencia. En un dibujo muy amplificado ($^{21}/_1$) del *Sympherobius angustus* Banks publicado en la revista « Pomona College Journal of Entomology », March 1910, p. 144, aparece la estriola con toda claridad ocupando un cuarto de la longitud del ala. En la descripción que hice de la *Psectra Buenoi* (1) (« Annales de la Société scientifique de Bruxelles », 1909) la menciono con el circumloquio de tiridio linear estrecho y la represento en las dos figuras que exhibo.

Se hace sensible sobre todo cuando la membrana del ala está manchada y cuando hay venillas intermedias ó que van del sector del radio al procúbito, pues aparecen en un sitio pálidas cual si estuviesen-cortadas por la estriola. En realidad la membrana

(1) Siendo esta especie el *Hemerobius delicatulus* Fitch y éste la forma alada de *Psectra diptera* Burm., en mi concepto podrá llamarse *Psectra diptera* Burm. var. *delicatula* Fitch.

misma forma un pliegue, bastante sensible en ejemplares grandes, tales como *Neuromus latratus* MAC LACHLAN var. *tonkinensis*.

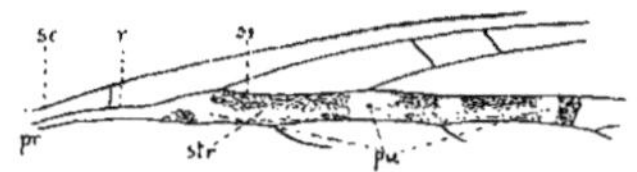

FIG. 3.

Neuromus latratus MAC LACHL. var. *tonkinensis* V. d. W.
Ala anterior, región de las pupilas.
str. estriola; *pu.* pupilas; *sc.* subcostal; *r.* radio; *sr.* sector del radio; *pr.* procúbito.

VAN DER WEELE (fig. 3), en la cual con mucho aumento se distingue convexa por la parte inferior del ala. Y donde quiera exista la estriola, mirando la membrana con grande aumento, se percibirá estriada en la misma dirección longitudinal de la línea pálida ó estriola.

Con esto se ve que tal modificación del ala no es debida al mero color, sino que es intrínseca á la membrana, ó sea es *órgano particular del ala anterior*.

Existe la estriola en muchas especies y géneros de Neurópteros pertenecientes á diversas familias, pero no parece exclusiva y propia de familia determinada. La he visto en muchos Hemeróbidos (*Boriomyia, Psectra, Sympherobius* etc.), en los Osmílidos (*Osmylus, Polystœchotes*), en los Neurómidos (*Neuromus, Chauliodes, Corydalis*).

Su *longitud* es varia, según las especies, mas puede en general decirse que ni llega á la base del ala, ni al margen externo. En el *Neuromus latratus* enlaza las tres pupilas. Su anchura siempre poca, de algunas centésimas de milimétro, rara vez décimas.

Se *diferencia* notablemente de un órgano análogo ó bolsa que se ha observado en algunos géneros de Tricópteros (*Catadice, Drusus, Eclisopteryx* de los Limnofílidos, *Silo* de los Sericostómidos) 1° en la figura, longitud, anchura, que son muy distintas; 2° la bolsa es propia de los ♂ ♂ de algunas especies, la estriola se halla en ambos sexos; 3° la bolsa está en el ala posterior, la estriola en la anterior. Finalmente la estriola es mucho más general, como se ha visto.

Su *importancia* en taxonomia es mucho menor que la de la pupila, por su mayor universalidad ó extensión. Podrá utilizars para caracterizar perfectamente algunas especies.

Qué *función* posea la estriola, falta averiguarlo. Es de creer que esté relacionada con la de las pupilas, pues en algunos géneros las enlaza ó pone en comunicación. Bien pudiera ser canal auditivo ú olfatorio. Como se encuentra mucho más extendida que la pupila, es de creer que en aquellos géneros y familias que cavecen de ésta hace sus veces.

3. Tiridio.

Este nombre se deriva sin duda del griego θυρίδιον, ventanilla. Con él se designa un pequeño espacio pálido, de ordinario semitransparente, generalmente sin pelo ó pubescencia (Mac Lachlan, A monographic revision and synopsis of the Trichoptera of the European fauna, p. 8, London, 1874-1880).

Este órgano es muy conocido entre los Tricópteros, tanto que se denomina celda ó área tiridial la que está detrás del tiridio ó sea entre las dos ramas del procúbito.

Pero no es exclusivo el tiridio de los Tricópteros, sino que se halla también en otras familias de Neurópteros.

Ya en 1903, en mi trabajo « Neurópteros prosostomios de la península ibérica » hice notar su presencia en los Panórpidos y denominé ramos tiridiales los que de él parten.

Su *posición* es bastante fija. En los Tricópteros y también en los Panórpidos (fig. 4ª) se halla precisamente en la axila de la horquilla

Fig. 4.

Thyridates chilensis Klug. Region tiridial del ala anterior.
r. radio; *sr.* sector del mismo; *pr.* procúbito; *to.* tiridio; *ta.* tiridiola.

ó primera bifurcación del ramo anterior del procúbito. En los Sócidos (Copeognatos) es frecuente ver un tiridio en ambas ramas del sector del radio, uno *anterior* en la axila de la horquilla apical y otro *posterior* en el peciolo de su correspondiente horquilla, en el sitio en que la venilla, cuando existe, cierra la celdilla discal. En ocasiones el tiridio anterior se extiende á la venilla radial y el posterior á las venas y venilla vecinas.

Lo hallo asimismo en los Osmílidos en la primera bifurcación del procúbito del ala anterior. En esto también conviene el *Polystœchotes punctatus* F., razón de más para incluirlo en la misma familia.

En la familia de los Panórpidos, por ser las dos alas de estructura muy semejante, el tiridio se encuentra en ambas.

Como ofrece particularidades notables en algunos géneros y especies, puede ser de mucha *utilidad taxonómica*. En diferentes ocasiones lo he utilizado en la descripción de varias especies de Panórpidos y aun, junto con otros caracteres, para la formación del género *Thyridates* (1) (fig. 4ª).

4. **Tiridiola.**

Como de la palabra núcleo en Citología se forma nucleola, así de la palabra tiridio derivo la de tiridiola.

La mencioné por primera vez en mi trabajo « Panorpides nouveaux du Japon » (Revue russe d'Entomologie, 1905, p. 275) al describir el *Bittacus nipponicus*. La he encontrado además en otros Panórpidos.

Se halla en la venilla que une ambas ramas del procúbito; ó si se quiere, en la axila de la horquilla segunda del ramo anterior del procúbito (fig. 4ª).

Se puede ver en ambas alas.

En muchas especies es casi invisible. En otras, sobre todo en el ala anterior, aparece doble : una mayor en el extremo anterior de dicha venilla (única en la fig. 4ª), invadiendo toda la axila de la referida horquilla, y otra menuda, en el extremo posterior de la misma venilla procubital.

5. **Ostiola** (latín *ostiolum*).

En el extremo del cúbito del ala anterior en muchos Tricópteros se encuentra una manchita pálida, á veces muy visible, que se ha denominado arquillo, *arculus*.

---

(1) Neurópteros nuevos, p. 14. (*Memorias de la Real Academia de Ciencias y Artes de Barcelona*, vol. VI, nº 29. Barcelona, 1908.)

En aquel sitio el cúbito se encorva en forma de un pequeño arco, por lo que la denominación de arquillo es muy propia aplicada á la porción final de la vena misma. No lo es tanto si se aplica á la mancha, en todo parecida al tiridio. Como por otra parte la palabra arquillo tiene una aplicación idéntica para una venilla de los Odonatos, es preferible usar otra palabra, para evitar confusiones.

A este fin, conservando la palabra arquillo para la porción terminal del cúbito, y si se quiere, para la porción adyacente del margen que se arquea, propongo la palabra *ostiola* (del latín *ostiolum*) para designar este órgano marginal del ala anterior de los Tricópteros.

Ni es la ostiola exclusiva de éstos, puesto que es fácil verla en muchas especies de Sócidos (Copeognatos) precisamente en la terminación marginal del cúbito. Su presencia podrá servir de guía para definir cuál vena sea el cúbito, á veces no muy patente en estos Insectos.

En los Sócidos la ostiola se corre á veces por la venilla intercubital, la que cierra la celdilla discal típica en estos Insectos, ó por la celdilla posterior, no menos típica en varios géneros de ellos.

Se ha utilizado la ostiola como elemento taxonómico en los Tricópteros, y lo mismo puede hacerse en los Sócidos y otros Insectos que la posean.

6. **Estigma.**

Los autores antiguos la llamaban *pterostigma* en latín ó en francés ptérostigme (de πτερόν, ala), y así prosiguen llamándolo algunos. Yo uso constantemente la palabra *estigma* por razón de la brevedad y por no ser posible la confusión, pues ya se entiende que se habla del estigma del ala y no de los torácicos ó abdominales.

Es órgano constante en muchas especies y aun familias y órdenes de Insectos (Neurópteros, Ortópteros, Hemípteros, Himenópteros, etc.), y ofrece las más variadas figuras, colores y consistencia. Unas veces es muy aparente, otras está casi desvanecido en cuanto al color, pero se ve mirando con fuerte aumento la región estigmática del ala.

Prescindiendo de su estructura y funciones, que no me incumben, advertiré que es órgano no despreciable en las descripciones.

Los estigmas de algunas especies de Sócidos (*Pterodela*, *Elipsocus*, *Peripsocus*, etc.) ofrecen una particularidad en su basse ó borde interno, que se prolongan hacia atrás en forma de talón ó muleta. Esta prominencia, muy característica en varios géneros, podría llamarse *cornisa*, ó en latín *crepido*, por su forma y hallarse en la base del estigma.

7. **Venillas discales.**

Parto del principio de ser mejor, por más exacto, llamar *venas* y *venillas* les líneas salientes que forman la reticulación ó malla de las alas : venas las longitudinales, venillas las transversales. Y con decir « venillas » no hay que añadir que son transversales, pues ya se entiende que van de una vena á otra.

Las venillas toman su denominación particular del nombre del campo en que se encuentran, ó de la vena que les antecede. Así se llamarán venillas *costales*, *subcostales*, *radiales* (entre el radio y su sector ó sectores), *intermedias* (entre el sector y el procúbito ó cúbito anterior), *intercubitales*, *cubitales*, etc. Las que se acercan á la base del ala llámanse en general *basilares* y las que caen hacia el medio *discales*.

Entre las discales son dignas de atención las *gradiformes*, dispuestas en escalera oblicua hacia el centro del ala. Cuando hay dos series, como con frecuencia sucede, su número en gracia de la brevedad lo expreso con una fórmula de quebrado. Así $^5/_7$ significa que hay 5 venillas en la serie gradiforme interna y 7 en la externa.

Tales venillas gradiformes pueden hallarse también en el campo costal en serie entre las venillas costales ordinarias, dividiendo á veces en dos el campo costal, como sucede en algunos géneros de Osmílidos y Mirmeléonidos. Con decir *venillas gradiformes* está quitada toda ambigüedad y no pueden confundirse con las propias costales que están entre dos venas, yendo de la costal á la subcostal.

Ueber Cymatophora or F. ab. albingensis Warn.

von Dr. K. HASEBROEK (Hamburg).

Im Brennpunkt aller descendenztheoretischen Ueberlegungen steht die Frage nach der Entstehung neuer guter Arten aus vorhandenen Abarten.

Der Melanismus der Falter ist schon seit langem für wert erachtet worden, in dieser Beziehung schärfer beobachtet zu werden. Es muss von fundamentaler Bedeutung sein, der Entwicklung einer neuen melanistischen Form in den Phasen ihrer Entstehung näher zu treten, zunächst einmal den *Beginn des Auftretens festzustellen*. Gerade der Beginn, die erstmalige Erscheinung ist möglichst genau festzulegen und zu analysiren.

In der *ab. albingensis* der *C. or* haben wir, wie es scheint, *eine überhaupt noch niemals und nirgends vorher beobachtete Form* des *Melanismus*. Wir haben ferner in ihr einen qualitativ intensiv ausgefärbten Melanismus, der so ausnahmslos weder bei *Amph. betularia ab. doubledayaria* noch *Ps. monacha ab. eremita* angetroffen wird. Das neue Tier steht einzig da.

Die Beschreibung der Type durch Herrn WARNECKE ist folgende : « Nigra, maculis albis », die Abbildung findet sich Intern. Entomol. Ztg. (Stuttgart) XXII, 1908, p. 126, und Intern. Entom. Ztg. (Guben), 1911, Nr. 6.

Diese Abart ist ausschliesslich bei Hamburg bisjetzt aufgetreten, zuerst 1904. Die ältesten Sammler haben sie bisher *niemals* trotz vieler Zuchten seit 50 Jahren beobachtet. Weitere Falter sind gefunden : 1905 = 1 Stück, 1906 = 1 Stück, 1907 = 2 Stück, 1908 = 2 Stück, 1909 = 10 Stück, 1910 = 30-40 Stück, letztere auch nunmehr in II. Generation gezogen und zwar aus Copula Abart × Stammform mit 50 % Stammform und 50 % Abart, ohne Uebergänge. Wir haben also eine ausgesprochen in *sich festliegende* melanistische Abart, *ohne Uebergänge zur Stammform.*

Das wichtigste Moment für die Entstehung der Abart ist die Oertlichkeit *Hamburg.* Hamburg ist auch sonst reich an Melanismen. Wir haben sie gefunden sowohl in Moorgegend als in der Heide.

Es handelt sich nun zunächst um die Frage, die ich hier auf dem Congress vorlege : *Ist irgendwo, sei es in früheren Zeiten oder in den letzten Jahren, unsere Abart* albingensis *gefunden worden? Ich knüpfe hier die Bitte an, im Fall sie gefunden wird, sobald als möglich darüber mir zu berichten.*

Da unsere Abart *albingensis* im Erscheinen begriffen ist, so sollten wir Lepidopterologen uns es besonders angelegen sein lassen, sofort die Weiterzucht zu erstreben : auch wenn es nur gelingen sollte, ihre *relative Festigkeit* jetzt, im Beginn ihres Auftretens, gegenüber späteren Jahren festzustellen, so wäre dies eine Tatsache, die im Lichte weiterer Erkenntniss von grösster Bedeutung werden könnte.

Aperçu sur la distribution géographique et la phylogénie des Fourmis,

par le Dr FOREL (Yvorne).

Le matériel de la famille des Formides, tel que nous le connaissons aujourd'hui, est considérable. Il se compose d'environ 4,877 espèces et sous-espèces ou races et de 1,200 variétés vivantes connues, ainsi que de 171 espèces et 18 variétés fossiles décrites, dont 64 de l'ambre baltique et sicilien.

Ces formes sont réparties dans environ 184 genres vivants et dans 12 genres exclusivement fossiles (6 de l'ambre et 6 autres). Je dis environ, les appréciations variant sur les genres et les sous-genres.

Parmi les 6 genres fossiles proprement dits (autres que ceux de l'ambre), 4 fondés par HEER sont plus que problématiques. 33 genres sont communs aux Fourmis vivantes et fossiles.

Les 196 genres de Formicides se répartissent en 5 sous-familles naturelles :

1. Ponerinæ.
2. Dorylinæ.
3. Myrmicinæ.
4. Dolichoderinæ.
5. Camponotinæ.

Si nous divisons la terre en ses grandes faunes, nous trouvons à peu près le groupement suivant :

		Espèces et races ou sous-espèces.	Variétés géographiques.
I. *Faune néotropique*	a) Amérique du Sud, sauf la Patagonie	961	208
	b) Amérique centrale	506	121
II. *Faune éthiopienne* (Afrique, au sud du Sahara). . . .		629	125
III. *Faune malgache* (Madagascar, Comores, Seychelles, Chagos, etc.)		230	63
IV. *Faune indo-malaise* (Inde, Indo-Chine, Andaman, Ceylan, îles de la Sonde, Philippines et une partie de la Chine et du Japon).		1,165	210
V. *Faune papoue et océanienne* (Moluques, Nouvelle-Guinée, Océanie)		335	42
VI. *Faune australienne* (Australie, Nouvelle-Calédonie, Tasmanie)		380	105
VII. *Faune paléarctique*	a) paléarctique proprement dite.	152	116
	b) méditerranéenne (avec N Afrique, Sahara, Asie Mineure, etc.)	294	158
VIII. *Faune néarctique* (Amérique du Nord)		352	105
IX. *Faune antarctique* (Nouvelle-Zélande, Patagonie) .		27	3
	Total. . .	5,031	1,256

L'excédent de 154 espèces et 56 variétés provient simplement de ce que ces 210 formes ont été comptées à double ou à triple, lorsqu'elles font partie de deux ou trois grandes faunes, surtout de la faune de l'Amérique centrale et de l'Amérique du Sud. On peut même à bon droit s'étonner de ce que le nombre des espèces communes à deux ou trois grandes faunes voisines n'apparaisse pas plus grand, ce qui provient sans doute de données faunistiques incomplètes dans les catalogues.

Il faut encore ajouter à la somme des 4,877 espèces et sous-espèces des grandes faunes 11 espèces cosmopolites transportées partout par les navires. Ce sont les suivantes :

Odontomachus hæmatodes L.
Monomorium destructor JERDON
— *floricola* JERDON
— *Pharaonis* L.
Solenopsis geminata F.
Pheidole megacephala F.
Tetramorium guineense F.
— *simillimum* SMITH
Tapinoma melanocephalum F.
Prenolepis longicornis LTR.
— *vividula* NYL.

Quatre espèces, dont deux d'origine indo-malaise et deux d'origine néotropique, sont actuellement en train de devenir cosmopolites par les transports. Ce sont :

Plagiolepis longipes JERDON
Triglyphothrix striatidens EMERY
Iridomyrmex humilis MAYR
Cardiocondyla Emeryi FOREL

Si nous ajoutons les 11 espèces décidément cosmopolites dont l'origine est incertaine, aux 4,877 espèces et races de grandes faunes, nous obtenons 4,888 espèces et races et 1,200 variétés, donc en tout 6,088 formes vivantes actuellement décrites. J'ajoute que ce compte a été fait en février 1910. Dès lors, environ 166 formes nouvelles ont paru dans divers travaux, ce qui porte le nombre des espèces et races vivantes décrites à 5,000 et celui des variétés à 1,250 en chiffre rond.

Ces chiffres bruts sont très relatifs et seulement approximatifs. Ils renferment des espèces douteuses. Mais ils donnent une vue d'ensemble sur ce qui est décrit et connu jusqu'aujourd'hui.

Les plus grands genres de Formicides sont *Camponotus* MAYR

avec 880 espèces, races et variétés, *Pheidole* WESTW. avec 505, *Polyrhachis* SHUCK avec 406 et *Cremastogaster* LUND avec 358.

Avant de passer aux grandes faunes et sous-faunes, je tiens à faire remarquer que le tableau ci-dessus est un peu artificiellement établi selon les besoins et le matériel. Si j'avais eu plus de matériel, j'aurais dû, par exemple, séparer de la faune néotropique la faune chilienne à l'ouest des Andes bien plutôt que celle de l'Amérique centrale, qui en est fort mal délimitée.

1. — **Faune néotropique et faune de l'Amérique centrale.**

Cette magnifique faune, relativement récente, se distingue par l'absence du genre *Polyrhachis* et de presque tous les genres archaïques constituant les reliquats de la faune de l'ancien monde, tels que les genres *Myrmecia*, *Harpegnathos*, *Mystrium*, *Myopopone*, etc. Elle contient néanmoins l'un d'eux, les *Stigmatomma*. La sous-famille des Dorylines y est représentée seulement par les genres *Eciton* et *Cheliomyrmex* qui sont exclusivement néotropiques. Mais la faune néotropique possède en outre des genres spéciaux très développés, tels que les *Cryptocerus*, *Procryptocerus*, *Pogonomyrmex*, *Pseudomyrma* et *Azteca*. Ces deux derniers sont spécialisés et adaptés à la vie arboricole. Ils fourmillent dans la forêt vierge américaine, où ils habitent les tiges et branches creuses, ou des nids en carton (les *Azteca*). D'une façon générale, dans la faune néotropique, une foule d'espèces de Fourmis habitent les cavités des plantes vivantes ou mortes. Les unes sont phylogéniquement adaptées à ces cavités, comme certaines *Azteca* et *Pseudomyrma* aux *Triplaris*, aux *Tococa*, *Coussapoa*, *Tachigalia*, *Duroia*, *Cecropia*, etc. Les autres habitent sans distinction les tiges sèches et creuses des broussailles ou autres cavités végétales; certaines même (*Pseudomyrma*) les tiges des graminées. Très intéressants sont les jardins de Fourmis dans les racines de touffes d'épiphytes (*Azteca Ulei*, *Camponotus femoratus*, etc.).

Les principaux genres spéciaux à la faune néotropique et à ses sous-faunes de l'Amérique centrale et du Chili sont *Typhlomyrmex*, *Cylindromyrmex*, *Belonopelta*, *Paraponera*, *Emeryella*, *Acanthoponera*, *Ectatomma* s. str. avec les sous-genres *Holcoponera* et *Gnamptogenys*, *Dinoponera*, *Neoponera*, *Thaumatomyrmex*, *Alfa-*

ria, Acanthostichus, Eciton, Daceton, Acanthognathus, Megalomyrmex, Ochetomyrmex, Allomerus, Wasmannia, Ischnomyrmex, Tranopelta, Forelius, Myrmelachista, Brachymyrmex, Gigantiops et le sous-genre *Dendromyrmex*. Le genre *Pogonomyrmex* lui est commun avec la faune néarctique et le genre *Dorymyrmex* avec la faune *antarctique*.

Mais le groupe le plus typique de la faune néotropique est le groupe ou la tribu des *Attii*, avec ses jardins de champignons. Le genre *Atta* en forme l'épanouissement le plus récent, avec les mœurs les plus spécialisées. C'est en coupant les feuilles des arbres de la forêt pour y établir leurs champignoneries que les *Atta* ont pu atteindre leur grande taille. On peut affirmer, je crois, en se basant tant sur leurs mœurs si spécialement adaptées que sur leur structure, qu'elles sont dérivées des genres de petits *Attii*, tels que les genres *Cyphomyrmex*, *Sericomyrmex* et *Apterostigma*, dont l'instinct horticole, comme l'a prouvé MÖLLER, est bien plus rudimentaire. Tous les *Attii* sont néotropiques. De même, les *Wasmannia* qui font des jardins temporaires de champignons. Je suis d'avis que tout le groupe néotropique des *Atti* dérive du groupe universel des *Strumigenii*, dont quelques genres surtout néotropiques, en particulier *Rhopalothrix* et *Ceratobasis*, se rapprochent beaucoup des *Cyphomyrmex*.

La faune chilienne est très curieuse, mais encore trop peu connue. Jusqu'ici on n'y a trouvé ni *Attii*, ni *Azteca*, ni *Cryptocerus*. Elle a des affinités marquées avec la faune antarctique, dont certaines formes semblent avoir immigré au Chili et dans les Andes, ainsi les *Melophorus*. Par contre, elle possède une *Pseudomyrma* et des *Myrmelachista* franchement néotropiques.

La faune de l'Amérique centrale n'est qu'une sous-faune de la faune néotropique. Elle se subdivise elle-même. Le Mexique et les grandes Antilles sont spécialement distincts du reste de la faune néotropique, tandis que du côté de Panama le mélange se fait. Les petites Antilles sont plus pauvres et en partie seulement spécifiées. Le genre *Macromischa* semble spécial à l'Amérique centrale, aux grandes Antilles et aux Bahamas.

En outre, on peut distinguer dans la faune néotropique celle de l'ouest des Andes de celle des grands bassins de l'Amazone, du Magdalena, de l'Orénoque et du Rio de La Plata. Les différences sont considérables.

II. — Faune éthiopienne.

Cette faune est, en somme, fort homogène et bien moins riche que la précédente. Elle se distingue par l'absence des *Ectatomma*, par sa pauvreté relative en *Dolichoderines* et par sa richesse en *Dorylines* de la tribu des *Dorylii*, c'est-à-dire en grandes Fourmis dites de visite (*Dorylus*, *Rhogmus*, *Anomma*). Les sous genres *Rhogmus* et *Anomma* sont propres à l'Afrique. Le genre *Ænictus* lui est commun avec la faune indo-malaise seule.

Elle offre une sous-faune tempérée, celle de l'Afrique du Sud, et de remarquables spécialisations désertiennes, dans le Kalahari surtout, mais aussi au Soudan et ailleurs.

Le singulier genre *Myrmicaria* lui est commun avec la faune indo-malaise, tandis qu'elle possède un *Atopomyrmex* qui la relie à Madagascar et à la faune indo-malaise. Le genre *Cremastogaster* y est d'une richesse et d'une complication remarquables. Par le bassin du Nil, elle vient rejoindre la faune méditerranéenne. L'Afrique tropicale est très riche surtout en grands Ponérines. Citons parmi eux les genres *Paltothyreus*, *Streblognathus*, *Plectroctena*, *Megaloponera*, *Ophthalmopone*, *Hagensia*, *Psalidomyrmex*, *Probolomyrmex* et *Escherichia* qui sont propres à sa faune. Parmi les Myrmicines, les genres *Melissotarsus*, *Diplomorium*, *Dicroaspis*, *Cratomyrmex* et *Ocymyrmex*, parmi les Dolichoderines, *Semonius*, *Engramma* et *Ecphorella* sont spéciaux à l'Afrique. Le genre *Aphomomyrmex* est presque identique aux *Myrmelachista* d'Amérique.

Les différences entre l'Afrique orientale et occidentale sont minimes, au contraire de l'Amérique du Sud, évidemment, parce qu'il n'y a pas de chaîne de hautes montagnes séparatrices comme celle des Andes.

III. — Faune malgache.

Cette singulière et antique faune de reliquats, faune dont j'ai fait une étude fort spéciale, s'étend aux Comores, aux Seychelles, aux Admirantes, aux îles Mascarènes et, enfin, aux îles Chagos, comme j'ai pu le prouver.

Elle se distingue par l'absence absolue de *Dorylines* et de *Polyrhachis*. Les *Dolichoderines* y sont aussi mal partagés qu'en Afrique. Par contre, le genre *Mystrium*, un reliquat dont on ne trouve qu'une autre espèce en Birmanie, et les genres *Simopone* et *Eutetramorium* lui sont spéciaux.

Les *Atopomyrmex* de Madagascar sont très voisins des *Podomyrma* d'Australie et de Nouvelle-Guinée. Le genre *Aëromyrma* de la faune malgache lui est commun avec l'ambre et avec l'Afrique. En somme, c'est une faune antique et spéciale. Ses affinités profondes la rapprochent de la faune papoue surtout. Elle offre avec la faune d'Afrique des échanges anciens et récents par immigration mutuelle de certaines espèces africaines immigrées à Madagascar et malgaches immigrées en Afrique. L'île intermédiaire « Europa » offre une faune plutôt malgache.

IV. — **Faune indo-malaise.**

Comprend l'Inde, Ceylan, l'Indo-Chine, le sud de la Chine, le sud du Japon, les Andamanes, les Nicobares, les îles de la Sonde et les Philippines. Elle possède des sous-faunes importantes dans chaque île et en outre une sous-faune de Ceylan, une autre du continent du Decan, une de Birmanie et d'Assam, une du nord-ouest et enfin une sous-faune de l'Himalaya.

D'après M. SARASIN, de Bâle, pour les Reptiles et les Amphibiens, Ceylan forme, avec la pointe sud-ouest de l'Inde (Kanara, Travancore), une faune spéciale bien plus riche que la faune continentale pauvre du Decan, et fort différente d'elle. Pour les Fourmis il n'en est pas tout à fait de même.

Sans doute le continent du Decan est relativement fort pauvre. Mais il n'est pas possible de trouver une limite nette entre le Kanara et le reste du continent, tandis que l'île de Ceylan offre des différences très marquées et ne peut être confondue avec le Kanara, dont la faune est très différente.

Je n'affirme pas cela à la légère, car je me base sur l'étude d'environ 650 formes de la faune de l'Inde et de Ceylan, que j'ai très spécialement étudiée.

J'en donne un exemple sur le genre *Polyrhachis*, si riche en espèces locales, et qui compte 61 espèces dans la faune de l'Inde et de Ceylan. Il existe 22 espèces de *Polyrhachis* à Ceylan, 42 dans la faune de Birmanie-Assam, 10 seulement dans le Kanara et à Travancore et 9 dans l'Inde centrale.

Sur ce nombre, 3 espèces sont communes aux quatre régions en question, 2 à Ceylan et à la Birmanie, 2 à Ceylan, Kanara et la Birmanie, 2 à Ceylan, Kanara et l'Inde centrale, 2 à la Birmanie et à l'Inde centrale, une enfin à Kanara et à la Birmanie, et 49 n'ont été trouvées que dans une seule des quatre régions. *Aucune espèce n'a été trouvée jusqu'ici exclusivement au Kanara et à Ceylan.*

Par contre, 13 espèces sont spéciales à Ceylan, 3 au Kanara et 2 à l'Inde centrale. Il me semble que c'est clair.

En outre, l'Inde continentale offre au nord-ouest une sous-faune désertienne (Rayputana) qui a de grandes analogies avec la partie orientale de la faune méditerranéenne, si bien qu'on peut la lui rattacher.

Par contre, je suis pour les Fourmis absolument d'accord avec les résultats de M. SARASIN en ce qui concerne la grande richesse de la faune d'Assam et de l'Indo-Chine comparée à celle du Decan. Cette magnifique sous-faune est bien distincte de celle du continent de l'Inde proprement dite et bien plus rapprochée de celle des îles de la Sonde. En particulier, les montagnes de Birmanie offrent des reliquats antiques tout à fait remarquables. Je cite le *Mystrium Camillae* EMERY, voisin des *Mystrium* malgaches, et le singulier genre *Myrmoteras* FOREL, qui représente un Camponotine tout à fait primitif avec certains caractères des Ponérines primordiaux.

Enfin, la faune de l'Himalaya offre un immense intérêt. Seule des faunes connues des montagnes, avec celle des Alpes birmanes, elle offre un nombre considérable d'espèces spéciales, c'est-à-dire d'espèces exclusivement himalayennes. Dans une étude spéciale que j'ai publiée, j'ai compté dans l'Himalaya 54 formes de Fourmis propres à l'Himalaya, 51 autres formes indo-malaises qui se trouvent aussi dans la plaine du Decan et sur ses collines, enfin 10 formes paléarctiques émigrées surtout dans l'Himalaya occidental. Les formes himalayennes propres laissent en bonne partie reconnaître une dérivation phylogénique indo-malaise ou paléarctique selon les espèces ou variétés.

La faune des îles Andamans et Nicobares a quelques formes spéciales; mais elle tient en somme de celles du Decan et de l'Indo-Chine.

La faune indo-malaise est extrêmement riche en *Polyrhachis*,

Ænictus, *Pheidole* et *Camponotus* ; elle ne possède par contre que 3 espèces de *Dorylus* (sous-genres *Alaopone* et *Typhlopone*); les sous genres *Anomma* et *Rhogmus* d'Afrique n'y existent pas. Parmi les reliquats il faut encore citer les genres *Harpegnathos* JERD. et *Myopopone* ROG. Le genre *Myrmicaria* lui est commun avec l'Afrique.

En propre, la faune indo-malaise possède encore les genres *Odontoponera*, *Cryptopone*, *Trigonogaster* et *Lophomyrmex* et *Rhopalomastix*, et le sous-genre *Stictoponera*, sans parler de ceux propres à Ceylan. Les genres *Echinopla* et *Diacamma* lui sont communs avec l'Australie, le genre *Vollenhovia* avec les faunes papoue, australienne et malgache, le genre *Pristomyrmex* avec les faunes papoue et australienne, enfin les genres *Pseudolasius* et *Liomyrmex* avec la faune papoue. Les genres *Dimorphomyrmex* et *Gesomyrmex* sont des reliquats spéciaux à Bornéo et à l'ambre.

L'île de Ceylan renferme un reliquat des plus curieux, le genre *Aneuretus* EMERY qui seul fait passage direct des Ponerines aux Dolichoderines. Puis elle possède, ainsi que Célèbes, un *Atopomyrmex* qui vient relier la faune malgache à la faune papoue et australienne. La *Mesoponera melanaria* de Ceylan a une sous-espèce australienne. Le sous-genre *Hemioptica* ROG. et quelques genres comme *Stereomyrmex* et *Acanthomyrmex* sont propres à Ceylan.

Enfin, le genre *Rhopalomastix* FOREL de Ceylan et de l'Inde centrale, tout en appartenant aux Myrmicines, offre de curieuses affinités avec les Ponerines, surtout par le ♂, et fait un peu transition entre les deux sous-familles (près de *Melissotarsus* d'Afrique).

V. — **Faune papoue.**

Cette belle et riche faune comprend les Moluques, la Nouvelle-Guinée, l'archipel de Bismarck et quelques autres îles. Elle offre des affinités plus grandes avec la faune australienne qu'avec la faune indo-malaise. Comme la faune malgache, elle ne renferme pas de Dorylines, mais par contre beaucoup de *Polyrhachis*, qui font défaut à Madagascar. Les *Podomyrma*, qui lui sont communes avec l'Australie, sont bien voisines des *Atopomyrmex* malgaches. Une foule d'espèces et de genres curieux, des *Pheidole* et des

Cremastogaster à épines ramifiées, etc., lui donnent un caractère très particulier. Les genres *Rhopalothrix, Rogeria, Prionopelta,* et le sous-genre *Rhizomyrmex* lui sont communs avec la faune néotropique, ce qui constitue un fait extrêmement curieux et remarquable; nous le devons aux recherches d'Emery qui a surtout étudié cette faune. Les genres *Podomyrma, Epopostruma, Orectognathus,* le sous-genre *Rhytidoponera,* etc., lui sont communs avec la faune australienne, les genres *Pheidologeton, Vollenhovia, Pristomyrmex,* etc., avec la faune indo-malaise. Sauf une espèce de Célèbes, le genre *Trapeziopelta* lui est propre. Le genre *Adelomyrmex* lui est spécial.

La sous-faune des îles océaniennes a été en grande partie envahie et détruite par les cosmopolites et les transports des faunes continentales, ainsi que celle des îles Sandwich. Néanmoins, dans diverses îles (Viti, Fidji, Salomon, etc.), il reste des formes spéciales et remarquables, parentes surtout de la faune papoue.

VI. — **Faune australienne.**

Cette faune contient sans contredit les plus belles espèces de Fourmis aux couleurs métalliques, etc. C'est, on le sait, une faune antique de reliquats, avec développements spéciaux. On peut la diviser en quatre parties principales sur le continent australien, sans parler des sous-faunes de la Nouvelle-Calédonie et de la Tasmanie. Ce sont :

1. La faune australienne proprement dite du sud-est et du sud;
2. La faune du Queensland, extrêmement riche, et faisant transition directe à la faune papoue par le cap York et les îles du détroit de Torres;
3. La faune désertienne de l'Australie centrale;
4. La faune assez distincte de l'Australie occidentale.

La faune australienne se distingue par l'absence des *Dorylinæ*, sauf deux espèces d'*Ænictus* si voisines d'espèces de l'Inde que leur importation relativement récente ne semble faire aucun doute.

Les Dolichoderines, extrêmement abondants, y offrent les genres spéciaux *Leptomyrmex, Froggattella, Turneria* et une grande richesse en *Iridomyrmex*. Ce dernier genre y présente des formes adaptées aux plantes et semblables aux *Azteca* d'Amérique.

Parmi les nombreux reliquats antiques, citons le magnifique genre *Myrmecia*, les genres *Prionogenys*, *Onychomyrmex*, *Notoncus*, *Opisthopsis*, les nombreux *Sphinctomyrmex*, *Amblyopone*, *Orectognathus*, etc. Les genres *Machomyrma*, *Mayriella*, *Dacryon*, *Lordomyrma*, *Myrmicorhyncha*, et le sous-genre *Stigmacros* lui sont aussi propres. Le genre *Melophorus* est propre aux faunes *antarctique* et *australienne*. Dans la faune désertienne de l'Australie centrale, il présente de nombreuses formes melligères, adaptées aux grandes sécheresses (provisions de miel dans le jabot). Il en est de même de certains *Camponotus* et *Leptomyrmex*.

VII. — **Faune paléarctique.**

La faune paléarctique proprement dite, celle des bois, des prairies et des montagnes de la région tempérée et froide, doit être distinguée de la sous-faune xérothermique et désertienne dite méditerranéenne. Cette dernière, bien plus riche, s'étend au nord de l'Afrique, à l'Asie Mineure, à l'Asie centrale et à une partie de la Chine, confinant aux faunes tropicales éthiopienne et indomalaise par des déserts ou de hautes montagnes (Himalaya, Birmanie).

La faune paléarctique proprement dite, relativement à son immense territoire la plus pauvre du monde, se distingue par les genres *Formica*, *Polyergus*, *Lasius*, *Myrmica*, *Harpagoxenus* et *Stenamma*, qui lui sont communs avec la faune néarctique, par les genres *Anergates* et *Formicoxenus*, qui lui sont propres, et par l'absence des Dorylines, la présence de deux seules espèces de Ponerines, l'absence de *Pheidole*, de *Cremastogaster*, de *Polyrhachis*, de *Monomorium*, de *Messor*, de *Prenolepis*, etc. Elle ne renferme que deux espèces de Dolichodérines. Le genre *Strongylognathus* lui est commun avec la faune méditerranéenne seule.

Cette dernière n'a que des Dorylines importés de la faune éthiopienne et une espèce de *Polyrhachis* (*simplex* MAYR) venue de l'Indo jusqu'en Asie Mineure. Elle est caractérisée par le genre *Myrmecocystus* (sous-genre *Cataglyphis* FORST.) et par de nombreux *Messor* et *Aphænogaster*, mais elle ne renferme guère que de petits Ponérines. Détruite par la periode glaciaire, la faune paléarctique n'a guère pu se repeupler dès lors que par les faunes désertiennes, ce qui explique sa pauvreté, surtout chez les Fourmis qui ont si peu de formes adaptées au froid. Le genre *Oxyopomyrmex* (avec le sous-genre *Goniomma*), ainsi que les genres parasites *Hagioxenus, Wheeleriella*, *Epixenus*, *Sifolinia*, *Phacota* et *Myrmoxenus* sont spéciaux à la sous-faune méditerranéenne. Il en est de même du sous-genre *Proformica*.

VIII. — Faune néarctique.

Bien plus riche que la faune paléarctique, elle possède en somme les mêmes genres, avec beaucoup plus d'espèces, dont une partie ne constitue que des sous-espèces ou variétés de leurs correspondantes paléarctiques. Il en est ainsi des *Formica rufa* et *fusca*, des *Camponotus herculeanus* et *fallax*, des *Lasius niger* et *umbratus*, de la *Myrmica rubra*, etc. Par contre, la faune néarctique présente les genres parasitaires spéciaux *Epœcus*, *Epipheidole*, *Sympheidole*, *Symmyrmica*. Un groupe de genres néotropiques possèdent des formes néarctiques dérivées d'eux, ainsi les *Pogonomyrmex*, les *Forelius*, les *Dorymyrmex* et surtout les *Atta* (*Trachymyrmex*) *septentrionalis*, *turrifex* et *arizonensis*, ainsi que l'*Atta* (*Moellerius*) *versicolor* des *Attii*.

Les formes tout à fait arctiques relèvent enfin de l'ancienne faune arctique commune aux deux hémisphères. Parmi elles, il faut surtout citer le *Camponotus herculeanus* L.

Quelques espèces paléarctiques ont été évidemment importées récemment dans l'Amérique du Nord, en particulier le *Tetramorium cæspitum* L.

Notons enfin les *Myrmecocystus* sens strict qui habitent la partie chaude et sèche des États occidentaux et qui s'étendent au Mexique dans la région de l'Amérique centrale.

IX. — Faune antarctique.

Comme je l'ai fait remarquer ailleurs, cette faune ne présente aucune parenté avec la faune arctique, mais seulement des convergences dues au froid. Elle possède un genre (*Huberia*) spécial à la Nouvelle-Zélande, et un autre (*Melophorus*) qui lui est commun avec la Patagonie, le Chili et l'Australie. Les Dorylines, les *Camponotus*, les *Polyrhachis*, les *Pheidole*, les *Cremastogaster* lui font absolument défaut. Le genre tropical *Monomorium* y a, par contre, développé toute une série de formes adaptées au froid, ce qu'il n'a pas fait dans le nord. Comme les *Melophorus* et les *Monomorium* antarctiques, le sous-genre *Acanthoponera* est commun à la Nouvelle-Zélande, à la Patagonie et au sud de l'Amérique du Sud. Deux *Amblyopone* y représentent les reliquats antiques (en Nouvelle-Zelande) et deux *Dorymyrmex* (en Patagonie et sur les Andes de l'Argentine) les Dolichodérines. La faune antarctique renferme en outre, en Nouvelle-Zélande, une *Ponera*, une *Euponera* (*Mesoponera*), une *Discothyrea*, deux *Orectognathus* et une *Strumigenys* ; c'est tout.

Je ne cite que pour mémoire le *Camponotus Werthi* FOREL trouvé dans la maison des explorateurs de Kerguelen, car c'est sans aucun doute une espèce importée. Aucun genre spécialement paléarctique ne se trouve dans la faune antarctique et vice versa. Les genres *Ponera* et *Strumigenys* sont les seuls genres tropicaux qui soient représentés dans les faunes arctique et antarctique, et encore le second ne l'est-il que dans la sous faune méditerranéenne; or ce sont deux genres universels et très répandus.

Ces faits généraux posés, comparons la faune fossile à la faune actuelle :

AMBRE.

A côté de genres actuels paléarctiques (*Formica*, *Lasius*, *etc.*) et tropicaux (*Iridomyrmex*, *Ectatomma*, *Podomyrma*, *Œcophylla*, *Leptomyrmex*, etc.), nous trouvons dans l'ambre baltique et sicilien les genres spéciaux de Myrmicines *Lampromyrmex*, *Stigmomyrmex* et *Enneamerus* situés près des *Aëromyrma* et des *Allomerus* actuels, *Hypopomyrmex* voisin de *Strumigenys*, le genre de Ponerine *Prionomyrmex* voisin de *Harpegnathos* et *Myrmecia*, enfin le genre de Camponotine *Rhopalomyrmex* voisin de *Gæsomyrmex* et de *Melophorus*. Les genres vivants *Aëromyrma*, *Gæsomyrmex*, *Œcophylla* et *Bradoponera* sont fortement représentés dans l'ambre, qui possédait donc une riche faune tertiaire tropicale en Europe. Quant aux *Macromischa* de l'ambre, elles me semblent aussi sujettes à caution que celles d'Afrique, qui sont des *Tetramorium*.

Fossiles proprement dits.

Les seuls genres spéciaux bien fixés sont *Attopsis*, voisin de *Cataulacus*, et *Lonchomyrmex*, voisin de *Myrmicaria*. Le genre *Poneropsis* est mal délimité et les autres sont douteux.

En somme, surtout dans l'ambre, les *Dolichoderines* et les *Camponotines* primordiaux (*Gæsomyrmex*, *Rhopalomyrmex*) prédominent.

Mais il ne faut pas oublier que l'ambre est relativement récent et que les vrais fossiles sont exceptionnels chez les Fourmis. Ce sont surtout les Fourmis ailées qui se trouvent comme fossiles. Elles ont évidemment été emprisonnées au moment où elles essaimaient, et c'est là ce qui explique l'abondance de certains genres qui essaiment en masse et l'absence d'autres genres où les accouplements sont plus individuels. On aurait donc tort de conclure de l'absence de ces derniers, dans le matériel dont nous disposons, à leur absence réelle à l'époque desdits fossiles. Il s'agit là de certains genres qui sans aucun doute constituent actuellement des reliquats fort primordiaux, tels que les genres *Mystrium*, *Amblyopone*, etc.

Phylogénie des Fourmis.

En effet, si nous cherchons à reconstruire l'arbre généalogique des Fourmis, nous en revenons toujours à leurs cousins les plus germains, les Mutillides et les *Apterogyna*, c'est-à-dire à des Hyménoptères porte-aiguillon à vie solitaire, sans pédicule abdominal, mais avec des ♀ aptères. Les rapports entre les ♂ de certains Ponérines et les ♂ des Mutillides sont frappants, si bien qu'Henri de Saussure m'affirmait que le ♂ d'un *Mystrium* récolté par Grandidier était un *Mutillide*. J'eus beaucoup de peine à le convaincre de son erreur. Ce qui m'engage à considérer la tribu des *Amblyoponii* comme le groupe le plus ancien des Fourmis connues, ce sont lesdites affinités et surtout le fait que le pédicule est si largement soudé au premier segment de l'abdomen proprement dit qu'il commence chez ces genres à prendre l'aspect d'un premier segment abdominal ordinaire, comme chez les Mutilles. Néanmoins les différences sont encore considérables, les ♀ sont ailées, etc.

EMERY croit que les ♀ des Fourmis descendent d'Hyménoptères à ♀ aptères, et que leurs ailes caduques se sont rééditées par suite des besoins de croisement dus à la vie sociale. C'est possible, mais ce sont là des hypothèses.

EMERY a fini par se ranger à mon opinion, adoptée par les autres Myrmécologistes, et qui consiste à rattacher aux *Ponérines* et non aux Dorylines les *Cerapachyi* et groupes voisins.

Cela dit, nous devons admettre pour les Formicides un ancêtre disparu, plus ou moins parent des Mutilles, des *Apterogyna* et des *Amblyoponii*. De cet ancêtre sont sortis d'abord les *Amblyoponii*, puis les autres Ponérines; cela me paraît hors de doute. Ceux-ci ont dû se diviser peu à peu en divers groupes. De l'un d'eux, les *Cerapachyi*, sont sortis les Dorylines. J'accorde la chose à EMERY ; cela me paraît hors de doute. Mais d'où sont sorties les trois autres sous-familles? Si nous prenons d'abord les Myrmicines, nous trouvons que leurs affinités sont surtout marquées avec des *Ponérines* plutôt primitifs, ainsi avec le genre *Myrmecia* et surtout avec les *Cerapachyi*. Le genre *Rhopalomastix* FOREL est fort instructif à cet égard, faisant transition entre les deux sous-familles. Je crois que les Myrmicines se sont détachés peu à peu desdits groupes des Ponérines chez lesquels le deuxième segment abdominal tend fortement, en se rétrécissant, à former un postpétiole. Donc les Myrmicines proviendraient à peu près du même groupe que les Dorylines. Mais ces derniers se différenciaient par les exigences de leur vie nomade et chasseuse, les Myrmicines, au contraire, par un plus grand développement de la vie sociale sédentaire, encore fort primitive chez les Ponérines primordiaux.

Nous avons vu le genre *Aneuretus* EMERY, grâce à son aiguillon, à la forme de son pédicule et aux caractères de son abdomen et de sa chitine, constituer une transition des Ponérines aux Dolichoderines. C'est là un anneau très intéressant de la chaîne phylogénique. Mais ce genre se rattache sans aucun doute à un tout autre groupe de Ponérines que les *Cerapachyi* et les *Myrmecia*. Il se rapproche bien plutôt des *Ponera* et formes analogues

Quant aux Camponotines, l'étude du gésier de leurs formes évidemment les plus primitives : *Melophorus, Gaesomyrmex, Dimorphomyrmex, Notoncus,* et surtout *Myrmoteras,* m'amène de plus en plus à comprendre qu'ils ne sont pas dérivés des *Dolichodérines*, comme je l'ai cru autrefois, mais directement des *Ponérines* par l'atrophie de l'aiguillon, le développement de la vie sociale, la transformation de l'appareil vénénifique et du gésier. Je crois que l'avenir éclaircira la question, mais, comme je l'ai déjà dit ailleurs, je considère maintenant les *Ponérines* comme la souche commune directe des quatre autres sous-familles.

AUTRES CONSIDÉRATIONS GÉOGRAPHIQUES.

A propos de la faune antarctique, nous avons parlé des phénomènes dits de convergence. Aujourd'hui, grâce aux recherches expérimentales relatives à l'influence des agents physiques et chimiques sur le développement des êtres vivants, nous possédons la clé de ces phénomènes. Immédiatement après DARWIN, on a cru pouvoir les expliquer par la sélection naturelle. Grâce aux travaux de SCHMANKEWITSCH, de MERRIFIELD, de STANDFUSS, de FISCHER, de Mlle DE CHAUVIN, de KAMMERER et PRZIBRAM, à Vienne, de SEMON et de bien d'autres, nous savons aujourd'hui que la couleur, la forme, bref tout l'organisme peut être directement modifié par l'influence engraphique lamarkienne des agents extérieurs, et que cette engraphie peut devenir héréditaire.

L'engraphie crée et la sélection trie. Il ne faut pas opposer ces deux agents transformateurs de la vie l'un à l'autre, comme on le fait si souvent par engoûment. Il faut saisir, au contraire, leur combinaison pour comprendre le résultat final.

C'est surtout à STANDFUSS que nous devons la preuve de l'influence de la température sur la transformation des espèces de Lépidoptères. Le froid et l'obscurité, joints en général à l'humidité, rendent les couleurs ternes; ils rendent aussi les téguments plus lisses, les formes plus arrondies. La lumière et la chaleur font le contraire.

Aussi voyons-nous dans la faune antarctique les espèces du genre *Melophorus*, par exemple, prendre tout l'aspect des *Lasius* arctiques, quoique la phylogénie de ces derniers soit tout autre. Nous-mêmes, les Myrmécologistes, nous nous y sommes laissé prendre dans le temps et nous avons considéré les *Melophorus* antarctiques comme des *Lasius*. EMERY a corrigé plus tard cette erreur par la dissection des *Melophorus* et montré qu'ils appartiennent à la tribu des *Plagiolepidii*. De même les *Huberia*, dérivées des *Monomorium*, ressemblent à des *Aphænogaster* et à des *Myrmica*. Et les *Monomorium* antarctiques eux-mêmes, avec leurs téguments lisses et leur couleur noire ou rousse, ont un autre cachet que les formes tropicales du genre.

Un phénomène analogue se produit pour la faune des montagnes; je signale en particulier celle de l'Himalaya dont j'ai déjà parlé. Dans nos Alpes suisses, la faune des Fourmis est la même que celle du nord de l'Europe; au contraire de celle de l'Himalaya, elle ne présente pas d'espèces propres, tout au plus d'insignifiantes variations. Cela tient probablement à ce qu'elle a été entièrement détruite à l'époque glaciaire, les Fourmis supportant fort mal le froid.

Ce dernier fait m'amène à une constatation intéressante d'ordre géographique, constatation que j'ai eu l'occasion de faire dernièrement.

J'ai pris des informations auprès d'explorateurs du Groënland et de l'Islande, après avoir longtemps et vainement cherché à obtenir des Fourmis de ces pays. Le résultat est qu'il n'y en existe aucune. Or, J. SPARRE SCHNEIDER a constaté la présence de plusieurs espèces de Fourmis : *Formicoxenus nitidulus*, *Leptothorax acervorum*, *Formica fusca* et *exsecta* au delà du 70° de latitude nord en Norvège arctique, ainsi que des *Formica rufa*, *Camponotus herculeanus*, *Myrmica lobicornis*, *sulcinodis* et *ruginodis* entre les 68° et 70°. Donc en tout neuf espèces arctiques, les mêmes que nous trouvons dans nos Alpes vers 2,000 mètres. Je suis certain, pour ma part, que ces espèces pourraient vivre en Islande et au sud du Groënland. Si elles n'y existent pas, c'est que, détruites à l'époque glaciaire, elles n'ont plus pu y rentrer faute de passage terrestre.

FAUNE DÉSERTIENNE.

Les régions très sèches et chaudes offrent des Fourmis fort spéciales, dont les mœurs se sont adaptées aux besoins du climat. Parmi ces adaptations, il en est de fort intéressantes. L'une est celle de l'hyperextension du jabot chez une partie des ♀ qui sont gavées de miel par les autres pendant la saison des pluies et de la végétation, pour servir ensuite, pendant la saison sèche et brûlante, de pots de conserves au reste de la communauté dans ses souterrains. Tels sont les *Myrmecocystus* du Mexique, du Texas et du Colorado, les *Melophorus* et certains *Leptomyrmex* de l'Australie centrale, les *Plagiolepis decolor*, *Trimeni* et *Jouberti* de l'Afrique du Sud, certains *Camponotus* des mêmes régions, etc.

Une autre adaptation est constituée par des rangées de longs cils chitineux raides, situés sous la tête et dirigés en avant comme une barbe, ainsi que devant l'épistome ou aux mandibules. Nous les trouvons dans le genre *Pogonomyrmex*, chez de nombreux *Messor*, chez les *Ocymyrmex*, divers *Holcomyrmex*, *Dorymyrmex*, *Oxyopomyrmex*, *Myrmecocystus*, *Camponotus*, etc., c'est-à-dire chez des genres et même des sous-familles très disparates, mais seulement et toujours chez des habitants du sable du désert. WHEELER, le premier, a été frappé de ce fait et a cru que ces longs cils servaient à transporter des gouttes d'eau ou à nettoyer le peigne des tibias du sable qui s'y attache. Mais c'est à SANTSCHI que nous devons d'avoir démontré par l'observation la vraie signification de ces longs cils qu'il appelle *psammophores*. Ils servent aux Fourmis du désert, obligées de creuser leur nid très profondément dans le sable, à transporter les boulettes de sable qu'elles accumulent et déplacent, sans que les grains de sable se détachent. Les cils sont assez rapprochés pour qu'un grain de sable ne passe pas entre deux d'entre eux, et cela permet à ces Fourmis de manipuler avec le sable comme aux nôtres avec la terre humide. Mais les boulettes de cette dernière n'ont pas besoin de treillis de cils pour demeurer cohérentes, tandis que les grains de sable — on le sait — n'adhèrent pas entre eux. C'est donc grâce à ces psammophores que les Fourmis du désert peuvent creuser leurs nids. SANTSCHI l'a observé directement.

Les Fourmis granivores ou moissonneuses, avec leurs épaisses mandibules et leurs individus à grosse tête, sont aussi, pour la plupart du moins, adaptées aux régions sèches, aux steppes ou aux déserts. Tels les *Messor*, les *Pogonomyrmex*, beaucoup de *Pheidole*, les *Holcomyrmex* et les *Oxyopomyrmex*. Leurs greniers les alimentent surtout pendant la saison torride, comme l'a montré EMERY.

AUTRES ADAPTATIONS.

Certaines Fourmis plates (*Cryptocerus*, *Cataulacus*) vivent dans des cavités végétales plates, et certaines espèces longues et cylindriques (*Pseudomyrma, Sima*), dans les tiges creuses à cavités cylindriques centrales. Ce genre d'adaptations fait partie de l'adaptation à la forêt, de même que les nids en carton sur ou dans les arbres, les jardins de Fourmis dans les touffes d'épiphytes, tels que les a découverts ULE, les mœurs des *Azteca* et des *Liomotopum*, etc.

Une autre faune de Fourmis, surtout de grands Ponérines, les *Leptogenys, Megaloponera, Dicamma, Paltothyreus*, etc., paraît adaptée à une vie *termitophage*. Ici, néanmoins, les observations sont encore trop sporadiques pour permettre une généralisation.

J'ai montré que la vie sous les pierres et dans des nids surmontés de dômes maçonnés est adaptée aux climats plutôt froids et humides, manquant de soleil, tandis que là où le soleil abonde et où le climat est sec, les Fourmis minent seulement la terre ou le sable, entourant leur porte d'un cratère, ou qu'elles vivent dans les végétaux. Dans ces pays-là on ne trouve que peu de nids sous les pierres. Ces faits jouent un rôle important dans la distribution géographique des espèces. Il faut les comprendre pour se rendre compte, par exemple, de la faune des îlots xérothermiques de l'Europe centrale.

Tous les phénomènes de convergence et d'adaptation que nous venons d'esquisser, et bien d'autres encore, s'enchevêtrent avec ceux des grands faits géologiques et avec les phénomènes de sélection. Ils rendent de mieux en mieux compte des particularités de la faune mondiale des Fourmis et des mœurs de ces insectes, à mesure que leur étude s'approfondit.

TRANSPORTS.

J'ai parlé plus haut des onze espèces cosmopolites et d'autres en train de le devenir. Les bateaux sont eux-mêmes envahis par ces espèces, en particulier par le *Monomorium Pharaonis*. Une quantité énorme de Fourmis, souvent de ♀ fécondes, sont transportées avec les végétaux. M. REH m'en a envoyé un nombre considérable, parmi elles même des espèces nouvelles ainsi arrivées à Hambourg. De cette façon, la faune des petites îles a été en grande partie détruite et remplacée par des cosmopolites ou d'autres espèces importées. Ainsi les faunes locales sont partout de plus en plus détruites par les cultures et les transports humains.

Tout dernièrement j'ai trouvé dans les rues de Smyrne ma *Pheidole teneriffana*. Mais je ne l'ai prise nulle part dans la brousse des environs de Smyrne. Je la crois donc importée. D'autres espèces, par contre, envahissent aussi la brousse, tels le *Monomorium floricola* et le *Tapinoma melanocephalum*.

Mon but a été de vous donner un simple aperçu de nos connaissances sur la distribution géographique des Fourmis sur le globe terrestre et en même temps de vous montrer les phénomènes de convergence, de climat, de phylogénie, d'habitat, etc., qui influencent cette distribution à côté de ses grandes causes géologiques. Il va sans dire que cet aperçu est très incomplet. Mon temps, c'est-à-dire mon surmenage perpétuel, ne m'a pas permis de faire les études nécessaires à un travail approximativement complet. Aussi je vous prie d'excuser cette simple causerie.

Une colonie polycalique de « Formica sanguinea » sans esclaves dans le canton de Vaud,

par le Dr A. FOREL (Yvorne).

La *Formica sanguinea* LATREILLE est, sans contredit, l'un des insectes les plus intéressants qui existent. Elle attaque des espèces plus faibles (*Formica fusca* et *rufibarbis*) et pille leurs nymphes qui, écloses chez elle, exécutent la plus grande partie du travail domestique. Mais malgré cela, la *Formica sanguinea* n'a pas perdu les instincts du travail. Elle est capable d'élever sa couvée, de traire les pucerons, de bâtir son nid, etc. Ses esclaves ne font que lui faciliter les travaux domestiques, en un mot, de la décharger. Dans mes « Fourmis de la Suisse », j'ai prouvé que de toutes nos Fourmis indigènes cette espèce possède les instincts les plus modifiables, les plus variables, les plus plastiques. Il n'existe en Europe qu'une seule variété définie de *Formica sanguinea* où, si l'on préfère, l'espèce ne varie pas sensiblement. Tout au plus certaines fourmilières ont-elles des individus tantôt plus grands, tantôt plus foncés, tantôt plus clairs, tantôt plus petits. Seulement en Espagne il existe une variété nettement plus claire. Les petits individus sont plus foncés que les grands, surtout sur le front et le vertex.

Nos fourmilières *sanguinea* en Suisse ont tantôt plus, tantôt moins d'esclaves. Seulement trois ou quatre fois dans ma vie j'ai trouvé autrefois des fourmilières tout à fait sans esclaves. La taille des individus, j'avais déjà cru le remarquer (« Fourmis de la Suisse », p. 359), était en somme fort petite chez elles. Mais on trouve aussi des *sanguinea* de petite taille avec des esclaves.

Dans l'Amérique du Nord, il existe plusieurs sous-espèces et variétés de la *Formica sanguinea*. Deux d'entre elles font toujours beaucoup d'esclaves. Une autre, que j'ai découverte moi-même au Canada, n'en fait jamais. C'est la race *aserva*. Cette *sanguinea aserva* établit des colonies composées de beaucoup de nids rapprochés les uns des autres et en communication les uns avec les autres. En cela, elle rappelle tout à fait les *Formica exsecta* et *pressilabris* dont les fourmilières constituent presque toujours des colonies considérables, composées de cinq, dix, trente, soixante et même jusqu'à deux cents nids.

Cela dit, je passe à l'observation entièrement nouvelle que je viens de faire au Chalet Boverat, près du Chalet à Gobet, dans le Jorat, au-dessus de Lausanne.

Je me trouvais dans une course avec les membres de la Maison du peuple de Lausanne. Assis par hasard près d'un nid de *Formica sanguinea* de petite taille, je voulus montrer aux personnes présentes leurs esclaves. J'eus beau chercher, je n'en trouvai point. Dépité, je me mis en route pour chercher un nid de *Formica fusca* avec nymphes, pour démontrer au moins le pillage esclavagiste. Mais j'eus beau chercher dans une grande prairie jusqu'à 150 mètres de distance environ, impossible de trouver un nid de *Formica fusca*. Seulement des *Lasius* et des *Tetramorium caespitum*. Cette prairie et quelques champs environnants sont entourés, de trois côtés, d'un grand bois de sapins très bien peigné et nettoyé. De guerre lasse, je revins avec trois ou quatre personnes qui m'accompagnaient, en suivant le bord du bois pour retourner au point du départ. Sur ce bord, je trouvai immédiatement un grand nid de *Formica sanguinea* dont les ♀ étaient de petite taille, comme dans le premier. J'eus beau creuser, impossible d'y trouver une seule esclave, ni une seule nymphe de *Formica fusca*. Quelques pas plus loin, je trouvai un second nid en communication avec le premier et ainsi de suite jusqu'au point de départ, sur une longueur de plus de 150 mètres. Je trouvai ainsi une quarantaine de nids, distants de 2 à 5 mètres les uns des autres, tous en communication les uns avec les autres par des chaînes de Fourmis et ne contenant aucune esclave.

En outre, quelques nids appartenant à la même colonie se trouvaient dans la prairie, un peu plus éloignés du bord du bois, et je suis certain que le nombre total des nids était de plus de quarante, mais je n'eus pas le temps de chercher partout. Je pris alors un petit sac rempli de Fourmis de l'un des nids (assez éloigné) pour les porter sur le premier. La confraternité entre les Fourmis du second et du premier nid fut immédiate. Aucun signe d'inimitié, ni même de méfiance. Il n'y a donc aucune erreur possible ; il s'agit d'une immense colonie polycalique de la *Formica sanguinea* européenne, absolument dépourvue d'esclaves. Ce fait n'a jamais été observé jusqu'ici, à ma connaissance. Comment l'expliquer?

On sait que WHEELER a découvert la manière dont se forment les fourmilières de la *Formica sanguinea*. Une femelle fécondée attaque à elle seule une petite fourmilière de *Formica fusca* ou *rufibarbis*, pille les nymphes, chasse les habitants et se procure ainsi des esclaves qui se chargent d'élever sa progéniture. Il est vrai que j'ai découvert, avec M. WHEELER lui-même, une fourmilière naissante avec deux femelles de *Formica sanguinea*, quelques ouvrières très petites et adultes de cette espèce et quelques nymphes et jeunes ouvrières de *Formica rufibarbis* Dans ce cas déjà, je soupçonnai que les femelles *sanguinea* avaient pu exceptionnellement élever elles-mêmes leur progéniture.

Il me semble évident qu'actuellement la colonie du Chalet Boverat a exterminé toutes les *Formica fusca* ou *rufibarbis* des environs et que c'est là probablement la raison pour laquelle elle n'a pas ou plus d'esclaves. C'est probablement aussi pour cela que les individus qui la composent sont si petits, — on peut dire rabougris, — faute d'avoir été bien nourris par de nombreuses esclaves dans leur état larvaire.

Mais comment cette colonie a-t-elle pu se former? A l'ordinaire la *sanguinea* ne possède qu'un, ou tout au plus deux, plus rarement trois nids, très rapprochés les uns des autres, et encore les habite-t-elle souvent alternativement plutôt que simultanément. Par contre, elle change assez souvent son domicile, probablement lorsqu'elle a trop pillé et affaibli les fourmilières environnantes des espèces esclaves.

Or, il est évident qu'une colonie de plus de quarante nids ne peut pas émigrer et changer de place comme une fourmilière n'ayant qu'un ou deux nids. Les fourmilières ordinaires de *Formica sanguinea* ne peuvent pas se recruter de nouvelles femelles fécondes de leur propre fourmilière lors de l'essaimage, par la simple raison que ces femelles ne viennent pas retomber sur leur petit domaine très local. Aussi les fourmilières de *Formica sanguinea* ne viventelles que quelques années. Au contraire, les fourmilières de *Formica rufa, pratensis, exsecta*, etc., avec leurs grandes colonies et leurs longues routes d'exploitation dans les environs, recueillent lors de l'essaimage les femelles de leur espèce et peuvent de ce fait exister sans changement pendant quarante-cinq ans, comme je l'ai observé moi-même, ou même pendant quatre-vingts ou cent ans, comme l'a démontré CHARLES DARWIN.

Il est évident pour moi qu'une colonie de *Formica sanguinea* de plus de quarante nids ne peut être entretenue par une seule ni même par deux femelles fécondes. Il me semble donc évident que dans ce cas particulier des circonstances spéciales ont produit dans cette colonie ce qui se passe dans les fourmilières de *F. rufa*, *exsecta* et autres, c'est-à-dire que les femelles sont retenues de temps à autre par les Fourmis allant et venant entre les nids de la colonie. La situation au bord intérieur d'un bois plus ou moins sémi-circulaire et dans une prairie en partie isolée semble propice à la chose, les femelles fécondées ayant beaucoup de chance de venir retomber parmi les ouvrières de leur grande fourmilière.

Mais tout cela n'explique pas l'origine première de cette colonie si curieuse et si exceptionnelle. Je laisse ce point en suspens, ne voulant pas faire de conjectures sur des bases insuffisantes. Il faut certainement qu'il y ait eu un concours exceptionnel de circonstances ayant fait revenir et retenir des femelles fécondes.

Ce qu'il y a de très remarquable, c'est de voir l'instinct polycalique (instinct qui pousse les individus d'une même fourmilière à fonder différents nids à une petite distance les uns des autres, tout en constituant une société commune), que nous trouvons normalement chez la race *aserva* de l'Amérique, venir remplacer accidentellement l'instinct esclavagiste chez des *Formica sanguinea* d'Europe.

Quoi qu'il en soit, cette observation montre une fois de plus la plasticité cérébrale de la *Formica sanguinea*.

Je ne crois pas me tromper en attribuant la petite taille de ces *sanguinea* à une dénutrition relative due à l'absence complète d'esclaves. Néanmoins, il existe des fourmilières dont les individus sont de petite taille et qui ont des esclaves. Je ne puis donc pas être trop affirmatif.

L'idée qu'il puisse s'agir d'une variété spéciale me semble devoir être écartée pour les raisons indiquées et d'après tout ce que nous savons sur la *F. sanguinea* d'Europe.

Der Begriff der Gattung in der heutigen Systematik,

von P. SPEISER (Labes).

Unsere ganze heutige Systematik baut sich gewissermassen auf den geistigen Grosstaten dreier Männer auf, die stets zusammen werden genannt werden müssen, wo systematische Zoologie getrieben wird. ARISTOTELES erkannte die innere Notwendigkeit eines zoologischen Systema; ihm verdanken wir die Erfassung des Gedankens, durch ihn wissen wir, *was* wir mit der Systematik wollen. LINNÉ hat uns dann gelehrt, *wie* wir es anzufangen haben, in das unübersehbar werdende Chaos der Erscheinungen Ordnung und Uebersicht zu bringen durch Schaffung der binären Nomenklatur. Endlich der dritte Grosse ist DARWIN, dessen Werk wir es verdanken, aufgewacht zu sein zu der kritischen Frage, *warum* wir ein solches System aufstellen müssen, und was dieses bedeutet. Drei geniale Männer sehen wir so an den Abschnitten der zoologischen Forschung stehen, von denen jeder einen tief einschneidenden Fortschritt herbeigeführt hatte. Aber es ist meines Erachtens das wahre Kennzeichen des wirklichen Genies, dass in seinen Werken sich mehr findet als das wirklich in dürren Worten ausgesprochene, dass vielmehr darüber weit hinaus unausgesprochene nur gewissermassen geahnte weitere Möglichkeiten sich wie Keime finden, die einer späteren Entwickelung entgegenharren.

Betrachten wir in diesem Sinne die Einführung der binären Nomenklatur durch LINNÉ, so will es ja auf den ersten Blick auch scheinen, als ob dieses *Systema naturae* nur eine bequeme Anordnung bieten will für die Masse der Einzelheiten. Dann war es auch gleichgiltig, ob dieses System ein künstliches oder ein

7*

natürliches war, und die immer wiederholten Streitereien über das Vorliegen eines solchen oder anderen wären überflüssig gewesen. Die damalige Zeit brauchte auch im letzten Ende kein natürliches System, denn auch die philosophische Betrachtung des Ganzen sah ja damals weiter nichts als ein Nebeneinander von Formen, die als *ab initio* unverändert gegeben betrachtet wurden. Dennoch konnte später, als man die Bedingungen erkennen und erschliessen lernte, nach denen man ein wirklich natürliches System aufstellen musste, namentlich durch DARWIN's Werk, LINNÉ's Methode glatt und mit Erfolg übernommen werden. Und die Bedingung hierfür war wieder in einer besonders genialischen Tat LINNÉ's gegeben, in der richtigen Konception des Gattungsbegriffes. Dessen strenge Durchführung war dasjenige, was allen älteren Versuchen eines Systemes gefehlt hatte, in der Einführung und Durchführung des Gattungsbegriffes ruht die Grundbedingung der binären Nomenklatur. Während die Nomenklatur aber eben nur ein Ausdrucksmittel ist, ist die Erfassung des Gattungsbegriffes und die Abgrenzung einer gegen die andere Gattung geradezu die grosse Tat, die der LINNÉ'schen Nomenklatur der dauernden Wert verleiht.

Denn wenn zunächst die Gattung wohl nichts weiter war als die Zusammenfassung des nächst Aehnlichen, so lernten wir doch bald und mindestens seit DARWIN's Werk verstehen, dass hier eine wirkliche Verwandtschaft vorlag und in der einfachsten Form nomenklatorisch zum Ausdruck gebracht war. Wichtiger als der wenig umstrittene Begriff der verschiedenen höheren Gruppen, wesentlicher fast als der schwankende Begriff der Art bleibt die Erfassung der Gattung als der kleinsten Zusammenfassung der nächst verwandten Formen. Die hohe Wichtigkeit dieser Begriffsbildung geht auch schon aus den mannigfachen Versuchen hervor, darüber klar zu werden, ob denn Gattungen in der Natur vorkommen, oder ob solche stets nur Abstraktionen menschlicher Systematik seien. Meiner Auffassung nach ist eine solche Fassung der Frage durchaus falsch, und es muss vielmehr gefragt werden, ob die Gattungen unserer heutigen Systeme den in der Natur vorhandenen Gattungen wirklich entsprechen, mit anderen Worten, ob unsere Systeme an der wichtigsten Stelle, in der Zusammenfassung der Gattungen, natürliche sind oder immer noch nur künstliche.

Die Antwort darf natürlich nicht fest umrissen und apodiktisch entscheidend lauten. Wir müssen uns vergegenwärtigen, dass alles menschliche Wissen und Einsehen notwendig stets Stückwerk ist und bleibt, und dass nur auf wenigen ganz kleinen günstigen Gebieten einigermassen vollkommenes erreicht werden kann und erreicht wird. Dann werden wir einsehen, dass wohl eine gewisse kleine Anzahl unserer Gattungen anscheinend allen Erfordernissen natürlicher Systematik entspricht, wir werden aber bei der weit überwiegenden Mehrzahl uns sagen müssen, dass da unser Wissen und Vermögen nicht ausreicht, und müssen uns damit begnügen das Beste zu wollen und das Erreichbare zu leisten. Wir dürfen dann aber nicht vergessen, dass wir nur einen augenblicklichen als mangelhaft empfundenen Zustand vor uns haben, an dessen Verbesserung zu arbeiten für uns Pflicht ist.

Wir erstreben ein natürliches System, d. h. ein solches, in dem die natürlichen verwandtschaftlichen Beziehungen der einzelnen Tiere zu einander zum Ausdruck kommen, in dem zum Ausdruck kommt, dass und möglichst auch wie sich die einzelnen heute von einander zu unterscheidenden Formen aus anderen Vorfahrenformen entwickelt haben. In meinen bisherigen Ausführungen habe ich schon zum Ausdruck gebracht, dass es meine Ueberzeugung ist, dass die Gattung, begrifflich richtig erfasst, in diesem natürlichen System ihre volle richtige Stelle hat, dass es mit anderen Worten natürliche Gattungen giebt, und nicht nur jeglicher Gattungsbegriff künstliche Trennung seitens des Menschen bedeutet.

Die Kriterien festzulegen, nach denen eine Gattung als natürliche wird anerkannt werden dürfen, macht bei diesem Begriff noch grössere Schwierigkeiten als bei dem Begriff der Art. Als Art fassen wir die Gesamtheit aller derjenigen Individuen zusammen, die eine eigentümliche Zusammenstellung von Eigenschaften gemeinsam haben, diese in gleicher Weise auf ihre Nachkommenschaft weitergeben und die unter einander selbst dann mit dauernd erhaltener Fruchtbarkeit und Generationstüchtigkeit gekreuzt werden können, wenn sie von den mittleren Werten jener Eigenschaften im einzelnen mehr oder weniger weit abweichen.

Gerade diejenigen Forscher, welche die Gattungen als rein künstliche Gebilde menschlicher Deduktion aufgefasst sehen wollen, sprechen den Gattungsbegriff an als eine blosse Parallele zum ebenso künstlichen Artbegriff, indem als Gattung ihrer Meinung nach solche Arten zusammengefasst werden, die einander näher zu stehen scheinen als den anderen Arten ihrer Familie, und die doch nicht genügend weitgehend abweichende Merkmale bieten um eine Abtrennung mit höherer Valenz zu rechtfertigen.

Ganz zweifellos ist die erste, ältere und in früheren Zeiten auch vollauf berechtigte Auffassung. Hat sie aber auch heute noch Berechtigung? Meiner Ansicht nach nicht! Seit uns DARWIN gelehrt hat die eigentliche Blutsverwandtschaft zwischen den einzelnen Arten zu begreifen, sind wir verpflichtet auch die höheren systematischen Kategorieen darauf hin zu prüfen, ob sie in ihrem Umfange solche Blutsverwandtschaft richtig zum Ausdrucke zu bringen vermögen. Hier sollte uns heute nur die Gattung beschäftigen, und wir haben zu fragen, wie wir imstande sind, eine solche Prüfung da vorzunehmen. Wieder müssen wir da zurückgreifen auf die Art: dort konnte uns das Experiment auf die Möglichkeit der Erziehung dauernd fruchtbarer Nachkommenschaft willkommene Hilfe bieten. Derartiges haben wir bei der Gattung nicht.

Ohne mich hier im Einzelnen auf die weiteren allgemeinen Fragen der sogenannten physiologischen Isolierung und ähnliches einlassen zu wollen, möchte ich doch die eine Seite einer solchen Definition des Artbegriffes betonen, die bei der Diskussion über die mögliche Artentstehung zu beachten ist, und uns hier wesentliche Gesichtspunkte zu bieten scheint. Entweder nämlich müssen wir annehmen und erschliessen, dass eine solche Art in der Weise entsteht, dass infolge irgend welcher Einflüsse alle oder ein mehr oder weniger hoher Procentsatz der Individuen einer bisher anders characterisierten Art in gleichem Sinne abändern, — oder wir müssen zugestehen, dass alle Individuen einer Art, die wir heute unter einem Artbegriff zusammenfassen, von einem einzigen zuerst in diesem Sinne abgeänderten Individuum abstammen. In Wirklichkeit scheint es, als ob *beide* Möglichkeiten vorkommen werden. Sobald wir das aber zugeben, werden wir etwas sofort bemerken, dass nämlich in dem ersten Falle die neu entstandene Art schon von vornherein ein grosses Areal bewohnen kann, während sie in dem zweiten Falle, eben an einem einzigen Punkte entstanden, von diesem Punkte aus ihre Ausbreitung nehmen muss. Darauf wollte ich hinaus.

Untersuchen wir dann ob denn auch bei der Gattung eine solche Teilung der Entstehungsweise denkbar ist, wie wir sie für die Arten besprochen haben, dass also einmal die wesentlichen Charactere gleichzeitig bei vielen einzelnen der nächstniederen Kategorie auftreten würden, oder andererseits von einer einzigen aus diese Charactere sich in der einmal erreichten Weise erhalten haben. Das führt uns zu der weiteren Frage, welche Charactere denn Gattungen characterisieren, welche anderen nur Arten. Es wird paradox klingen, wenn ich behaupte, eine derartige Scheidung ist unmöglich, und doch gleichzeitig behaupte, es giebt natürliche Gattungen. Ein sehr bequemes Schlagwort fasste die Frage so, dass « Anpassungscharactere keine Gattungscharactere » sein könnten. Ich gehe nun nicht so weit wie gewisse andere Zoologen, die eben alle und jegliche Charactere nur Anpassungscharactere sein lassen, ich gebe zu, dass wahrscheinlich viel mehr als gewöhnlich angenommen wird, korrelative Charactere sein werden, damit ist aber nichts gewonnen, weil diese Korrelationen eben zu den Anpassungen gehören. Auch mit der Auffassung als Ergebnisse innerer Wachstumsgesetze ist viel in dieser Frage nicht anzufangen. Vielmehr wird man meiner Meinung nach am ehesten der wahren Sachlage nahe kommen, wenn man sagt, ältere Charactere sind Gattungsmerkmale, jüngere oder die jüngsten die der Arten. Damit ist, so höre ich schon die Kritiker rufen, der Spekulation Tür und Tor geöffnet. Und in der Tat, so sehr ich in der Einzelforschung dafür eintreten muss, die Spekulation zu Hause zu lassen, in diesen allgemeinen Fragen kann man sie nicht ganz entbehren. Nur wenn man z. B. strenge die durch LAMARCK, DARWIN und ihre Nachfolger erarbeitete Erkenntnis festhält, dass jegliches Organische aus einer weniger vollkommenen Vorstufe hervorgegangen ist, nur dann, meine ich, kann man in sorgfältiger Fortbildung dieses Gedankens zu einer der Wahrheit wahrscheinlich nahekommenden Auffassung des Gattungsbegriffes kommen, der sich dem natürlichen System als nicht mehr künstliches Gebilde einfügt. Dann sehen wir im Geiste ein Merkmal auftreten oder einen Complex von Merkmalen, die gegenseitig in Abhängigkeit stehen, und die in ihrer Gesamtheit, sagen wir, zunächst eine Art characterisieren. Indem nun dieses Merkmal unverändert festgehalten wird, findet sich Veranlassung, dass eine Gruppe von Individuen noch neue Eigenschaften erwirbt, eine andere Gruppe kann andere Eigenschaften erreichen, eine dritte wieder andre, alle aber halten das erstgemeinte Merkmal aufrecht. Somit ist dieses das ältere und damit Gattungsmerkmal.

Wenn wir die Frage so anfassen, dass wir für Gattungsmerkmale erfordern, sie sollen ältere, schon fester als andere der Organisation mehrerer Arten inhärente Charactere sein, dann vermeiden wir eine Klippe. Dann kann nämlich niemals eine Gattung aufgestellt oder gehalten werden, die aus dem Grunde unnatürlich ist, und einem natürlichen, auf phylogenetischen Beziehungen fussenden System nicht mehr genügt, weil da convergente Entwickelungsreihen von verschiedener Grundlage aus gleiches erreicht haben. Infolge solcher Convergenzen würde eine polyphyletische Gattung resultieren, und solche müssen in einem natürlichen System vermieden werden.

Wie schon einmal gesagt, und wie aus den hier aufgestellten Forderungen abermals hervorgeht, kommt man bei dem Versuch einer kritischen Beurteilung der bestehenden und zu schaffenden Gattungsbegriffe nicht aus ohne Speculation. Dann aber hat man die Pflicht, von jeder nur erdenklichen Seite her das Erreichte zu beleuchten. Und da will ich heute zu allen sonstigen schon mehr oder weniger lange beachteten und befolgten Grundsätzen den einen als meines Erachtens recht besonders wichtigen hervorheben : Es muss meiner Meinung nach ganz besonders die Geographie, die Tatsachen und Möglichkeiten der Verbreitung ganz massgebend auch in dieser Frage der natürlichen Berechtigung des Gattungsbegriffes wesentlich in Rücksicht gezogen werden.

Ich habe es vorher ausgeschlossen, dass aus mehreren Arten her durch Auftreten von Characteren, die dann eben ja jüngere als die Artcharactere wären, eine natürliche Gattung sich rekrutieren könnte. Es muss also auf stets nur eine Art zurückgegangen werden, aus der sich jede Gattung phylogenetisch entwickelt hat. Nun giebt es aber nur wenige Arten, die primär ein wirklich besonders weites Verbreitungsgebiet haben, und in früheren Erdperioden ist das kaum anders gewesen. Auf specielle Ausführungen im Einzelnen glaube ich hier verzichten zu müssen, um nicht zu lang zu werden. Ich darf nur auf dem Entomologenkongress hervorheben, dass jene besonders weit verbreiteten Arten zumeist Wassertiere sind, die also unter ganz wesentlich anderen Bedingungen leben, als die allermeisten Insekten. Jedenfalls, und darauf will ich hinaus, verpflichtet uns die so eben entwickelte Erkenntnis dazu, die vorhandenen wie die neu zu errichtenden Gattungen darauf hin zu prüfen, ob sie in ihrer geographischen Verbreitung dem Gedanken entsprechen, dass sie aus einer einzigen Art durch Aufspaltung entstanden sind. Ein schönes Beispiel bietet die Staphylinidengattung *Dinarda*, bei welcher in exactester Weise nachgewiesen ist, wie ihre Arten sich in einem einheitlichen geographischen Areal so anordnen, dass die älteste, erste Art das Centrum innehat, die von ihr abgeleiteten an verschiedenen Stellen der Peripherie sich dann angliedern. So klar werden die Verhältnisse nur selten liegen, und es wird immer der mühevollen Arbeit eines geeigneten Specialforschers bedürfen, um aus der Erscheinungen Fülle solche Klarheit zu erarbeiten. Bei der Umschau aber in der grossen Reihe der übrigen Gattungen werden wir schon oft genug auf Tatsachen stossen, die nach dem von mir entwickelten Gesichtspunkten erhebliche Zweifel zulassen daran, ob die bestehenden und bekannten Merkmale sie wirklich als natürliche characterisieren. Wir werden solche grossen und nahezu über den ganzen *orbis terrarum* verbreitete Gattungen wie z. B. *Lycoria*, *Culex*, und *Tipula* unter den Mücken, *Musca* und *Lauxania* unter den Fliegen, *Pieris* unter den Tagfaltern, *Agrotis* unter den Eulenfaltern u. a. m. mit gewissem Vorbehalt betrachten lernen. Sicherlich werden sie, wenn auch vorerst die morphologische Einheitlichkeit vorzuherrschen scheint, zerlegbar sein und zerlegt werden müssen, wobei dann der Gesichtspunkt der geographisch-historischen Entstehungsmöglichkeit der Gruppen wohl in Berücksichtigung gezogen werden kann. Und andererseits rechtfertigen sich gerade durch die hier angeregte Betrachtungsweise die vielfachen kleinen Genera, die auch heute noch vielen Systematikern überflüssige und unerwünschte Haarspaltereien zu sein scheinen. Eine neu gefundene Art unter irgend welchem Zwang an eine schon bestehende Gattung anzuschliessen, bedeutet in dieser Beleuchtung immer das Aussprechen einer Theorie über ihre Entstehung, und dieses auszusprechen, dürfte doch bei nur einigermassen eigentümlich gestalteten Tieren nicht so ganz ratsam sein. Es ist daher durchaus zu empfehlen, solchen einigermassen merkwürdigen Tieren allemal den Rang einer besondere Gattung zuzuerkennen; meistens geschieht das ja auch. Lernen wir dann später neue Formen kennen, die eine Verknüpfung rechtfertigen, so ist dann immer noch Zeit, den Anschluss an eine schon bestehende Gattung auszusprechen, und dieses Vorgehen ist viel richtiger, wissenschaftlich viel besser begründet, als wenn etwa infolge solcher späteren Funde eine erstmal der Gattung A eingereihte Art dort herausgeholt und nun zur Gattung B gezogen werden muss. Ganz in demselben Sinne ist die Aufspaltung älterer Sammelgattungen in noch so viele kleinere Gattungsgruppen zu betrachten. Sobald, möchte ich sagen, die morphologischen Anhaltspunkte für eine Aufteilung durch geographische Tatsachen unterstützt werden, sind wir zu einer solchen Aufteilung nicht nur berechtigt, sondern gewissermassen verpflichtet, da so eine viel klarere und durchsichtigere Anordnung des Ganzen ermöglicht wird. Als Beispiel möchte ich auf die von mir vorgenommene

Aufteilung der Nycteribiidengattung *Penicillidia* verweisen, wo die Tatsachen der geographischen Verbreitung die Abgrenzung zweier Untergattungen gewissermassen forderten, welche sich morphologisch als jünger differenziert dem verbleibenden Stamm der Gattung mit einer weiter umfassenden Verbreitung gegenüberstellen.

Oftmals werden gerade bei solcher Betrachtungsweise die Rätsel, die uns unsere immer unvollkommene Kenntnis des Wirklichen und Vorhandenen aufgiebt, nur rätselhafter. Dahin ist z. B. die Tatsache zu rechnen, dass von der ganz zarten, sehr wenig flugtüchtigen, gut characterisierten Mückengattung *Idiophlebia* (*Limoniidae*) eine Art auf den Karolinen im Stillen Ozean, die bisher einzige bekannte andere Art in Kamerun vorkommt; beide so ähnlich, dass sie ganz ausschliesslich durch die Gestalt der Genitalien zu unterscheiden sind. Ein jeder wird aus der von ihm vorzugsweise studierten Gruppe parallele Beispiele anzuführen wissen.

Aber die Rätselfragen werden zu vertieften Aufgaben für den denkenden Geist, wenn sie weiter ausschauend erfasst werden. Die verpönte Spekulation hat einen Inhalt bekommen, wenn die Frage der natürlichen Gattung von der Seite der geographisch-historischen Möglichkeiten her beleuchtet wird. Ich bin der Ueberzeugung, dass dieses Hilfsmittel, dieses Adjuvans, bei bewusster und kritischer Anwendung sehr wesentlich dazu beitragen kann und wird, die vielfältigen heutigen Genera mit dem von uns allen erstrebten Erfolge durchzuarbeiten, dass unser *System* mehr und mehr ein *natürliches* wird.

Atavistische Erscheinungen im Bienenstaat (« Apis mellifica L. »). *Müssen wir dem Bienenei zwei oder drei Keimesanlagen zuschreiben? Entdeckung der « Sporen » (Calcaria) bei der Honigbiene;*

von Profr.-Dr. H. v. BUTTEL-REEPEN (Oldenburg i. Gr.).

Unter den sozialen *Apidæ* nimmt die Staatenbildung der Honigbiene (*Apis mellifica* L.) die höchste Stufe ein. Diesen recht verwickelten Komplex von Anpassungen synthetisch zu erforschen, gewährte einen besonderen Reiz, und ich habe vor einigen Jahren unter eingehendster Würdigung aller einschlägigen Verhältnisse versucht, die phylogenetische Entstehung des Bienenstaates in ihren Grundzügen darzulegen (1). Es ergab sich hierbei auch das Vorhandensein einiger, offenbar aus uralter Vergangenheit isoliert festgehaltener resp. wieder rückgebildeter, morphologischer und anatomischer Bildungen, sowie biologischer Verhältnisse, die damals aber nur sehr flüchtig und zum Teil gar nicht berührt wurden. Ich möchte diese Erscheinungen hier unter Zugrundelegung einer unlängst veröffentlichten sehr interessanten Hypothese von DEMOLL einer Betrachtung unterziehen, ohne hierbei alles Einschlägige berühren zu können.

Sehr auffällig ist die Bauart der Königinzelle. Neben den

---

(1) Die stammesgeschichtliche Entstehung des Bienenstaates, sowie Beiträge zur Lebensweise der solitären und sozialen Bienen (Hummeln, Meliponinen, etc.), Leipzig, 1903.

bekannten sechseckigen wagerecht angeordneten Zellen der Arbeiterinnen und Drohnen sehen wir sie bedeutend grösser in Gestalt einer Eichel senkrecht an der Wabe herabhängen. Diese rundliche Form zeigt uns zweifellos eine altertümliche Bauweise, die uns bei näherer Betrachtung der ganzen Verhältnisse, den Gedanken nahe legt, hier noch ein Stück aus fernster Vergangenheit zu besitzen. Aber auch das Verfahren der Bienen mit dieser Zelle ist ein an alte Zeiten erinnerndes. Wir sehen u. a. bei den Hummeln (*Bombinæ*) und bei den stachellosen Bienen (*Meliponinæ*), also bei primitiveren Staatenbildungen, dass die Brutzellen nach einmaliger Benutzung abgerissen werden. In der erwähnten Arbeit wies ich nun darauf hin, dass bei der *Apis mellifica* nur die runden Königinnenzellen (Weiselzellen) diesem phyletisch ursprünglichen Triebe verfallen, während die anderen Zellen fortdauernder Benutzung unterliegen. Also nur bei dieser vielleicht phyletisch ältesten Bauart offenbart sich noch dieser alte Trieb.

Aber auch die Insassin der Zelle giebt uns noch durchaus altertümliche Züge neben ausserordentlich weit vorgeschrittenen einseitigen Vervollkommnungen. So wies ich erstmalig darauf hin (p. 158) (2), dass gewisse Gliedmaassen der Königin — die Metatarsen — « rückgebildet » seien und Anklänge an eine von mir im Bernstein des Samlandes entdeckte Zwischenform, die einerseits an Apis aber auch an *Melipona* erinnert, aufweisen. Ich gab dieser Bernsteinbiene daher den Namen *Apis meliponoides* (2). Selbstverständlich erstreckt sich diese Rückbildung bei der Königin nicht nur auf die Metatarsen sondern wir sehen derartige Rückschläge sich ja meist in der ganzen Organisation äussern. Ich betonte dieses, wenn ich so sagen darf, phylogenetische Herabsinken der Königin auch schon früher, indem ich darauf hinwies, dass im stammesgeschichtlichen Aufsteigen die Hauptveränderung auf Seiten der Königin liegt, « die von ihrer Höhe als frühere Allesschafferin (*) herabsinkt, fast alle die ihr eigentümlichen

(2) *Apistica.* Beiträge zur Systematik, Biologie, sowie zur geschichtlichen und geographischen Verbreitung der Honigbiene (*Apis mellifica* L.), ihrer Varietäten und der übrigen Apis-Arten. (« Mitt. Zool. Museum zu Berlin », III. Bd. 2., Heft, 1906, pp. 119-201.)

(*) Die Hummelkönigin z. B. ist solch eine Allesschafferin, da sie alles im Neste selbst beschafft, den Bau: das Sammeln des Nektars, des Pollens und der Baustoffe; die Eiablage, Fütterung, Verteidigung, u. s. w.

Instinkte verliert und nur noch Eierlegemaschine ist, während die Arbeiterinnen alle Instinkte ihres früheren Weibchentums behalten, also die Bau- und Fütter- resp. Sammelinstinkte etc. und nur den Begattungstrieb einbüssen, dafür aber einige neue Instinkte hinzugewinnen u. s. w. (p. 49) (1). Es steht diese Ansicht im Gegensatz zu der WEISMANN'schen, nach der die Arbeiterinnen die meisten Instinktsveränderungen zeigen, eine Ansicht, die sich aber, so glaube ich, nicht aufrecht erhalten lassen dürfte.

Dieses Herabsinken zeigt sich uns übrigens auch in ausgesprochener Weise bei den Schmarotzerhummeln (*Psithyrus*) (vgl. p. 7) (1), wie auch bei den solitären Schmarotzer- oder Kukuksbienen, die vielfach sogar fast ganz den Habitus der Wespenvorfahren wieder annehmen (p. 96) (1).

Unter dem Titel : « Die Königin von *Apis mell.*, ein Atavismus » (3), gab nun Dr. R. DEMOLL eine sehr anregende, gedankenreiche Studie heraus, die nachzuweisen versucht, dass, weil die Königin eine atavistische Stufe sei, die auch heute noch von den Arbeitsbienen durchlaufen werden müsse, die Annahme von drei Keimesanlagen im Bienenei (WEISMANN) überflüssig sei. Es handele sich sehr wahrscheinlich nur um zwei Keimesanlagen, eine männliche und eine weibliche, die vermeintlich dritte sei nur ein Atavismus, der aus der weiblichen Keimesanlage hervorginge und kein « vollständig neuer Determinantenkomplex ».

Hat man nun aber bisher die Anlage der Arbeiterin als einen « vollständig neuen Determinantenkomplex » betrachtet ? Ich glaube nicht. WEISMANN sagt in der jüngsten Publikation über diese Verhältnisse (4) Folgendes : « Wenn es bei den Bienen ausser Weibchen und Männchen noch sog. Arbeiterinnen giebt, so wird das nur auf einer besonderen Art von Iden beruhen können, die ursprünglich echt weibliche waren, dann aber für den Bestand der Art vorteilhafte Abänderungen vieler ihrer Determinanten eingingen und nun zu « Arbeiterin-Iden » sich umgestalteten ». Hier wird ein « vollständig neuer Determinantenkomplex » nicht angenommen. Auf Grund unserer biologischen und stammesgeschichtlichen Erforschungen ist eine derartige Auffassung wohl auch nicht gut möglich.

---

(3) « Biolog. Centralbl. », Bd. XXVIII, Nr. 8, Erlangen, 1908, pp. 271-278.
(4) « Vorträge über Deszendenztheorie », 2. Aufl., Iena, 1904, Bd. I, p. 319.

Seit langem wissen wir, wie sehr eng die körperliche Organisation der Königin und die der Arbeiterin zusammenhängt, wie eine Organisation, wenn man so sagen darf, in die andere übergeführt werden kann, so dass schliesslich eine Mischung beider Charaktere stattfindet und wir Individuen erhalten, *die weder eine normale Königin noch eine normale Arbeiterin darstellen*.

Entnimmt man z. B. einem Volke die Königin, so legen die Arbeitsbienen über Arbeiterinnenlarven die sog. Nachschaffungszellen an, also Königinzellen, die einfach durch Erweiterung und rundliche Verlängerung der engen Arbeiterinzellen geschaffen werden. Je nach dem Alter der Larven schlüpfen nun mehr oder minder « gute » Königinnen heraus; je älter die Larve war, je kleiner und arbeitsbienenähnlicher werden die Königinnen. Das sind seit Jahrzehnten bekannte auch in der bienenwirtschaftlichen Litteratur betonte Verhältnisse (5), die aber oft übersehen oder gar als « widersinnig » (H. v. IHERING) (6) bezeichnet werden. IHERING kommt zu diesem Schlusse auf Grund einseitiger Studien an stachellosen Bienen,—den Meliponen,—da bei diesen, und das ist hier wichtig zu betonen, alle Zellen gleich sind, d. h. alle drei Bienenwesen entstehen in ganz denselben gleichgrossen Zellen, eine Tatsache, mit der uns schon FRITZ MÜLLER 1879 bekannt gemacht hat (vgl. p. 53) (1). IHERING erwähnt, dass die Nahrungsmenge in den Zellen der Königinnen und Arbeiterinnen die gleiche sei (p. 283) (6). Ich komme auf diese Verhältnisse zurück.

Bei dem erwähnten Ignorieren oder Verkennen jener Verhältnisse im Staate der *Apis mellifica* möchte ich vorerst noch gründlicher an dieser Stelle auf dieses sehr interessante Ineinandergehen der beiden Formen hinweisen.

So sehr vorsichtig man sich im Allgemeinen der bienenwirtschaftlichen Litteratur gegenüber zu verhalten hat, so kann ich hier auf die einschlägigen zuverlässigen Untersuchungen eines Bienenzüchters, des Pfarrers KLEIN (7), hinweisen. Seine Fest-

---

(5) « Bienenzeitung », 1859, p. 8. (Dr. DÖNHOFF : « Ueber die künstliche Erziehung von Zwergköniginnen »), ferner ebenda, 1884, p. 234, etc.

(6) Biologie der stachellosen Honigbienen Brasiliens. (« Zool. Jahrb. » Abt. Syst. 19. Bd., Heft 2 u. 3, Iena, 1903, p. 283.)

(7) Futterbrei und weibliche Bienenlarve, in « Die Bienenpflege », Jahrg. 26, Heft 5, Ludwigsburg, 1904.

stellungen sind dem Resultat nach nicht neu, aber sie erledigen dieses Thema so vielseitig unter teilweise neuer Versuchsanordnung und unter Beigabe von Abbildungen (s. Tafel IV), dass sie hier des Näheren erwähnt werden sollen. Ich bemerke, dass ich aus eigenen Beobachtungen das Wesentliche zu bestätigen vermag.

Zur besseren Klarlegung der Verhältnisse schicke ich eine Uebersicht der Entwicklungsstadien voraus, die uns Durchschnittswerte angiebt.

		Königin. Tage.	Arbeiterin. Tage.	Drohne. Tage.
Dauer der Eientwicklung	ca.	3	3	3
Dauer der Larvenernährung	ca.	6	6	6
Einspinnen und Ruheperiode	ca.	2	4	7
Umwandlung zur Puppe und Imago	ca.	5	8	8
Zusammen		16	21	24

Entnimmt man nun eine $^1/_2$—1 oder 1 $^1/_2$ tägige Larve einer Arbeiterinzelle und bringt sie in eine Königinzelle, so entsteht daraus eine *vollkommene* Königin (Tafel IV, Fig. 1), weil bis dahin auch die Arbeiterlarve offenbar denselben gleichartigen Futterbrei erhält. Aber schon bei Larven von 2 $^1/_2$—3 $^1/_2$ Tagen zeigt sich bei derselben Prozedur ein gewisser kleiner Unterschied, die Königinnen bleiben in der Körpergrösse etwas zurück und die Beine der Königinnen, die stets heller gefärbt sind als die der Arbeiterinnen, zeigen besonders an den Hinterschienen, wo sich bei den Arbeiterinnen eine Vertiefung — das sog. Körbchen — befindet, dunklere Flecke. Auf der Zeichnung (Fig. *b.* und *c.*) finden sich schon schwache Andeutungen der Körbchenvertiefung. Ich kann diesen Befund nicht bestätigen. Vielleicht sollen die Schattenstriche auch nur die erwähnten Flecken andeuten, was auch mit sonstigen Angaben KLEIN's harmonieren würde. Aber bei Larven, die schon ca. 4 $^1/_2$ Tage in der Arbeiterzelle gewesen und dann erst umgelarvt wurden, haben wir bereits einen erheblichen Unterschied (Fig. *d*) durch die erfolgte Einwirkung des Arbeiterfutters, das wohl schon vom 2. Tage an eine andere Zusammensetzung erfährt.

Nach v. PLANTA (8) ist die Zusammensetzung des Futters, wie ich hier einfügen möchte, die folgende :

ANGABEN in Procenten.	Königinlarve im Durchschnitt.	Drohnenlarven			Arbeiterlarven		
		unter 4 Tagen.	über 4 Tagen.	im Durchschnitt.	unter 4 Tagen.	über 4 Tagen.	im Durchschnitt.
Eiweisskörper . .	45,14	55,91	31,67	43,79	53,38	27,87	40,62
Fett	13,55	11,90	4,74	8,32	8,38	3,69	6,03
Zucker	20,39	9,57	38,49	24,03	18,19	44,93	31,56

Den Drohnen- und Arbeiterlarven wird nach dem 4. Tage vorverdauter Pollen und vom 5. Tage an Honig und Pollen gereicht.

Die Figur *e* zeigt uns ein Resultat, das KLEIN folgendermaassen erhielt. Eine $\frac{1}{2}$-1 $\frac{1}{2}$ tägige Arbeiterlarve wurde auf 2. Tage in eine Weiselzelle gebracht und dann wieder in eine Arbeiterzelle, wo sie noch 1-1 $\frac{1}{2}$ Tag als Arbeiterin ernährt wurde. Der Erfolg ist auffälliger Weise der, dass die Arbeitereigenschaften stark hervortreten, nur die Körbchenhaare (s. Hinterbein *e*) sind kürzer als bei *f*. Offenbar ist die Zehrung in der Weiselzelle wohl eine kärgliche gewesen infolge der Störung. Ich habe bei meinen Versuchen ein derartiges Stadium bis jetzt wenigstens auf diese Weise nicht erzielen können.

Doch nun zurück zu der Studie DEMOLL's. DEMOLL sagt : « Nach der bisherigen Auffassung musste man notwendig im Bienenei mit WEISMANN drei Keimesanlagen annehmen; eine männliche, eine weibliche und die einer Arbeiterin. Nun ist es aber auffallend, dass eine dritte Art von Individuen und mithin eine dritte Keimesanlage in ein und demselben Ei nur da auftritt, wo die Auswahl des Nährmaterials, das während der Entwickelung aufgezehrt wird, in der Hand der Fütternden liegt, also von diesen

(8) « Zeitschr. für phys. Chemie », von HOPPE-SEYLER, XII, Heft 4, p. 327-354, u. XIII, Heft 6, p. 552-561.

eventuell willkürlich geändert werden kann ». Ist diese Angabe DEMOLL's richtig? Ich glaube nicht. Ich verweise auf die Verhältnisse bei den Meliponen bei denen alle drei Bienenformen, wie soeben erwähnt, in gleich grossen Zellen entstehen und bei denen, wie ich hier hinzufüge, die Fütterung, soviel wir wissen, nicht « willkürlich geändert » werden kann, denn die Zellen werden *bevor* die Königin die Eier ablegt mit Pollen und Honig gefüllt und unmittelbar nach der Eiablage *geschlossen*, wie ich das schon in der « Stammesgeschichte » (1) betonte (pp. 34, 47, 54), und keinerlei Beobachtung weist bisher darauf hin, dass die Zellen während des Larvenlebens wieder geöffnet werden, um eventuell eine Aenderung des Futters zu bewirken. Die Nahrungsmenge ist überdies, wie erwähnt, der Quantität nach die gleiche (*). Andererseits sehen

(*) Auf das « Meliponastadium » bei *Apis* wies ich in der « Stammesgeschichte » hin (1, p. 59). Wir finden dieses Stadium bei der indischen Riesenbiene *Apis dorsata* F. Auch dort sind alle Zellen gleich. Bestimmt bekannt ist es von den Arbeiterinnen- und Drohnenzellen. Da aber noch niemals Weiselzellen beobachtet wurden, während sie uns von den anderen indischen Apis-Arten bekannt sind, so zweifle ich persönlich nicht daran, dass auch die Königinnen in denselben Zellen entstehen wie die Arbeiter und Drohnen. Ob aber diese Gleichheit der Zellen eine Gleichheit des Futters bedingt, wie DEMOLL allerdings nur für die Drohnen und Arbeiter annimmt (die Königin wird von ihm nicht erwähnt), ist immerhin doch fraglich. Die Quantität mag vielleicht die gleiche sein, aber die chemische Zusammensetzung dürfte vielleicht durch Zusatz besonderer Drüsensecrete eine gewisse Aenderung erfahren, denn die Königinnen (als auch die Drohnen) entstehen nur zu gewissen Zeiten, die sich biologisch von den anderen unterscheiden, sofern es gestattet ist, von den Verhältnissen bei *Apis mellifica* L. zu urteilen. Es scheint mir, dass wir — namentlich im Hinblick auf die Verhältnisse bei *Melipona* (sofort nach der Eiablage geschlossene Zellen) — gezwungen sind, irgend welche äusseren Reize, die im Innern der Wachszelle — also doch wohl im Futter — vorhanden sein müssen, anzunehmen. Dass die Königin der *Dorsata* ebenfalls in den kleinen Zellen entsteht, scheint mir ferner auch aus den stark differierenden Grössenangaben hervorzugehen. Die Länge der Arbeiter ist 16-18 mm., die Länge der Königin nach BINGHAM 18-21 mm., die Länge der lebenden Königin nach DATHE 23 mm. (vgl. 2) (p. 160). Nach BENTON ist sie « etwas grösser als die Arbeiter » i. l. (« Bienenw. Centralblatt », Hannover, 1881, p. 222). Die jungfräulichen Königinnen dürften, wie bei *Melipona*, die Arbeiter anfänglich an Grösse nicht oder unwesentlich übertreffen und erst später durch das Anschwellen der Ovarien beträchtlich grösser werden.

wir bei den Hummeln diese Möglichkeit einer willkürlich zu verändernden Fütterung, da nachweislich die Zellen beim Füttern wieder geöffnet werden und doch haben wir hier nach meiner Auffassung nur Männchen und Weibchen und brauchen daher (wie übrigens auch bei den sozialen Vespiden unserer Breiten) nur *zwei* Keimesanlagen im Ei anzunehmen, wie ich das in der « Stammesgeschichte » (1) näher ausführte (p. 37). Die sog. Arbeiter bei den Hummeln dürfen, so glaube ich, nicht mit den Bienenarbeitern verglichen werden, denn sie sind nur schlechter ernährte aber in jeglicher Hinsicht vollkommene Weibchen (*). Ich muss auf die erwähnte Arbeit verweisen.

(*) In einer nach Abhaltung des vorliegenden Vortrages (Aug. 1910) erschienenen sehr interessanten Arbeit : « Zur Phylogenie des Hymenopterengehirns » (« Ienaisch. Zeit. » Nat., Bd. XLVI N. F. XXXIX, Heft 2, 1910), sagt VON ALTEN : « ... ich möchte noch darauf hinweisen, dass bei den nicht perennierenden Hummel- und Wespenstaaten das Weibchen (in Bezug auf das Gehirn) am höchsten entwickelt ist, worauf die Arbeiterinnen und schliesslich die Männchen folgen, während bei *Apis mellifica* die Arbeiterinnen höher stehen als Weibchen und Männchen. Dieses Resultat scheint mir deshalb von einigem Interesse zu sein, als von einigen Seiten, im besonderen VON BUTTEL-REEPEN, die Ansicht vertreten wird, dass die Hummel- und Wespenarbeiterinnen nur kleine, mangelhaft ernährte, im übrigen aber morphologisch und anatomisch vollkommene Weibchen seien, also nicht direkt mit den Arbeiterinnen von *Apis mellifica* verglichen werden könnten. Demgegenüber glaube ich behaupten zu können, dass die Hummelarbeiterinnen sich von den Weibchen durch eine nicht nur absolut, sondern auch relativ geringere Ausbildung der pilzhutförmigen Körper und damit der Instinkte unterscheiden; dass wir also demnach auch bereits im Hummel- und Wespenei, wie nach WEISMANN im Bienenei, drei getrennte Anlagen für die drei verschiedenen Formen annehmen müssen ».

Da die vortrefflichen Untersuchungen VON ALTEN's sonst ausgezeichnet mit den biologischen Befunden zusammengehen, sind seine Ausführungen auch in diesem Punkte sehr beachtenswert, aber aus welchem Grunde bin ich zu der im Text ausgesprochenen Ansicht gelangt? Nun, weil eben, soweit unsere Erfahrungen reichen, *keine* « geringere Ausbildung der Instincte » zu entdecken ist oder doch nur eine so unbeträchtliche Abweichung, wie das Ausbleiben der Begattungstriebes, dass durch die so gut wie vollkommene Gleichheit der Instincte eben die Ansicht entstehen *musste*, es hier hinsichtlich der Keimesanlage mit *gleichartigen* Wesen zu tun zu haben. Wenn man bedenkt, dass auch die Bienenkönigin, wenn sie unbefruchtet in die Eiablage eintritt, — wie das ja die Hummelarbeiterinnen etc. ständig tun, — *keinen Begattungstrieb* zeigt, so wird man dem fehlenden Begattungstrieb bei den Hummel- oder Wespenarbeitern auch keinen Wesensunterschied zuschreiben.

Die Fütterungsart bei den *Meliponinæ* ist also dieselbe wie bei vielen *solitären* Hymenopteren, wo nur Weibchen und Männchen auftreten. Da wir nun aber auch zahlreiche *solitäre* Hymenopteren kennen, die die Zellen nicht schliessen, sondern die Jungen andauernd füttern (also wie bei *Apis mellifica*), z. B. *Cerceris*, *Bembex rostrata*, *Bembex spinulæ*, *Lyroda subita*, *Monedula punctata*, etc. (vgl. 1)(p. 34) und doch nur Weibchen und Männchen entstehen, so wird auch durch diese Tatsache der allgemein gehaltene Ausspruch DEMOLL's widerlegt, dass « nur da wo die Auswahl des Nährmaterials in der Hand der Fütternden liegt, eine dritte Art von Individuen auftritt ». Die Fütterung bei diesen andauernd fütternden solitären Wespen [wir kennen auch eine solitäre Biene, — *Allodape* —, welche dieselbe Erscheinung zeigt (9)], ist nun zweifellos je nach der Beute und der Witterung eine wechselnde aber das Mehr oder das Minder oder auch die qualitative Verschiedenheit, die jedenfalls eine unwillkürliche ist, ergiebt nur Variationen in der Grösse der späteren Imagines, eine dritte Art von Individuen tritt nicht auf, weil eine *dritte* Keimesanlage im Ei nicht existiert. Nur dort wo eine solche vorhanden, können Reize irgend welcher Art eine dritte Form bewirken.

Wie geschieht denn nun aber die Differenzierung der drei Formen bei den Meliponen? Ich gab vorhin einige Mutmaassungen, aber wir wissen es eben nicht und ich kann hier nur das wiederholen, was ich schon in der « Stammesgeschichte » (1) hierüber sagte : « Es lassen sich allerhand Theorien aufstellen, aber die ganzen Verhältnisse sind so wenig bekannt, dass man nur sagen kann, dass hier die Forschung kräftig einzusetzen hat, bevor eingehendere Erklärungsversuche, die einigermaassen Hand und Fuss haben, beginnen können ».

Aber mit dieser diffizilen Frage haben wir uns hier nicht zu beschäftigen, wir wollen hier nur feststellen, dass tatsächlich eine dritte Keimesanlage im Bienenei vorhanden sein muss. DEMOLL wendet nun aber u. a. weiterhin dagegen ein, dass es wahrscheinlicher sei, im Bienenei nur zweierlei Keimesanlagen anzunehmen, weil durch eine Fütterung der Bienenkönigin, die einer phylogenetisch früheren Fütterungsweise entspricht, in einer phylogenetisch älteren Zelle die Entwickelung eines befruchteten Eies

---

(9) FRIESE, H., Die Bienen Afrika's, Iena, 1910.

in der Weise modifiziert wird, dass der Embryo nicht mehr die ganze Entwickelung durchläuft, sondern schon auf einem früheren Stadium diese abbricht und so als atavistische Form die Zelle verlässt. »

Angenommen diese Angaben seien richtig, so muss man unter Heranziehung alles bisher Vorgebrachten sagen, dass diese Modifikationsmöglichkeit eben doch nur auf einer besonderen Keimesanlage möglich erscheint, zumal die Königin nicht einfach als eine Vorstufe der Arbeiterin zu betrachten ist; sie zeigt uns eine solche Fülle von Instinktsverlusten und Organisationsänderungen und eine solche Anpassung an die einseitig ausgebildeten Bedürfnisse des *heute* bestehenden Bienenstaates, die ohne Zweifel erst spätere Neuerwerbungen resp. Neuausgestaltungen sind, dass zu ihrer Hervorrufung durch äussere Reize eine ganz besondere Keimesanlage notwendig angenommen werden muss. Die Königin in ihrer heutigen Gestalt ist nur lebensfähig im Volksverbande, sie hat, wie erwähnt, alle Sammel-, Bau- und Fütterinstinkte verloren, ihr fehlen das erste Speicheldrüsensystem, die Sammelapparate und die Wachsdrüsensecretion, der Stachel ist anders geformt ebenso die Mundgliedmaassen, etc. Eine solche atavistische Vorstufe wäre als solche ohne den complicierten Volksverband, wie wir ihn heute sehen, nicht lebensfähig. DEMOLL wendet dagegen ein, dass aber die *Anlagen* zu allem diesem Notwendigen bei der Königin vorhanden seien, und das sei das Wesentliche. Ich habe schon in der « Stammesgeschichte », im Gleichklang mit WEISMANN, betont, dass Arbeiterin *plus* Königin — *cum grano salis* — früher *eins* gewesen seien oder mit anderen Worten : die Keimesanlage war *früher* dieselbe, jetzt aber geht sie in vielen Dingen weit auseinander und bleibt ständig so. Diese fest bestehenden auseinandergehenden Anlagen sind heutzutage eben differente Keimesanlagen. Ich wenigstens kann mir eine andere einigermassen befriedigende Erklärung nicht machen.

DEMOLL sieht die der Königin fehlende Futtersaftdrüse (Speicheldrüse) als eine Neuerwerbung der Abeiterinnen an. Nehmen wir diese Ansicht als richtig an, so muss man fragen, wie soll mit dieser erst neu erworbenen Anlage eine « phylogenetisch frühere Fütterungsweise » bewirkt werden, denn grade diese Futtersaftdrüse (System I) erzeugt nach allgemeiner Annahme den milchigen der Königinlarve ausschliesslich gereichten Futtersaft. DEMOLL meint aber, dieser Futtersaft sei *flüssig*, er enthielte keine festen Bestandteile (nämlich Pollen) wie bei den Arbeiterinnen und Drohnen und eine *flüssige* Fütterung sei zweifellos die phylogenetisch ältere. Das erscheint aber doch als eine wohl etwas weit hergeholte Analogie. Und es ist noch die Frage, ob die frühere « flüssige » *animalische* Kost, wie sie — DEMOLL zieht hier die Wespen heran — auch heute noch den Wespen gereicht wird, wirklich nur flüssig war. Meine Beobachtungen an den heutzutage mit animalischer Kost genährten Hymenopteren sprechen dagegen. Zur Bekräftigung weist DEMOLL aber auf die *Polistes*-Arten hin und sagt : « Ferner sei darauf hingewiesen, dass da, wo wir heute einen Uebergang von animalischer zu vegetabilischer Kost konstatieren können, wie bei *Polistes*-Arten, nicht Pollen » (also nicht feste Bestandteile) « sondern Honig gereicht wird. Interessant hierbei ist, dass solche honigfütternde Wespen erst dem dem Ausschlüpfen nahen Insekt die vegetabilische Nahrung darbieten, während die Larve noch mit animalischer Kost versorgt wird ».

Hier bleibt etwas unverständlich. Dem « Ausschlüpfen nahe Insekten » werden nicht gefüttert, fressen auch nicht, da sie als Puppen resp. als unausgeschlüpfte Imagines, wie natürlich DEMOLL selbst weiss, im geschlossenen Kokon ruhen. *Nur* die Larve frisst. Bei *Polistes* verzehren aber nur die *Imagines* Honig. V. SIEBOLD weist ausdrücklich hierauf hin (10), dass der Honig *nicht* verfüttert wird, und ich konnte diese Beobachtung bestätigen (1) (p. 60).

Aber DEMOLL meint, auch die Instinkte, die wir jetzt bei den Arbeiterinnen sehen, ruhen nur bei der Königin, sie seien in der Anlage vorhanden. Er bekräftigt seine Ansicht durch die Beobachtung, dass eine junge Königin « direkt nach dem Ausschlüpfen, also bevor sie von Arbeiterinnen gefüttert werden konnte, in einen Zwinger gebracht, sofort sich an die Blumen machte, die in ziemlicher Entfernung in dem Gehäuse auf den Boden gelegt wurden und dabei ihre Zunge eifrig betätigte ».

Wenn man aber weiss, dass es die erste Beschäftigung einer im Stock auskriechenden Königin ist — ich habe das vielfach beobachtet — sofort zu einer der nächsten Honigzellen zu eilen, um dort

(10) Beiträge zur Parthenogenesis der Arthropoden. Leipzig, 1871, p. 31.

sehr ausgiebig ihren Hunger zu stillen, so wird man in dieser DEMOLL'schen Beobachtung nichts Besonderes finden, zumal wenn in Betracht gezogen wird, dass die DEMOLL'sche Königin « schon einen Tag vorher in Gefangenschaft verbracht » hatte, also wohl sehr hungrig war. Der Nektargeruch musste sie sofort zu den Blumen leiten. Wenn aber DEMOLL ferner meint : « Das Anfliegen von Blumen ist aber weiterhin wieder notwendig, um die Instinkte der Brutversorgung (Zellenbau, Fütterung) auszulösen » und weil dieses bei der Königin normalerweise unterbleibt, dass nur deshalb ein Ruhen dieser Instinkte bei ihr eintritt, so scheint mir DEMOLL zu übersehen, dass die jungen Arbeitsbienen, die niemals den Stock verlassen haben (denn erst gegen den 12.-14. Tag ihres Imagineslebens werden sie « Feldbienen »), die Instinkte der Brutversorgung, etc., ganz besonders gut zeigen, nach allgemeiner Annahme sogar besser als die blütenbesuchenden Feldbienen, die sich normaler Weise dann fast nur den Aussenbeschäftigungen hingeben. Hier werden also bei den sog. Hausbienen (im Gegensatz zu Feldbienen), die auch *Brutammen* genannt werden, diese Instinkte ohne Blumenbesuch und zwar besonders gut betätigt. Ich wies hierauf in meiner Arbeit « Sind die Bienen Reflexmaschinen » ausgiebig hin (11).

Es scheint mir also doch wohl an verschiedener Erbmasse zu liegen, wenn die Instinkte so differieren.

Ich möchte hier noch folgende Differenz zwischen Königin und Arbeiterin betonen. Die Königin ist sehr viel grösser als die Arbeiterin. Das ist anscheinend leicht zu erklären, da sie in einer sehr viel grösseren Zelle unter so starker Fütterung erzogen wird, dass stets nicht verzehrter Futterbrei am Grunde der Zelle zurückbleibt, während sich die Arbeiterin mit der kleinen engen Zelle und spärlicherer Fütterung begnügen muss. Erzieht man aber Arbeiterinnen in den nicht unwesentlich grösseren Drohnenzellen (p. 71) (10), so behalten sie ihre normale Grösse trotz reichlicher Nahrung; auch Drohnen, die man hin und wieder in Weiselzellen findet, in denen sie aber meistens absterben, haben die normale Grösse. Die enge Arbeiterinzelle, die man so häufig als Zwangsjacke ansieht, entspricht also den normalen Bedingungen, wie das auch biologisch verständlich und eigentlich selbstverständlich ist.

---

(11) Leipzig, 1900.

In dieser Grössendifferenz zwischen der Königin und den Arbeiterinnen finden wir also wiederum eine heute bestehende beträchtliche Verschiedenheit. Die sich hieraus weiter ergebenden Schlüsse werde ich gleich näher berühren.

Diese Grössenunterschiede erwecken nun die Frage : War die Königin der *A. mellifica* in vergangenen Zeiten eben so gross wie die Arbeiter, wie wir es heute noch, wie vorhin betont, bei der *Apis dorsata* und auch bei den Meliponen finden? Oder war sie von Anfang an stets etwas grösser? Die Frage ist nicht so leicht zu entscheiden, wie z. B. H. v. IHERING vom Standpunkt des Meliponen-Forschers anzunehmen scheint. Ich habe mich auf Grund vergleichender Betrachtungen unter Berücksichtigung aller bekannten Staatenbildungen für die Annahme entscheiden müssen, dass die Königin bei *Apis mellifica* wohl stets grösser gewesen sein dürfte als die Arbeiterinnen, und nur auf Grundlage dieser Ansicht ist ja auch die Auffassung der grossen Weiselzelle als einer atavistischen Bauform besonders verständlich.

Der Gedanke liegt freilich nahe, anzunehmen, dass es im Interesse des Volksverbandes lag, allmählich in der Eiablage leistungsfähigere Weibchen zu erzielen, und dieser Gedanke ist zweifellos ein sehr richtiger, aber es ist nicht notwendig, dass sich die *ursprüngliche* Grösse der (jungfräulichen) Weibchen durch das Uebergehen zu sehr volkreichen Staatenbildungen wesentlich veränderte (in einigen Ausnahmefällen ist es vielleicht geschehen), denn wir sehen, dass sich die Weibchen bei den Termiten ebenso wie bei den Meliponinen je nach den Bedürfnissen des Staates erst secundär, d. h. stets erst im individuellen Leben, durch das Anschwellen der Ovarien zu enormer Grösse ausdehnen. Hier ist anscheinend trotz riesiger Leistungsfähigkeit (Termiten) die ursprüngliche Grösse wohl kaum verändert worden, doch können Zweifel hierüber vorhanden sein. Vergegenwärtigt man sich aber nun, dass die Königin von *Apis mellifica* früher, d. h. in jenem Stadium als die Staatenbildung einsetzte, Allesschafferin war und daher nach *jeder* Richtung hin ausserordentlich leistungsfähig sein musste, so müssen wir hierfür starke, kräftige Weibchen postulieren, die dem Kampf ums Dasein gewachsen und vor allen Dingen fähig waren, die Trockenperiode oder den Winter *einsam* zu überstehen.

So finden wir auch heute noch die stets noch *einsam* überwinternden Hummeln- und Wespenköniginnen, deren Staaten uns, so glaube ich, Uebergangsbildungen zeigen (1) *beträchtlich grösser* als die resp. « Arbeiterinnen », obgleich die Ansprüche an die Eiablage im Verhältnis zur *Mellifica*-Königin *gar nicht in Vergleich gezogen werden können*, da sie so sehr viel schwächere sind. Aber es ist noch ein anderer Punkt heranzuziehen, der vielleicht zu Gunsten meiner Ansicht sprechen dürfte.

Worin besteht der Vorteil der Staatenbildung im Lebenskampfe? In der Vielheit der Individuen also der Brutenährerinnen etc. Je volkreicher sich im Durchschnitt ein Staatsverband halten kann, je überlegener wird er meistens im Kampf ums Dasein sein. Es würde mich zu weit führen, diesen Gedanken hier ausführlicher zu begründen (vgl. 1) (p. 102). Auf welchem Wege konnte Naturzüchtung diese Vielheit erreichen, dadurch dass jeder Volksgenosse, also jeder Arbeiter mit dem zulässig geringsten *Minimum* an Nahrung, Pflege und Raum auskam. Je genügsamer die Arbeiter bei voller Leistungsfähigkeit in dieser Hinsicht wurden, je kräftiger stand das Volk da und konnte im Vorzug vor anderen von einer phylogenetischen Stufe zur anderen emporsteigen. Dieses Minimum bedingte aber meines Erachtens bei den Arbeitern der *Apis mellifica* ein gewisses Abnehmen der Körpergrösse. Der starke Fettansatz, den sich die Königin früher heranmästen musste, um die Zeiten der Not zu überdauern resp. um winterständig zu sein, war für die Arbeiter nicht mehr vonnöten, ebenfalls nicht mehr der Körperraum für volltätige Ovarien, hierdurch musste naturgemäss die *Körpergrösse abnehmen* (vielfach wohl auch durch besondere biologische Bedingungen resp. Anpassungen). Die nötige Reservenahrung wurde nicht mehr in Gestalt von Fett im Körperinnern deponiert, sondern in Gestalt von Honig und Pollen in die leeren Brutzellen oder ursprünglich vielleicht auch in besondere dafür geschaffene Zellen.

Diese ganzen Verhältnisse haben wohl sicher auch ein Kleinerwerden der Eier bewirkt, da die Selektion auch nach dieser Richtung vielfach tätig gewesen sein dürfte (*), naturgemäss wuchs

(*) Man vergleiche z. B die winzig kleinen Eier der *A. mellifica* mit den riesigen einer solitären Biene z. B. *Halictus quadricinctus* F. Es ist mir sehr wohl bekannt, dass die Meliponen im Verhältnis zur *A. mellifica* sehr grosse Eier aufweisen, aber die Meliponen haben auch nicht die hohe Stufe der *A. mellifica* erreicht.

hiermit auch die Leistungsfähigkeit der Königin, ohne ein Zunehmen der Körpergrösse zu bedingen. Dass diese Wege unter besonderen biologischen Bedingungen nicht immer beschritten wurden, ist sicher. Im Allgemeinen aber lässt sich wohl der Satz — namentlich auch im Hinblick auf die Ameisen — aufstellen, dass nicht die jetzige Körpergrösse der Arbeiter bei den sozialen Insekten der *ursprünglichen* Grösse der phylogenetischen Gründer der betr. Staatsgebilde am nächsten steht, sondern dass in vielen Fällen wohl die Grösse der jetzigen *jungfräulichen* Weibchen der der Gründer näher gestanden haben dürfte.

Dieser Grössenunterschied zwischen der Königin der *A. mellifica* und den Arbeitern beruht nun m. E. ebenfalls auf differenter Keimesanlage. DEMOLL hat diesen Punkt nicht in den Kreis seiner Betrachtungen gezogen, wohl aber die sicher erst im Staatsverbande erfolgte Grössenzunahme des Ovariums. Er meint, das alles spräche nicht gegen seine Ansicht, denn die *Anlage* wäre auch bei der Arbeiterin vorhanden.

Es würde mich zu weit führen auf diesen und auf manchen anderen Einwand DEMOLL's einzugehen, zumal auch diese m. E. nicht stichhaltig sind. Die WEISMANN'sche Ansicht dürfte daher wohl zu Recht bestehen bleiben, nur müsste auf Grund unserer heutigen Kenntnisse eine andere Formulierung vielleicht — wie nachstehend — erfolgen :

Bei den Bienen finden wir ausser den Männchen stark einseitig ausgebildete fast aller weiblichen Instinkte verlustig gegangene Weibchen (Königinnen), die nur noch die Eiablage besorgen und ferner Arbeiterinnen, die alle weiblichen Instincte bewahrten und nur die Begattungsfähigkeit einbüssten. Aus stammesgeschichtlichen Gründen müssen wir annehmen, dass beide Formen *einem* echt weiblichen Determinantencomplex entsprungen sind, der aber für den Bestand der Art vorteilhafte Abänderungen vieler seiner Determinanten einging und sich zu Königin-Iden und zu Arbeiterin-Iden umgestaltete, so dass wir jetzt drei Keimesanlagen annehmen müssen.

Ich möchte jetzt nur noch kurz auf einige andere atavistische Erscheinungen aufmerksam machen.

Geht die Königin in Stocke verloren und ist keine Brut zur Nachzucht vorhanden, so schreiten die normaler Weise ganz sterilen Arbeiterinnen infolge dieser besonderen biologischen Verhältnisse durch die unterdrückten Fütterinstincte etc. und auch wohl infolge der stärkeren Ernährung, da der im Körper sich vorbereitende Futterbrei keine Ablagerungstätte findet und wieder resorbiert wird, zur Eiablage. Es tritt da nun nicht etwa eine Ersatzkönigin auf, sondern schliesslich wird die *Mehrzahl* der Arbeiter im Stock eierlegend. Also auch ein Anklang an stammesgeschichtlich ältere Verhältnisse.

Es ist nun interessant zu konstatieren, dass hierbei offenbar sehr alte Instinkte erwachen, die sich in der merkwürdigen Art und Weise der Eiablage äussern. Wir sehen noch heute bei den Hummeln stets mehrere Eier in den Zellen, oft eine ganze Anzahl, und seltsamer Weise finden sich bei der Eiablage durch Arbeiter der *A. mellifica* auch stets mehrere Eier in den Zellen, während die Königin normaler Weise stets nur je ein Ei in eine Zelle deponiert. Dann ist es seltsam, dass von den legenden Arbeitern nicht die in der weitaus grössten Mehrzahl im Stock vorhandenen Arbeiterinnenzellen im Vorzug benutzt werden, sondern mit besonderer Vorliebe die grösseren Drohnenzellen. Da bei den Wespen (12) und Hummeln (13) auch heute noch die Arbeiter stets an der Erzeugung der Männchen beteiligt sind, so liegt auch hier vielleicht ein atavistischer Zug vor und besonders in der von den legenden Arbeitsbienen erfolgenden Benutzung der *Pollenzellen*. Der in den Arbeiterzellen aufgespeicherte Pollen füllt die Zelle stets nur dreiviertel aus, es ist daher die Möglichkeit gegeben, in diesen noch verbleibenden kleinen Raum Eier abzulegen. Wenn man bedenkt, dass in früheren Zeiten wohl zweifellos die Ernährung so vor sich gegangen ist, wie wir es heute noch bei fast allen solitär lebenden Bienen und auch noch bei den stachellosen Bienen (*Meliponen* und *Trigonen*) sehen, nämlich erst Ansammlung von honigdurchtränktem Pollen und dann darauf Ablage des Eies, so wird man kaum bezweifeln können, dass wir es hier mit einem phylogenetisch altem Instinkt zu tun haben.

Das ganze Verstörtsein des Volkes hervorgerufen durch die Weisellosigkeit und den Mangel an Brut dokumentiert sich in diesem Durcheinander alter Instinkte, die uns aber wertvolle Fingerzeige in längst dahingegangene Zeiten geben.

---

(12) MARCHAL, PAUL, Comptes rendus 1893.

(13) LEUCKART, Zur Kenntniss des Generationswechsels und der Parthenogenesis bei den Insekten, Frankfurt 1858, p. 428

Das Volk der egyptischen Biene, *Apis fasciata* LATR. hat insofern altertümliche Züge bewahrt, als eierlegende Arbeitsbienen dort sogar beim Vorhandensein einer eierlegenden Königin vorkommen sollen. Ich kann diese Angaben aus der bienenwirtschaftlichen Literatur nach meinen neuerlichen Beobachtungen an der *A. fasciata* nicht bestätigen, womit nicht gesagt sein soll, dass diese Angaben unrichtig sind, denn im heissen Klima ist manches möglich, was im kühlen Norden nicht oder nicht so leicht möglich ist. Jedenfalls entstehen eierlegende Arbeiterinnen sehr viel schneller bei der *A. fasciata* als bei der *A. mellifica*. Auch das Auftauchen zahlreicher jungfräulicher Königinnen in demselben Stock neben der Stockmutter bei der *A. fasciata* und der kaukasischen Biene (*Apis mellifica* var. *remipes* PALLAS) weist auf frühere Zeiten hin, denn bei der *Apis mellifica-mellifica* ist das ganz ausgeschlossen. Bei den stachellosen Bienen begegnen wir derselben Erscheinung (*).

Ich kann mich hier auf die Aufzählung weiterer Atavismen nicht einlassen. Ich möchte nur noch die Aufmerksamkeit auf eine sehr interessante Beobachtung lenken, die ich vor einigen Jahren machte, als ich mich eingehender mit der Systematik der Honigbienen (2) beschäftigte.

Bekanntlich nimmt die *Apis mellifica* unter allen europäischen Bienen durch den Mangel der Sporen (Calcaria) an den hintersten Schienen eine ganz auffällige, seltsame Ausnahmestellung ein.

Wozu die Sporen dienen, ist nur zum Teil bekannt. Nach allgemeiner Annahme sollen sie Stützorgane beim Laufen und besonders beim Kriechen in engen Nestgängen sein (14), doch befriedigt diese Erklärung nicht völlig, da wir sie bei Insekten sehen, die keine Nestgänge bauen z. B. bei *Polistes* und sie andererseits bei ungemein gewandten Insekten, die sich oft überdies

(*) Diese jungfräulichen Königinnen bei den Meliponen waren übrigens der « europäischen Bienenforschung », die bei H. v. IHERING (6) — zum Glück unbegründeter Weise — nicht gut wegkommt, längst vor der Entdeckung durch IHERING bekannt. DRORY beschrieb sie schon 1872. Auch die Weiselzellen bei *Trigona* kannten wir schon vorher durch SILVESTRI.

DRORY, E., Quelques observations sur la Melipone scutellaire. Bordeaux 1872; SILVESTRI, FILIPPO, Contrib. alla Conosc. dei Meliponidi (Riv. Pat. Veget. Anno X, Portici 1902). Vgl. a. (1), p. 52 u. 54.

(14) FRIESE, H., Monographie der Bienengattung *Ceratina*. (« Term. Füzetek. », vol. XIX, parte 1. Budapest, 1896, p. 16.)

ange Nestgänge anlegen wie die stachellosen Bienen, nicht vorkommen. Jedenfalls dienen die Sporen nach meiner Beobachtung jetzt vielfach doppelten Zwecken, z. B. besitzt *Vespa crabro* L. an den *drei Beinpaaren*! Putzscharten, denen die Sporen, die ebenfalls einen mehr oder weniger starken Bürstenbesatz tragen, so zugeordnet sind, dass sie das zu reinigende Glied in die Putzscharte hineindrücken resp. umschliessen. Sämmtliche solitäre Bienen tragen in den bis jetzt bekannten ca 8000 Arten (mit weinigen Ausnahmen *Ceratina œnea*, etc.) alle die doppelten typischen Sporen an den Schienen der Hinterbeine, nur bei *Samba calcarata* FRIESE sind sie — als Unicum — in der Einzahl vorhanden (15).

Offenbar sind die Sporen ein uralter Besitz, der nach meiner Ansicht vielleicht auf schnellaufende, räuberische, in engen Gängen wohnende Vorfahren hindeutet. Schon sehr früh, ich beschränke mich hier auf die Hymenopteren, verbreiterte sich wohl der Sporn der Vorderbeine zu dem bekannten Deckel der Putzscharte und sind dann die Doppelsporen der beiden anderen Beinpaare, je nach den sich wandelnden biologischen Verhältnissen, umgeformt worden und teilweise oder ganz weggefallen. Vielleicht war bei vielen Hymenopteren an den ersten beiden Beinpaaren von Anfang an nur je *ein* Sporn vorhanden.

Da nun die *Mellifica* zweifellos von sporentragenden Vorfahren abstammt, so forschte ich nach diesen Sporen intensiver und ich fand sie denn auch allerdings nicht bei der Imago sondern bei der *Puppe!*

Wenn man sich vergegenwärtigt, ich ziehe hier nur die Holometabolen heran, dass die Larve und demgemäss auch die Puppe stammesgeschichtlich *jünger* ist als die Imago, ohne dabei präimaginale Formen ganz einzubüssen (16), so war die Aussicht, dort phylogenetisch ältere Spuren zu finden recht schwach (*), zumal wir grade bei den Jugendstadien der Holometabolen eine so weitgehende Fähigkeit treffen, alles umzumodeln und das nicht mehr Brauchbare zu beseitigen. Ich war daher sehr überrascht, die

---

(15) FRIESE, H., Neue Bienenarten aus Ostafrika. (« Deutsch. Ent. Zeitschr. », 1908, pp. 568-569.)

(16) DEEGENER, P., Die Metamorphose der Insekten. Leipzig, 1909.

(*) Phylogenetisch ältere Spuren bei Puppen sind bis jetzt *sehr selten* konstatiert worden.

Sporenrudimente in recht kräftiger Anlage zu entdecken. Bedenkt man, dass diese weichen Hautausstülpungen für die Puppe keinerlei Bedeutung haben, so ist es auf 's Höchste erstaunlich, dass sich dieser Atavismus durch so viele Jahrhunderttausende hindurch erhalten konnte.

Ich verdanke das auf Tafel IV, Figur 2, reproduzierte Microphotogramm der Freundlichkeit des hiesigen Directors am Naturhistorischen Museum, Profr.-Dr. Martin, dem auch an dieser Stelle verbindlichster Dank ausgesprochen sei. Der gänzlich durchsichtige Sporn erforderte zur besseren Sichtbarmachung eine Färbung des Hinterbeins. Borax-Carmin erwies sich als geeignet.

Auf der Abbildung sehen wir zu äusserst die Contouren der Puppenhaut, darunter die Umrisse der Imago, am Gelenk gegenüber dem Sporn die Chitinborsten des sog. Wachskammes und tiefer die Reihen der « Bürste », etc. Der zweite sehr viel kleinere Sporn ist auf der Abbildung nicht sichtbar. Die Lage des zweiten Sporns sei aus der nachstehenden schematischen mit dem Zeichenprisma (Abbe) entworfenen Zeichnung Figur 1 ersichtlich.

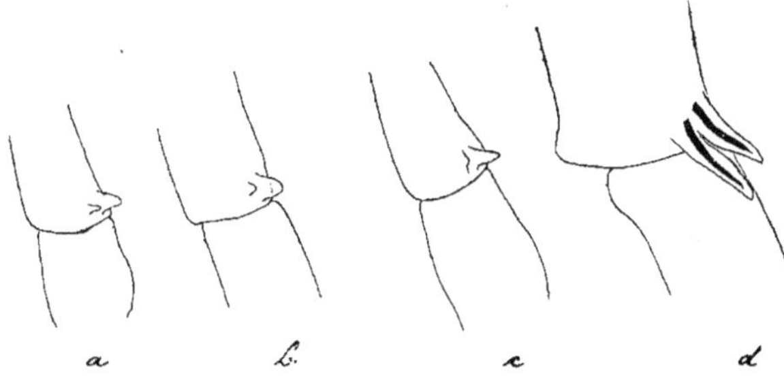

Fig. 1.

Die Sporen (Calcaria)- Rudimente in Gestalt von Hautausstülpungen an den Hinterschienen der Puppen von *Apis mellifica* L.
a = Arbeiterin, *b* = Königin, *c* = Drohne. Daneben zum Vergleich *d* = die Sporen bei der Puppe von *Bombus distinguendus* Mor.

Die Sporen finden sich bei den Puppen aller drei Bienenwesen in der typischen Form, d. h. der eine Sporn ist länger als der andere. Die Grössenunterschiede sind hier allerdings auffällig stark. So markiert sich bei der Drohnenpuppe der kleinere Sporn nur noch als ganz schwache Erhebung, dafür ist der andere im Verhältnis zu dem der Königin und der Arbeiterin der längste.

Sehr interessant ist, dass alle drei Formen hinsichtlich der Sporenausgestaltungen konstant von einander abweichen. Es wird hierdurch der Schluss nahe gelegt, dass schon in jenen fernen Zeiten als die *Apis mellifica* noch Sporen als Imago trug, bereits eine Differenzierung in Arbeiter- und Königinform oder doch in biologisch resp. in den Instinkten von einander abweichende Formen durchgeführt war !

Die Bedeutung dieses an und für sich geringfügigen Befundes erhöht sich vielleicht dadurch, dass auf Grund dieser Feststellung ein Insekt, das durch seine Staatenbildung und durch seine volkswirtschaftliche Bedeutung das Interesse weitester (Laien) Kreise in Anspruch nimmt, *im Gegensatz zu früher* ein *besonders gutes* und nicht wegzudisputierendes *Beweisstück* geworden ist *für die Descendenztheorie.* Auf die « Unveränderlichkeit der Biene » ist schon so oft von Gegnern der Entwicklungslehre gepocht worden, dass die gefundenen kleinen Sporen die Gedanken nun möglicherweise eher in andere Richtung lenken werden. Für den Naturforscher bilden sie freilich nur ein kleines unbedeutendes Glied mehr in einem überreichen Gefüge.

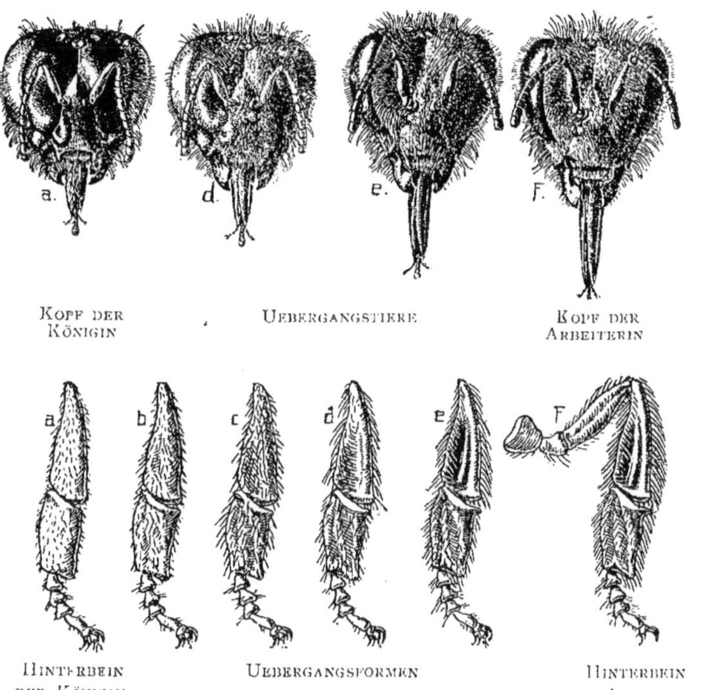

Fig. 1. — *Köpfe* und *Hinterbeine* weiblicher Bienen nach vorangegangener Bebrütung als :

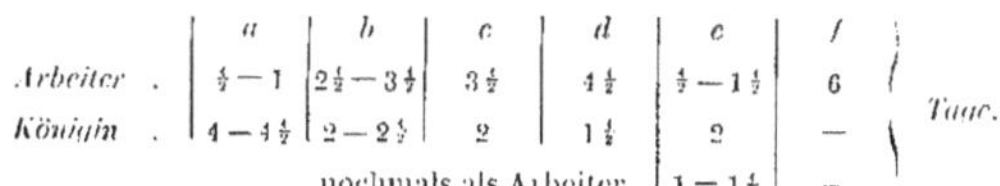

	a	*b*	*c*	*d*	*e*	*f*	
Arbeiter .	½ — 1	2½ — 3½	3½	4½	½ — 1½	6	*Tage.*
Königin .	4 — 4½	2 — 2½	2	1½	2	—	
nochmals als Arbeiter					1 — 1½	—	

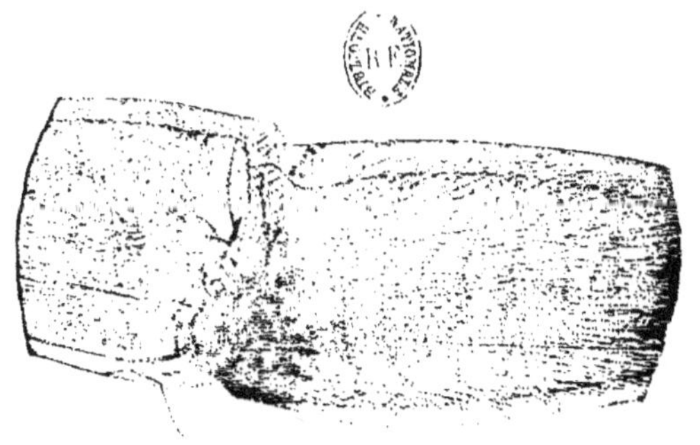

Fig. 2. — Sporn des Hinterbeins der Puppe der Arbeiterin von *Apis mellifica* L.
24 × Phot. Vergrösserung.

H. von BUTTEL-REEPEN. — ATAVISTISCHE ERSCHEINUNGEN.

Recherches physiologiques sur la coloration des cocons de certains Lépidoptères,

par I. DEWITZ (Metz).

Quelques auteurs, et plus particulièrement POULTON, ont signalé des cas où des cocons de Lépidoptères font voir, d'après la couleur de leur support, une teinte variant du blanc ou du brun clair au brun foncé.

Pour se former une opinion sur cette adoption, il serait nécessaire de connaître l'origine du pigment et les facteurs qui exercent une influence sur l'apparition de ce dernier.

RÉAUMUR avait déjà fait quelques observations sur la coloration du cocon de certains Lépidoptères; il cite notamment l'exemple du *G. neustria*.

Les cocons mentionnés chez RÉAUMUR se laissent diviser en deux groupes. Dans le premier groupe, la chenille confectionne un tissu parfois peu serré, qu'elle imprègne à l'aide de ses mandibules d'une pâte sortie de l'anus et qui représente le contenu de ses tubes de Malpighi (*G. neustria*, *L. salicis*). Ou bien elle fait de cette pâte une coque dont elle entoure son léger cocon (*B. lanestris*). D'après mes informations, la couleur du cocon ne varie que dans le dernier cas.

Au second groupe appartiennent les cocons de certaines espèces de *Saturnides* et peut-être aussi ceux d'autres *Bombycides*. Lorsque la chenille de ces espèces cesse de manger, elle se débarrasse du contenu de son tube digestif et fait sortir de l'anus un liquide qui est brun au début, mais qui devient incolore vers la fin de son évacuation. Après s'être vidée, la chenille file un volumineux et solide cocon, qui est parfaitement blanc, et, vingt-quatre heures après, elle se vide une seconde fois. Le cocon devient alors mou et flasque comme si on l'avait trempé dans l'eau, et, chose particulièrement curieuse, il change de couleur et devient brun. C'est le liquide incolore que la chenille, enfermée dans le cocon, a expulsé et qui contient beaucoup de cristaux des tubes de Malpighi qui a provoqué ce changement. Quant aux échanges intimes qui se produisent lorsque le liquide incolore du tube digestif est mis en contact avec le cocon incolore, LEVRAT et CONTE croient que le liquide n'est autre que la partie fluide du sang de la chenille qui s'est accumulée dans l'intestin. Il s'oxyderait, brunirait à l'air et communiquerait à la soie la teinte brune que tout le monde connaît. La soie elle-même ne participerait donc pas au changement de couleur; ce serait au contraire un corps étranger, couvrant les fils de soie, qui deviendrait brun. Suivant mes recherches personnelles, la solution du problème me paraît être toute différente. Mes observations se rapportent surtout aux cocons du *S. pavonia*.

Pour obtenir des cocons blancs, indemnes de toute souillure du liquide intestinal, j'attendis que le ver se fût vidé pour la première fois et commençât à filer. J'obstruai alors l'anus de la chenille en laissant tomber sur lui une goutte de vernis séchant rapidement ou en faisant une ligature à l'extrémité du corps, de sorte que les cocons ainsi obtenus se composèrent uniquement de produits des glandes séricigènes ou d'autres sécrétions sortant de la bouche.

La plus curieuse des réactions chimiques de ces cocons est celle que détermine l'eau. Des morceaux de cocons blancs mis dans l'eau deviennent vite bruns. La chaleur (45°) accélère beaucoup ce changement de couleur. En même temps que les morceaux, l'eau reçoit une teinte rouge, ce qui indique que des parties solides du tissu se sont dissoutes. Par contre, le pouvoir du cocon blanc de brunir au contact de l'eau est détruit lorsqu'on le plonge pendant quelques instants dans l'eau bouillante. Ceci paraît indiquer que nous nous trouvons en présence d'un enzyme, bien que ce fait n'exclue pas une autre interprétation. Car l'eau bouillante dissout partiellement le grès et l'extrait du cocon.

Des corps réducteurs tels que l'hydroxylamine et le cyanure de potasse sont très nuisibles au changement de couleur des morceaux blancs mis dans l'eau, tandis que des corps oxydants comme l'acide chromique ou le permanganate de potasse déterminent promptement l'apparition du pigment.

Le dernier corps se décolore en même temps.

Les alcalis donnent d'intéressants résultats. Une solution de carbonate ou de bicarbonate de soude brunit en peu de temps les morceaux blancs et devient elle-même brune, ce qui est d'autant plus intéressant, que plusieurs auteurs ont signalé, dans le liquide que le ver prêt à se chrysalider évacue, la présence de bicarbonate de potasse. Quand on verse une solution étendue de soude caustique sur les morceaux, le liquide est tout de suite rose et devient plus tard brun ; mais les morceaux restent blancs. L'hydroxylamine, qui est d'une grande puissance réductrice, s'oppose à ce changement de couleur.

Je suis en mesure d'ajouter, à ces réactions des cocons blancs, d'autres non moins intéressantes. C'est ainsi qu'on obtient avec des cocons blancs et aussi avec des cocons normaux de couleur brune les réactions de guajac et de guajacol qui sont propres aux enzymes oxydants (oxydases). Les deux sortes de cocons décomposent aussi l'eau oxygénée, et cela prouve que nous avons ici encore un autre enzyme, c'est-à-dire celui qui porte le nom de catalase. A ce point de vue, il importe de dire que les cocons appartenant aux femelles paraissent avoir un pouvoir un peu plus grand de décomposition que les cocons des mâles.

Je crois pouvoir résumer de la manière suivante les faits que je viens d'énumérer :

Avant de faire son cocon, la chenille de certaines espèces de *Saturnides* rejette le contenu de l'intestin dans lequel ne se trouve à la fin qu'un liquide incolore. Elle file alors et, vingt-quatre heures après, emprisonnée dans son cocon, qui est parfaitement blanc, elle se vide de nouveau, et ce liquide incolore, qui contient du bicarbonate de potasse, fait brunir le cocon incolore. Le corps qui subit ce changement, le chromogène, se trouve dans le tissu même et il donne naissance au pigment brun sous l'influence de l'oxygène de l'air et du liquide alcalin sorti de l'anus de la chenille.

Malheureusement, je ne suis pas suffisamment documenté pour pouvoir dire si la lumière peut modifier ou supprimer le changement de couleur des cocons blancs. Je suis cependant à même d'affirmer que des chenilles dont l'anus n'a pas été obstrué peuvent fournir des cocons blancs aussi bien à un endroit obscur ou mal éclairé qu'au soleil, pourvu que l'air et le voisinage contiennent très peu d'humidité. D'autre part, des morceaux de cocons blancs exposés dans les environs de Nice, en juillet, au soleil, en présence de l'eau ou d'une solution de carbonate de soude, se colorèrent très bien.

La coque du cocon du *B. lanestris*, qui se compose uniquement du contenu de l'intestin et surtout de celui des tubes de Malpighi, m'a donné des résultats à peu près analogues à ceux que j'ai obtenus en me servant des cocons de *S. pavonia*. On se rappellera que la chenille de *B. lanestris* mâche la pâte destinée à la confection de la coque. Je suppose donc que le changement de couleur des cocons de certains Lépidoptères est dû à des sécrétions sortant de la bouche de la chenille.

Note on the chemical composition of the red-coloured secretion of «Timarcha tenebricosa»,

by E. WACE CARLIER, M. D, F. R. S. E., F. E. S., etc., Professor of Physiology, and C. LOVATT EVANS (Birmingham).

In April of the year 1910, on the last day of my holiday in Norfolk, I gathered 16 examples of this bloody-nosed Beetle, and observing how copiously they poured forth their characteristic secretion on the slightest provocation, it occurred to me that enough might be available for chemical analysis if they were kept alive, properly fed, and irritated at regular intervals, say every 24 hours.

It was quite noticeable that some individuals produced more secretion than others, the females, which are larger than the males, usually yielding most, but even amongst these the quantity varied considerably, and though they were kept on a moist grass sod, in conditions as nearly natural as possible, the amount yielded gradually diminished in quantity as time went on, and practically ceased after about a week or ten days, the last specimen dying some three weeks after the date of capture. I was not successful in obtaining a second supply of these Beetles after my return here, as they seem to have emerged unusually early this year, and are not common in the immediate neighbourhood of Birmingham.

Mr. C. LOVATT EVANS, an expert chemist working in my laboratory, undertook to analyse this secretion for me, and devised some ingenious methods for dealing with the minute quantity of secretion daily available. His report is appended.

As far as I am aware, no attempt to gain an insight into the chemical characters of this secretion has been undertaken in this country, therefore I have the less hesitation in bringing before you the somewhat imperfect analysis here detailed in the hope that others more fortunately situated may be induced to repeat the experiments on a larger scale, as doubtless the part played by this discharge, which must be of an exhaustive nature to the Beetle, derived as it is either from the salivary glands or the crop, may be thereby better understood. The secretion is poured forth, not from the joints or neck as in oil Beetles, but from between the jaws as in *Carabidæ*.

Some work of this nature has I believe been done on the Continent many years ago, but unfortunately I have been unable either to procure a copy or find the name of its author, and I must therefore apologise if the work done here has been previously published abroad.

Mr. C. Lovatt Evans' Report.

Methods. — The Beetles, irritated by handling, yielded their secretion at intervals in droplets which were collected as soon as exuded in a fine capillary tube applied to them, and thence transferred by blowing into a larger one. The total amount obtained was, however, small, amounting only to a few cubic centimetres, and the tests necessary for its analysis had to be carried out on a very small scale, and for this on many occasions the following device was used :

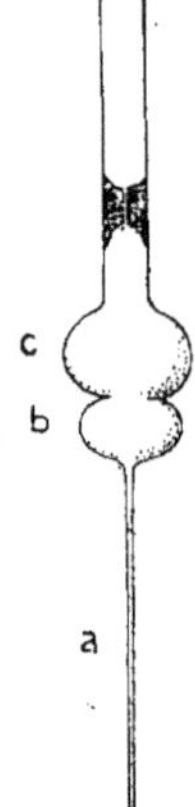

Fig. 1.

A piece of apparatus shown in the accompanying sketch was made, consisting of two superimposed perforated glass bulbs of small size surmounted by a sucking tube partially blocked immediately above the upper bulb, to prevent splashing. The lower bulb communicated with a long capillary tube into which a certain amount of the secretion could be aspirated, followed by the reagent to be used, their mixing occurring in bulb (*b*); when well mixed, the contents of (*b*) were blown back again into the capillary tube (*a*), and examined there for turbidity, etc., with the naked eye and under the microscope.

To filter off such precipitates the following device was had recourse to : A piece of glass tubing of small calibre was drawn out at one end into a capillary tube as indicated n figure 2, and a small plug of filter-paper pulp was firmly pressed into the neck of the capillary tube (*c*) at (*a*). The wide part of the tube was then filled with water, which was forced through the plug (*a*) by means of a cyclepump attached by pressure tubing to

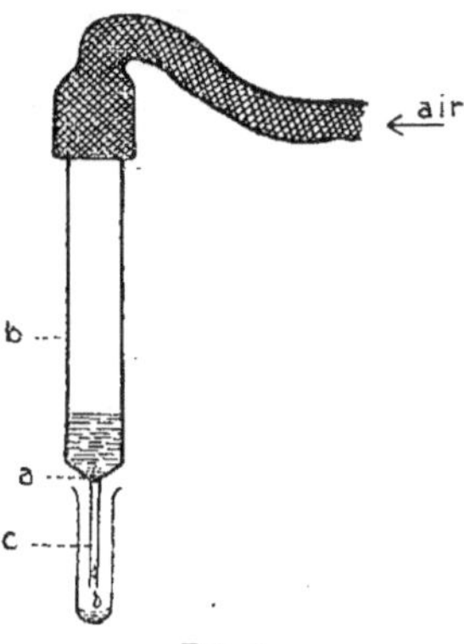

FIG 2.

the wide end of (*b*). Moderate pressure was all that was needed and the plug was firmly fitted into the neck of the capillary tube by the time the water in (*b*) had been driven through it.

A small amount of the fluid to be tested was then placed in (*b*) and forced through the filter plug in the same manner, driving the water in tube (*c*) before it. On emerging from (*c*) the filtrate was collected in another capillary tube and was then available for further testing.

Properties of the fluid. — The fluid when first exuded is of a clear reddish-brown colour, perfectly mobile, and mixible without change with small quantities of water. It has a somewhat bitter taste, but no smell and gave with litmus an alkaline reaction. Small living tadpoles placed in it for some hours showed no signs of distress, and therefore to them at any rate it is non-toxic. This fluid contains proteins, pigments, inorganic salts, and other constituents. A full determination of these was impossible owing to the small quantity secretion obtainable.

Inorganic constituents. — On evaporation and ignition the fluid yielded a fair amount of light friable ash, almost insoluble in distilled water, though readily soluble in dilute hydrochloric acid. No quantitive estimation of ash was undertaken, as it would require more than was available to do so with any accuracy. When the secretion is allowed to stand overnight, it becomes turbid, owing to crystals visible to the naked eye, making their appearance in it. Under the microscope these crystals were seen to consist of colourless tufts, similar in appearance to those found in certain urines and termed « stellar phosphates », which is another term for calcium phosphate. The deposited crystals were carefully filtered off, washed, dissolved in dilute hydrochloric acid, and subjected to the usual analysis. On evaporation and incineration a white mass with no charring was obtained, in which the ions of calcium and phosphoric acid were easily revealed by the usual tests. Therefore the crystals were undoubtedly of calcium phosphate.

Search for the other inorganic constituents was then made, a quantity of 2 c.c. being sacrificed for this purpose. The organic matter was first removed by treatment with aqua regia and subsequent evaporation three times with hydrochloric acid. The solid thus obtained taken up in the smallest amount of dilute hydrochloric acid and subjected to the usual course of inorganic analysis with presence of phosphates. Owing to the considerable amount of calcium phosphate present the chief constituents were Ca'' and PO_4'''. Somewhat considerable amounts of iron were, however, found, and also magnesium, potassium, and sodium. These various metals seemed to be chiefly present in the form of phosphates and chlorides, as no other acids could be detected.

The proteins. — The fluid gives all the protein reactions very readily : thus alcohol, mercuric chloride, picric acid, nitric acid and other protein reagents give very bulky and solid precipitates. The Xanthoproteic and Millon's reaction are both obtained, as is also Piotrowsky's reaction, but the colour is masked by the pigmented nature of the liquid itself. Heat coagulates the proteins, the greater part being thrown down at 75-80° C.; the coagulum thus formed carries down most of the pigment with it.

When the secretion is added to several times its bulk of water, a bulky light precipitate is obtained. When this precipitate is filtered off and the filtrate concentrated in vacuo over sulphuric acid to a small bulk, it still gives the Piotrowsky's reaction, though more feebly. The presence of both albumins and globulins is therefore indicated. Their presence is proved further by the separation with ammonium sulphate. Half saturation (in a capillary tube) gives a precipitate which when filtered off and drawn into a wider capillary tube and mixed with about six times its bulk of saturated ammonium sulphate solution yields a further precipitate. After this is removed by filtration, the filtrate on concentration and refiltration is found to contain neither propeptones nor peptones, which are therefore absent from the original fluid.

Pigment. — When the liquid is added to about twelve times its bulk of alcohol, the whole of the proteins are precipitated as a white coherent mass, and the pigments are retained in the alcoholic supernatant fluid, which becomes yellow. Similar results are given with ether, except that the proteins are only partially coagulated, and form with the water and inorganic constituents a pasty mass at the bottom of the narrow tube.

The alcoholic solution, when spectroscopically examined, exhibits no definite bands in the visible spectrum; a diffuse absorption of the violet end, and to a less extent of the extreme red, being all that can be observed. In all probability, therefore, the colouring matter is a lipochrome. This is borne out by two other characters, viz. :

1. The ethereal solution is bleached when exposed to sunlight;
2. When excess of strong nitric acid is added to the fresh secretion, the colour changes to an evanescent azure blue, turning to a dirty greenish grey.

Lipochromes are therefore at all events the chief pigments.

It seems possible also that other pigments may be present. Thus if the secretion be dropped upon a filter-paper and dried at the ordinary temperature, a brown stain is left; if the stained paper be now extracted with dry ether, the lipochromes are removed and dissolved in the ether, but the stain remains and shows a greenish tinge, which on exposure to light for some days vanishes.

Other constituents. — When a column of the liquid is mixed in a capillary tube with sodium hypobromite solution, bubbles of gas very soon make their appearance, suggesting the presence of urea in the fluid; this was not further investigated. Small amount of some copper-reducing substance seems to be present in the solution. This may be a sugar, but its amount is very small indeed and no osazone could be obtained from it.

Careful search was made for the presence of digestive enzymes, but none were found. Proteolytic enzymes cannot be present in it, for if such were the case auto-digestion of the proteins present in it must necessarily occur, yielding propeptones, etc., and we have already seen that these are absent. But if this fluid is derived from the salivary glands as has been suggested, diastatic ferments ought to be present in it, as these beetles are phytophagous. Nevertheless when a 2.5 °/₀ solution of soluble starch is mixed with 1/5 of its bulk of the secretion and placed in a thermostat at 20° C. gradually raised to 35° for 30 minutes, no reducing sugar is obtained, and therefore probably no enzyme is present in the original fluid.

No further investigation could be pursued owing to the supply of secretion having completely failed.

Les accouplements anormaux chez les Insectes,

par E. OLIVIER (Moulins).

Il a été constaté chez les Insectes des cas d'accouplement anormal entre individus du même sexe et aussi entre individus d'espèces différentes.

MULSANT (« Bull. Soc. Ent. France », 1848, p. LX) a observé assez souvent des *Melolontha vulgaris* accouplés avec des *M. hippocastani*, mais il ne précise pas le sexe des sujets.

LABOULBÈNE (« Ann. Soc. Ent. France », 1859, p. 567) a reçu du Dr PUTON deux *Melolontha vulgaris* accouplés, présentant les attributs extérieurs du sexe masculin : il les a disséqués et acquit la certitude que ces deux Insectes possédaient le même appareil générateur et étaient bien réellement des mâles.

C. BERG, professeur à l'Université de Buenos-Ayres, avec lequel j'étais en relations, m'écrivait qu'il capturait fréquemment des *Photinus fuscus* GERM. ♂ accouplés entre eux. Les ♂ de cette espèce étaient très communs aux environs de la ville, et les ♀ beaucoup plus rares, et cette disproportion entre les sexes pouvait motiver, disait-il, l'aberration de ces unions.

Ce sont là des faits monstrueux analogues à celui relaté par PERAGALLO, qui est encore plus extraordinaire, puisqu'il s'agit d'acteurs appartenant à deux espèces différentes (« Ann. Soc. Ent. France », 1863, p. 661). Cet entomologiste a capturé à plusieurs reprises, aux environs de Menton, des *Rhagonycha melanura* ♂ accouplés avec des *Luciola lusitanica* ♂, ce rapprochement contre nature ayant toujours lieu entre mâles et les Lucioles y faisant toujours fonction de femelles.

Il serait moins étrange, en effet, d'admettre une sorte de croisement hybride, si un *Rhagonycha* ♂ s'unissait à un *Luciola* ♀ ou un *Luciola* ♂ à un *Rhagonycha* ♀; mais il n'en est pas ainsi, et l'union a toujours été constatée entre les mâles des deux espèces.

J'ai pu voir, dans la collection de M. VON HEYDEN, un *Phosphœnus hemipterus* ♂ accouplé avec un *Lampyris noctiluca* ♀ (« Rév. des Lampyrides », p. 41, in « Abeille », 1884).

Ces unions illégitimes et contre nature n'ont été observées jusqu'à présent, à ma connaissance, que chez des Coléoptères de mêmes espèces ou d'espèces voisines.

Le fait que je viens faire connaître est bien plus étrange, puisqu'il concerne des Insectes appartenant à des ordres bien différents.

Le 6 juin dernier, j'ai capturé, aux environs de Moulins, un *Telephorus bicolor* ♂ accouplé avec un Diptère *Ephippium thoracicum* LATR. ♀.

Le *Telephorus* était cramponné sur le dos de l'*Ephippium*, qui courait sur le sol, essayant de s'envoler, mais retombait après avoir franchi quelques centimètres, le poids qu'il portait l'empêchant probablement de prendre franchement son vol.

Les deux Insectes étaient tellement bien unis que j'ai pu les saisir et les asphyxier dans un flacon à cyanure sans qu'ils se détachent. Je les ai conservés ainsi et j'ai le plaisir de les soumettre aux membres du Congrès, qui pourront se convaincre *de visu* de ce cas extraordinaire d'aberration génésique.

The Distribution of the Yellow Fever Mosquito (*Stegomyia fasciata* FABRICIUS) and general notes on its bionomics,

by FRED. V. THEOBALD, M. A., F. E. S., etc.

The theory instituted by FINLAY, that the Tiger, Brindled, Spotted Day or Striped Mosquito, then known as *Culex tæniatus*, was the carrier of yellow fever, has now become a proven fact. The importance of this insect has in consequence become very great.

The *Stegomyia fasciata* (pl. V) of FABRICIUS is the type of a new genus which I created in 1901. Its structural and bionomical features differing very much from true *Culex*.

The main characters are the flat cephalic and scutellar scales, which alone will separate the genus from other allied Culicine groups. The insect is a very marked one, and at present there is no known species with which it can be confused. The peculiar thoracic adornment, together with the leg and abdominal banding, will at once identify this insect.

The synonymy has already been referred to (« Monograph of Culicidae of the World », vol. I, III, IV and V, 1901-1910).

One point only in this paper need be mentioned, namely, the reasons I have retained FABRICIUS's name *fasciata*. Firstly, because medical men over most of the world now know the insect by this name, and any alteration seems to me under such circumstances to be likely to lead to needless confusion. Secondly, there is a possibility that MEIGEN's *Culex calopus* described from Portugal is not this species. His name is considered by many entomologists to be correct on account of the fact that in 1789 VILLIERS described a *Culex fasciatus*. No type of this exists and the description is far too vague, as far as I can see, for any one to identify an insect from it. In fact no one has attempted to use the name for VILLIERS's insect as far as I am aware.

In 1804 MEIGEN described a *Culex fasciatus* which is undoubtedly only *Culicada nemorosa* MEIGEN 1818, and it may thus be said that that name should be adopted for *nemorosus* (1818), and so in any case our yellow fever carrier cannot be called *fasciata*, but prior to this he had described the same insect as *Culex reptans* (1804). Personally I do not feel at all sure that MEIGEN's *calopus* does refer to this insect, and think FICALBI was quite right in separating his *elegans* from it, because MEIGEN says the thorax of *calopus* was marked with dark lines. If it is not *calopus,* then DESVOIDY's name *frater* (1827) would be correct instead of *fasciata*.

However, as long as the insect is known and can be identified, it does not matter whether we call it *calopus, frater* or *fasciata* for practical purposes.

VARIATION. — This Mosquito is subject to much variation, which one would naturally expect when we consider its wide range and its great abundance.

In Bonny and Old Calabar many specimens show the abdominal banding, which is normally basal, involving both sides of the segments. Similar variations also occur in Australia. I described this variety from Brisbane, Queensland, as var. *Queenslandensis* (vide p. 19). Now and again this form becomes almost entirely covered on the abdomen with creamy scales. There is also some variation in the leg banding. One marked variety *Luciensis* THEOB. has the tip of the last hind tarsal black. This variety occurs in the West Indies, Brazil, and British Guiana, and I have seen it from India.

In rarer cases the median thoracic lines are absent. This is ARRIBALZAGA's *Culex mosquito* (1891). These well marked varieties of the yellow fever carrier are referred to later.

In one lot I hatched in England the variety *mosquito* occurred from the same batch of eggs with typical specimens. There is also some variation in size and colour of the thoracic scales, in *queens-*

landensis they are sometimes almost golden brown. Besides variation in markings there is considerable variation in size. From Australia I have noticed that the females are larger than those from S. America and the West Indies, but the males are of normal dimensions. From Bonny and Old Calabar the specimens of both sexes seemed to be small.

Very small specimens occur in parts of India. GILES says : « A dwarf variety from Calcutta is the smallest true gnat I have met with ».

The only two Portuguese specimens I have seen were abnormally large. I have seen all sizes from various parts of the world varying between 2^{mm}50 and 5^{mm}. Some which I bred in England from dried ova, and which fed irregularly when larvæ, reached barely 2^{mm}5. Size is evidently a question of nourishment in the larval stage.

Habits and food.

The adult *S. fasciata* feeds largely upon the blood of man. It is a most vicious and persistent biter. DURHAM (1901) has shown that it also feeds on the blood of dogs, and he also found it feeding on an agouti and on bats.

It is practically a Day Mosquito. In the West Indies it bites mostly between 1 and 3 p. m. But Dr. GRAY, in a letter to Mr. AUSTEN, says : « You will observe that I have caught them at all hours of the day, and even late at night ». LOW also noticed this in the West Indies. He wrote me as follows : « Whilst in Trinidad, unless I went to bed under a mosquito net, I found it impossible to rest and sleep in the heat of the day owing to the continual bites of this pest. It used to bite one on the feet and ankles in the mornings, and again, as GRAY says, bites all night as well ».

In Para DURHAM (1901) tells us it is solely a day Gnat. It will bite soon after sunrise, its time of chief activity in Para being in the middle of the day, from 12 to 2, and on till 5.

In Gambia DUTTON (1903) found it biting at 4 p. m. Practically we may say it may bite at any time during light, but that one is generally free from its attacks at night.

The adults breed the first day after their emergence from the pupæ. A single male will fertilize a number of females, and Low has observed in the West Indies that they copulate by night as well as during the day.

The adults under certain conditions can live for a considerable time. They have been known to live on board ship for forty-seven days, and BANCROFT in Queensland has kept females in confinement for two months. This longevity of the adults is very important, and explains how several outbreaks of yellow fever have occurred where the species is not indigenous.

REED has pointed out that this insect dies within five or six days when deprived of water; it is not likely therefore that it can be imported in trunks or crates of bananas, etc. I think however that this statement wants verifying, as I remember having kept one alive for ten days in a glass top pill box.

The general facts of the life-history have been traced out by GOELDI (1905), GRABHAM (1905) and others.

The ova are laid separately, not in rafts as in the genus *Culex*. According to GRABHAM, they may be laid in chains, about a quarter of an inch between each egg.

BANCROFT (1908) has also noticed the same in Queensland, stating that they are laid « singly on water in chains with an interval between each egg of a quarter of an inch or more ». Those figured by GOELDI (1905) do not show this chain-like appearance.

In colour the ova are black, elongated oval in form and completely covered with a reticulated membrane; with air in some of the reticulated cells.

The incubation period varies, but is somewhere between 6 and 20 hours.

NEWSTEAD (1910) found it to vary between 6 and 12 hours. The larval stage lasted 9 days. The pupal stage 3 days. The complete life-cycle thus possibly taking only 12-13 days. DURHAM (1901) found in Para that the first pupa occurred on the 8th day, and hatched into a male on the 10th day, and others on the 12th day.

A technical description in English of the larva and pupa may be found in the « Bulletin of Entomological Research (1) ».

---

(1) W. WESCHÉ, « On the larval and pupal stages of West African Culicidæ », vol. 1, pt. 1, p. 25, 1910.

The chief and first complete description was given by GOELDI in « Os Mosquitos No Pará » (pp. 95-100).

The adults are especially attracted to dark surfaces for a resting place, and it has been noticed that they are especially attracted to dark clothing. LOW wrote me from the West Indies in 1902 that « black and blue clothes are where they especially choose to rest, and it is really astonishing, if one has none of those hanging about, the reduction in their numbers ».

How far this species can fly we cannot at present say, but it has been stated that if a ship does not anchor within 200 fathoms, it is safe from becoming invaded by these Mosquitoes.

Vitality of the ova.

Two cases show very clearly the length of time the ova may remain dormant, an important point in considering the artificial distribution of this Mosquito.

Dr. FINLAY, the founder of the Yellow Fever Mosquito theory, sent me ova from Cuba, dry, in a test-tube. From neglect these eggs after their journey were left for two months. As an experiment they were placed in tepid water, and in less than 24 hours there were larvæ swimming in the jars. How soon the ova hatched I do not know as they were put in at night, and being away early next day observations were not made.

They developed very slowly, and many died. None pupated until three weeks after, one female hatched and five males (1). One of the males being the distinct *Mosquito* variety.

NEWSTEAD (1910) checked my observations regarding the long vitality of the ova, when dried, with the following results :

Two lots of eggs were dried in air, one subsequently placed in a desicator with chloride of calcium for 24 hours, and sent to England in tightly corked glass tubes.

The following dates give the result of this experiment :

September 10th-11th. Eggs laid at Manáos, Amazon.

October 26th. Reached England. Placed in water at 23° C.

---

(1) THEOBALD, « Monograph of Culicidæ », vol. 3, p. 143, 1903.

October 27th. 12 larvæ hatched during the previous night and one after 12 hours immersion.

October 28th. Larvæ began moulting.

October 30th. All completed first moult.

November 4th. Larvæ pupated.

November 7th. First imago (a male) hatched.

November 8th. A male and female hatched.

The breeding jar was kept at a temperature of 23° C. It was thus seen in this case that eggs, which had remained dormant for from 45 to 47 days, incubated.

Those I hatched had remained dormant for a longer period. How long this period of rest can remain when the eggs are desiccated we cannot at present say, but sufficient has been shown to prove that eggs laid in a small collection of water on a ship which may suddenly dry up, may remain dormant for at least two months, and this alone would enable the species to be carried great distances in dried-up tanks of old sailing ships, much further now by rapid ocean transit

The larvæ.

The larval form of *Stegomyia fasciata* can at once be told from that of its common campanion (*Culex fatigans* WIEDEMANN). In fact it has no general resemblance to a typical Culex, and the behaviour is quite different. The Stegomyias pass most of their time at the bottom of the water and slowly come to the surface to take air, and then go down again. Their colour also differs, being greyish white, whilst in Culex the colour is generally some shade of brown. Nevertheless they come to the surface to breathe, and so are as easily killed by paraffin as other Culicines.

Time of appearance of the adults.

The period over which this Gnat extends varies. For instance in Brisbane we learn from BANCROFT (1908) that it is present and bites throughout the year. In the Philippine Islands, according to LUDLOW (1908), « in places where the rainy season begins in showers in April, and it is distinctly rainy in June, *S. calopus* apparently appears in June or July, increases greatly in numbers and is most noticeable in July, August, September and October, and then gradually diminishes again ».

In Cuba it is specially troublesome in the rainy season, where, according to DESVOIDY, it was called « Mosquito » by the natives.

From different parts of the world it has been sent me every month of the year. In some places it always seems to occur, in others a certain seasonal prevalence occurs, but at present this has not been satisfactorily worked out.

Pairing of *Stegomyia fasciata*.

From general observations it seems a single male may fertilize many females. Dr. ST. GEORGE GRAY found in St. Lucia that the males fly from one female to another, apparently never tiring.

The method of copulation, as first pointed out by GRAY, and later admirably figured by GOELDI (1902), is for the male to get under the female, who may or may not carry him off and complete the process in the air.

Breeding places of *Stegomyia fasciata*.

There is no doubt that, with a few notable exceptions, this insect is a purely domesticated Gnat. With the exception of one being found in the scrub of South Queensland by Dr. BANCROFT, and one or two other casual notes sent me, the general consensus of opinion and observation show that the Yellow Fever Mosquito is a companion of man.

BANCROFT, writing in his work on the Mosquitoes of Queensland (1908), however, says : « It is an introduced species and *never goes wild* ». The single specimen sent me from the scrub was evidently exceptional.

NEWSTEAD (1907) records the following breeding places in the Congo region : « Adults were hatched from larvæ and pupæ taken from water collected in old pots and tins and from a foul smelling pool used for steeping manioc ».

Again NEWSTEAD (1910) records as follows from the Amazon region : « Great numbers were found in large puddles in the open, in association with Anophelines and other Mosquito larvæ, although there were houses with suitable receptacles within 50 feet of these habitats. They occurred also in large, deep collections of terrestrial water, averaging 3 feet in depth. On one occasion swarms of larvæ were found in a barrel of water containing the macerated and putrid remains of a number of frogs, upon whose carcasses they seemed to have been thriving ».

In Para, DURHAM (1901) records its habitats as follows : « Casual water in vessels, etc., in and about houses, such as buckets, tins, wash-tubs, rain-gutters, ant-guards, larger and deeper collections of water as casks and hogsheads full of rain water. Also in the bilge water of barges and lighters, and on the S. S. Viking ». Further DURHAM says it is « not found in cesspools, stable runnings or natural puddles in forest or streets ».

LOW, writing from the West Indies, states as follows : « It is essentially a domestic species, breeding in water barrels, tanks, tubs, fountains and other collections round houses. It is very commonly found with *Culex fatigans*. I have never found it in any other situations such as in the country collections of water favoured by *Anopheles* and other Mosquitoes, which may be termed wild Mosquitoes in contradistinction to the domestic (1) ».

In Gambia DUTTON (1903) found this Gnat breeding in a tub of water, in wells and in a canoe.

In Jamaica GRABHAM (1905) found it in tanks, tubs, wells, fountains, empty tins and calabashes.

In New Orleans it was found most abundantly in the large round water cisterns used at the back of every house.

BALFOUR has noticed that they will breed in the bilge water and tanks on the Nile steamers.

Speaking generally this insect breeds in any small or moderately small collection of water near man's habitations, especially artificial collections of water, such as in jars, tins, calabashes, cisterns, vats and tubs and on board ships. But at the same time we see it may go to larger and natural collections of water, but that such is unusual.

---

(1) THEOBALD, « Mono. Culicidæ of the World », III, 142 (1903).

Such small collections of water as are held up *pro tempore* in gutters have also been shown to be breeding places.

All these breeding places must be looked to if we wish to stamp out this pest near habitations.

Do the males bite?

Ficalbi I believe has stated that the male *C. elegans* (i. e. *S. fasciata*) bites. Writing from Mombasa, a correspondent, Mr. McKay, also stated that he found the males bit now and again.

Bancroft (1900) finds in Queensland that the females only bite. This biting by the male is however unusual, judging from the consensus of opinion, and the absence of biting habits in all other male Mosquitoes.

The possibility of *S. fasciata* being split up into several so-called species.

If we take the minute larval characters of the Culicidæ to be specific differences (when they vary) then I feel sure we shall have this insect split up into a number of species.

Let us take two examples of the larvæ of this insect and their characters of siphon spines and combs.

Goeldi (1905) figures the siphon spines as 12 in number, Wesché (1910) 10 (the latter remarks that their number and shape are not reliable characters); if we take the comb scales, we find Goeldi figures them with a long terminal and two to three small basal lateral processes, Wesché with one basal lateral process on each side, which seem to vary from 8 to 9 in number.

The difference in the siphon spines appears marked, hence we might assume the African specimens belong to a different species to the American.

I have found the same variations in those from any country when numbers are examined, and for this reason consider the minute larval characters of no taxonomic value as far as our present knowledge goes.

Transportation by ship and train.

From various sources we learn of this insect being found both on ships and in trains. It is certain that by this means it has gained its wide distribution. DURHAM (1901), writing in connection with this insect in Pará, reports : « All ship's captains are agreed that Gnats come on board with the advent of lighters, and the men on the tenders complain of the abundance at night when lying close to the wharfs. The species found on the lighters were town kinds : *Stegomyia fasciata* and *Culex fatigans* ».

« Investigation of a number of lighters at Para showed that *S. fasciata* larvæ were present here and there in the bilge water of the covered lighters. Open lighters were free, but this was due to their being used for coal, and having a thin tarry film over the water ».

BALFOUR (1909) says concerning Nile steamers : « No sooner did a steamer with its barges tie up, and the hatches were removed than a black cloud of Mosquitoes emerged and sought breeding places ashore. The majority proved to be *Stegomyia calopus* ».

This is speaking of the river steamers on the Nile coming from 240 miles south of Khartoum on the White Nile, and similar steamers from the Blue Nile were also found to be laden with *S. fasciata*.

Another important piece of evidence is that of Dr. CUMMINGS of the United States Marine Hospital Service; he reported that a Spanish barque, 65 days from Rio de Janeiro, arrived at the South Atlantic Quarantine Station carrying a veritable plague of Mosquitoes, most all of which were *Stegomyia fasciata*. According to the Captain's statement, they had been on board only 43 days, quantities of larvæ were found in the tanks.

Pregnant females are now known to travel long distances.

I have received specimens taken on liners in the Suez Canal, other cases of this insect being found on board ships in the East are quoted in the distribution lists.

In fact it seems quite a common thing for this domesticated Gnat to travel with man on board ship. This is a very important matter to notice, for in this way the Yellow Fever Mosquito in still being spread from place to place. It has been through this means of transport, namely ships, that this insect has assumed such a worldwide distribution, as shown in the following list of localities where it has been recorded.

This method of transportation was great in the old days of sailing ships, for the Mosquitoes we know breed in the water in the tanks carried for drinking purposes. At the present day distribution is going on, but perhaps not to the same extent, as there is little opportunity for them breeding, and the adults, although they can live for considerably over a month, are much more susceptible to climatic changes than the ova, larvæ and pupæ. That the adults can travel long distances we know, as long as they are in a warm climate, but if they have to round Cape Horn or through any cold area, there is not the same possibility of their surviving in the adult stage as in the larval, pupal or desiccated egg stage.

It is in this respect that the opening of the Panama Canal may mean a serious menace to life and trade in the East.

We see from what BALFOUR and others say, that swarms of *Stegomyia fasciata* fly on board ship, and can be carried great distances in warm climates. The opening of the Panama Canal will bring ships direct from the chief endemic center of yellow fever and the chief habitat of *Stegomyia fasciata* (its only carrier) along a line favourable to the insects (fig. 1).

It will be only 12 days for a ship leaving the Panama Canal to get to the Sandwich Islands, and 23 days on to the Philippine Islands. A *Stegomyia fasciata* may live 47 to 50 days, and may thus reach the Philippines direct from the endemic centre of yellow fever alive.

A case of yellow fever occurs on board ship 3 to 4 days after the patient is bitten by the Gnat. In 12 days the organism of yellow fever migrates from the Stegomyian stomach to the salivary glands. Then the Gnats can inoculate at intervals of 3 days for a considerable time. When these infected Gnats come to port they fly on land, bite people, and in this way yellow fever breaks out. The insect is perhaps already living there, probably in numbers, and thus an epidemic of yellow fever is brought about. It may not occur as a native insect, but flies to land and lives a certain time.

That this has been the cause of outbreaks of yellow fever we know from the outbreak at Swansea, when there were 30 cases and 17 deaths in 1875. The same boat that brought the yellow fever patients brought the *Stegomyia fasciata*; some must have flown on land and inoculated people (1), and then, as the insect can not live here, died out, and so the fever ceased. We have seen the same in New York, in Philadelphia and even in New Haven and Providence. *Stegomyia* do not breed there, but during the epidemic they must have been imported and feasted off the inhabitants' blood, and so caused the epidemics, and then died out naturally (2).

There is, as was pointed out by Sir PATRICK MANSON (1903), a very grave danger of yellow fever appearing in the East. We have in many or, we may say, most ports the *Stegomyia fasciata*; it only wants the fever introducing, and then it would run wild in the overcrowded and unsanitary oriental cities, where it could not be coped with so easily as in America and the West Indies, for one would be dealing with a very different type of people.

All the more reason therefore for stamping out as far as we can this insect wherever it occurs.

Parasites of *Stegomyia fasciata*.

A certain number of parasites have been found on this cosmopolitan insect, but they do not seem to be of any economic importance. MARCHOUX, SALIMBENI and SIMOND (1903) refer to a kind of *Mucor* occurring in great abundance on this Gnat at certain periods of the year (DYÉ).

The same authors found at Rio de Janeiro certain *Myxosporidia* which they placed in the genus *Nosema*.

At Vera Cruz *Myxococcidium Stegomyiæ*, PARKER, BEYER, POTHIER (1903) has been found as a parasite.

Whilst LOW and others of course refer to *Filaria bancrofti* in this Gnat at Saint-Lucia, Trinidad and on the Niger, but the evolution is not complete in this Mosquito, and it is difficult to see that it can in any way affect the host.

---

(1) Most of the cases died on board ship, but some occurred on land and these only about 200 yards from the ship which came from the West Indies.

(2) It has been suggested that the range of this *Stegomyia* was formerly wider than now. There is no evidence whatever to show this, rather the contrary.

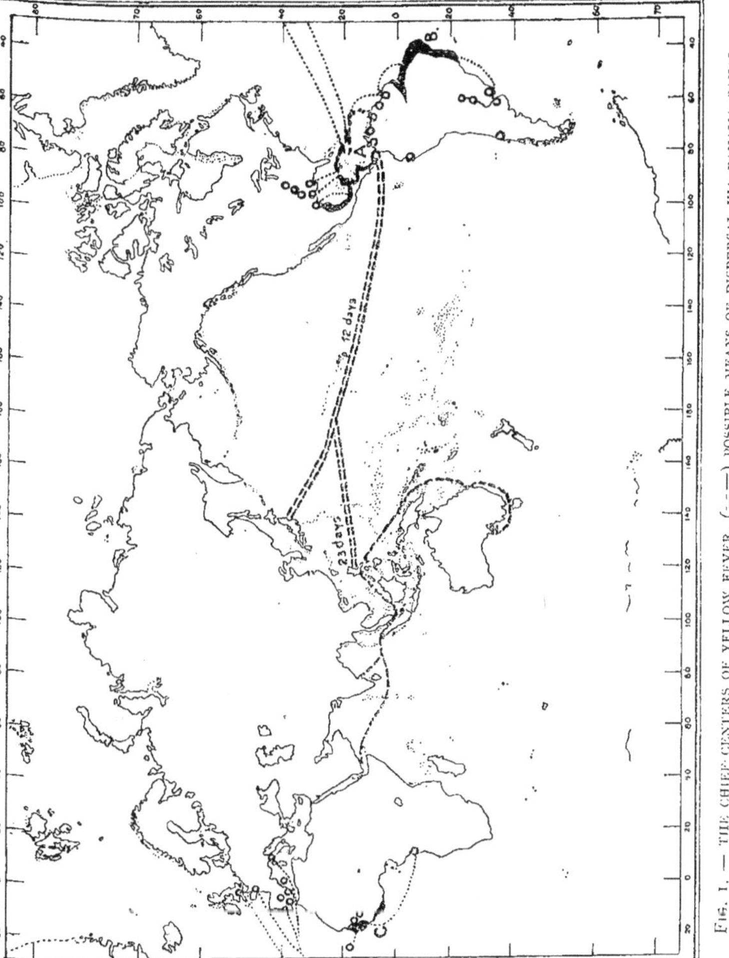

FIG. 1. — THE CHIEF CENTERS OF YELLOW FEVER. (=·=) POSSIBLE MEANS OF DISPERSAL VIA PANAMA CANAL.

The possibility of their introduction into Britain.

It has been stated that we may receive in England numbers of this Gnat from ships, carrying bananas from the West Indies and South America. It is quite possible that some may at any time come over in any ships, not only those carrying bananas. It is quite possible, if introduced, that they may live for a little while in such ports as Bristol, Swansea, etc. It is even possible that infected specimens may land, just as they must have done at Swansea during the outbreak of yellow fever there. It is extremely improbable from what we know of its distribution and its likes and dislikes that it could establish itself here and breed.

All cases so far that I have investigated of Mosquito plagues supposed to be due to importation by ships have proved to have been caused by one or other of our twenty-two native species of *Culicidæ*.

The distribution of *Stegomyia fasciata*.

Howard, writing on this subject in 1905, says: « We may expect to find this species everywhere in the moist tropical zone, or at all events, when introduced at any point within the low moist tropics, it may be expected to establish itself ». Besides the tropics we know it extends well into subtropical even temperate regions.

The most northernly line seems to run no further than between 45° and 44° (at Ravenna, Ficalbi) and the most southernly the uplands of Victoria, say 38°.

I do not think its range will be found to extend further north, but it may extend further south, down to about 40° to 43° in South America.

Anywhere between these limits we may expect to find in *favourable localities* this yellow fever carrier, or may expect to have it become established if once introduced.

The localities from which it has so far been recorded are given in the following pages.

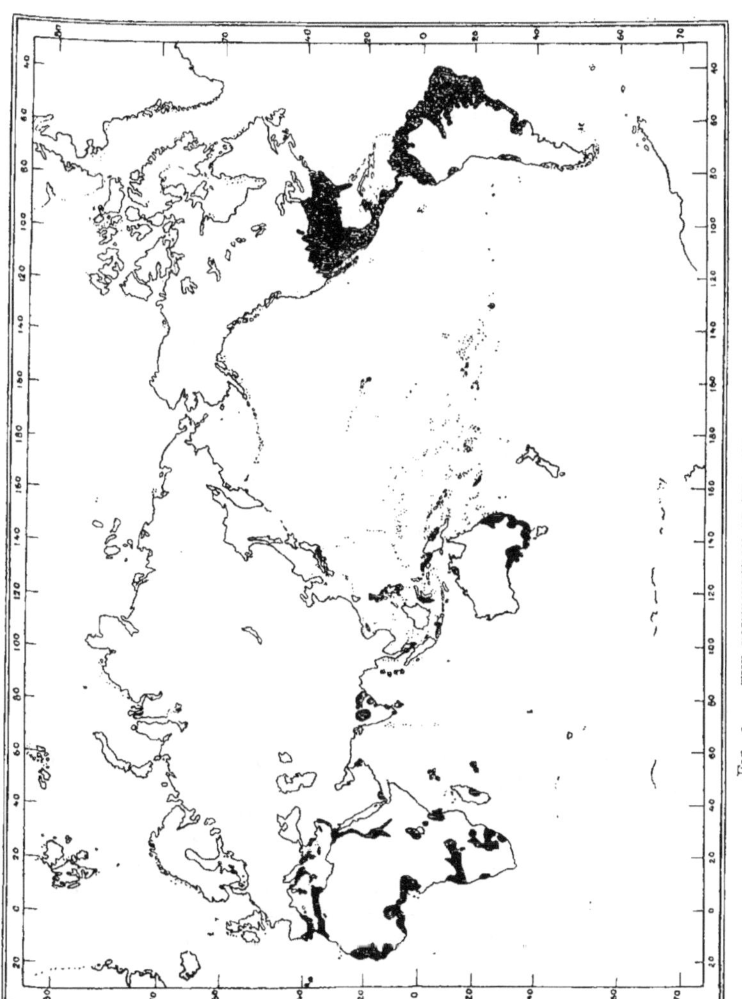

FIG. 2. — THE DISTRIBUTION OF THE YELLOW FEVER MOSQUITO.

Distribution of *Stegomyia fasciata* FAB. (Fig. 2).

Africa. — Natal, Durban (CHRISTOPHERS, 7-11.I.99); Orange River Colony; Salisbury, Mashonaland (MARSHALL, IV); Rhodesia; Transvaal, Komatipoort (BOSTOCK), Leysdorp (COPLAND), Pretoria (THEILER and SIMPSON); Delagoa Bay (JOSÉ SANT'-ANNA); Zanzibar (MACKINDER); Nairobi; Mombasa (MC KAY); Uganda (DYÉ); Madagascar, Tanarive (VENTRILLON); Reunion (DYÉ); Mauritius (MACQUART, D'EMMEREY DE CHARMOY)(common near the sea shore, but more scarce in the high parts of the Island); Seychelles, Victoria (DENMAN); Zomba and British Central Africa (DANIELS); Congo region, Matadi, Wathen, Leopoldville (XII. 03), Sendive, Kasongo (V.03), Tshofa, Lusambo; Cameroons, Cap Blanc (DYÉ); Calabar (ANNETT); Bonny (ANNETT); Lagos (AUSTEN, VIII); Accra (GRAHAM); Ashanti (GRAHAM); N. and S. Nigeria (HANLEY, ANNETT, JEMMETT); Ivory Coast (DYÉ and BLANCHARD, common); Sierra Leone, Freetown (WALKER and AUSTEN) (1); Gambia (BATHURST, common), McCarthy Island (VI) (BURDETT and DUTTON); Senegambia (REES); Senegal, Dakar (DYÉ); Morocco (DYÉ), Algeria, Tunis (SERGENT); Egypt, Cairo (WILLCOCKS), Port Said (ROSS), Alexandria (ANDRES); Ismailia (ROSS and GORGAS); Sudan, Khartoum and Nile generally (BALFOUR); Pibor (KING).

Asia. — *Turkey* in Asia, Smyrna (BIGOT in MACQUART). *Palestine,* Tyre and Sidon (CROPPER); *Arabia,* Muscat (BRUNETTI), Aden.

India. — Quilon (JAMES, 2-7.III.1900); Madras (GOODRICH 12.XII.1899); Madras town, Puri, Orissa Coast (ANNANDALE); Calcutta (DANIELS, GILES, ANNANDALE, BRUNETTI, PAIVA, in May, July, August, September and October); Central Provinces; Eastern Bengal, Chittagong (HALL, 21.IX.03; Bhim Tal, Kumaon, 4800 feet, ANNANDALE, IX); N. Bengal, Purneah (PAIVA, VIII); Oude, Lucknow (PAIVA, XI); Punjab, Ferozepore district (ADIE); Assam, Lushai Hills (MACLEOD). On board ship, Bay of

---

(1) WALKER described this insect from Sierra Leone as *Culex formosus*

Bengal, between mouth of the Hooghly River and Rangoon (22-23.II.08, ANNANDALE); on board ship 10 miles off Coconada on Madras Coast (15.IV.08, PAIVA); common on board ship all the way from Calcutta to Rangoon in February (ANNANDALE).

Burma. — Mandalay (ANNANDALE); Upper Burma (WATSON); Rangoon (ANNANDALE).

Siam. — Phrapatoon (WOOLEY 15-17-26-28-29.III.07), and Siam (SKEAT).

Malay Peninsular, ports along coast (DANIELS); Perak (DURHAM); Singapore (BIRO and P. DE FONTAINE, 2-8-12-30.VII.99).

French Cochin China. — Saigon (BROUQUET).

Islands. — Andaman Islands (RAY WHITE); Nicobar Islands; Ceylon (GREEN).

Philippine Islands. — Aparri, Cagayan; Antimonan, Tayabas; Balayan, Batangas; Balinag, Bucalan; Bamban, Tarlac; Binan, Laguna; Binangenan, Rixal; Boac, Marinduque; Bonongon, Samar; Bongabong, Nueva Ecija; Bulalacao, Mindoro; Bulam, Sorsogon; Camp Bumpus, Leyte; Cayagan, Mindano; Calamba, Laguna; Calavag, Tayabas; Culapan, Mindoro; Catubig, Samar; Camp Connell, Samar; Corregidor Islands; Cottabato, Mindanao; Daet, Ambos Camerines; Camp Daraga, Albay; Desmarinas, Cavite; Dumaguete, Negros; Camp Eldridge, Laguna; Camp Gregg, Pangasinan; Hagonoy, Bulacan; Camp Hartshorne, Samar; Ilo-ilo, Panay; Infanta, Tayabas; Jolo-jolo; Lucena, Tayabas; Macabebe Pampanga; Majayjay, Laguna; Manila; Mariquina, Rizal; Meriveles, Bataan; Camp Mc Grath, Batangas; Naga, Cebu; Naic, Cavite; Nasugbu, Batangas; Orion, Bataan; Ormac, Leyte; Camp Overton, Mindanao; Panique, Tarlac; Pasig, Rizal; Polo, Bulacau; Puerto Princess, Paragua; Romblon, Romblon; Salomaque, Ilocos, Sur; Samal, Bataan; Santa Cruz, Laguna; San Francisco de Malabau, Cavite; San Isidro, Nueva Acija; San Jose, Abra; Santa Maria, Bulacan; San Mateo, Rizal; Camp Stotsenberg, Angeles, Pampanga; Santo Tomas, Batangas; Taal, Batangas; Tanay, Rizal; Tarlac, Tarlac; Tiaong, Tayabas; Tobaco, Albay; Tuguegaras, Cayagan; Camp Wallace, San Fernando de Union; Camp Ward, Cheney, Cavite; Warwick Barracks, Cebu, Fort Wm. Mc Kinley, Rixal, Camp Wilhelm, Tayabas (LUDLOW).

Japan. — Tokyo (WOOD, 3.VIII.1899).

Malay Archipelago and East Indies. — Java, Batavia, Soekaboemi, Garvet (MARLATT); Samarang (JACOBSON, I-08); Makessar, Celebes (WALKER) (1); New Guinea, Dorey (WALKER) (2); Friedrich Wilhelmshafen (BIRO).

Australia. — Port Darwin, Northern Territory; Brisbane (SKUSE) (3); Bupengary, Queensland (SKUSE and BANCROFT); New South Wales (SKUSE); Victoria, from the Malarious uplands (FRENCH); South Australian.

New Zealand. — No records have come from here.

Europe and Mediterranean Islands. — Portugal [MEIGEN (4), MACQUART]; Spain, Malaga, (AINSWORTH) and SCHIENER); Italy, Pisa, Leghorn, Florence, Naples, Calabria, Spezia, Siena, Ravenna, Maggio, Sardinia, Sicily [RONDANI (5), FICALBI]; Greece, Kobus, Poros (KRÜPER); France, Saint-Louis (DYÉ).

Cyprus, Larnaka (BORDAN); Crete; Malta; Gibraltar (BIRT. 22. IX.09).

England? (STEPHENS) (6).

---

(1) Described by WALKER as *Culex impatibilis*, 1860. When I examined the remnants of WALKER's type in the British Museum in 1899. enough remained to show that it was undoubtedly this species.

(2) WALKER described this as *Culex zonatipes*. (« Proc. Linn. Soc. », V. 229)

(3) This is SKUSE's *Culex Bancroftii* (1889).

(4) Described by MEIGEN from an ill-preserved specimen, the locality only being given as Portugal.

(5) Although there is no trace of this in RONDANI's collection, we may be sure his original identification was correct, as FICALBI has shown it to be common in Italy.

(6) VERRALL records *Culex calopus* MEIGEN as British in his first List of British Diptera, but in the second edition I left this species out in the list of British *Culicidæ* I prepared for him, and the name does not occur. It may have been found alive, or even we may say must have been during the Swansea yellow fever attack and probably this accounts for the record, but cannot trace it, and Mr. Austen tells me he knows of no authentic British record.

South America. — Panama (MALLET, 22.IX.99) Colon, Panama; Ecuador, Guayaquil (LUDLOW); Peru, Callao (LUDLOW); Chile, (GUADICHAUD, MACQUART) (1); Chile, Valparaiso (LUDLOW); Argentina, Buenos Ayres (ARRIBALZAGA); Rio de Janeiro (PERYASSU); Brazil generally (except Therezina, Pientry), Amazon region, present all the year round, no marked seasonal prevalence; Para, the common household Mosquito about the city and outskirts, as at the Yellow Fever Hospital, Hospital Santa Casa, Marco da legua, Asilo dos alienados (Leper asylum), at Pinheiro and Onteiro; in vessels and lighters off Para (DURHAM); Manáos, common in houses about the city and in the suburban region of 'Cachoeirinha' (DURHAM), Caupinas (HOWARD), Bahia, São Paulo (LUTZ); French Guiana, Cayenne (DYÉ and NEVEU LEMAIRE); British Guiana, New Amsterdam (ROWLAND, 4.IV. 1899); Demerara (QUELCH, 16.VI.99); Colombia (2).

Central America. — British Honduras, Belize (IX), Ceiba, Puerto Cortez; Guatemala, Livingston, Puerto Barrios; Nicaragua, Bluefields; Costa Rica, Limon, Bocas del Tors (HOWARD).

Mexico. — Jampico, Acapulco, Quanajuato, Frontera, Vera Cruz, La Paz (Lower California), Coatzocoalcos, Pachuca, Tuxpan, Nantla, Ilacotalpam, Maxatlan, San Blas, Carmen, Cozumel, Champoton, Perihuete, Las Penas, Tepic, Poekutla, Progress, Monterey, Cordoba, Orizaba, Saluia Cruz, Saltillo, Cindad Victoria, Linares, Merida, Tonala, Rincon Antonio (HOWARD).

North America. — *Texas* (GLENN HERRICK), E. Texas (WOLDERT), Galveston, Fort Sam Houston (LUDLOW), Victoria, San Diego, Tyler (HOWARD), Laredo, Austin, San Antonio, Corsicana, Brownsville, Alice, Fort Mc Intosh (LUDLOW), Colorado, Dallas, Paris, Edna, Fort Bliss, Fort Ringgold, Fort Brown (LUDLOW).

Louisiana (GLENN HERRICK); Natchitoches (JOHNSON, 6.X.),

(1) This refers to MACQUART's *Culex toxorhynchites*. The remnants of the type in the Jardin des Plantes at Paris evidently show this species, but the locality may be Brazil (« Dip. Exot. », 1 p. 25).

(2) GILES records a specimen so named in the British Museum collection. I was unable to trace this.

Ruddock; New Orleans (VEAZIE, HOWARD, BOYCE); Baton Rouge, Napoleonville, Hammond, Shreveport, Franklin (HOWARD), New Theria, Patterson, Jackson Barracks (LUDLOW).

Mississipi (GLENN HERRICK), Pass Christian, Summit, Quarantine Station, Biloxi (HOWARD), Vicksburg, Clarksdale, Tutwiter, Belzoni, Holly Springs, Jackson, Winona, West Point, Tupelo, Corinth, Morgan City.

Alabama. — Mobile, Decatur, Auburn, Tuscumbia, Huntsville, Yazoo City.

Georgia (GLENN HERRICK, COQUILLETT), Atlanta, Pelham, Augusta, Savannah (1), Brunswick (HOWARD), Augusta Arsenal and Fort Mc Pherson (LUDLOW).

Florida. — Fort Barrancas (LUDLOW), Ley West (HOWARD and LUDLOW).

S. Carolina. — Charleston, Sullivan's Island (HOWARD), Columbia, Fort Freemont, Fort Moultrie (LUDLOW).

N. Carolina. — Beaufort, Winston (HOWARD), Raleigh, Greensboro, Charlotte, Salisbury (LUDLOW).

Tennessee. — Savannah (WIEDEMANN) (2), Nashville (HOWARD), Knoxville, Clarkesville, Chatanooga, Memphis, Columbia, Dechard, Athens, Bristol (LUDLOW).

Arkansas (GLENN HERRICK), Hot Spring (HOWARD and LUDLOW), Helena (LUDLOW).

Kentucky. — Lexington, Middlesboro, Louisville, Richmond (HOWARD and LUDLOW).

Illinois. — Cairo (HOWARD).

Missouri. — St. Louis (HOWARD).

Indiana. — Jeffersonville.

Virginia (GLENN HERRICK), Virginia Beach (HOWARD), Norfolk (3), Lynchberg, Danville, Richmond (HOWARD).

Maryland. — Baltimore (CARTER) found breeding in fresh water on fruit wharf.

Arizona. — Nogales (HOWARD).

California. — San Diego, Angel Island (CARTER).

---

(1) WIEDEMANN described *Culex tæniatus* from Savannah, but whether the Savannah in this State or not we do not know.

(2) WIEDEMANN merely says Savannah, which we do not know.

(3) In former years this place suffered much from yellow fever (HOWARD).

West Indies. — Cuba, Habana, Quantanamo, Daiquiri, Baracoa, San Antonio de los Banos, Cayamas, « Yaquaramoa », Santiago, Caimanera, Batabano, Santiago de los Kegas, Quemados, Isle of Pines (SCUDDER, ROBINEAU DESVOIDY, MACQUART, GUÉRIN, THEOBALD, HOWARD), Porto Rico (FABRICIUS).

St. Kitts; Nevis; Antigua; Carriacon, Grenadine Islands (LOW); St. Vincent (LOW, 25.V.1899); St. Lucia (GRAY, 26.VI.99, GALGAY, 21.IX, and 21.XII.99); Montserrat (LOW); Dominica (LOW); Grenada (LOW and BROADWAY, 14.II.1900); Trinidad (LOW, HEWLETT); Barbados (LOW); Jamaica, Kingston (JOHNSON, POWELL, CRUNDALL, 13.VII, and IX.99), common in Jamaica (GRABHAM, 7-12-15.IX, and 24.XI.99).

Bahama Islands. — Narsan, Spanish Wells, Current, Harbor Island, Tarpon Bay, San Salvador, Long Island, Government Harbour (HOWARD).

Oceanic Islands. — Azores, Teneriffe, Santa Cruz (GRABHAM); Samoa, Apia; Sandwich Islands, Honolulu, Hilo (TERRY and HOWARD); Fiji Islands (HEWLETT, JEPSON, IV-V-VI).

Pitcairn Islands (LORD CRAWFORD).

Bermuda (HARVEY, VII & 6.VIII.99).

Varieties of *S. fasciata* FABRICIUS.

1. Var. *Luciensis* THEOBALD.

« Mono. Culicid. », 1, 287 (1901).

Differs from the type only in having a clearly defined black band at the tip of the last hind tarsal segment.

Localities. Demerara, Georgetown (QUELCH, 16.VI.1899); St. Lucia (LOW); Trinidad; Philippine Islands (LUDLOW); Seychelles.

2. Var. *Queenslandensis* THEOBALD.

« Mono. Culicid. », 1, 297 (1901).

The mid lobe of scutellum has deep purple scales, instead of silvery white, and the abdomen has basal and apical pale bands, and an irregular broad median dorsal line of the same. In some cases the pale scales cover the whole of the abdomen. Queensland, Brisbane (BANCROFT, XI); Nigeria (HANLEY, VII & VIII), males and females being very white; Seychelles.

3. Var. *Mosquito* R. DESVOIDY.

Culex mosquito R. DESVOIDY.

« Dipt. Argentina », p. 60 (1891); Mono. Culicid., 1, p. 295 (1901). From DESVOIDY'S description it seems that in the insect he described as *Culex mosquito* the thorax had only silvery spots and lateral semi-lunar silvery spots. Had the median pale parallel lines been present, he would have been sure to have mentioned them, unless he described a worn specimen.

There exists, however, a distinct variety in which the two median lines are absent, and for this variety I retained DESVOIDY'S name *Mosquito*.

VON RŒDER (« Stett. Ento. Zeit. », p. 388, 1885) referred to this as a synonym of *fasciatus*.

It has been sent me from St. Lucia, Cuba, Jamaica (GRABHAM), Brazil and the Argentine; Calcutta; Singapore (BIRO); Philippine Islands (LUDLOW); Cyprus, Larnaka (BORDAN), and Greece, Poros (KRÜPER).

4. Var. *persistans* BANKS.

« Philippine Journal of Science », 1, 9, 996 (1908).

BANKS pointed out that in all the Philippine specimens he had seen there were additional white lines at the sides outside the curved silvery lateral lines. He mentioned these specimens as being a distinct species.

This variation is also quite common in N. and S. American forms, also in those from the West Indies, and I have seen it in specimens from India and Fiji.

Writing of this variation Miss LUDLOW (1908) says the differences pointed out by BANKS in the Philippine specimens show merely on account of inaccurate descriptions of *fasciata*. I do not think this is necessarily so, for I have seen far more specimens without the extra lines than with.

Miss Ludlow has compared large numbers of Philippine *Stegomyia* with those from the Southern States of America, and finds them the same. All the Philippine specimens I have seen have been true *Stegomyia fasciata* or its varieties.

There is no doubt all four are merely varieties and founded on purely colour and marking differences. There is no structural difference between them and the typical form.

REFERENCES.

1805. Fabricius, Systema Antliatorum, 36, 13.

1818. Meigen, J.-W., Systematische Beschreibung der bekannten europäischen zweiflügeligen Insekten, 1, 3.

1827. Robineau-Desvoidy, J.-B., Essai sur la tribu des Culicides (« Mémoires de la Société d'histoire naturelle de Paris », III, 390)

1828. Wiedemann, C.-R., Aussereuropäische zweiflügelige Insekten. (Als Fortsetzung des Meigen'schen Werkes, H. 2, T. 1, S. 10 u. 13.)

1831 (? 35). Guérin-Méneville, F. E., et Percheron, A., Genera des insectes. Paris. Pl. II, fig. 1.

1831. Brullé, A, Insectes. Expédition scientifique de Morée, Section des sciences physiques. Paris, 1re partie, Zoologie.

1834. Macquart, Suites à Buffon. (« Diptera », I, 35, 8.)

1838. Idem, Diptera exotica, I.

1848. Walker, F., List of Diptera in the British Museum, I, 4.

1860. Idem, Journal of the Proceedings of the Linnæan Society of London, IV, 91.

1861. Idem, Ibidem, V, 229 *(C. zonatipes)*.

1878. Osten-Sacken (baron), Catalogue of North American Diptera, 2nd Edit., 19.

1889. Skuse, F. A. A., Diptera of Australia, Pt. V, *Culicidæ*. (« Proceedings of the Linnæan Society of N. S. Wales », 2nd S., III, 1717.)

1889. Ficalbi, E., Notizie preventive sulla zanzare italiana V. Nota preventiva. Descrizione di une specie nuova *(Culex elegans)*. (« Estratto dagli Atti della R. Accademie dei Fisiocritici ». Se. IV, vol. 1.)

1891 Arribalzaga, F. L., Dipterologia argentina. (« Revista del Museo de la Plata », p. 60.)

1896. Filcalbi, E., Revisione sistematica della famiglia delle *Culicidæ* europee (Gen. *Culex*, *Anopheles*, *Aedes*). Firenze, 241, 5, 6 et 246, 7. (« Extracts Bollettino della Soc. Entom. Ital. », XXI-XXII, 1889-1890, etc.)

1896. HOWARD, L. O., Mosquitoes. (« Bulletin nº 4. N. Se. U. S Department of Agriculture, Division Entomology ». 22.)

1899. GILES, G. M., « Journal of Tropical Medicine », p. 64 *(Culex rossii)*.

1899. IDEM, A description of the *Culicidæ* employed by Major R. Ross, I. M. S., in his investigations on Malaria. (« Journal of Tropical Medicine », II, 64) *(Culex rossii)*.

1899. FICALBI, E, Venti specie di Zanzare *(Culicidæ)* italiane, classate descrite e indicate secondo la loro distribuzione corografica. (« Bolletino della Soc. Entomologia italiana », XXXI, 203, 12. Firenze.)

1900. GILES, G. M., A plea for the investigation of Indian *Culicidæ*. (« Journal Bombay Natural History Society », XIII, 606.)

1900. HOWARD, L. O., « Bulletin nº 25, new series. U. S. Department of Agriculture », 30, 31.

1900. GILES, G. M., A Handbook of Gnats or Mosquitoes. London. 216, 13; 220, 14; 224, 16; 264, 15; 230, 22; 231, 24; 232, 25; 235, 25; 237, 30; 244, 38; 255, 48; 263, 57; 283. 78; 286, 85; 287, 86.

1901. THEOBALD, F. V., Liverpool School of Tropical Medicine. (« Mem. » IV, App. III, Notes on a collection of *Culicidæ* from West Africa.)

1901. IDEM, A Monograph of the Mosquitoes or *Culicidæ* of the World, I, 289.

1901. IDEM, Notes on a collection of Mosquitoes from West Africa and descriptions of new species. Reports of the Thompson Yates Laboratories. Report of the Malarial Expedition to Nigeria. (« Liverpool School of Tropical Medicine », IV, app. III.)

1901. GILES, G. M., Notes on Indian Mosquitoes. (« Journal of Tropical Medicine », IV, 159.)

1901. MAC DONALD, I., La propagation du paludisme par les Moustiques, avec une note sur leur rôle à Rio-Tinto (sud d'Espagne). (« Thèse de Paris ».)

1901. DURHAM, H. E., Liverpool School of Tropical Medicine. (« Mém. » VII, p. 55. Report of the Yellow Fever Expedition to Para.)

1901. HERRICK, GLENN. W., Mississippi Agricultural Station. (« Bulletin » 74. Some Mosquitoes of Mississipi and how to deal with them, p. 17.)

1901. HOWARD, L. O., Mosquitoes. How they live, etc., p. 136.

1902. GILES, G. M., A Handbook of Gnats or Mosquitoes, 2nd edition, 372, 4.

1902. THEOBALD, F. V., A short description of the *Culicidæ* of India, with a description of the new species of *Anopheles*. (« Proceedings of the Royal Society », LXIX, 383.)

1902. GOELDI, W. E., Os Mosquitos No Pará.

1902. NEVEU-LEMAIRE, M, Description de quelques moustiques de la Guyane. (« Archives de parasitologie », VI, nº 1, p. 5.)

1902. DYÉ, LÉON, Notes et observations sur les Culicides. (« Ibid. », VI, nº 3, p. 367.)

1902. Vincent. G. A., Observations on human filariasis in Trinidad. (« British Medical Journal ». January 25th.)

1903. Theobald, F. V., Monograph of the *Culicidæ* of the world, III, 141.

1903 Dutton, J. E., and Theobald, F. V, Report of the Malarial Expedition to the Gambia, 1902 (« Mem. », X. p. 6. Appendix. Liverpool School of Tropical Medicine.)

1903. Howard, L. O., Concerning the geographical distribution of the Yellow Fever Mosquito. (« Public Health Reports », vol. XVIII, n° 46, November.)

1903. Marchoux. Salimbeni et Simond, La fièvre jaune. (« Annales de l'Institut Pasteur », 665.)

1903. Blanchard, R. et Dyé. L., Notes sur les Moustiques de la Côte d'Ivoire. (« Comptes rendus des séances de la Société de biologie ». IV. 570.)

1903. Theobald, F. V., Note on the genus *Stegomyia* and its distribution. (« Transactions of the Epidemiological Society of London », N. S., XXII, pp. 91-97, map.)

1903. Manson, Sir Patrick, The relation of the Panama Canal to the introduction of yellow fever into Asia. (« Transactions Epidemiological Society of London », N. S., XXII. pp. 60-72.)

1905. Theobald. F. V., et Grabham, M. The Mosquitoes of Jamaica, p. 19.

1905. Theobald, F. V., A catalogue of the *Culicidæ* in the Hungarian National Museum. (« Annales Musei Nationalis Hungarici », III, 73.)

1905. Howard. L. O., Concerning the geographi al distribution of the Yellow Fever Mosquito. (« Public Health Reports », XVIII, No 46, 03. Revised, Sep. 10. 1905.)

1905. Dyé, Léon, Les parasites des Culicides. (« Archives de parasitologie », IX, n° 1. 5.)

1905. Goeldi, E. A., Os Mosquitos no Pará, pp. 96-106. (« Memorias do Museu Goeldi.)

1906. Ludlow, C. S., The distribution of Mosquitoes in United States, as shown by collections made at Army Posts, 1904-1905, p. 6.

1906. Banks, C. S., A list of Philippine *Culicidæ* with descriptions of some new species. (« Philippine Journal of Science », 1, No 9, p. 984.)

1906. Theobald, F. V., Some notable instances of the distribution of injurious Insects by artificial means. (« Science Progress ». No I, p. 14.)

1906. Idem. Report of the injurious Insects for the year ending April 1st, 1906, p. 106.

The possibility of the Introduction of the Yellow Fever Mosquito by banana ships into England.

1907. Newstead. R., Insects and other Arthropoda collected in the Congo Free State. (« Annals of Tropical Medicine and Parasitology », I, No I, Se. T. M., 14.)

1908. THEOBALD, F. V., Records of the Indian Museum, II and III, No 30, 291.

1988. ROSS, RONALD, Report on the prevention of malaria in Mauritius. Addenda I, p. 133 The Mosquitoes of Mauritius.

1908. LUDLOW, CLARA, The Mosquitoes of the Philippine Islands. The distribution of certain species and their occurrence in relation to the incidence of certain diseases, p 8.

1908. PERYASSU, A. S., Os Culicideos do Brasil. (« Traballo do Instituto de Manguinhos ». 63. map.)

1908. BANCROFT, T. L., List of the Mosquitoes of Queensland. (« Annals of the Queensland Museum », No 8.)

1909. BALFOUR. A., Mosquitoes with reference to immigration and horse sickness. (« Cairo Scientific Journal », III, No 37, p. 1.)

1910. THEOBALD, F. V., Monograph of the *Culicidæ* of the world, V. 188.

1910. NEWSTEAD, R., The Amazon Yellow Fever Expedition. (» Annals of Tropical Medicine and Parasitology », IV, No 1, Se. T. N., 143.)

1910. WESCHÉ, W., On the larval and pupal stages of West African *Culicidæ*. (« Bulletin of Entomological Research », I, pt. 1, p. 25.)

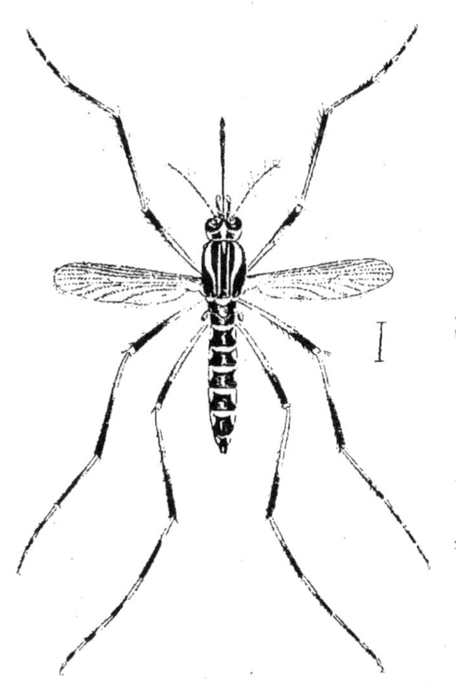

Stegomyia fasciata Fabr. Female.

FRED. V. THEOBALD. — YELLOW FEVER MOSQUITO.

Destruction of Mosquitos by small fish in the West Indies,

by Sir DANIEL MORRIS, K. C. M. G., D. Sc. F. L. S.

As a supplement to the interesting information placed before the Congress by Professor F.V. THEOBALD, F. E S., it would be useful to place on record the valuable services rendered by a small fish (*Girardinus pœciloides*), popularly known as « millions ». In 1905 Mr. C. KENRICK GIBBON suggested that probably the absence of malaria at Barbados was due to the presence in the ponds in that island of the small fish above referred to. The Mosquito usually regarded as causing malaria is a species of *Anopheles*. The breeding places of this are shallow streams, pools and swamps. So far, this Mosquito is unknown at Barbados, and notwithstanding the frequent and intimate communication between Barbados and the neighbouring islands where malarial fever is more or less endemic the *Anopheles* has not established itself. Three inferences have been drawn from this fact : 1° That the *Anopheles* has failed to obtain a footing at Barbados because it breeds in localities where the *Girardinus* abounds; 2° that in consequence Barbados continues free from malarial fever; 3° that if the *Girardinus* were introduced into countries where the *Anopheles* is abundant, both the Mosquito and malaria would probably be greatly diminished, or eradicated. In a recent publication issued by the Imperial Department of Agriculture in the West Indies Mr. BALLOU gives a brief account of the habits of *Girardinus*, its relation to Anopheles, and the manner in which the fish may be successfully introduced to other portions of the tropics (1). *Girardinus pœciloides* measures about 35 m. in length; it frequents water too shallow for larger fish; is viviparous, and increases very rapidly. It is closely allied to *G. versicolor* of San Domingo and G. *formosus* of South Carolina and Florida. Mr. BALLOU, after carrying on investigations extending over a long period, is of opinion that *Girardinus* is a very efficient enemy of Mosquitos. He has observed it devouring the eggs, larvæ and pupæ of *Culex* and *Stegomyia* as well as of *Anopheles*. Instances are quoted of the success of the introduction of *Girardinus* to other parts of the West Indies. More recently small consignments of the fish have been forwarded to West Africa and India. Somewhat similar small fish known as « top minnows » have been utilised to control Mosquitos in the United States. In 1905 the U. S. Fish Commission was instrumental in introducing a large number of « top minnows » with good results from Texas to the Hawaiian Islands. It is probable that many other small fish that multiply rapidly would prove of value in controlling Mosquitos. The subject is one which is deserving of careful attention in all tropical countries.

---

(1) *Millions and Mosquitos.* Pamphlet Series (Imperial Department of Agriculture), N° 55. London, Dulau and C°, 1908. Price : 3 pence.

Ueber zweigbewohnende Cicindelinenlarven und ihre Entdeckung in Brasilien durch Herrn Jos. ZIKAN,

von WALTHER HORN (Berlin).

Im Jahre 1897 hat KONINGSBERGER im botanischen Garten von Buitenzorg (Java) Cicidelinenlarven entdeckt, welche ihre Gänge in den dünnen Zweigen der Kaffeebäume anlegen. Seitdem ist eine ziemlich umfangreiche Literatur über diesen Gegenstand erschienen, deren Zusammenstellung hier folgt :

KONINGSBERGER : « Mededeelingen uit 's Lands Plantentuin », 1897, n° 20, pp. 58-59 ; 1898, n° 22, p. 45 ; 1901, n° 44, p. 113 ; « Mededeelingen uitgaande van het Departement van Landbouw », 1908, n° 6, p. 92.

R. SHELFORD : « Journal Royal Asiat. Sol. », Straits Branch, 1906, p. 283 ; « Proc. Ent. Soc. London », 1905, p. 72 ; « Idem. », 1908, p. 43 ; « Trans. Ent. Soc. London », 1907, p. 83.

W. HORN : « Deutsche Ent. Zeit. », 1908, p. 125.

Van Leeuwen : « Mededeel. van het Algemeen Proefstation Java », Salatiga 1909, IIe Ser. n° 15; « Cultuurgids 1909, II, n° 3 »; « Tijdschrift voor Entomologie », 1910, LIII, pp. 18-40.

Obwohl manche genaueren Einzelheiten über die Herstellungsweise der Gänge, die Beziehung zu den einzelnen Arten und selbst Gattungen etc. noch nicht oder nur lückenhaft bekannt sind, lässt sich doch wohl der allgemeine Satz jetzt schon aufstellen, dass die Larven aller *Collyris*- und *Tricondyla*- Arten ihre Gänge in frische Zweige der verschiedensten Pflanzenarten bohren und stellenweise dadurch geradezu als Massenschädlinge auftreten.

Nach dem von mir in letzter Zeit ausgearbeiteten System der Cicindelinen zerfallen diese in 2 Hauptgruppen : Alocosternale und Platysternale Phylen. Zur ersteren Gruppe gehören nur baumbewohnende, zur zweiten meist erdbewohnende Imagines. Die Kenntnis der Larven ist nun zwar heutzutage noch sehr gering, immerhin war zu beachten, dass von den alocosternalen Arten *nur* zweigbewohnende, von den platysternalen Arten *nur* erdbewohnende Larven bekannt waren. Dies brachte mich auf den Gedanken, ob nicht alle alocosternalen Gattungen (es kommen ausser den orientalischen, *Collyris* und *Tricondyla*, noch das madagassische Genus *Pogonostoma* und das neotropische Genus *Ctenostoma* in Frage) im Gegensatz zu allen platysternalen zweigbewohnende Larven haben könnten. Meine Versuche, Sammler in Madagaskar für diese Frage zu interessieren, scheiterten völlig ; dagegen hatte ich das Glück, in Brasilien einen intelligenten böhmischen Volksschullehrer, Herrn Jos. F. Zikan, in Mar de Hespanha (Minas Geraes) brieflich anleiten zu können. Natürlich richtete er nach meiner Weisung sein Augenmerk lange Zeit vor allem auf lebende Zweige, Büsche und Stämme. Ein volles Jahr verging, ohne irgend welche Erfolge, obwohl der fleissige und sorgfältige Beobachter stellenweise dort, wo Imagines von *Ctenostoma* vorkamen, den Boden des Urwaldes auf den Knieen rutschend absuchte. Schliesslich entdeckte er 1908 durch Zufall ein winziges, etwa 1 mm. im

Umfang zeigendes Loch im *Zentrum der frei nach oben ragenden Bruchstelle* eines dünnen *in der Erde steckenden* morschen Zweigstückes : Es war ein Cicindelinenlarvengang ! Im darauf folgenden Jahre fand Herr ZIKAN an derselben Lokalität ein anderes morsches Zweigstück, welches *frei am Boden lag* und eine *seitliche* Oeffnung mit einem Cicidclinenlarvengang à la *Collyris* aufwies. Seitdem glückte es ihm, noch ein halbes Dutzend weiterer Zweiggänge vom Typus des ersten Fundes zu entdecken. Nur das 1908 gefundene Zweigstück ist bisher in meine Hände gelangt, ebenso stehen noch von Herrn ZIKAN die genauen Aufzeichnungen aller Fundnotizen etc. aus; immerhin kann ich schon hier bekannt geben, dass Herr ZIKAN durch recht mühsame Zuchtversuche festgestellt hat, dass 2 *Ctenostoma*-sp. solche Holzgänge anlegen, und zwar legt *Cten. rugosum* Klug seine Eier in alte Bohrlöcher anderer Insekten, während *Cten. unifasciatum* DEJ. dieselben in das obere Ende von morschen, in der Erde steckenden Zweigstücken in selbstgebohrten Löchern ablegt. Es gebraucht ½ - 1 Stunde zum Ablage eines Eies. Folgende Schlussfolgerung ist also wahrscheinlich schon jetzt aufzustellen :

Im Gegensatz zu den in zahlreichen Individuen vorkommenden und in *lebenden* Zweigen etc. ihre Larvengänge bohrenden orientalischen alocosternalen Cicindelinen-Genera *Collyris* und *Tricondyla*, bewohnen die spärlichen Larven der neotropischen Gattung *Ctenostoma* Gänge in *morschen* abgefallenen Zweigen. Stets verlaufen die Gänge in der Längsachse des Holzes. Die Larvengänge der madagassischen alocosternalen Cicindelinen-Gattung *Pogonostoma* wird man mit einer gewissen Wahrscheinlichkeit gleichfalls in Zweigen etc. zu suchen haben. Zweigbewohnende Larven platysternaler Cicindelinen sind bisher noch nicht bekannt. Die Larven der neotropischen Genera *Euprosopus* und *Odontochila* leben in der Erde, wie Herr ZIKAN durch Züchtung festgestellt hat. Die Seltenheit der *Ctenostoma*-sp. dürfte in den Schwierigkeiten ihre Erklärung finden, welche die in den losen, morschen Zweigstücken lebenden Larven haben, bevor sie glücklich am Ende ihrer Entwickelung anlangen. Nur ein winziger Prozentsatz der Brut wird voraussichtlich das Imaginalstadium erreichen.

Ueber fossile Insekten,

von A. HANDLIRSCH (Wien).

An der Hand von 75 Rekonstruktionen (1) versuchte der Vortragende, der Versammlung ein Bild der Evolution vorzuführen, welche die Insekten vom Oberkarbon bis zur Tertiärzeit durchgemacht haben. Von den primitiv organisierten Urinsekten oder *Paläodictyopteren* (Pl. VI, Fig. 1, 2) ausgehend, können wir mehrere scharf getrennte Entwicklungsreihen verfolgen, deren Endglieder zum Teile wieder erloschen sind, zum Teile aber in der jetzt lebenden Insektenwelt vor uns liegen.

Eine dieser Entwicklungsreihen führt über die *Protorthopteren* (Pl. VI, Fig. 3, 4, Pl. VII, Fig. 5) zu den echten Orthopteren (Pl. VIII, Fig. 6) (Locustoiden und Acridioiden) und Phasmoiden, eine andere über die *Protoblattoiden* (Pl. VIII, Fig. 7, Pl. IX, Fig. 8) zu den *Blattoiden* (Pl. VII, Fig. 9).

In beiden Reihen hat sich ein Uebergang von der ursprünglich jedenfalls amphibiotischen Lebensweise der Paläodictyopteren zu einer rein terrestrischen Lebensweise vollzogen und gleichzeitig

---

(1) Diese Rekonstruktionen sind keine Phantasiegebilde sondern das Resultat sorgfältiger Präparationsarbeit, ergänzt durch Schlussfolgerungen morphologischer Natur. Nach den Zeichnungen HANDLIRSCH's wurden von Dr. LÜLKEMÜLLER in Baden die Diapositive hergestellt, wofür ihm hier herzlichster Dank ausgedrückt sei.

haben die ursprünglich horizontal ausgebreiteten Flügel die Fähigkeit erlangt, sich nach hinten über das Abdomen zurückzulegen. Es ist kaum zweifelhaft, dass diesen Entwicklungsreihen auch die hochspezialisierten Physopoden, Hemimeriden, Dermapteren einerseits und die Psociden mit den Mallophagen und Pediculiden, die Termiten und Mantiden anderseits angehören. Höchst wahrscheinlich sind auch die Coleopteren und Hymenopteren (Pl. IX, Fig. 10) holometabol gewordene Abkömlinge der mit den Protoblattoiden beginnenden Reihe.

Selbstständige Entwicklungsreihen führen über die *Protodonaten* zu den echten Odonaten (Pl. VII, Fig. 11) oder Libellen und über die *Protephemeriden* (Pl. VIII, Fig. 12) zu den Plectopteren oder Ephemeriden (Pl. VIII, Fig. 13); in beiden wurde die amphibiotische Lebensweise beibehalten und es kam weder zu einer Holometabolie noch zum Zurückdrehen der Flügel. Eine fünfte Reihe beginnt bei den merkwürdigen *Megasecopteren* (Pl. IX, Fig. 14) des Karbon, bei welchen noch die Flügel in ursprünglicher Weise horizontal ausgebreitet sind, im übrigen aber bereits mancherlei Spezialisierungen aufweisen. Zweifellos waren auch diese Formen noch nicht holometabol und höchst wahrscheinlich amphibiotisch. Nach Ansicht des Vortragenden bilden sie den Ausgangspunkt für die bereits holometabolen Panorpaten, von denen sich sowohl die Phryganoiden (Trichopteren) als die Dipteren mit Einschluss der Puliciden und die Lepidopteren (Pl. VIII, Fig. 15) ableiten lassen.

Selbstständige Entwicklungsreihen bilden offenbar auch die *Megalopteren* oder Sialiden, die *Raphidiiden*, bezw. die *Neuropteren* (Pl. IX, Fig. 16, Pl. X, Fig. 17) im engsten Sinne (Hemerobiden und Verwandte), doch fehlen uns in diesen Reihen noch die paläozoischen, den Anschluss an die Paläodictyopteren vermittelnden Uebergangsformen, die offenbar noch amphibiotisch und heterometabol gewesen sind, in Bezug auf den Verlauf des Flügelgeäders aber jedenfalls nicht sehr verschieden von den uns bekannten mesozoischen Typen der Sialiden und Prohemerobiden.

Mit dem merkwürdigen *Eugereon Böckingi*, dem einzigen bisher bekannt gewordenen Vertreter der *Protohemiptera*, beginnt die Reihe der heterometabol gebliebenen Hemipteroiden oder Rhynchoten (Pl. VII, Fig. 18), bei welchen die Mundorgane in so überaus charakteristischer Weise umgewandelt sind, und selbstständige Reihen bilden jedenfalls auch die Embioiden, die vielleicht mit Hilfe der *Hadentomoiden* (Pl. VII, Fig. 19) an die Paläodictyopteren angeschlossen werden können, endlich noch die bis zum Ende des Paläozoikum zu verfolgenden *Perlarien*, so dass die noch heute lebend erhaltenen Insektengruppen sich auf wenigstens 9, vielleicht aber 11, falls Sialiden, Raphidien und Neuropteren nicht gemeinsame Stammformen haben sollten, oder sogar auf 13 Reihen verteilen, wenn sich die Ableitung der Hymenopteren und Coleopteren von blattoiden Typen als irrig erweisen sollte. Alle diese Reihen laufen schliesslich an der Basis in den Paläodictyopteren zusammen.

Es liegt in der Natur der Sache, dass die uns bekannt gewordenen fossilen Insekten, die bereits die stattliche Zahl von je 1000 für das Paläozoikum und Mesozoikum und von etwa 6000 für das Kainozoikum erreicht haben, nicht eine ununterbrochene Entwicklungsreihe bilden, sondern dass darunter nur weinige sind, welche direkt als Stammformen neuer Gruppen betrachtet werden können. Alles andere sind Seitenäste, in irgend einer Richtung spezialisiert, mehr oder minder lang erhalten, aber schliesslich wieder erloschen. Bereits aus den Paläodictyopteren sind sicher zahlreiche solcher blind endender Reihen hervorgegangen wie z. B. die *Sypharopteroiden* (Pl. IX, Fig. 20), *Mixotermitoiden*, *Hapalopteroiden*, und auch unter den Protorthopteren, Protoblattoiden und anderen Ordnungen finden wir derartige Elemente, wie z. B. die langhalsigen Gerariden u. a.

Das Studium der fossilen Insekten hat ausser den oben skizzierten rein phylogenetischen Resultaten auch schon so manche für die Tiergeographie, Geologie, Klimatologie und allgemeine Biologie wichtige Tatsache ergeben; es hat uns neuerdings gezeigt, dass ein, wenn auch langsames, Fortschreiten in der Spezialisierung stattfindet, dass nichts durch allzu lange Zeit ganz unverändert bleibt und dass die Evolution einer Tiergruppe vielfach in Wechselbeziehungen zur Evolution anderer Organismen und zu den allgemeinen Lebensbedingungen steht. Wir sehen, dass die Insektenwelt des Paläozoikums von jener des Mesozoikums und diese wieder von jener des Kainozoikums eben so verschieden ist, wie die Floren oder die Vertebraten-, Mollusken-, Echinodermenfaunen dieser Perioden von einander verschieden sind.

Wenn uns die fossile Pflanzenwelt gestattet, auf ein feuchtwarmes frostfreies Klima der Steinkohlenzeit zu schliessen und auf eine wesentliche Verschlechterung dieses Klimas im Perm, so finden wir auch in der Paläoentomologie neuerdings eine Bestätigung dieser Hypothesen, denn die vorwiegend riesigen Karboninsekten mit ihrem ausgesprochen thermophilen Habitus weichen einer viel unansehnlicheren Fauna, in welcher, analog den Jahresringen der Bäume, in dem gleichzeitigen heterophyletischen Auftreten der Holometabolie wohl mit Recht Anzeichen für das Auftreten von kalten oder trockenen Jahreszeiten erblickt werden dürfen Auch der auffallende Grössenunterschied zwischen Lias- und Malminsekten, in unseren Breiten, legt umsomehr den Gedanken an Klimaschwankungen nahe, als er mit einem — und + in der Entwicklung der Riffkorallen — dieses geohistorischen Termometers par excellence — zusamenfällt.

Wie enorm wichtig aber die Erwerbung der Holometabolie für die gesammte Evolution der Insekten war, erhellt aus der Tatsache, dass unsere heutige Insektenwelt mindestens siebenmal so viele holometabole als heterometabole Formen enthält, und demgemäss werden wir nicht irren, wenn wir selbst das Auftreten der angiospermen Pflanzen in der Kreidezeit, welches erwiesenermassen einen ganz eminenten formbildenden Einfluss auf die Insekten ausübte, an Bedeutung noch geringer einschätzen als jene Faktoren, welche die Holometabolie bewirkt haben.

Eine genaue Analyse der Formen lehrt uns, dass die primäre Ernährungsweise der Insekten höchst wahrscheinlich eine karnivore war, denn es zeigen nicht nur sehr viele alte Fossilien im Bau ihres Kopfes, ihrer Vorderbeine, etc. Anhaltspunkte für eine solche Annahme sondern es sind auch noch heute die tiefststehenden Elemente der meisten Gruppen oder Reihen nicht typisch phytophag.

Schon jetzt lässt sich aus dem Studium der fossilen Insekten so manches wichtige Detail für eine allgemeine Organologie ziehen, in erster Linie natürlich für die Flügelmorphologie, denn wir können an der Hand der Fossilien nicht nur den strikten Beweis erbringen, dass die beiden Flügelpaare ursprünglich homonom und keineswegs, wie man eine Zeitlang meinte, fächerartig waren, sondern wir können sehen, wie sich die uns heute vorliegenden weitgehenden Differenzierungen doch alle aus einem einzigen Urtypus herausgebildet haben. An der monophyletischen Ableitung der gesammten heute so enorm artenreichen und an Mannigfaltigkeit der Formen alles übertreffenden Klasse der pteryzogenen Insekten ist schon aus diesem Grunde nicht mehr zu zweifeln.

Wenn nun schon die relativ geringe Zahl der bisherigen fossilen

Insektenfunde (1) manchen Lichtstrahl in dunkle Ecken der Naturgeschichte geworfen hat, so sollten wir mit allem Eifer bestrebt sein, dem Schosse der Erde noch mehr von diesen kostbaren Urkunden zu entreissen, denn wir sind noch immer sehr weit von unserem Ziele entfernt : von einer exakten Biologie.

BEMERKUNGEN ZU DEN BEIGEFUEGTEN PROBEBILDERN.

Pl. VI, Fig. 1. — *Stenodictya lobata* BRONGN. — Ein Paläodictyopteron aus dem Oberkarbon von Commentry. Man beachte die homonomen horizontal ausgebreiteten Flügel mit ihrem höchst primitiven Geäder, ferner die flügelartigen Anhänge des Prothorax, die Seitenlappen der Hinterleibessegmente und den einfachen Bau des Thorax. Ergänzt sind die Fühler, Ocellen und Tarsen ($^2/_3$ natürliche Grösse).

Pl. VI, Fig. 2. — *Eubleptus Danielsi* HANDL. — Ein Paläodictyopteron aus dem Oberkarbon vom Mazon Creek, Illinois, Nordamerika. Man beachte die Homonomie der Flügel und Segmente sowie die habituelle Aehnlichkeit mit Ephemeriden. Ergänzt sind die Ocellen, Fühler und Tarsen (× 2).

Pl. VI, Fig. 3. — *Dieconeura arcuata* SCUDDER. — Ein Protorthopteron aus dem Oberkarbon vom Mazon Creek in Nordamerika. Man beachte die Stellung der Flügel in der Ruhelage, die Vereinigung der Medialader des Vorderflügels mit dem Sector radii, die Spezialisierung des Cubitus, des Analfeldes der Vorderflügel, das Auftreten eines faltbaren vergrösserten Anallappens der Hinterflügel, den verlängerten Prothorax, den prognathen Kopf, die Gonapophysen und die noch wenig verlängerten Hinterbeine. Die Fühler sind nach verwandten Formen ergänzt, die Tarsen und Ocellen auf Grund morphologischer Schlussfolgerung (× 2).

Pl. VI, Fig. 4. — *Gerarus longicollis* HANDL. — Ein aberranter Typus der Protorthopteren vom Oberkarbon des Mazon Creek in Nordamerika. Auffallend durch die enorme Verlängerung des Prothorax und den kleinen prognathen Kopf. Beine und Fühler sind teils nach Analogie mit verwandten Formen, teils auf morphologischer Grundlage ergänzt (× 2).

---

(1) Vergl. HANDLIRSCH, « Die fossilen Insekten und die Phylogenie der rezenten Formen ». Leipzig, 1908.

Pl. VII, Fig. 5. — *Œdischia Williamsoni* BRONGNIART. — Ein zu den echten Locustoiden hinüberleitender Typus der Protorthopteren aus dem Oberkarbon von Commentry in Frankreich. Man beachte die Verbindung von Medialis und Sector radii im Flügel, die langen Fühler und die zu typischen Sprungbeinen umgewandelten Hinterbeine (natürliche Grösse).

Pl. VIII, Fig. 6. — *Elcana Geinitzi* HEER. — Eine Locustoide aus dem Lias von Dobbertin in Mecklenburg. Man beachte die blattartigen Anhänge der Hinterschiene, welche es dem Tiere ermöglichten, sich auf nassem Schlamm oder sogar auf der Oberfläche des Wassers fortzubewegen Das Geäder gleicht auffallend jenem der Acridier, aber die langen Fühler und die Legescheide verweisen das Tier in die Gruppe der Locustoiden. · erci, Tarsen und Ocellen ergänzt (× 5).

Pl. VIII, Fig. 7. — *Protophasma Dumasi* BRONGNIART. — Eine Protoblattoide aus dem Oberkarbon von Commentry in Frankreich. Man vergleiche dieses Bild mit den von BRONGNIART, SCUDDER und anderen hergestellten Rekonstruktionen! Zu beachten ist der scheibenförmige Prothorax, der freie Kopf, das breite Costalfeld der Vorderflügel, der Analfächer der Hinterflügel und das netzartige Zwischengeäder. Fühler, Ocellen, Tarsen und Cerci durch morphologische Schlussfolgerung ergänzt (²/₃ natürliche Grösse).

Pl. IX, Fig. 8. — *Eucænus ovalis* SCUDDER. — Eine Protoblattoide aus dem Oberkarbon vom Mazon Creek in Nordamerika. Auffallend ist der coleopterenähnliche Habitus, der durch die derbe Beschaffenhei der gewölbten Flügeldecken erhöht wird, der freie prognathe Kopf, der scheibenförmige Prothorax, das kleine Analfeld der Vorderflügel, der faltbare Analfächer der Hinterflügel, das Vorhandensein der Cerci und die Tibienrinne an den kräftigen Vorderschenkeln. Tarsen, Fühler und Ocellen ergänzt (× 2/3).

Pl. VII, Fig. 9. — *Aphthoroblattina Johnsoni* WOODWARD. — Eine der ältesten und primitivsten echten Blattoidenformen aus dem Oberkarbon Englands. Man beachte den über den Kopf ausgedehnten scheibenförmigen Prothorax, das Costal- und Analfeld des Vorderflügels, die noch deutliche Gliederung des Radius in Stamm und Sektor und die deutlichen Queradern. Ergänzt nach verwandten Formen sind Fühler und Beine (× 1/7).

Pl. IX, Fig. 10. — *Pseudosirex* sp. — Ein siricidenähnliches Hymenopteron aus dem Malm von Solnhofen in Bayern. Das Geäder erweist sich als primitiv im Vergleiche mit jenem der lebenden Siriciden, mit denen das Tier habituell weitgehende Uebereinstimmung zeigt. Ergänzt sind nur die Ocellen und die Gliederung der Tarsen (natürliche Grösse).

Pl. VII, Fig. 11. — *Tarsophlebia eximia* HAGEN. — Eine Odonate aus dem Malm von Bayern. Auffallend durch die Kombination von Charakteren der Anisopteren und Zygopteren im Bau des Kopfes und der Flügel, ferner durch die auffallend langen nach vorne gerichteten Beine (× 1'7).

Pl. VIII, Fig. 12. — *Triplosoba pulchella* BRONGN. — Eine Protephemeride aus dem Oberkarbon von Commentry in Frankreich. Zu beachten sind die 3 langen Anhänge am Hinterende des auffallend homonom segmentierten Körpers und der einfache Kopf, ferner die homonomen Flügel mit den deutlichen Schaltsektoren. Fühler, Füsse, Ocellen und Genitalfüsse morphologisch ergänzt (× 2).

Pl. VIII, Fig. 13. — Larve von *Phthartus rossicus* HANDLIRSCH, einer Ephemeroide aus dem Perm Russlands. Sehr bemerkenswert ist das Vorhandensein von Extremitätenkiemen auf dem 9. Segmente (× 2).

Pl. IX, Fig. 14. — *Mischoptera Woodwardi* BRONGN. — Ein Megasecopteron aus dem Oberkarbon von Commentry in Frankreich. Die Flügel sind noch homonom und in der Ruhe horizontal ausgebreitet. Die Zahl der Queradern ist limitiert und das Geäder im Vergleich zu jenem der Paläodictyopteren hochspezialisiert. Der Körper zeigt noch sehr homonome Segmentierung. Auffallend sind die langen Cerci und die Kopfform Ergänzt sind die Ocellen und die Tarsen (3/4 natürliche Grösse).

Pl. VIII, Fig. 15. — *Eocicada Lameerei* HANDL. — Eine Lepidopterenform aus dem Malm von Solnhofen in Bayern. Erweist sich durch die weitgehende Heteronomie der Flügel und durch den dicken kurzen Leib als bereits hochspezialisiert, obwohl die Erhaltung des Stammes der Medialis noch ein primitives Merkmal darstellt. Ergänzt sind die Fühler und Ocellen (2/3 natürliche Grösse).

Pl. IX. Fig. 16. — *Prohemerobius prodromus* HANDL. — Ein primitives Neuropteron aus dem Lias von Dobbertin in Mecklenburg. Beachtenswert sind die vielen Gabelzinken der Längsadern und der ursprüngliche Bau des Radius. Rekonstruiert sind die Körperanhänge und Ocellen (× 11).

Pl. X, Fig. 17. — *Kalligramma Haeckeli* WALKER. — Ein riesiges Neuropteron aus dem Malm von Solnhofen, auffallend durch die Augenflecken aller Flügel und durch die eigentümliche Anordnung der Adern. Beine und Fühler sind rekonstruiert (2/3 natürliche Grösse).

Pl. VII, Fig. 18. — *Mesobelostomum deperditum* GERM. — Eine Belostomide aus dem Malm von Bayern. Zeigt alle Charaktere der rezenten Belostomen (natürliche Grösse).

Pl. VII, Fig. 19. — *Hadentomum americanum* HANDLIRSCH. — Einziger Vertreter der Gruppe Hadentomoidea aus dem Oberkarbon des Mazon Creek in Nordamerika. Auffallend durch die homonomen nach hinten faltbaren Flügel, deren Geäder in eigentümlicher Weise durch Reduktion spezialisiert ist. Körperanhänge rekonstruiert (× 1'4).

Pl. IX, Fig. 20. — *Sypharoptera pneuma* HANDL. — Einziger Vertreter der Gruppe Sypharopteroidea aus dem Oberkarbon des Mazon Creek in Nordamerika. Auffallend durch die noch sehr homonome Segmentierung des Körpers und die homonomen wenigstens teilweise zurücklegbaren Flügel, deren Geäder in ähnlicher Weise spezialisiert erscheint wie bei den Megasecopteren. Fühler, Ocellen und Tarsen sind ergänzt (× 5).

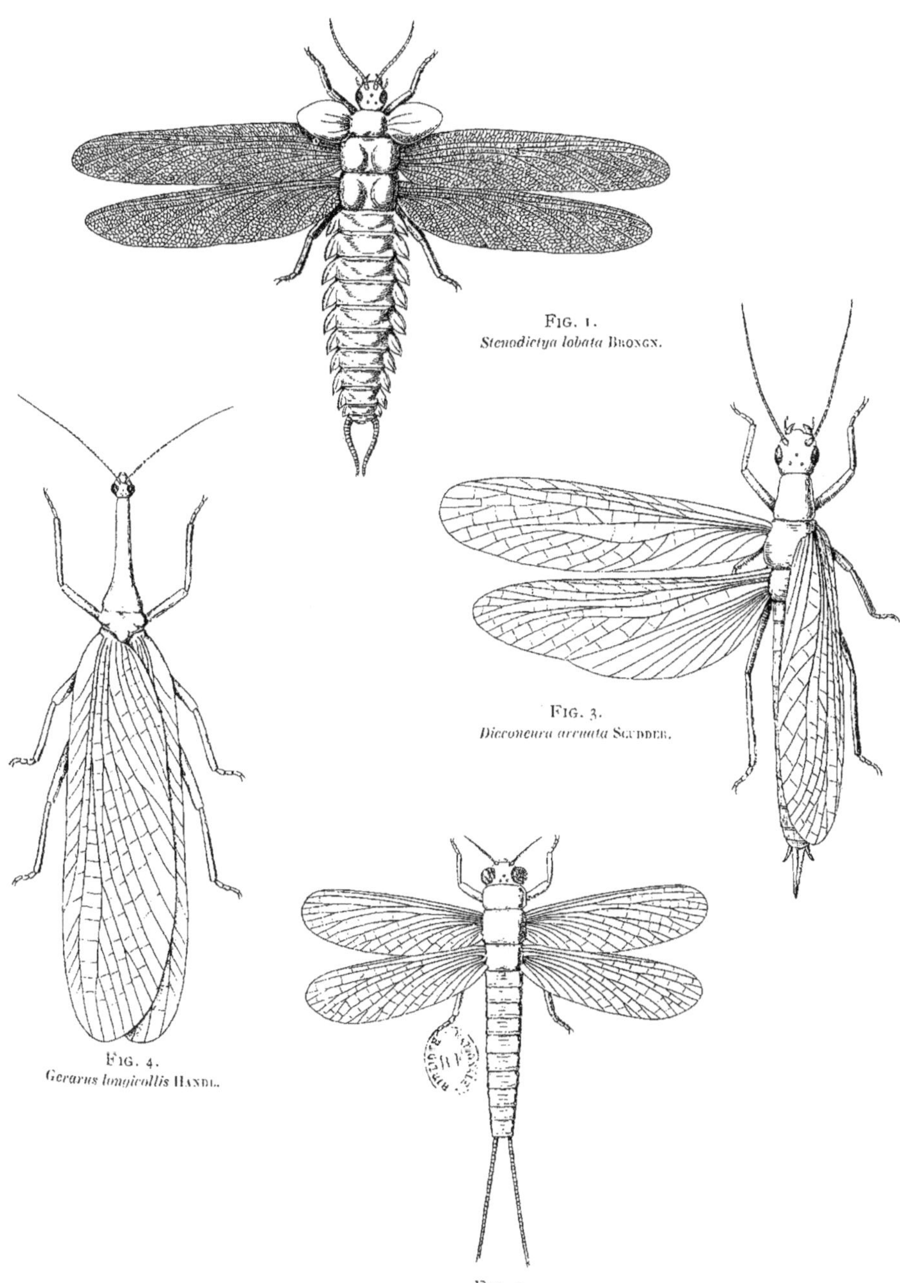

Fig. 1.
Stenodictya lobata Brongn.

Fig. 3.
Dicconeura arcuata Scudder.

Fig. 4.
Gerarus longicollis Handl.

Fig. 2.
Eubleptus Danielsi Handl.

A. HANDLIRSCH. — UEBER FOSSILE INSEKTEN. — 1.

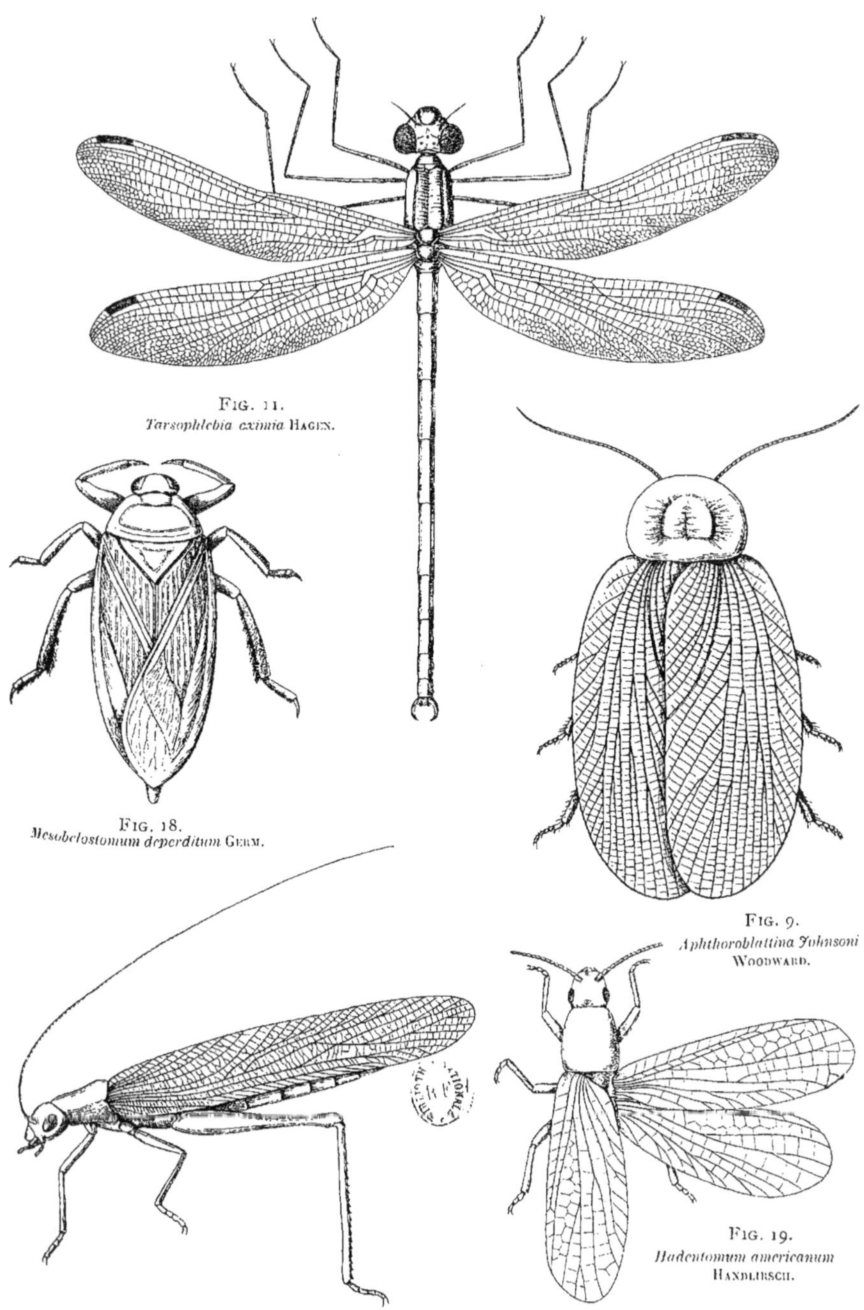

Fig. 11.
Tarsophlebia eximia Hagen.

Fig. 18.
Mesobelostomum deperditum Germ.

Fig. 9.
Aphthoroblattina Johnsoni Woodward.

Fig. 19.
Hadentomum americanum Handlirsch.

Fig. 5.
Œdischia Williamsoni Brongn.

A. HANDLIRSCH. — UEBER FOSSILE INSEKTEN. — II.

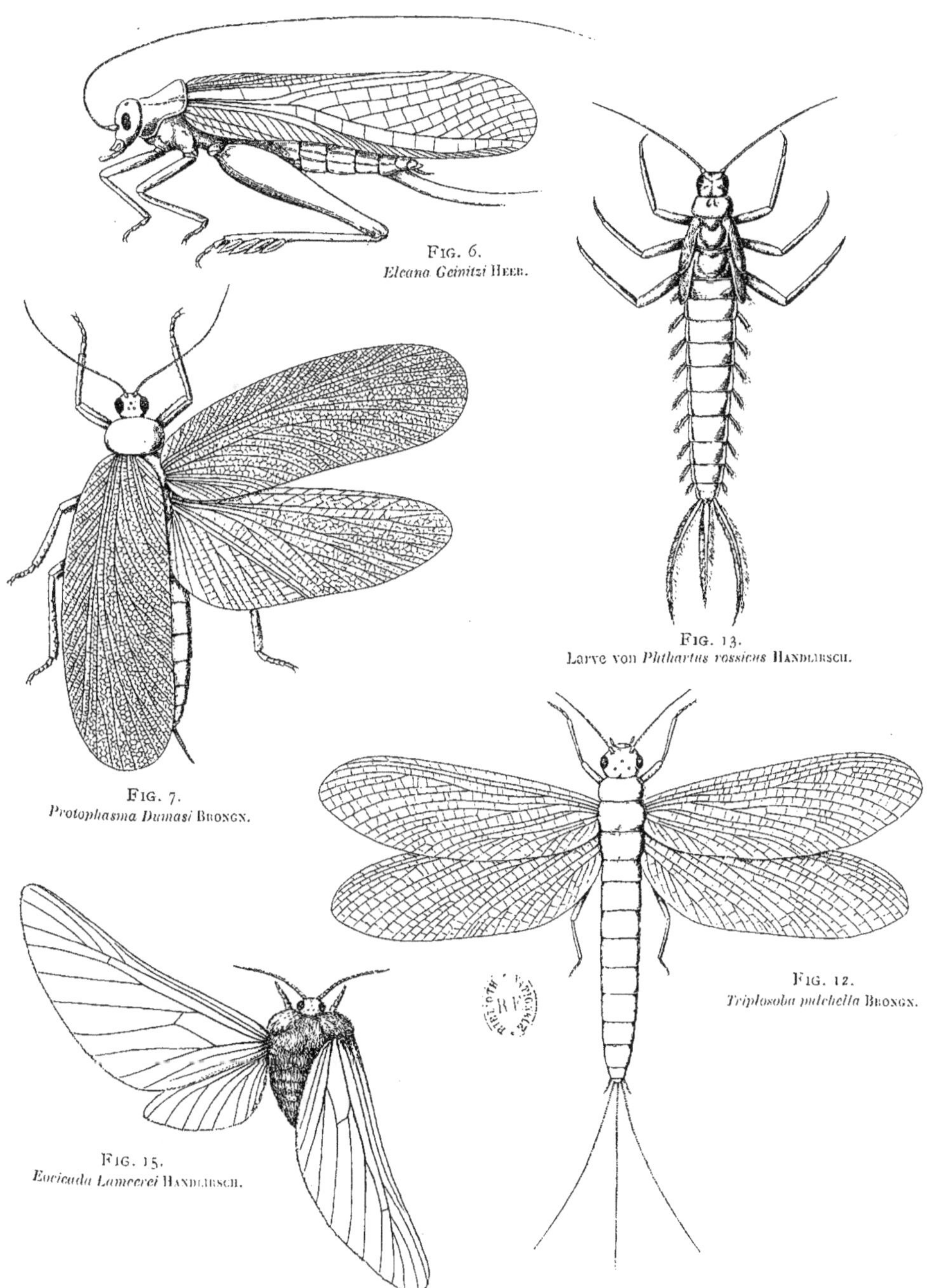

Fig. 6.
Elcana Geinitzi Heer.

Fig. 13.
Larve von *Phthartus rossicus* Handlirsch.

Fig. 7.
Protophasma Dumasi Brongn.

Fig. 12.
Triplosoba pulchella Brongn.

Fig. 15.
Eocicada Lameerei Handlirsch.

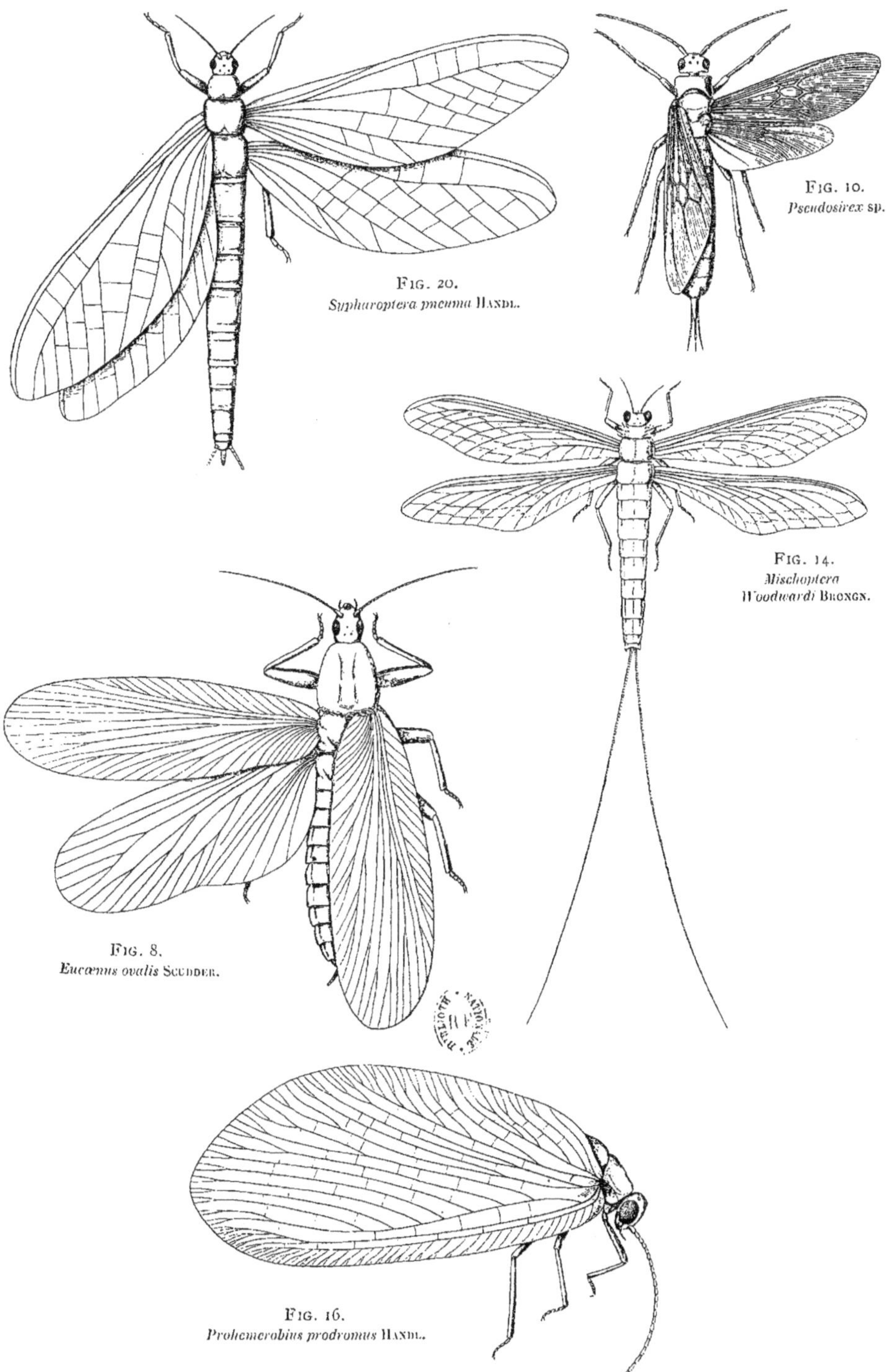

Fig. 20.
Sypharoptera pneuma Handl.

Fig. 10.
Pseudosirex sp.

Fig. 14.
Mischoptera Woodwardi Brongn.

Fig. 8.
Eucœnus ovalis Scudder.

Fig. 16.
Prohemerobius prodromus Handl.

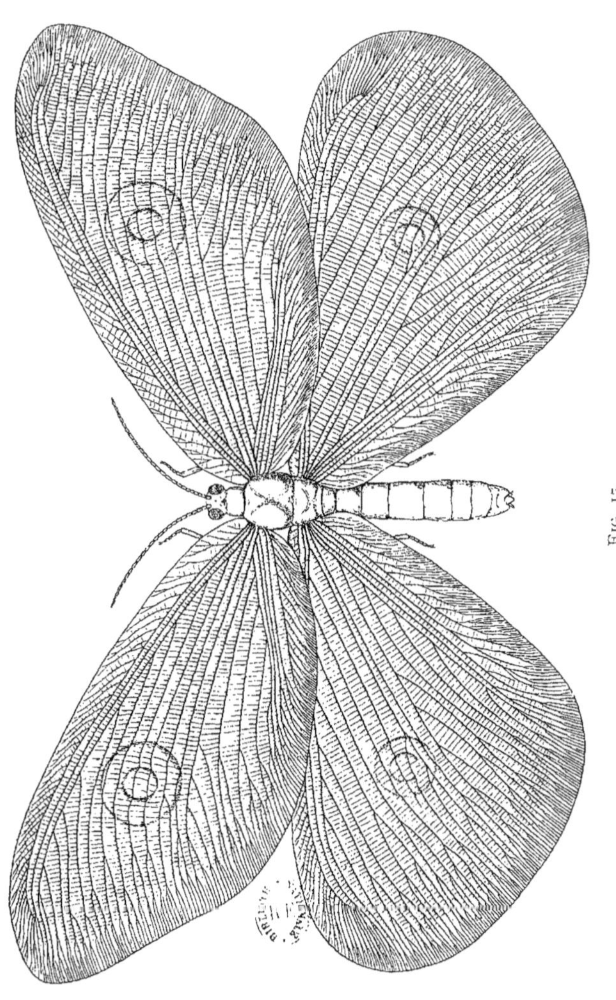

Fig. 17.
Kalligramma Haeckeli Walter.

A. HANDLIRSCH. — UEBER FOSSILE INSEKTEN. — V.

Notizie preventive e informazioni sulla « Sphenoptera lineata F. » *(geminata* ILL.) *(Coleottero buprestide)* e la larva di un Lepidottero che attaccano la Sulla *(Hedysarum coronarium* L.) della Tunisia e della Sicilia,

T. DE STEFANI PEREZ (Palerme).

A dare ragione del perchè questa mia comunicazione vien fatta nell' odierno Congresso entomologico, mi corre l'obbligo di dichiarare che non sono le poche e incomplete notizie biologiche che posso fornire sugli Insetti che gravemente danneggiano la Sulla che voglio far conoscere, ma i rapporti invece che intercedono tra questi Insetti e la coltivazione, per essi compromessa, di una preziosissima pianta che, per molti paesi, costituisce l'unica risorsa come foraggera.

L'egregio Dott. FRANCESCO TUCCI, Direttore dell' Istituto Zootecnico di Palermo, che in Italia e all' estero gode giustamente fama non solo di valente zootecnico, ma ancora di valentissimo agricoltore, fu quello che richiamò la mia attenzione sui danni da questi Insetti arrecati alla Sulla.

Egli, con lettera del 21 dicembre 1909, veniva informato e consultato dall' Ispettore di Agricoltura a Tunisi Sig. VERRY sul proposito di alcune larve che danneggiavano la Sulla nel protettorato, attaccandone il fittone.

Per essere più preciso nella storia di questo soggetto, credo miglior partito riportare integralmente la lettera del Sig. VERRY gentilmente communicatami dal Dott. TUCCI :

« Tunis, le 21 décembre 1909.

» Monsieur,

» Je me permets de vous adresser des échantillons de Sulla attaqués par un ou plusieurs Insectes qui causent actuellement les plus grands ravages sur plusieurs points de la Tunisie où ce fourrage est cultivé.

» Ces échantillons proviennent du domaine de Saint-Cyprien (Société des Fermes françaises, près de Tunis) où déjà le mal avait été signalé l'an dernier, mais sur une échelle beaucoup plus faible que cette année. La Sulla attaquée provient de semis faits l'année dernière en même temps que le blé précédent. Cette Sulla, qui avait végété dans la céréale, s'était mise à repousser depuis les dernières pluies.

» Presque aussitôt le mal s'est manifesté. Les feuilles se flétrissent et se dessèchent progressivement. Les pieds les plus faibles meurent. Les plus forts, ceux possédant un gros pivot radiculaire, ne meurent pas, mais restent languissants, n'émettant pas de belles pousses vigoureuses comme les pieds sains. Le mal se présente par taches, comme s'il s'agissait du ver blanc. Parfois dans certaines taches, souvent très étendues, pas un seul pied n'est resté sain.

» Le terrain est formé d'un sol convenant à merveille à la Sulla : terre argileuse ou argilo-calcaire, fertile et fraîche. Sur quelques points, l'égouttement du terrain serait peut-être insuffisant, mais ces points ne paraissent pas être plus éprouvés que le reste.

» Quand on arrache un pied et qu'on examine le pivot de la racine, on trouve, le plus souvent, ce dernier rongé extérieurement en présentant des sillons irréguliers qu'il faut, peut-être, attribuer au ver blanc dont on trouve de place en place quelques exemplaires paraissant cependant en trop petit nombre pour justifier le caractère assez général de cette sorte de dégât. Parfois on trouve à la surface externe du pivot de petits trous paraissant être dus au passage d'une larve.

» Si on sectionne le pivot suivant son axe vertical médian, on trouve à l'intérieur la moelle rongée et remplacée par des débris pourris et des excréments de couleur jaunâtre. C'est du collet de la racine, de haut en bas par conséquent, que le mal paraît se propager, l'extrémité du pivot étant quelquefois saine. De nombreux exemplaires examinés et ceux envoyés vous permettront, sans doute, de faire une partie de ces observations, qui m'ont permis de constater la présence :

» 1° Et c'est là le cas le plus fréquent, d'une petite larve cylindrique blanche à tête rousse;

» 2° D'une larve « fil de fer » à différents états de développement, depuis le bâtonnet jaune luisant et rigide, toujours implanté perpendiculairement à l'axe de la racine, isolé ou rarement par groupes de deux ou trois, jusqu'à la forme complète, ayant 2 centimètres de longueur.

» Il ne m'a pas été possible d'en conserver d'exemplaires;

» 3° Cette larve, beaucoup plus rare que la précédente, paraît correspondre à l'Insecte parfait, de couleur mordorée, que vous trouverez à l'intérieur de plusieurs racines et qui paraît appartenir au genre *Elater* (Taupin).

» Je serais porté à croire que l'on est en présence de deux Insectes différents, la larve n° 1° correspondant à un Insecte autre que le Taupin, probablement à un Lepidoptère. Je vous serais reconnaissant des renseignements que vous pourriez me fournir sur cette question : nom du ou des Insectes, leur biologie et moyens pratiques de lutter contre eux. A priori, il me paraît qu'il sera bien difficile de lutter, car comment atteindre ces larves à l'intérieur de la racine? Je me propose néanmoins d'essayer le sulfure de carbone.

» La question présente le plus grand intérêt pour l'agriculture tunisienne. En effet, en dehors de la perte, que je crois certaine, des fourrages de la Sulla actuellement attaqués, l'extension de cette intéressante plante, considérée comme seule fourragère, pourrait être compromise. Les agriculteurs paraissent s'être rendu compte de la nécessité de faire entrer une légumineuse dans leurs assolements. La Sulla, dans les terres où elle est spontanée, donne les plus brillants résultats. On se demande par quoi il faudrait la remplacer, si la culture en devenait impossible à cause des parasites dont il vient d'être question. C'est dire toute l'importance que les indications que vous pourrez me donner — et pour lesquelles je vous adresse à l'avance mes plus vifs remerciements — pourront avoir pour nos agriculteurs.

» Un de mes correspondants me prie de demander s'il existe en Sicile des variétés de *lupin* poussant dans les sols argileux et, dans l'affirmative, s'il serait possible de s'en procurer de la graine. Ce lupin servirait comme engrais vert dans les vignes. Je vous serais reconnaissant si vous vouliez bien me fournir quelques indications à ce sujet.

» Veuillez agréer, Monsieur, l'assurance de mes sentiments très distingués et dévoués.

» J. VERRY. »

Da questa lettera ci si può già formare il concetto di quale importanza sia la Sulla nei paesi caldi; essa è la sola che vegeta rigogliosa in terreni poco fertili, che non richiede irrigazione e che immagazzina nel terreno una grande quantità di azoto che lo fertilizza e lo rende adatto, più di qualunque concimazione artificiale, ad una buona produzione di grano.

Terreni, che poco prima dell' avvicendamento della Sulla, producevano tre o quattro sementi, hanno dato otto ed anche sedici sementi dopo.

Questi pregi della Sulla, che d'altronde, come vedremo, non sono oli, sono valsi a far grandemente modificare l'agricoltura in non poche contrade italiane e di altri paesi. Si tratta di una foraggera di primissimo ordine formendo essa un ottimo alimento verde al bestiame, un apprezzatissimo fieno allo stato secco, e quel che conta molto, non richiede quasi nessuna coltura; di fatti, basta affidarne il seme, sgusciato o no, al terreno, spargendovelo a spaglio e sinanco senza la cura d'interrarlo, per avere un discreto prato al primo anno, splendido al secondo. Questa pianta, che è spontanea in Algeria, nella Grecia, nella Spagna, nelle Isole Baleari e italiche e nella Calabria da dove si estese al nord italico sin sulle colline dell' astigiano, è una di quelle che può risolvere le sorti dell' agricoltura di un paese; così in Tunisia, dove essa è del pari spontanea, e come forse lo è in tutte le coste settentrionali africane, non domanda che di essere appena carezzata per acquistare quel valore che gl' indigeni non hanno mai saputo darle.

Ma io non devo parlare della coltivazione di essa; chi ne volesse sapere di più può consultare con vantaggio la monografia dello SBROZZI « La Sulla » scritta per la Biblioteca agraria Ottavi (1);

(1) SBROZZI, Dino-La Sulla (Biblioteca agraria Ottavi, vol. XIX), Casale, 1899.

solo dirò di più che questa leguminosa, coi suoi fiori leggiadramente scarlatti, fornisce abbondante miele all' industre ape, e infine ch'è stato mio precipuo scopo quello di far notare a Voi, valenti Colleghi, l'importanza di una pianta sotto tutti i riguardi degnissima delle maggiori cure.

Così, allorquando essa è minacciata da una malattia o da un qualunque parassita che ne attenta la vita, diviene un dovere degli studiosi prendere in essame il malanno e fare di tutto per debellarlo, e nel caso specifico degli Insetti che l'attaccano spetta oggi agli entomologi un tale compito.

Ecco perchè mi rivolgo a Voi, Onorevoli Colleghi, e vi prego di interessarvi dell' argomento, tanto più che oltre alla speranza di poter venire in aiuto ad una pianta di tanta importanza, ci si offre il destro di studiare alcuni Insetti la cui biologia, i di cui costumi sono punto noti. Lo Sbrozzi già citato dice che, pregio immenso della Sulla è quello di essere pianta rusticissima e quindi poco pensierosa del clima e non infestata mai da parassiti dannosi come quelli che rovinano altre foraggere.

Questa affermazione reggeva una volta, ora non più perchè, da quanto oggi siamo venuti a conoscere, si deve credere che gl' Insetti, dei quali ci occuperemo, sono capacissimi di annientare un sulleto sul campo.

Altri Insetti veramente dannosi la Sulla non ne ha o almeno non li conosciamo; ma è accaduto in Sicilia che i sulleti, in alcune annate, sono stati sofferentissimi senza che se ne potesse comprendere la causa; oggi tali deperimenti, invero saltuari nel tempo, dopo la conoscenza dei gravi danni che vi apportano la larva di un Coleottero e quella di un Lepidottero, possono attribuirsi ad essi : Ma sono poi veramente questi i soli Insetti a danneggiare la Sulla?

Comunque, io per ora sottopongo agli Onorevoli Congressisti quel tanto che attribuisco all' azione dannosa esercitata sopra l'importante Edisarea dai due Insetti da me osservati.

Permettetemi intanto un pò di cronaca :

Da che la Francia estese il suo protettorato alla Reggenza della Tunisia, ha fatto di tutto per migliorare le condizioni di quel paese. L'agricoltura è stata curata in modo speciale, e mi piace quì ricordare come in questo ramo non bisogna disconoscere quale valevole contributo hanno apprestato gl' italiani o meglio i siciliani, e in modo specialissimo i coloni dell' isola di Pantelleria. La Repubblica ha ben riconosciuto la nota sobrietà e la tenacia nel lavore del contadino siciliano per mezzo del quale ha potuto rendere alcune zone, dall' accidia araba lasciate in abbandono, plaghe ubertosissime.

Tra le colture molto rimunerative curate in Tunisia troviamo la Sulla la quale, nei terreni argillosi e freschi, assicura un reddito molto importante e vegeta rigogliosa.

Intanto, da qualche anno a questa parte un grave malanno è sopravvenuto alla preziosa foraggera; due robusti Insetti si insinuano nel fittone della pianta e vi apportano guasti molto gravi. Il danno è evidentissimo, e quei coloni ne sono grandemente preoccupati, anzi cominciano ad essere convinti di dover rinunziare alla coltivazione della Sulla che era per loro di grande risorsa. Questa convinzione dei coloni tunisini nasce dal fatto dello espandersi del malanno perchè, mentre pochi anni addietro esso non si era avvertito, ora, di anno in anno, è andato sempre più estendendosi su vasta scala.

Le piante invase dalle larve di questi Insetti si marcano subito per le loro foglie appassite in sul principio dell' invasione e dissecate più tardi; le più tenere muiono ben presto, quelle più robuste languiscono, cestiscono poco e non emettono più getti vigorosi come le piante sane. Il male si presenta per chiazze, spesso molte estese, così che nessuna pianta resta indenne in quel dato spazio.

Il male è grave, molto grave ed è necessità correre alla non facile ricerca di un rimedio prima che esso possa divenire più generale, prima che la disillusione possa giungere a fiaccare la buona volontà di tanti agricoltori. Aggiungo che il male, che si credeva limitato alla Tunisia, lo abbiamo anche nei sulleti siciliani; si tratta di Insetti che si trovano in diversi paesi e che un aumento di prolificazione potrebbe domani renderli dovunque nefasti.

Uno degli Insetti danneggiatori à la *Sphenoptera lineata* F. (*geminata* ILL.) specie ben nota e di cui l'Insetto adulto è di un leggiadro color marrone dorato, glabro, lucentissimo, lungo da 10-14 mill., con testa relativamente piccola, incassata nel corsaletto, finemente punteggiata come il torace; questo più grande della testa, largo quanto la base delle elitre : quest' ultime costate e come tutto il corpo, attenuantesi gradatamente verso la parte posteriore; spazio intercostale con due linee longitudinali di punti impressi; parti inferiori del corpo brevemente pubescenti. Antenne serrate, brune. Piedi brevissimi. Tegumenti del corpo molto resistenti. La larva, che ho trovato nella Sulla in contrada Principato presso Petralia

Soprana (Sicilia) in maggio, discretamente sviluppata, ma non adulta, ha nulla di particolare, essa rassomiglia ad una qualunque altra di buprestide.

La femmina di quest' Insetto, all' epoca delle prime pioggie, sorvola sui campi di Sulla, si ferma sulle incipienti piantine e tra le tenere foglie, che appena cominciano a scoppiare, depone uno o due uova in vicinanza del colletto. Dopo pochi giorni, da queste uova schiudono le larvette le quali immediatamente attaccano il fittone lateralmente al colletto e vi scavano un piccolo cunicolo sino a raggiungere il midollo; esse, una volta giunte in questa parte della pianta, vi si stabiliscono nutrendosi del midollo che corrodono avanzando sempre più in basso; colà compiono il loro accrescimento e le loro metamorfosi sino a quella di Insetto adulto, e in questo stadio vi passano l'inverno immobili e senza più nutrirsi.

Come si vede, il leggiadro e dannoso buprestide vive completamente a spese della radice della Sulla, divorandone il midollo e anche le altre parti tenere circostanti e vi scava così una larga galleria, apportando alla pianta guasti considerevoli e irreparabili.

I detriti e gli escrementi della larva vengono accumulati e pressati verso l'alto della galleria come un turacciolo; il piccolo buco di entrata resta così tappato in modo che tra l'ambiente esterno e la larva ogni comunicazione resta chiusa.

Questa disposizione, che in natura vediamo ripetersi per altri Insetti, in questo caso della *Sphenoptera lineata* ha hanche grande importanza perchè serve a preservare la larva dall' umidità, diversamente le pioggie, immettendosi nella galleria scavata dall' Insetto nel fittone, potrebbero apportare grave danno alla larva perchè difficilmente si evaporerebbero o verrebbero assorbite e quindi ristagnando far marcire la radice; questi accidenti in tal modo vengono evitati e l'Insetto, in quella specie di astuccio, vi dimora riparato dagli agenti esterni.

L'Insetto, compi tele sue metamorfosi durante i mesi invernali, resta nicchiato, allo stato perfetto, nella sua galleria e solo ne vien fuori appena si eleva la temperatura; allora risale lungo il fittone spostando il turacciolo di detriti, buca il colletto della pianta alla sua estremità cioè, tra i getti che formano il cesto, e viene fuori, probabilmente per rintracciare altre piante (leguminose?), a ripetere il suo ciclo biologico.

Devo inoltre portare a vostra conoscenza che la Sulla, al primo anno dopo la semina, non viene mai attaccata dalla *Sphenoptera*, forse perchè i giovani fittoni ancora esili non sono adatti alla vita dell' Insetto, e questo fatto porta di conseguenza che tutto lo svolgimento, dall' uovo ad Insetto perfetto, nella *Sphenoptera lineata*, deve necessariamente avvenire da settembre od ottobre a maggio e giugno cioè, in nove mesi circa. Nè può essere altrimenti perchè appunto in maggio e primi di giugno del secondo anno, la Sulla viene falciata e più tardi, in està, le sue radici distrutte con la zappa onde preparare il maggese per la prossima semina del grano.

In tal modo, l'aver trovata in maggio dentro i fittoni di Sulla qualche rara larva ancora giovane di questo buprestide, come è a me successo in Sicilia, non infirma per nulla la biologia della specie, dovendo tali larve ritenersi come ritardatarie che con la zappatura del terreno e la conseguente distruzione delle radici di Sulla, sono condannate a perire.

Se si seziona un fittone di queste piante attaccate dall' Insetto nel senso della sua lunghezza, si trova all' interno il midollo corroso e come ho detto, rimpiazzato da detriti ed escrementi di color giallastro; seguendo allora il cunicolo verso l'alto (questo cunicolo scavato dalla larva non giunge mai all'estremità più bassa del fittone) si arriva al piccolo foro di entrata, non sempre chiaramente visibile perchè gli elementi della pianta, durante lo sviluppo, sono giunti qualche volta a turarlo e farlo scomparire.

Della biologia di quest' insetto si conosceva quasi nulla, come si conosce ben poco dei buprestidi in genere, tanto più che la specie in discorso credo che sia la prima volta che viene notata come dannosa ad una pianta coltivata di gran valore.

Il Sig. VERRY nella sua lettera accenna di avere osservato altre larve attorno e dentro il fittone della Sulla, e le sue esattissime osservazioni posso confermarle alla mia volta per averle fatte anche in Sicilia; così le larve che egli indica col nome di *fil de fer* non sono che quelle di Coleotteri *Elateridæ* del genere *Agriotes*; *vers blancs*, o larve di *Scarabæidæ*, in un terreno sullato se ne trovano parecchie, ma contrariamente a quanto vediamo avvenire in un campo di grano dove le larve e l'Insetto perfetto dell' *Hybosorus arator* ILL. possono recare gravi danneggiamenti, la Sulla dalle larve di scarabeidi ha poco da temere, come non ha nulla a temere dalle larve di *Agriotes;* invece *la petite larve cylindrique blanche à tête rousse*, di cui anche parla l'egregio Sig. VERRY, è realmente da riferirsi ad un Lepidottero e concorre grandemente a danneggiare la Sulla.

Questa larva, che si trova isolata o a coppie dentro il fittone della Sulla spesso con la *Sphenoptera lineata*, tanto il Dott. TUCCI, quanto io, l'abbiamo trovato frequente in un sulleto del Barone SABATINI in contrada Principato nel mese di maggio, e le piante che l'albergavano erano in stato molto sofferente, tanto che il contadino incaricato di estirparle, già sapeva riconoscere dall' aspetto quelle che ne erano affette.

E' a credersi quindi che i sulleti sofferenti o improduttivi altre volte osservati in Sicilia, siano stati ridotti a mal punto dalle larve della *Sphenoptera* e da quelle del Lepidottero sino ad oggi specificamente non accertato.

Allo scopo appunto di conoscere la specie, nella speranza di potere ottenere l'Insetto perfetto, ho stabilito alcuni allevamenti artificiali allogando in vasi e tenendo in casa alcune radici di Sulla invase da queste larve; esse sin oggi sono vive e prospere e non dispero quindi di ottenere l'Insetto perfetto.

Come vedete, egregi Signori, si tratta di un simpatico problema entomologico-agrario da risolvere, ed io mi rivolgo a Voi, che potendovi trovare in condizioni favorevoli, vogliate contribuire alla sua soluzione.

Ecco intanto la descrizione di questa larva :

Bianca o bianco-carnicina, lunga in media un centimetro mezzo, più ingrossata verso la parte anteriore, composta apparentemente di 13 segmenti oltre la testa, nodosi, cioè ben distinti uno dall' altro per un marcatissimo restringimento. Fori tracheali piccoli e fulvicci. Piedi anteriori armati di piccole unghie fulve.

La testa, poco più piccola degli anelli toracici, è chitinosa, di color baio, con grandi macchie oculari dello stesso colore e con le parti orali nere.

Primo segmento toracico-dorsale leggermente chitinoso, di color fulviccio come la testi. Pochi e sparsi peli cenerini brevissimi si notano su tutte le parti del corpo.

E' mia opinione che l'intromissione di questa larva, nel fittone della Sulla, deve avvenire dopo quella della *Sphenoptera*, approfittando della via tracciata dal buprestide. Dico ciò perchè questa larva, nei campioni di Sulla avuti dalla Tunisia, l'ho spesso trovata in Compagnia della *Sphenoptera* in dicembre e sempre soprastante al Coleottero in stato perfetto.

Queste poche e incomplete notizie ho voluto dare a Voi, egregi Colleghi, perchè rivolgiate la vostra attenzione sopra tali Insetti dannosissimi alla più preziosa foraggera che abbia l'agricoltura di motti paesi.

Voler tentare la distruzione di queste larve, colpendole direttamente nel terreno, annidate come sono nell' interno delle radici, è cosa certamente molto difficile; forse si potrà giungere a garentire la Sulla dai loro attacchi, con misure di difesa preventiva cioè, intese a tenerli lontani da essa all'epoca in cui gli Insetti depongono le uova.

Da questo punto di vista non è difficile riuscire ad un risultato positivo; ma per raggiungerlo è necessario conoscere minutamente la biologia dei due Insetti che oggi conosciamo in parte; a tale scopo ho voluto dare la presente comunicasione, nell' odierno Congresso, perchè gli entomologi volessero prendere in considerazione questo sogetto di importanza massima per la scienza e per l'agricoltura.

La risoluzione di questo problema non darebbe solo la soddisfazione che ogni studioso prova nello strappare un velo alla natura, ma esso verrebbe inoltre coronato dalla riconoscenza di tutti gli agricoltori meridionali, di tutti gli agricoltori di quei paesi dove la Sulla costituisce una delle colture importantissime, anzi la più importante di tutte nell' avvicendamento granario.

Ueber die Einwirkung der Röntgenstrahlen auf die Entwicklung der Schmetterlinge,

von Dr. K. HASEBROEK (Hamburg).

Zusammenfassung.

I.

1. Die 3-5 Tage alten Puppen von *Van. urticæ, atalanta* und *io* reagiren nicht nachweisbar auf die Röntgenbestrahlung. Auch die Winterpuppe von *D. euphorbiæ* ist trotz der sehr häufigen Bestrahlung während des Winters nicht beeinflusst worden.

2. Die Raupen von *Van. urticæ* und *io* scheinen, abgesehen davon, dass sie bei den stark beeinflussten *Van. urticæ* etwas kleiner blieben, nicht zu leiden. Diesem gegenüber kann ein Sterben von *Char. jasius* Raupen, das beobachtet wurde, nicht auf eine Bestrahlung bezogen werden, um so weniger, als hierzu keine Kontroltiere vorhanden waren.

3. Der Vorgang der Puppenbildung an sich wird bei *Van. urticæ* und *io* nicht gestört. Eine Störung der Puppenbildung bei *Char. jasius*, welche in Platzen der Haut bestand, ist jedoch auffallend, da erstens weder mir noch anderen Züchtern ähnliche Vorgänge vorgekommen sind, zweitens die Störung mit einer Beobachtung unter Röntgenstrahlen von BORDIER an Seidenraupen sich scheinbar deckt.

4. Die Zeit der Puppenruhe wird weder bei der Bestrahlung der Raupe noch der Puppe von *Van. urticæ* und *io* beeinflusst.

5. Die Bestrahlung über das letzte Raupen- und das erste Puppenstadium bewirkt bei *Van. urticæ* tiefgreifende Veränderungen, welche einerseits für den Aufbau der Epithelialgebilde, der Schuppen und Haare, degenerativer Natur sind, andererseits in der Vermehrung der schwarzen Pigmentirung bestehen. Ob die allgemeine Vitalität des Gesammtorganismus leidet, ist nicht mit Sicherheit zu sagen, da das vorliegende Zeichen hierfür, die Abnahme des Flugvermögens, — die Falter konnten nur flatternd sich von einem erhöhten Punkt auf den Boden herablassen, — auch wohl auf die mangelhafte Schuppenbildung zurückgeführt werden kann.

6. Es ist sehr wahrscheinlich, dass für die Wirkung der Bestrahlung ein bestimmtes, kurz begrenztes Stadium der Entwicklung des Falters in Frage kommt, das getroffen werden muss. Vielleicht fällt dieses Stadium mit der Zeit des Hervorspriessens der Schuppen aus ihrer Basalanlage zusammen, welche nach den Beobachtungen SEMPER's eine sehr kurze ist.

II.

Der folgende Versuch mit *Plusia moneta* ergab ein so auffallendes Resultat hinsichtlich feinerer Veränderungen der Haargebilde, dass ich ihn für wert halte, hier im Auszuge mitgeteilt zu werden.

Vier erwachsene Raupen wurden im Holzgazekästchen von 8 × 12 × 6 cm, mit der Futterpflanze versehen, vom 21. Mai bis 28. Mai 1907 vormittags und nachmittags 4 — 15 Minuten lang im Abstand von 35 cm mit Blendenvorrichtung bestrahlt.

Die Raupen verpuppten sich in typischen Cocons resp. am 18., 19., 20., 22. Mai 1907. Die Bestrahlung traf die Tiere also dreimal im ersten Puppenstadium, einmal die noch umherlaufende Raupe. Die Falter schlüpften am 15., 16., 20. Juni 1907. Sie sind sämtlich, wenn auch in verschiedenem Grade, verändert. Ich beschreibe den am stärksten beeinflussten Falter :

Der Falter sieht wie abgeflogen aus; seine Zeichnung erscheint verwischt. Die Flügelsäume erscheinen defekt. Es besteht Transparenz der Flügel durch zu geringe Beschuppung. Die Behaarung ist spärlich, die Schöpfe am Leibe sind nicht mehr vorhanden.

1. **Oberseite :** Vorderflügel : Zeichnung nur noch an der Spitze im Winkelstrich und in den Wurzelquerlinien angedeutet. Mittel-

schatten verschwunden. Silbermakel nur noch in Spuren vorhanden. Farbenton im übrigen wie beim normalen Falter.

Hinterflügel : Einfarbig, doch ist die dunklere Randtönung nicht mehr vorhanden.

2. **Unterseite** : Vorderflügel und Hinterflügel einfarbig, Querlinien und Mittelschatten sind so gut wie ganz ausgelöscht.

Somit haben wir also im grossen und ganzen dieselbe Art der Beeinflussung durch die Bestrahlung, wie ich sie bei *V. urticæ* fand : Bei Fehlen jeden Einflusses auf Puppenbildung und Zeit der Puppenruhe Störung im Aufbau von Schuppen und Haaren.

Die 60-fache lineare Vergrösserung ergab nun aber folgende auffallende Veränderungen :

1. Die normal silberweisse Fläche der Makel ist unregelmässig mit braunen Schuppen bestreut, wie wenn ein Wirbelsturm die Schuppen der braunen Umgebung auf die Makel hinübergeweht hätte. Die Schuppen selbst sind kleiner als normal und haben die Zähnung zum Teil verloren (Fig. 1).

TEILE DES MAKELRINGES.

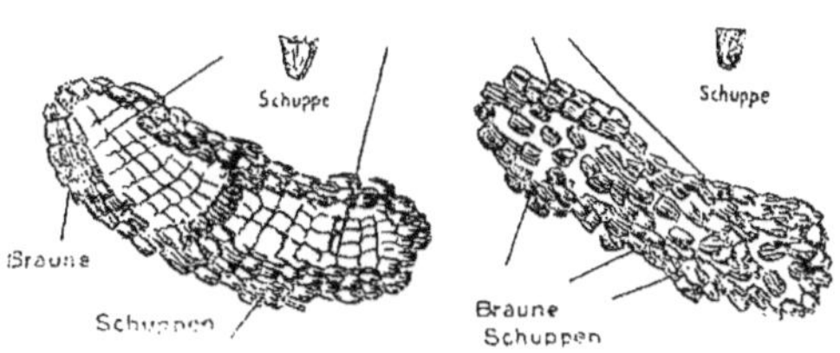

FIG. 1.
a) Normal. b) Bestrahlt.

2. Die Fransen sind an ihrem Ansatzsaum nicht mehr scharf begrenzt, spärlich, verschieden lang, schmäler und an den Enden fast ausnahmslos zweizackig, vereinzelt lanzettförmig gegenüber der meistens vierzackigen normalen Form (Fig. 2).

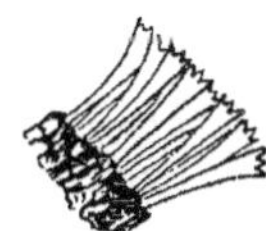

FIG. 2.
a) Normal. b) Bestrahlt.

3. Die Schopfhaare sind weniger lang, schmäler, am Ende meistens zweizackig und durchweg ohne braune Scheckung gegenüber den vierzackigen und stark gescheckten normalen Haaren (Fig. 3).

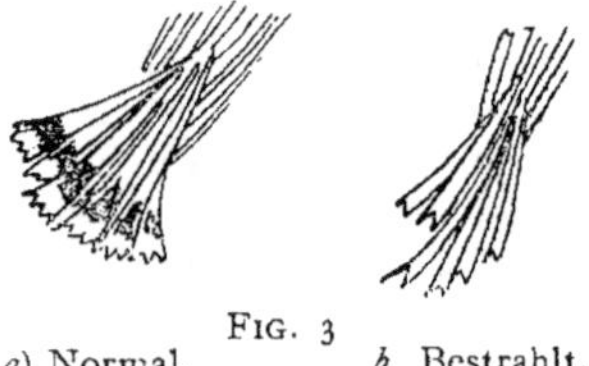

FIG. 3
a) Normal. *b* Bestrahlt.

Was also diesen Versuch ausgezeichnet, ist die durchgreifende Reduktion der Breite und der normalen 4 Zacken auf 2 und 1 Zacken an den Haarenden. Vielleicht ist dies für die mit starker Behaarung augestatteten Nachfalter charakteristisch. Weitere Versuche mit den stark behaarten Spinnern sind in dieser Hinsicht von mir in Aussicht genommen.

Fourmis et leurs hôtes,

par H. Donisthorpe (Londres) (1).

Tous les ordres dont se compose la classe des Insectes ont des représentants dans les fourmilières, de même que les Arachnides, les Isopodes et les Myriapodes. On peut approximativement partager les hôtes des fourmilières en les quatre catégories suivantes, proposées par le R. P. Wasmann :

1° Myrmécophiles proprement dits;
2° Hôtes tolérés par les Fourmis;
3° Hôtes traités en ennemis par les Fourmis;
4° Parasites.

I. — Myrmécophiles proprement dits.

Ce sont peut-être les plus intéressants de tous, en ce que, d'une part, ils sont élevés et nourris par les Fourmis, et qu'en revanche ils sécrètent une liqueur sucrée que les Fourmis sucent avec avidité. Certains Coléoptères portent en différents points de leurs téguments des mèches de poils dorés sous lesquelles se produit cette exsudation. Leur languette est large et courte, et leurs palpes labiaux sont rudimentaires, caractères parfaitement adaptés au fait d'être nourris par les Fourmis. J'en citerai quelques exemples :

Le plus caractéristique est un petit Coléoptère aveugle, qui a reçu

(1) Traduction faite par M. le cap. J. Sainte-Claire-Deville.

le nom de *Claviger* en raison de ses antennes renflées en massue. En Angleterre, il a été observé dans les nids de la petite Fourmi jaune (*Lasius flavus*), de même que dans ceux de la petite Fourmi des jardins (*Lasius niger*) et d'une race particulière de celle-ci (*Lasius alienus*). Sur la face dorsale de son abdomen se trouve une dépression où se produit la sécrétion recherchée par les Fourmis. Le *Claviger* est élevé et nourri par elles, bien qu'à l'occasion il dévore leurs nymphes. Lorsque la fourmilière est troublée, les Fourmis entraînent les *Claviger* et les mettent en sûreté dans les galeries intérieures du nid. Les *Claviger* sont parfois très abondants dans la même fourmilière : j'en ai capturé plus de soixante-dix, à l'île de Portland, dans un seul nid de *Lasius flavus*.

Passons maintenant aux Coléoptères du genre *Atemeles*. Voici côte à côte nos deux espèces britanniques, *A. paradoxus* et *A. emarginatus*, le premier beaucoup plus rare que le second. Ils ont chacun deux hôtes différents, appartenant à deux genres de Fourmis différents, un *Formica* et un *Myrmica*. Vers le mois de février, les *Atemeles* quittent les *Myrmica* pour les *Formica*, dans les nids desquels sont élevées leurs larves ; en été ou en automne, l'*Atemeles*, fraîchement éclos, abandonne les colonies des *Formica* pour celles des *Myrmica*, dans lesquelles il passe l'hiver.

L'hôte dont je vais maintenant parler est un des plus intéressants parmi les habitants des fourmilières. C'est encore un Staphylinide, le *Lomechusa*, dont, jusqu'en 1906, deux exemplaires seulement avaient été observés en Angleterre. Le premier avait été capturé en 1710, à Hampstead Heath, par Sir HANS SLOANE ; le second par le Dr. LEACH, vers 1820, entre Cheltenham et Gloucester, au cours d'un voyage en mail-coach. En mai 1906, j'eus la bonne fortune de le découvrir à Woking, en explorant les colonies du *Formica sanguinea*, grande Fourmi d'un caractère belliqueux et d'une couleur rouge-sang. Je l'y ai retrouvé depuis en nombre. Cette Fourmi est une des plus curieuses de nos espèces indigènes, en raison de ses mœurs intelligentes et du fait qu'elle pratique l'esclavage. Elle exécute des raids au détriment des colonies de *Formica fusca* et rapporte comme butin les nymphes de cette dernière ; lorsque ces nymphes éclosent, elles jouent le rôle d'esclaves et travaillent pour leurs vainqueurs.

Voici l'esclave, le *Formica fusca* ; voici maintenant le *Lomechusa* (pl. XI, fig. 1) lui-même en train de recevoir sa nourriture de son hôte le *Formica sanguinea* ; voici les palpes labiaux avortés du

Lomechusa; voici enfin la Fourmi léchant la sécrétion produite par le Staphylin au moyen des premiers tergites de son abdomen, lesquels sont pourvus de mèches de poils dorés (pl. XI, fig. 2).

Un problème des plus intéressants auquel ait donné lieu le *Lomechusa* est celui de la production, dans les nids envahis par lui depuis plusieurs années, d'une certaine proportion de « *pseudogynes* » ou fausses femelles. Ce phénomène extraordinaire a été constaté par le R. P. WASMANN. La larve de *Lomechusa* ressemble à une nymphe de Fourmi; elle a six pattes, mais elle n'en fait aucun usage, et ses mouvements imitent ceux de la nymphe de Fourmi. Non seulement les Fourmis alimentent ces larves par la bouche, mais encore elles les placent au milieu de leur propre progéniture. Les larves de *Lomechusa* en profitent pour détruire de grandes quantités de larves d'ouvrières. A la longue, ce fait est la cause d'un appauvrissement du nid en ouvrières. On sait que chez les Fourmis les reines sont le produit d'une nourriture spéciale; en présence de la pénurie d'ouvrières, les Fourmis essayent donc d'y remédier en changeant le régime d'une partie des larves qu'elles avaient destinées à faire des reines, de manière à les transformer en ouvrières. Mais leur développement est déjà trop avancé, et le résultat est d'en faire des *pseudogynes*. Voici une reine, une ouvrière et une *pseudogyne*. Cette dernière, comme vous pouvez le voir, a le thorax conformé à peu près comme la reine; elle est à la fois incapable de travailler et de combattre comme une ouvrière, et inapte à s'acquitter des fonctions de reine en pondant des œufs; c'est une non-valeur pour la fourmilière. Voilà comment peut éventuellement se produire la destruction d'une colonie. Des centres ainsi dépeuplés le *Lomechusa* passe à d'autres nids.

Les Fourmis trouvent un aliment à leur goût chez d'autres Insectes, tels que les Aphides ou Pucerons, et les poursuivent sur les plantes qu'ils fréquentent. Elles recueillent les œufs de certaines espèces et, lorsqu'ils éclosent, les transportent hors du nid et les placent sur les plantes qui leur conviennent. Plusieurs espèces d'Aphides ont des mœurs souterraines. Voici une petite espèce, *Tychœa setariæ*, qui se trouve sur les racines des plantes dans les nids du *Lasius flavus* et d'autres Fourmis. Voici une autre espèce : c'est une forme ailée, qui a été décrite par le regretté M. BUCKTON sur un individu découvert par moi-même, à Oxshott, dans une fourmilière en dôme de *Formica rufa*; le nom de cet Aphide est *Lachnus formicophilus*.

Les Coccides, si curieux par leur forme, qui rappelle celle d'une petite Tortue, sont aussi l'objet des recherches des Fourmis qui les lèchent avec avidité. Plusieurs espèces n'ont été trouvées que dans les fourmilières. Voici le *Ripersia Tomlini*, découvert par moi-même en Angleterre. Il vit sur des racines souterraines, dans les colonies du *Lasius niger* et d'autres Fourmis. Quand une fourmilière est inquiétée, les Fourmis se hâtent de mettre en sûreté les Coccides en les emportant dans leurs galeries. L'espèce que voici est le *Ripersia Donisthorpei*, décrit sur un individu découvert par moi-même, à Charing, dans un nid de *Ponera contracta*, et qui m'a été dédié par l'auteur.

Toutes les chenilles de *Lycénides* ou « Papillons bleus » sont pourvues, sur le septième segment abdominal, d'une glande que les Fourmis sollicitent en la frappant de leurs antennes jusqu'à ce qu'elle sécrète une goutte de liquide qu'elles s'empressent de lécher. Les Fourmis protègent les chenilles de Lycénides contre les Ichneumons et contre leurs autres ennemis, et tolèrent la présence de leurs chrysalides dans la fourmilière. Voici le vulgaire *Corydon* de nos collines calcaires; le cliché suivant représente sa chenille. Les glandes sont situées sur le septième segment abdominal.

Enfin, un petit Coléoptère clavicorne, l'*Amphotis marginata*, qui passe toute son existence dans les colonies de *Lasius fuliginosus*, peut être considéré comme un Myrmécophile vrai, car il est élevé et nourri par les Fourmis; toutefois on n'a jamais observé qu'il produisît une sécrétion.

II. — **Hôtes tolérés par les Fourmis.**

Les hôtes de cette catégorie comprennent un grand nombre de types très variés. Plusieurs sont simplement tolérés ou passent inaperçus grâce à leur petite taille ou à leur vivacité; d'autres sont protégés par l'épaisseur et la dureté de leur carapace ou par leurs grandes dimensions. Quelques-uns ne passent dans les fourmilières que les premières phases de leur existence et sont indépendants des Fourmis une fois parvenus à l'état parfait. Enfin, certains Insectes ressemblent absolument aux Fourmis et se trouvent ainsi protégés contre les ennemis du dehors lorsqu'ils poursuivent leur proie autour des fourmilières.

Un petit Cloporte blanc, *Platyarthrus Hoffmannseggi*, se trouve dans les colonies de toutes nos espèces de Fourmis, de même qu'un Collembole, *Beckia albina*, ce dernier parfois en grand nombre. Les Fourmis ne font aucune attention à ces deux animalcules, et tout se passe absolument comme s'ils étaient invisibles. Ce sont les balayeurs de la colonie, fonction qui leur est commune avec plusieurs animaux de la même catégorie, tels, par exemple, que les Acariens myrmécophiles. Voici un Acarien, le *Trachyuropoda Bostocki*, considéré jusqu'ici comme très rare aussi bien sur le continent qu'en Angleterre. Il vit avec *Lasius umbratus*; je l'ai pris l'an passé en nombre avec cette Fourmi à Whitsand Bay.

La toute petite espèce que voici est inféodée au *Tetramorium cæspitum*; elle a reçu le nom de *Lælaps equitans*, parce qu'elle se sert de la Fourmi comme d'une monture et grimpe sur son dos pour se faire transporter dans les nids.

Les petites espèces de Coléoptères *Histerides* sont aussi chargées de la propreté des nids. Leur carapace extrêmement épaisse et dure les met à l'abri des morsures des Fourmis. L'*Hetærius ferrugineus* vit avec *Formica sanguinea* et *fusca*, et avec d'autres Fourmis. C'est presque un Myrmécophile vrai, car les Fourmis le lèchent souvent,

Les deux Staphylins que vous voyez sont deux de nos quatre espèces ou races de *Dinarda*, qui vivent chacune avec une Fourmi différente. L'un, *Dinarda dentata*, vit avec *Formica sanguinea*; l'autre, *Dinarda Hagensi*, vit avec une Fourmi rare et localisée, *Formica exsecta*. Je l'ai découvert, il y a deux ans, à Bournemouth; il était alors nouveau pour la Grande-Bretagne. Les *Dinarda* partagent la nourriture des Fourmis, dont elles dévorent parfois les œufs; ils dévorent aussi les Acariens, de même que les cadavres d'Insectes qui peuvent se trouver dans les galeries. Lorsqu'ils sont attaqués par les Fourmis, ils se défendent en dirigeant sur elles un jet de vapeur, moyen de défense qu'ils partagent avec tous les Staphylins myrmécophiles. J'en parlerai avec plus de détails lorsque nous en viendrons aux hôtes traités en ennemis par les Fourmis.

Ces deux petites Fourmis parasites, *Formicoxenus nitidulus* et *Solenopsis fugax*, vivent à l'intérieur des fourmilières occupées par des Fourmis de plus grande taille. La première est parfois abondante dans les nids de *Formica rufa*; une de ses particularités est que le mâle, à l'encontre de ceux de la plupart des autres Fourmis, est aptère. Quant à la seconde, je l'ai observée une fois à Sandown, formant une petite colonie isolée et indépendante de toute autre Fourmi.

Une petite Araignée, *Thyreosthenius biovata*, que j'ai capturée le premier en Angleterre, à Guestling, peut se trouver dans les nids du *Formica rufa* partout où existe cette Fourmi. La Fourmi ne fait aucune attention à elle, mais l'Araignée, lorsqu'elle rencontre une Fourmi, s'enfuit rapidement d'un autre côté. Elle fait surtout sa proie des petites Mouches qui vivent dans les fourmilières.

Voici un petit Diptère nouveau pour la science, *Phyllomyza formicæ*; je l'ai élevé en nombre dans mes nids d'observation de *Formica rufa*.

L'Araignée que voici, *Tetrilus arietinus*, était aussi nouvelle pour la faune britannique. Je l'ai capturée à Oxshott, aussi bien avec *Formica rufa* qu'avec *Lasius fuliginosus*. Chez le mâle, les palpes sont extrêmement développées.

Encore une Áraignée : c'est le *Micaria scintillus*. Celle-ci, lorsqu'elle est vivante, ressemble beaucoup à une Fourmi et en reproduit les allures. Elle n'habite pas les fourmilières, mais on la trouve courant autour en compagnie du *Formica fusca*, avec lequel elle a une ressemblance superficielle.

Quelques Hémiptères ont, spécialement pendant leurs premiers états, un facies qui rappelle celui des Fourmis. Voici une forme jeune de l'*Alydus calcaratus*, lequel, à toute époque de son existence, court au milieu des Fourmis (pl. XI, fig. 3).

L'espèce que vous allez voir, *Nabis lativentris*, se prend aussi dans sa jeunesse dans le voisinage des fourmilières. La ressemblance de cette espèce avec une Fourmi tient à ce que la base de son abdomen est blanche, avec une tache médiane foncée qui imite le pédicelle d'une Fourmi. Vue de profil, sa silhouette rappelle aussi celle d'une Fourmi.

Un très petit Hémiptère, *Piezostethus formicetorum*, dont vous avez sous les yeux le mâle et la femelle, se trouve en Écosse avec *Formica rufa*. On n'en connaissait chez nous qu'une seule capture à Braemar, lorsque l'année dernière j'en ai pris un certain nombre à Rannoch.

Un beau Coléoptère, *Cetonia floricola*, est aussi chez nous une espèce propre à l'Écosse. Il ne passe que ses premiers états dans les fourmilières; à l'état parfait, on le trouve sur les fleurs ou autour des plaies d'arbres. La Cétoine femelle cherche un nid en dôme du

Formica rufa et y dépose ses œufs. La larve fait sa nourriture des matières végétales qui composent le nid; lorsqu'elle est près de se transformer, elle se construit avec de la terre et de menus débris une loge que représente la figure ci-contre. L'Insecte, au moment de son éclosion, crève sa loge, sort de la fourmilière et s'envole.

Une « Bête-à-Bon-Dieu », *Coccinella distincta*, se trouve aussi dans les dômes de *Formica rufa* ou dans leur voisinage. Elle fait sa proie des Pucerons ou Aphides, qui vivent eux-mêmes dans la fourmilière. Lorsqu'elle est attaquée par les Fourmis, elle contracte les pattes et les antennes, et fait la morte; les mandibules des Fourmis n'ont alors aucune prise sur ses élytres lisses.

Passons maintenant aux métamorphoses d'un autre Coléoptère, le *Clytra quadripunctata*, lequel passe aussi sa première existence dans les nids des *Formica rufa*. J'ai étudié sa biologie dans mon nid d'observation. La femelle cherche un arbre ou un buisson situé au-dessus d'une fourmilière de *Formica rufa*; elle laisse tomber sur le nid ses œufs, qu'elle a recouverts de ses propres excréments à l'aide de ses pattes postérieures. L'œuf ainsi habillé ressemble absolument à une petite bractée où à l'extrémité d'un chaton de bouleau. Les Fourmis s'en saisissent et le transportent à l'intérieur de la fourmilière. La larve se construit, à l'aide de ses excréments, un fourreau qu'elle agrandit au fur et à mesure de sa croissance, en faisant passer peu à peu à l'extérieur les matériaux de l'intérieur; elle vit des débris végétaux de l'intérieur du nid. Lorsqu'elle est près de se transformer, elle fixe son fourreau à une brindille ou à un autre objet. Elle se retourne alors dans le fourreau et opère sa nymphose, la tête dirigée vers le gros bout. L'Insecte, une fois éclos, brise le fourreau et s'échappe du nid. Le *Clytra*, comme vous le voyez, offre une ressemblance superficielle avec une Coccinelle, et précisément avec le *Coccinella distincta* que nous venons d'étudier tout à l'heure. Je considère cette ressemblance comme un cas de « *mimétisme mulérien* »; j'ai en effet observé que le *Clytra* est laissé de côté par tous les insectivores auxquels je l'ai offert; or on sait depuis longtemps qu'il en est de même des Coccinelles, probablement à cause de leur odeur désagréable.

Un Microlépidoptère, *Myrmecozela ochraceella*, vit également dans les nids du *Formica rufa*; chez nous on ne le trouve qu'en Écosse. Il dépose ses œufs sur les monticules élevés par la Fourmi. Sa chenille, pour se protéger des Fourmis, se construit une petite loge faite de brindilles et de feuilles liées ensemble par de la soie;

Elle se transforme en chrysalide dans cette loge. J'ai capturé le Papillon à Rannoch, au mois de juin, en agitant le dessus du nid pour le contraindre à prendre son vol.

Quant au beau Diptère que voici, le *Microdon mutabilis*, je l'ai élevé de larves recueillies, à Porlock, dans des nids de *Formica fusca*. La larve, extrêmement curieuse, ressemble à une petite Limace. Elle vit dans les galeries vides des fourmilières, et on dit que les Fournis la nourrissent comme elles nourrissent les Coccides. J'ai vu, en effet, dans mon nid d'observation les Fourmis la caresser affectueusement. Elle se transforme en pupe dans l'intérieur du nid; voici la loge nymphale. La Mouche, une fois éclose, ouvre son enveloppe et s'envole.

III. — Hôtes traités par les Fourmis en ennemis.

Nous arrivons maintenant à la catégorie des hôtes traités en ennemis et pourchassés par les Fourmis. La plupart appartiennent à différents genres de *Staphylinidæ*. Ce sont des assassins et des bêtes de proie qui sèment la mort dans les nids. Leur taille fait qu'ils ne peuvent passer inaperçus des Fourmis. J'ai vu un *Myrmedonia* attaquer une Fourmi, la saisir par le dessus du cou et la tuer. L'espèce que vous avez sous les yeux, *Myrmedonia humeralis*, vit à la fois avec *Formica rufa* et avec *Lasius fuliginosus*; il en est de même du *Quedius brevis* que je vous présente comme second exemple. Le *Lamprinus saginatus* vit avec diverses Fourmis; il s'introduit dans les nids, emporte les œufs et les dévore.

Voici maintenant un *Myrmedonia funesta* aux prises avec un *Lasius fuliginosus* (pl. XI, fig. 4), Fourmi avec laquelle il vit. A l'état vivant, ce Staphylin offre une ressemblance superficielle avec la Fourmi; tous deux sont d'un noir de jais et brillants. Quand le Coléoptère est attaqué par la Fourmi, il relève son abdomen au-dessus de son corps et en dirige l'extrémité vers la face de la Fourmi. Celle-ci recule, et le Staphylin reprend sa course. Quand j'ai forcé la Fourmi à entrer en lutte avec lui, elle lâche rapidement prise et souvent se met à courir en cercle comme si elle était folle; parfois, l'une de ses antennes reste un certain temps raidie dans une position contractée. Ces Staphylinides exhalent une odeur forte et pénétrante, très appréciable quand on les touche, et c'est la sécrétion qu'ils projettent qui les protège des Fourmis. J'ai disséqué sous le microscope les glandes qui sécrètent cette vapeur et suis en train d'en rechercher la composition chimique.

IV. — **Parasites.**

Nous arrivons à la dernière catégorie, qui est celle des parasites au sens propre du mot. Ils se divisent en Endoparasites et Ectoparasites, et comprennent, avec plusieurs petits Hyménoptères appartenant aux Proctotrypides, Chalcidides et Braconides, quelques représentants de l'ordre des Acariens.

Voici le *Pezomachus macropterus*, parasite du *Formica rufa*. Le petit Proctotrypide dont vous voyez maintenant le mâle, *Tropidopria fuliginosa*, était nouveau pour la faune britannique quand je l'ai obtenu d'éclosion, dans mes nids d'observation, de *Formica rufa* et de *Lasius fuliginosus*. J'ai aussi obtenu dans les mêmes nids un parasite de la famille des Cynipides. Il était nouveau pour la science, et le Dr KIEFFER l'a décrit sur mes exemplaires sous le nom de *Kleditoma myrmecophila*.

Voici un petit *Bracon* également nouveau pour la science et non encore décrit. J'en ai capturé des femelles à Weybridge. Elles volaient comme des oiseaux de proie au-dessus d'un nid de *Formica rufa*, s'abattant sur les Fourmis pour déposer leurs œufs sur leurs téguments. J'ai obtenu le mâle d'éclosion dans mon nid d'observation.

L'Acarien que voici appartient au genre *Antennophorus* (pl. XI, fig. 5), littéralement « porteur d'antennes », ainsi nommé parce que ses pattes frontales sont dirigées en avant et ressemblent à des antennes. L'espèce s'appelle *Antennophorus grandis*; elle était nouvelle pour la faune britannique. Je l'ai prise en nombre dans les colonies de *Lasius fuliginosus* à Wellington College, Sherwood Forest, etc. Il se tient sur le menton de la Fourmi et, lorsqu'il a besoin de nourriture, il sollicite les parties buccales de la Fourmi avec ses pattes frontales, jusqu'à ce que celle-ci laisse échapper une goutte de liquide qu'il s'empresse de sucer. Souvent la Fourmi essaie en vain de s'en débarrasser en se frottant le menton sur le sol, pendant que l'Acarien passe sur le dessus de sa tête. La figure que voici montre le parasite installé sur la tête de la Fourmi (pl. XI, fig. 6).

La très petite espèce qui suit, *Laelaps oophilus,* se tient sur les paquets d'œufs de Fourmis. Il ne se nourrit pas des œufs, mais trouve sa nourriture dans le fait que les Fourmis lèchent leurs œufs. C'est un cas de ce que l'on appelle la « syntrophie ». Il était nouveau pour la faune britannique lorsque je l'ai capturé, l'année dernière, à Porlock, sur les paquets d'œufs du *Formica fusca.*

Une autre espèce, *Cilibano comata,* était également nouvelle pour la faune britannique. Je l'ai trouvée en nombre à Whitsand Bay, fixée sur les larves de la petite Fourmi jaune (*Lasius flavus*).

FIG. 1. — *Lomechusa* alimentée par la Fourmi.

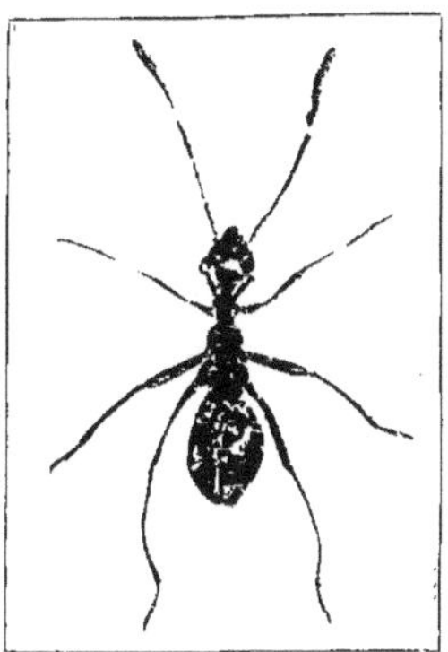

FIG. 3. — *Alydus calcaratus.*

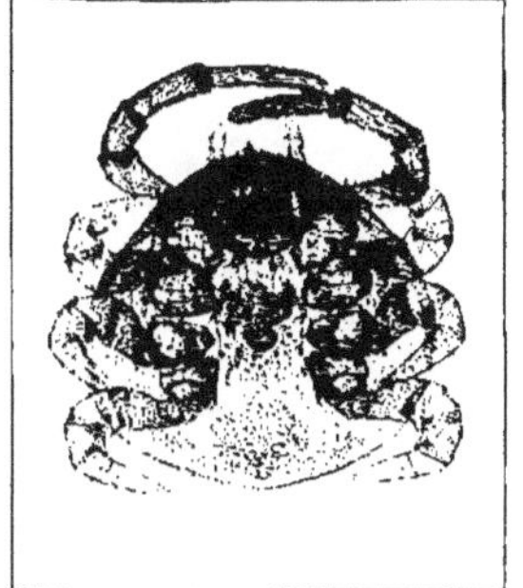

FIG. 5. — *Antennophorus grandis.*

FIG. 2. — Fourmi léchant le *Lomechusa.*

FIG. 4. — *Myrmedonia* aux prises avec la Fourmi.

FIG. 6. — *Antennophorus* et Fourmi.

H. DONISTHORPE. — FOURMIS ET LEURS HOTES.

Die Ameisen und ihre Gäste,

von P. E. WASMANN, S. J. (Luxemburg) (1).

Hundert Jahre biologischer Ameisenforschung sind verflossen, seitdem der Genfer PETER HUBER im Jahre 1810 seine klassischen « Recherches sur les mœurs des Fourmis indigènes » veröffentlichte. Wir feiern daher in diesem Jahre ein Centenarium der Ameisenbiologie, und da wollen wir zuerst einen kurzen Rückblick werfen auf die seitherige Entwicklung der Myrmekologie. Ihr Charakter ist ein echt *internationaler*, indem Forscher der verschiedensten Länder und Nationen an ihr sich beteiligten.

Die schon von LATREILLE begründete *Systematik* der Ameisen erhielt durch GUSTAV MAYR um die Mitte des vorigen Jahrhunderts einen neuen Aufschwung. Für ihren weiteren Ausbau haben sich AUGUST FOREL, CARLO EMERY, ERNEST ANDRÉ, WHEELER, RUZSKY und Andere verdient gemacht, so dass wir gegenwärtig bereits über 5000 lebende und fossile Arten und Unterarten in dieser Familie kennen. Die *Anatomie* der Ameisen ist nach den älteren Arbeiten von MEINERT, FOREL, u. s. w. namentlich durch die zahlreichen Publikationen von CHARLES JANET mächtig gefördert worden. Auch dem mikroskopischen Studium der Richtungskörperbildung im Ameisenei hat man sich neuerdings zugewandt (W. SCHLEIP). Was uns hier jedoch besonders interessiert, das ist die Entwicklung der *Bionomie*, der Kunde von der *Lebensweise* der Ameisen.

Das Werk des Vaters der biologischen Ameisenkunde, PETER

---

(1) Nur die Einleitung ist unverkürzt wiedergegeben; der an die Lichtbilder sich anschliessende Inhalt des Vortrages ist kürzer zusammengefasst.

HUBER, wurde erfolgreich weitergeführt von AUGUST FOREL in der Schweiz, von CARLO EMERY in Italien, von SIR JOHN LUBBOCK (Lord AVEBURY) und neuerdings von HORACE DONISTHORPE in England, von GOTTFRIED ADLERZ in Schweden, von ERNEST ANDRÉ und ganz besonders von CHARLES JANET in Frankreich, in Nordamerika von MC COOK und seit zehn Jahren durch zahlreiche hervorragende Arbeiten von WILLIAM MORTON WHEELER, in Russland von KARAWAIEW, u. s. w. Die Namen derjenigen Forscher — welche sich für einzelne Zweige der Ameisenbionomie besondere Verdienste erworben haben, für die Kenntnis der Beziehungen der Ameisen zu ihren Gästen, für das Studium der Gründungsweise der Ameisenkolonien und der Entwicklung des sozialen Parasitismus und der Sklaverei, für die Erforschung der Pilzgärten der Blattschneiderameisen, des Nestbaues der hochinteressanten Weberameisen, die ihre Larven als Webschiffchen benutzen, u. s. w. — sind viel zu zahlreich, als dass sie bei der Kürze der Zeit hier ausdrücklich genannt werden könnten.

Durch ihre raschen und allseitigen Fortschritte ist die moderne Ameisenkunde einerseits zu einer so *reichentwickelten* und *reichverzweigten Spezialwissenschaft* geworden, dass es dem einzelnen Forscher gar nicht mehr möglich ist, das ganze Gebiet vollständig zu beherrschen. Die Arbeitsteilung musste daher immer mehr einsetzen, besonders auch auf dem Gebiet der Myrmekophilenkunde, dessen Bearbeitung das Zusammenwirken von Spezialisten der verschiedensten Klassen und Ordnungen der Arthropoden erheischt.

Andererseits ist aber namentlich die bionomische Ameisenkunde aus dem Rahmen des spezialwissenschaftlichen Interesses *herausgetreten : die vergleichende Psychologie* hat der psychologischen Bewertung der Ameisentätigkeiten in erhöhtem Masse ihre Aufmerksamkeit zugewandt; die *Deszendenztheorie* hat unter den Ameisengästen eine Fülle interessanter Belege für die Bildung neuer Arten, Gattungen und Familien der Insekten durch Anpassung an die myrmekophile Lebensweise gefunden; sie hat zugleich auch in der hypothetischen Stammesgeschichte des sozialen Parasitismus und der Sklaverei bei den Ameisen eines der lehrreichsten Beispiele der Instinktentwicklung entdeckt; ja, die *Sozialwissenschaft* hat sogar den Versuch gemacht, in den Ameisenstaaten Vorbilder für menschliche Staatseinrichtungen zu finden. Allerdings ist dabei zu berücksichtigen, dass die Ameisen trotz der hohen Analogie, welche zwischen vielen Betätigungen ihrer sozialen Instinkte und den menschlichen Intelligenzhandlungen sich zeigt, keine Miniaturmenschen sind. Die wissenschaftliche Ameisenforschung ist von der Romantik der Vermenschlichung des Ameisenlebens längst zurückgekommen und sieht in den wunderbaren Leistungen des kleinen Ameisengehirns *instinktive* Tätigkeiten, die jedoch innerhalb gewisser Grenzen durch die Sinneserfahrung des Individuums plastisch modifizierbar sind. Sie kann daher die Ameisen ebensowenig als blosse Reflexmaschinen wie als intelligente Miniaturmenschlein gelten lassen; die Wahrheit bezüglich des Seelenlebens der Ameisen liegt vielmehr in der Mitte zwischen jenen beiden Extremen.

Wegen der Kürze der Zeit muss ich es mir leider versagen, alle diese hochinteressanten Beziehungen der Ameisenbiologie näher zu entwickeln. Ich muss mich darauf beschränken, an der Hand von *Lichtbildern* (1) Ihnen einige besonders fesselnde Hauptpunkte aus dem Leben der Ameisen und der Ameisengäste vorzuführen.

I. — Die Organisation der Ameisenstaaten.

Die einfache Ameisenkolonie stellt eine Familie im engeren oder weiteren Sinne dar. Sie umfasst nämlich eine oder mehrere Generationen von Abkömmlingen eines oder mehrerer Weibchen derselben Ameisenart. Die Stammutter ist die befruchtete Königin, welche die Kolonie gegründet hat. Die Abkömmlinge sind teils flügellose Formen des weiblichen Geschlechtes, die sogenannten Arbeiterinnen, teils junge geflügelte Männchen und Weibchen, teils auch noch junge, aber bereits befruchtete und entflügelte Königinnen. Die Arbeiterkaste kann sich wieder in verschiedene Formen gliedern, nämlich in echte Arbeiterinnen und in Soldaten, welche durch ihre mächtige Kopf- oder Kieferbildung von den Arbeiterinnen sich unterscheiden. Soldaten kommen unter unseren paläarktischen Ameisen nur bei wenigen Gattungen vor (*Colobopsis*, *Myrmecocystus*, *Pheidole*). Die Arbeiterinnen selber können wieder in grosse und kleine Individuen zerfallen, von denen erstere manchmal, z. B. bei *Camponotus*, wahre Riesen sind im Vergleich zu den letzteren; noch weit stärker entwickelt ist dieser Dimor-

(1) Von den 40 photographischen Lichtbildern des Vortrages ist nur ein Teil hier reproduziert.

phismus bei exotischen Gattungen wie *Pheidologethon*. Bei manchen Ameisenarten finden sich neben den geflügelten Weibchen, welche erst nach der Paarung ihre Flügel ablegen, auch flügellose echte Weibchen, sogenannte ergatoide Königinnen. Ein typisches Beispiel hierfür, das schon PETER HUBER bekannt war, bietet die Amazonenameise (*Polyergus rufescens*) (vgl Fig. 8, a). Bei den tropischen Wander- und Treiberameisen (*Eciton* und *Dorylus*) kommen sogar blos flügellose Weibchen, und zwar von relativ riesiger Grösse, vor. Bei manchen Ameisenarten findet sich selbst ein mannigfaltiger Pleomorphismus der Weibchen, der in verschiedenen Uebergängen zwischen Weibchen und Arbeiterin zum Ausdruck kommt. Viel seltener sind flügellose, und dann meist arbeiterähnliche (ergatoide), Männchen; sie sind von wenigen Ameisenarten bekannt, und zwar entweder neben den normalen geflügelten Männchen oder als einzige männliche Form. Ein Beispiel letzterer Art zeigt uns die glänzende Gastameise *Formicoxenus nitidulus* (Fig. 1), wo die Männchen wegen ihrer hohen Arbeiterähnlichkeit acht und dreissig Jahre lang unbemerkt blieben, bis ADLERZ sie 1884 entdeckte.

Bei vielen Ameisenarten kann man mehrere Königinnen zugleich in einer Kolonie finden. Bei unserer Waldameise *Formica rufa* kann ihre Zahl sogar gegen hundert in einem Neste erreichen. Ferner kann eine Ameisenkolonie mehrere Nester besitzen, die gleichzeitig oder abwechselnd bewohnt werden. Sogenannte Saisonnester, die je nach den Jahreszeiten wechseln, sind z. B. bei *Formica sanguinea* und bei *Prenolepis longicornis* beobachtet worden. Durch die Mehrzahl der Königinnen in einer Kolonie unterscheiden sich die Ameisenstaaten auffallend von den Staaten der Honigbiene. Letztere tragen vergleichsweise mehr einen monarchischen, erstere einen republikanischen Charakter, zumal die Königin bei den Ameisen in viel geringerem Grade den Mittelpunkt der instinktiven Tätigkeiten der Arbeiterinnen bildet als bei den Bienen. Die grössere individuelle Selbständigkeit der Ameisenarbeiterinnen im Vergleich zu jenen der Bienen beruht wohl grossenteils auf ihrer längeren Lebensdauer, welche bei den *Formica*-Arten meist drei Jahre beträgt. Die Lebensdauer der Königinnen kann sogar zwölf Jahre überschreiten.

Der Polymorphismus bildet zwar die organische Grundlage der Ameisenstaaten, indem durch ihn die Arbeitsteilung zwischen den Mitgliedern derselben Kolonie bedingt wird. Aber die Auslösung der im Ei schlummernden organisch-psychischen Anlagen erfolgt durch die Brutpflege von Seite der Arbeiterinnen. Was ein Männchen oder ein Weibchen werden soll, scheint, ähnlich wie bei den Bienen, schon im Ei vorherbestimmt zu sein (1); aber auf die Differenzierung der verschiedenen Formen des weiblichen Geschlechtes übt die Brutpflege einen bestimmenden Einfluss. Aus den befruchteten Eiern einer und derselben *Formica*-Königin können entweder geflügelte Weibchen oder flügellose Arbeiterinnen oder Zwischenformen erzogen werden. Namentlich die als Pseudogynen bezeichneten Mischformen, die wir als eine Wirkung der *Lomechusa*-Pflege kennen lernen werden, bieten hierfür einen hohen Wahrscheinlichkeitsbeweis.

2. — Sozialer Parasitismus und Sklaverei bei den Ameisen.

Gehört die Bewohnerschaft eines Ameisennestes einer einzigen Ameisenart an, so nennt man sie eine *einfache* (ungemischte) *Kolonie*. Setzt sie sich dagegen aus verschiedenen Ameisenarten zusammen, so spricht man nach FOREL's Vorgang entweder von *zusammengesetzten Nestern* oder von *gemischten Kolonien* : in ersteren leben die fremden Arten nur räumlich nebeneinander, in letzteren verbinden sie sich zu einer Haushaltung mit gemeinschaftlicher Brutpflege. Dies sind die zwei Unterabteilungen, in welche die *soziale Symbiose* zwischen Ameisen verschiedener Arten sich gliedert. Wir wollen hier blos auf die gemischten Kolonien unser Augenmerk richten, und zwar mit besonderer Berücksichtigung der Entwicklung *des sozialen Parasitismus* und der *Sklaverei*. Es ist dies eines der interessantesten Kapitel in der phylogenetischen Entwicklung der Instinkte im Tierreich.

Schon PETER HUBER hatte entdeckt, dass in den Kolonien der blutroten Raubameisen (*Formica sanguinea*) und der Amazonenameisen (*Polyergus rufescens*) neben der Herrenart noch die Arbeiterinnen einer Sklavenart leben, welche von ersterer aus den Sklavennestern als Puppen geraubt und dann zu Hilfsameisen erzogen werden; das sind also sklavenhaltende (dulotische) Kolonien. Aber es gibt noch andere gemischte Kolonien, in denen die

(1) Dass aus parthenogenetischen Eiern nicht blos Männchen sondern auch Arbeiterinnen entstehen (REICHENBACH, etc., bei *Lasius*), ist jedenfalls eine seltene Ausnahmeerscheinung.

Hilfsameisen nicht durch Sklavenraub in die Gesellschaft der Herrenart gelangen; hier spricht man von sozialem Parasitismus der letzteren. Ueber den Ursprung der Sklaverei sind seit CHARLES DARWIN (1859) verschiedene Hypothesen aufgestellt worden. Indem man in neuerer Zeit die Erscheinungen des sozialen Parasitismus zum Vergleiche heranzog, ist über jenes interessante Problem wenigstens einiges Licht verbreitet worden. Darüber sind alle Forscher einig, dass die Stammesgeschichte des sozialen Parasitismus und der Sklaverei bei den Ameisen keine einfache Entwicklungreihe bildet, sondern eine Anzahl verschiedener, von einander unabhängiger Parallelreihen; aber über die näheren Beziehungen zwischen sozialem Parasitismus und Sklaverei gehen die Ansichten auseinander. Sie werden es deshalb entschuldigen, wenn ich hier nur meinen eigenen Gedankengang kurz verfolge, wie ich ihn namentlich im « Biologischen Centralblatt » 1909 (1) näher begründet habe.

Beginnen wir mit der *abhängigen Koloniegründung* und ihrer Beziehung zum sozialen Parasitismus und der Sklaverei bei *Formica*. Als ursprüngliche Form der Koloniegründung ist auch hier — wie überhaupt bei den Ameisen — die *selbständige* Koloniegründung anzusehen, wie wir sie z. B. in der *fusca*- Gruppe heute noch vorfinden. Hier vermögen die Weibchen (Fig. 5, *a*) nach dem Paarungsfluge ihre neuen Kolonien selbständig, d. h. ohne Hilfe von Arbeiterinnen, zu gründen. Wie ist von hier aus der soziale Parasitismus und die Sklaverei entstanden? Den ersten Schritt bildete wahrscheinlich der Uebergang der Arbeiterinnen zur *acervicolen Lebensweise*, durch welche die Kolonien individuenreicher wurden und einen grösseren Nestbezirk beherrschen konnten, wie wir es in der *rufa*-Gruppe sehen. Hiedurch war aber den Weibchen die Gelegenheit geboten, ihre neuen Niederlassungen mit Hilfe von Arbeiterinnen der eigenen Art zu gründen. Als zweiter Schritt in der parasitischen und der dulotischen Richtung folgte dann eben hierdurch bei den Weibchen der *rufa*-Gruppe, dass sie die selbständige Koloniegründung aufgaben und auf die Mithilfe von Arbeiterinnen bei derselben angewiesen waren, also zur *abhängigen Koloniegründung* übergingen. Es ist jedenfalls eine merkwürdige Erscheinung, dass alle parasitischen *Formica*-Formen der

---

(1) Ueber den Ursprung des sozialen Parasitismus, der Sklaverei und der Myrmekophilie bei den Ameisen.

alten wie der neuen Welt acervicol sind und zur *rufa*-Gruppe gehören oder in nächster Verwandtschaft zu ihr stehen; letzteres gilt auch für die dulotische *sanguinea*-Gruppe, welche durch morphologische Uebergänge mit der *rufa*-Gruppe verbunden ist. In der *rufa*-Gruppe haben wir ferner biologische Uebergänge vom fakultativen sozialen Parasitismus zum obligatorischen. *F. rufa* (Fig. 2.) und *pratensis* gründen ihre Kolonien meist mit Hilfe von Arbeiterinnen der eigenen Art, nur fakultativ mit fremden Hilfsameisen (*F. fusca*, Fig. 4, *b*). Bei *F. truncicola* (Fig. 3) und *exsecta* (Fig. 4, *a* und 5, *b*) in Europa sowie bei *F. consocians* und einer Reihe anderer, von WHEELER entdeckter nordamerikanischer Formen ist die letztere Koloniegründung bereits obligatorisch. Bei mehreren von WHEELER beschriebenen parasitischen *Formica*-Formen Nordamerika's sowie bei unserer *F. exsecta* (Fig. 5, *b*) fällt die Kleinheit der Weibchen auf, die bereits einen Fortschritt in der parasitischen Anpassungsrichtung darstellt, während z. B. bei *F. rufa*, *pratensis* und *exsectoides* die Weibchen sehr gross sind und noch keine Spur einer parasitischen Umbildung zeigen. Alle die ebenerwähnten *Formica*-Formen, deren Weibchen ihre Kolonien fakultativ oder obligatorisch mit den Arbeiterinnen einer fremden Hilfsameisenart gründen, bilden nur *zeitweilig* (*temporär*) *gemischte* Kolonien. Nach dem Aussterben ihrer ursprünglichen Hilfsameisen — ca drei Jahre nach der Gründung — werden diese Kolonien wieder *einfache*, ungemischte Ameisenkolonien und können als solche noch Jahrzehnte lang weiterwachsen. *Dauernd* gemischte Kolonien finden wir dagegen bei den *dulotischen Formica*-Formen, welche nach dem Aussterben der ursprünglichen Hilfsameisen sich durch Puppenraub neue verschaffen. Wie schliessen sich diese Formen an die vorhergehenden, und speziell an ein *rufa*-ähnliches Anfangsstadium des sozialen Parasitismus, an?

Wenn bei einer grossen, starken acervicolen *Formica*-Art, welche bereits von der selbständigen zur abhängigen Koloniegründung übergegangen war, ein Wechsel in der *Ernährungsweise* der Arbeiterinnen, durch klimatische Aenderungen veranlasst (1), eintritt, so dass sie immer ausschliesslicher von Insektenraub, und zwar besonders von fremden Ameisenpuppen, sich ernähren, dann

---

(1) Wie der Uebergang des Waldklimas in ein Steppenklima diese Veranlassung bieten kann, habe ich speziell für *Formica sanguinea* gezeigt.

ist die Grundlage geboten *für die Entstehung der Sklaverei;* denn dafür, dass unter den geraubten fremden Puppen diejenigen der Hilfsameisenart *aufgezogen* werden, ist bereits durch die abhängige Koloniegründung gesorgt. Selbst *Formica truncicola* und *exsecta*, welche in freier Natur keine Sklavenräuber sind, behalten in ihren bereits wieder einfach gewordenen Kolonien die Neigung bei, die Arbeiterpuppen ihrer ehemaligen Hilfsameisenart zu erziehen, wenn man ihnen in Beobachtungsnestern solche gibt, während sie die Puppen anderer fremder Arten auffressen oder wenigstens die aus denselben entwickelten Arbeiterinnen töten. *Die Entstehung der Sklaverei* bei einer *Formica* Form wie *sanguinea* (Fig. 6), beruht also auf *zwei Faktoren : 1. auf der abhängigen Koloniegründung ihrer Weibchen mit Hilfe einer fremden Ameisenart; 2. auf der Neigung der Arbeiterinnen, fremde Puppen als Beute zu rauben.* Nach dieser Ansicht ist somit der hypothetische Ursprung der Sklavenzucht in die Gattung *Formica* auf eine gemeinschaftliche Wurzel mit der Entstehung des sozialen Parasitismus in derselben Gattung zurückzuführen, nämlich auf ein Anfangsstadium der abhängigen Koloniegründung, das dem heutigen *rufa*-Stadium einigermassen vergleichbar ist. Jedenfalls genügt die Neigung zum Puppenraub *allein* nicht, um die Entstehung der Dulosis bei *Formica* oder bei irgend einer anderen Ameisengattung zu erklären; denn es gibt viele Ameisenarten, namentlich in der Unterfamilie der Dorylinen, welche fremde Ameisenpuppen als Beute rauben, und doch keine Sklavenzüchter sind, weil eben der erstere der beiden obigen Faktoren, die von einer fremden Hilfsameisenart abhängige Koloniegründung, fehlt. Die Hauptfrage bei der Erklärung der Sklaverei ist nicht : Warum *raubt* die betreffende Ameisenart fremde Puppen als Beute? sondern : Warum *erzieht* sie *Hilfsameisen* aus denselben? Diese zweite Frage bleibt auch ungelöst, wenn man (mit Emery und Viehmeyer) die Dulosis direkt aus einem « primitiven Raubweibchenstadium » abzuleiten versucht, für welches uns überdies alle tatsächlichen Anhaltspunkte fehlen; denn, wie Emery selbst zuerst gezeigt hat, sind die heutigen parasitischen und dulotischen Ameisen von der Gattung ihrer heutigen Hilfsameisen stammesgeschichtlich herzuleiten; dort aber finden wir *nirgendwo* solche primitive Raubweibchen, wohl aber mannigfaltige Abhängigkeitsverhältnisse in der Koloniegründung einer Art von derjenigen einer anderen verwandten Art. Es sei übrigens ausdrücklich bemerkt, dass der hypothetische Ursprung der Dulosis bei *Formica* nicht schlechthin auf andere dulotische Ameisengattungen z. B. unter den Myrmicinen ausgedehnt werden darf. Auch andere Wechselbeziehungen als jene des fakultativen sozialen Parasitismus können dort zur Entstehung der Sklavenzucht geführt haben (*Harpagoxenus-Leptothorax*). Durch das zufällige Ueberleben geraubter Ameisenpuppen in einem fremden Neste (CH. DARWIN) lässt sich der Ursprung der Sklaverei jedoch nicht erklären.

Wahrscheinlich von einem *sanguinea*-ähnlichen Stadium ausgehend, jedenfalls aber mit der Entwicklung der Dulosis in der Gattung *Formica* stammesgeschichtlich zusammenhängend, stellt die Gattung *Polyergus* den Kulminationspunkt des Sklavereiinstinktes in der Unterfamilie der *Camponotini* dar. Vergleichen wir die Kieferbildung der europäischen Amazonenameise *Polyergus rufescens* mit derjenigen unserer *Formica sanguinea* (Fig. 7), so zeigt sich eine auffallende Verschiedenheit. *Formica sanguinea* (Fig. 7, *a*) hat normale dreieckige Oberkiefer mit einem gezähnten Innenrand (Kaurand), *Polyergus* dagegen (Fig. 7, *b*) schmale, scharfspitzige Sichelkiefer. In diesen morphologischen Unterschieden spricht sich auch die Verschiedenheit des Sklavereiinstinktes beider aus : *Formica sanguinea* steht auf einer primitiven Stufe der Entwicklung jenes Instinktes, und derselbe ist sogar bei den verschiedenen nordamerikanischen Rassen dieser Art wiederum verschieden weit entwickelt, wie namentlich WHEELER gezeigt hat. *Sanguinea* hält relativ wenig Sklaven, kann dieselben sogar ganz entbehren und ist nicht von ihnen abhängig. Die Amazonenameise dagegen steht in ihrer europäischen und in ihren nordamerikanischen Rassen auf dem Höhepunkt der Dulosis, lebt nur noch vom Sklavenraub und entfaltet hierbei das glänzendste Kriegertalent, das wir im Tierreiche kennen. Ihre Kiefer sind einseitig zu Mordwaffen ausgebildet und zu den häuslichen Arbeiten ungeeignet; ja sie hat sogar den Instinkt der selbständigen Nahrungsaufnahme verloren und muss aus dem Munde ihrer Sklaven gefüttert werden : die Ueberentwicklung der Dulosis ist bereits mit deutlichen Merkmalen der parasitischen Degeneration verbunden. Ihre Kieferbildung bringt beides zum Ausdruck, die Licht- und die Schattenseiten ihrer organisch-psychischen Entwicklung. Das häufige Auftreten einer flügellosen Weibchenform, sogenannter ergatoider Königinnen

(Fig. 8, *a*), deutet ferner auf den Beginn der Inzucht hin, obwohl die normale Weibchenform bei ihr immerhin noch zahlreicher ist als die ergatoide.

Mit der Gattung *Polyergus* ist die Entwicklung der Dulosis in der Unterfamilie der *Camponotinen* abgeschlossen. Wenden wir uns deshalb jetzt zu den *Myrmicinen*. Eine ganz isolierte Stellung nimmt die europäisch-nordamerikanische Gattung *Harpagoxenus* (*Tomognathus*) ein. Sie ist sehr wahrscheinlich von ihrer Hilfsameisengattung *Leptothorax* stammesgeschichtlich abzuleiten; die Männchen sind nur schwer zu unterscheiden von denjenigen der Hilfsameise. Der europäische *Harpagoxenus sublævis* (Fig. 9, *a*), der früher für hochnordisch galt, ist von VIEHMEYER auch in Sachsen gefunden worden; er entdeckte hier neben der schon bekannten ergatoiden Weibchenform (Fig. 9, *a*), welche in Skandinavien die einzige zu sein scheint, auch noch geflügelte, normale Weibchen. Wahrscheinlich ist *Harpagoxenus* ursprünglich entstanden durch eine Mutation der Weibchenformen in einer zu *Leptothorax* gehörigen Stammart. Hiermit ist nicht ausgeschlossen, dass sie später mit ihren heutigen Hilfsameisen in zusammengesetzten Nestern lebte (WASMANN und VIEHMEYER), bevor sie zur Dulosis gelangte.

Einen anderen Stamm der Entwicklung des Sklavereiinstinktes unter den Myrmicinen bildet die Gattung *Strongylognathus*, die wahrscheinlich von ihrer Hilfsameisengattung *Tetramorium* herzuleiten ist. Die südlichen Arten dieser mediterranen Gattung sind noch kräftige und volkreiche Sklavenräuber, die sich die Puppen ihrer Hilfsameisenart (*Tetramorium cæspitum*) mit Gewalt zu verschaffen vermögen. Die nördliche Art, *Strongylognathus testaceus* (Fig. 10), die in Mitteleuropa bis Holland vorkommt, ist dagegen zum permanenten sozialen Parasitismus übergegangen, indem ihre Kolonien ausser den Arbeiterinnen auch ein Weibchen von *Tetramorium* beherbergen, das ihnen die neuen Hilfsameisen liefert. Die Arbeiterinnen von *Strongylognathus testaceus* unternehmen keine Sklavenjagden mehr; sie sind auch zu klein und zu wenig zahlreich dazu. Wahrscheinlich waren es die nordischen Klimaverhältnisse, welche hier den äusseren Anlass boten für den Uebergang von der Dulosis zum permanenten sozialen Schmarotzertum; denn, wenn eine südliche sklavenraubende Art nach Norden vordringt, werden ihre Sklavenjagden, deren Ausübung an ein bestimmtes Temperaturoptimum gebunden ist, immer seltener erfolgen und schliesslich ganz aufhören. Bei *Strongylo-*

gnathus testaceus ist, in Verbindung hiermit, auch die Körpergrösse und die Individuenzahl der Arbeiterinnen bedeutend herabgesunken, lauter Anzeichen einer parasitischen Degeneration der Art. Die Säbelkiefer dieser kleinen Ameisen sind gleichsam nur noch ein stammesgeschichtliches Andenken an ihre glänzende dulotische Vergangenheit. Die Vorgeschichte der Gattung *Strongylognathus* bis zu jenem Stadium, auf welchem die südlichen sklavenraubenden Arten heute noch stehen, ist übrigens ebenso unbekannt wie die künftige Weiterentwicklung des sozialen Parasitismus bei der nördlichen Art; beide können wir nur vermutungsweise ergänzen, erstere durch den Vergleich mit *Polyergus* unter den Camponotinen, letztere durch den Vergleich mit den arbeiterlosen Schmarotzerameisen, zu denen wir jetzt übergehen.

Wenn bei einer ehemals dulotischen Art wie *Strongylognathus testaceus* die parasitische Degeneration weiter fortschreitet, so wird endlich die eigene Arbeiterform ganz aussterben und durch jene der Hilfsameisen ersetzt werden, so dass die ehemalige Herrenart nur noch aus Männchen und Weibchen besteht. Wir kennen eine ganze Anzahl solcher arbeiterloser Schmarotzerameisen aus dem paläarktischen und nearktischen Gebiet, neuerdings auch eine aus Ostindien (*Wheeleriella Wroughtoni* For.). Eine der paläarktischen Arten, *Wheeleriella Santschii* (Fig. 11), ist in Tunesien von SANTSCHI in den Kolonien von *Monomorium Salomonis* entdeckt worden und besitzt, wie auch die nordamerikanischen Gattungen *Epoecus*, *Sympheidole* und *Epipheidole*, geflügelte, noch ziemlich normale Geschlechtstiere. Auf der tiefsten Stufe der Degeneration steht dagegen unsere kleine bei *Tetramorium* lebende schwarze Schmarotzerameise *Anergates atratulus*, deren Männchen (Fig. 12) puppenähnlich sind, und deren befruchtete Weibchen, um den Untergang der Art durch ihre Fruchtbarkeit aufzuhalten, eine riesige Physogastrie besitzen. Aber für keine dieser arbeiterlosen Schmarotzerameisen können wir mit Bestimmtheit nachweisen, dass ihr Parasitismus aus einer ehemaligen Dulosis hervorgegangen ist. Es gibt nämlich noch drei andere Wege, welche theoretisch zu demselben Ziele führen : die Weiterentwicklung eines ehemaligen temporären Parasitismus, die parasitische Degeneration eines ehemaligen Gastverhältnisses und endlich das relativ plötzliche (mutationsartige) Auftreten eines neuen Dimorphismus der weiblichen (und später auch der männlichen) Form in der ehemaligen Stammart und jetztigen Hilfsameisenart. In jenen Fällen, wo z. B. wie bei *Sym-*

pheidole, *Epipheidole* und *Epixenus* die parasitische Gattung den Geschlechtstieren der Wirtsgattung sehr ähnlich ist, dürfte die letztere Erklärung sogar die wahrscheinlichste sein. Dies wird auch durch die Entdeckung einer neuen parasitischen *Pheidole*-Art, *Ph. symbiotica*, in Portugal bestätigt, deren Männchen und ergatoide Weibchen in den Nestern von *Pheidole pallidula* leben. In letzteren Fällen haben wir vermutlich eine relativ rasche Entstehung der arbeiterlosen Schmarotzerart anzunehmen, da dieselbe wahrscheinlich niemals (seit ihrer Abtrennung von der Stammart) eine eigene Arbeiterform besass, also auch keiner Zeit bedurfte, um dieselbe. zu « verlieren ». In anderen Fällen dagegen, wo die Schmarotzerameise morphologisch sehr weit von ihrer jetztigen Hilfsameisenart und mutmasslichen ehemaligen Stammart abweicht, war wohl ein längerer Entwicklungsgang erforderlich, der mit einem wirklichen Aussterben der eigenen Arbeiterform verbunden war. Dies gilt z. B. für *Anergates atratulus*, für welche Gattung wir nur vermutungsweise *Tetramorium* als Stammform annehmen können, und für welche es gar nicht so unwahrscheinlich ist, dass sie aus einer ehemals dulotischen Form durch Vermittlung eines dem heutigen *Strongylognathus testaceus* ähnlichen Stadiums hervorging, zumal auch *Strongylognathus* bei *Tetramorium* lebt und von dieser nämlichen Gattung abzuleiten ist. Aber mehr als Vermutungen stehen uns für die Stammesgeschichte von *Anergates* einstweilen nicht zu Gebote.

Die bisherigen Forschungen über die hypothetische Stammesgeschichte des sozialen Parasitismus und der Sklaverei haben jedenfalls zu der Erkenntnis geführt, dass dieselbe nur eine *ideale Einheit* bildet, in Wirklichkeit aber *aus einer Menge real verschiedener Entwicklungsreihen sich zusammensetzt*, die bei verschiedenen Gattungen und Arten in verschiedenen Unterfamilien des Ameisenstammes zu verschiedenen Zeiten begonnen haben und bis heute verschieden weit fortgeschritten sind. Je mehr es gelingt, mittelst neuer Beobachtungen und Versuche diese *einzelnen* Stammesreihen aufzuhellen, desto mehr werden wir auch fortschreiten in der *allgemeinen* Erkenntnis des phylogenetischen Zusammenhanges zwischen Parasitismus und Sklaverei bei den Ameisen. Ebenso wie auf morphologisch-paläontologischen Gebiete, so ist auch hier eine wahre Bereicherung unseres Wissens nicht von theoretischen Reflexionen sondern von kritischen Einzeluntersuchungen zu erwarten.

3. — Das echte Gastverhältnis (Symphilie).

Während das Zusammenleben von Ameisen verschiedener Arten unter den Begriff der *sozialen Symbiose* fällt, bezeichnet man als *individuelle Symbiose* ihre Vergesellschaftung mit nichtsozialen Tieren, insbesondere mit anderen Arthropoden. Wir haben uns also hier mit den sogenannten *Ameisengästen* aus anderen Familien und Ordnungen der Insekten und der übrigen Gliederfüsser zu befassen. Die Zahl der *gesetzmässigen Myrmekophilen* betrug schon 1894 gegen 1200; heute dürfen wir sie auf circa 2000 veranschlagen. Ihre Beziehungen zu den Ameisen sind sehr mannigfaltige und lassen sich in fünf Hauptklassen einteilen, die jedoch durch vielfache Uebergänge verbunden sind. Wir unterscheiden *Symphilen* oder echte Gäste, *Synöken* oder indifferent geduldete Einmieter, *Synechthren* oder feindlich verfolgte Einmieter, ferner *Parasiten* (Ento- und Ectoparasiten) und endlich *Trophobionten* oder Nutztiere der Ameisen. Ich werde hier nur auf die erste dieser Klassen kurz eingehen, weil wir manche interessante myrmekophile Staphyliniden noch später bei den Wanderameisen kennen lernen werden, und namentlich auch deshalb, weil Herr DONISTHORPE einen eigenen Vortrag über die einheimischen Ameisengäste halten wird.

Die *echten Ameisengäste* (Symphilen) werden von den Ameisen gastlich gepflegt wegen bestimmter *Exsudate,* welche flüchtige Produkte des Fettgewebes (bei den Lomechusini) oder eines adipoiden Drüsengewebes (Clavigerinæ, Paussiden, etc.) darstellen, während bei den physogastren Termitengästen das Blutgewebe das hauptsächliche Exsudatgewebe ist. Sehr mannigfaltig entwickelt sind die äusseren Exsudatorgane, gelbe Haarbüschel, Hautporen, Hautgruben, etc., an denen die Gäste von ihren Wirten beleckt werden. Ein Nahrungsmittel — wie z. B. die zuckerhaltigen Ausscheidungen der Aphiden — scheinen die Exsudate der echten Gäste für die Ameisen nicht zu sein, sondern nur ein angenehmes Reizmittel.

Unter den myrmekophilen *Koleopteren* sind es hauptsächlich drei Gruppen, die durch ihr echtes Gastverhältnis zu den Ameisen hervorragen : die *Lomechusini* unter den Staphyliniden, die *Clavigerinæ* unter den Pselaphiden, und endlich unter den *Paussiden* weitaus die meisten Gattungen von *Pleuropterus* bis

Paussus. Die übrigen Symphilen unter den Käfern nenne ich hier nicht.

a. Bei den *Lomechusini* ist das echte Gastverhältniss insofern am höchsten entwickelt, als diese Käfer nicht nur von ihren Wirten beleckt werden (erste Stufe), sondern auch aus dem Munde derselben regelmässig gefüttert werden (zweite Stufe), und endlich auch die Larven dieser Käfer von den Ameisen gleich der eigenen Brut erzogen werden (dritte Stufe). Der grösste Vertreter der *Lomechusini* ist unsere europäische *Lomechusa strumosa* (Fig. 13), welche einwirtig bei *Formica sanguinea* lebt und daselbst auch ihre Larven (Fig. 14) erziehen lässt. Letztere ahmen in ihrer Haltung die unbeweglichen Ameisenlarven nach, obwohl sie sechs Beine besitzen, und werden gleich den Ameisenlarven, ja sogar noch viel eifriger als diese, von den Wirten gefüttert. Ueberdies aber nähren sie sich, namentlich in der ersten Jugend, von den Eiern und jungen Larven der Ameisen und fressen dieselben massenhaft auf; dadurch sind sie tatsächlich die schlimmsten Feinde ihrer Wirte. Die Arten der Gattung *Atemeles* sind nicht einwirtig wie die *Lomechusa*, sondern regelmässig doppelwirtig. Die Käfer leben im Herbst und Winter bei den kleinen roten Knotenameisen *Myrmica* und gehen dann im Frühling zur Fortpflanzungszeit zu *Formica* über, wo sie ihre Larven erziehen lassen; und zwar hat jede *Atemeles*-Art oder -Rasse eine bestimmte *Formica*-Art oder -Rasse als Larvenwirt (1). Die Doppelwirtigkeit der *Atemeles* bedingt einen viel höheren Grad der Initiative dieser Käfer gegenüber den Ameisen, als wir sie bei *Lomechusa* finden. Die *Atemeles* ahmen durch « aktive Mimikry » das Benehmen der Ameisen in hohem Grade nach, besonders in der Aufforderung zur Fütterung (Fig. 15). Der Schaden, den ihre Larven der *Formica*-Brut zufügen, ist ähnlich wie bei *Lomechusa*. In Nordamerika sind die Lomechusini durch die Gattung *Xenodusa* (Fig. 16) vertreten, deren Arten doppelwirtig sind wie die *Atemeles*, aber statt *Myrmica Camponotus* als zweiten Wirt haben. Ihre Larven werden bei *Formica* erzogen auf Kosten der Ameisenbrut wie bei den obigen Gattungen.

Wie sehr die *Formica*-Arten durch die Larven der Lomechusini geschädigt werden, zeigt sich auch darin, dass durch ihre

(1) Vgl. hierzu auch « Die Anpassungscharaktere der *Atemeles* » unter den Vorträgen in der Sektion II. für Bionomie in diesen Congressverhandlungen.

andauernde Erziehung der normale Brutpflegeinstinkt der Ameisen pathologisch verändert wird ; statt echter Weibchen erziehen sie nämlich krüppelhafte Zwischenformen zwischen Arbeiterin und Weibchen, die sogenannten *Pseudogynen*, welche für den Ameisenstaat völlig nutzlos sind. Es sind arbeiterähnliche Formen mit bucklig aufgetriebenem, weiblichen Mittelrücken. Am häufigsten kommt es zur Pseudogynenerziehung bei unserer *Formica sanguinea* (Fig. 17, *a.* Arbeiterin, *b.* Pseudogyne), welche die Larven der einwirtigen *Lomechusa strumosa* züchtet. Hier konnte ich den Zusammenhang zwischen der Zucht jener Adoptivlarven und der Pseudogynenbildung auch auf statistischem Wege nachweisen. Minder häufig führt die Erziehung der Larven von *Atemeles* und *Xenodusa* zur Entstehung von Pseudogynen, weil diese Käfer infolge ihrer Doppelwirtigkeit nicht stets in dieselben individuellen *Formica*-Kolonien zurückgelangen, in denen sie selber erzogen worden waren. Durch das Ueberhandnehmen der Pseudogynen in einem Neste wird schliesslich der Untergang der Wirtskolonie herbeigeführt. Die angeblich so « intelligenten » *Formica* züchten also in den Larven der *Lomechusini* tatsächlich ihre schlimmsten Feinde.

Wenn wir aber die stammesgeschichtliche Entwicklung der Symphilie verfolgen, in welcher die Amikalselektion, d. h. die instinktive Zuchtwahl, welche die Ameisen gegenüber ihren Gästen ausüben, eine grosse Rolle spielt, so müssen wir sogar sagen : *die Ameisen haben sich in den Lomechusini ihre schlimmsten Feinde selber herangezüchtet !* Näher auf die psychologische und die phylogenetische Seite dieses interessanten Problems hier einzugehen, verbietet leider die Kürze der Zeit (1).

b. Viel harmloser sind die Beziehungen der *Keulenkäfer* (*Clavigerinæ*) zu den Ameisen. Die Käfer werden von ihren Wirten eifrig beleckt und aus ihrem Munde gefüttert, tun aber der Ameisenbrut keinen Schaden, obwohl sie manchmal an kranken oder verwundeten Larven zehren. Wir kennen bereits, hauptsächlich durch RAFFRAY's Arbeiten, 40 Gattungen der *Clavigerinæ* mit weit über 100 Arten. Die Lebensweise unseres kleinen gelben Keulenkäfers ist schon seit 1818 recht gut bekannt, und doch sind

(1) Vgl. hierüber « Die psychischen Fähigkeiten der Ameisen » (« Zoologica », Heft 26), 2. Aufl., 1909, p. 149; « Ueber Wesen und Ursprung der Symphilie » (« Biolog. Centralbl. », 1910, Nrn. 3-5).

die *Larven* sämtlicher Clavigerinæ noch unentdeckt. Ein Bild von den Anpassungscharakteren dieser Käfer bietet ein 4 mm. langer Riesenkeulenkäfer aus Madagaskar, *Miroclaviger cervicornis* (Fig. 18), der ausser einer grossen Abdominalgrube reich entwickelte gelbe Haarbüschel an verschiedenen Köperstellen zeigt.

c. Die an Formenmannigfaltigkeit so reiche Familie der *Fühlerkäfer* (*Paussiden*) ist für das Studium der myrmekophilen Anpassungen sehr fruchtbar, kann aber hier nur ganz kurz behandelt werden. Schon im unteren Oligocän des baltischen Bernsteins finden wir drei Gattungen, von denen zwei (*Paussoides* und *Paussus*) wahrscheinlich damals schon zu den echten Ameisengästen gehörten, während die dritte (*Arthropterus*) noch in ihren heutigen Vertretern den primitivsten Trutztypus aufweist. Unter den gegenwärtigen Gattungen treffen wir symphile Charaktere schon bei *Pleuropterus* (Fig. 19) trotz der noch zehngliedrigen Fühler, indem die Gruben des Halsschildes und der Flügendeckenbasis zu Exsudatorganen dienen. Mit der weiteren Entwicklung der Symphilie ist bei den Paussiden eine fortschreitende Reduktion der Fühlergliederzahl vorhanden. Auf der höchsten Stufe, bei *Paussus,* sind die Fühler nur noch zweigliedrig, und die Fühlerkeule nimmt die mannigfaltigsten Gestalten an (1), unter denen die Muschelform in der innigsten Beziehung zum echten Gastverhältnisse steht. Ein schönes Beispiel hiefür bietet *Paussus howa* (Fig. 20) aus Madagaskar, der bei *Ischnomyrmex Swammerdami* lebt. Im Innern des muschelförmigen Fühlerbechers liegt ein mächtiges Drüsenzellenlager, wie es der beifolgende Fühlerschnitt von *Paussus cucullatus* zeigt (Fig. 21). Aber auch unter den Stirnporen, unter den Halsschildgruben und in der Pygidialregion findet sich adipoides Drüsengewebe in grossem Umfange, und bei vielen Paussus sind auch die symphilen Trichombildungen (rotgelbe Haarbüschel, etc.) in der mannigfaltigsten Form und an den verschiedensten Körperstellen reich ausgebildet (2). Trotz der

---

(1) Arten mit linsenförmiger Fühlerkeule wie *P. arabicus* und *Favieri* stehen nach Escherich's Beobachtungen auf einer niedrigeren Stufe der Symphilie als *P. turcicus,* der eine muschelförmige Fühlerkeule hat.

(2) Ueber die Exsudatorgane und Exsudatgewebe der Paussiden siehe : « Zur näheren Kenntnis des echten Gastverhältnisses » (« Biolog. Centralbl. », 1903) und « Die moderne Biologie und die Entwicklungstheorie », 3. Aufl., 1906, p. 370 ff.

hohen Entwicklung ihrer Exsudatorgane werden die *Paussus*, soweit bisher bekannt, von den Ameisen nur beleckt, nicht aus ihrem Munde gefüttert; sie leben vielmehr als Räuber von der Ameisenbrut. Die durch Böving zum erstenmal sicher beschriebenen *Paussus*-Larven (von *P. Kannegieteri*) sind ebenfalls carnivor; ihre Physogastrie und die Drüsengewebe ihrer Hinterleibspitze deuten allerdings nebenbei noch auf ein echtes Gastverhältnis hin.

4. — Der Nahrungserwerb der Ameisen.

Der Nahrungserwerb der Ameisen ist ein mannigfaltiger und bietet mehr Analogien mit menschlichen Erwerbszweigen, als wir sonst im Tierreiche finden. Für viele unserer einheimischen Ameisen ist die Hauptnahrungsquelle *Viehzucht*, d. h. die Beschäftigung mit Blatt- und Schildläusen, welche sie teils ausserhalb der Nester besuchen, teils in ihren Nestern halten, und denen sie die zuckerhaltigen Exkremente durch Streicheln mit den Fühlern (« Melken ») entlocken. Auch die Honigraupen mancher Schmetterlinge, besonders aus der Familie der Lycæniden, werden von einheimischen und tropischen Ameisen als Melkvieh benutzt, und auch tropische Homopterenlarven liefern ein reiches Kontingent hierzu. Gewisse einheimische Ameisen (*z. B. Lasius flavus*) bewahren sogar die Eier der Blattläuse während des Winters in ihren Nestern auf. Innerhalb sechs verschiedener Gattungen verschiedener Erdteile gibt es ferner *Honigameisen*, welche eine Arbeiterkaste zu lebendigen « Honigtöpfen » heranfüttern, aus deren Kropf die Kolonie zur Zeit der Dürre ihre Nahrung bezieht. Das klassische, schon von Mc Cook beschriebene Beispiel ist die Honigameise des Göttergartens in Colorado (*Myrmecocystus hortideorum*). Andere Ameisen wiederum sind *Körnersammler*, welche Vorräte von Sämereien eintragen und in Kornkammern ihrer Nester aufspeichern. Ueber ein Jahrhundert lang hatte man den Bericht der heiligen Schrift über die getreidesammelnden Ameisen Palästinas für eine Fabel gehalten, bis er durch die Beobachtungen über die Gattung *Messor* im Mittelmeergebiet glänzend bestätigt wurde. Neuerdings hat Neger sogar gefunden, dass *Messor barbarus* die Sämereien zu einer Art Ameisenbrot verarbeitet. Die klassischen « Ackerbauameisen » Nordamerika's gehören zur Gattung *Pogonomyrmex*, und wenngleich die übertriebene Romantik des « Ackerbaues » von *Pog. barbatus* durch WHEELER's kritische Untersuchungen zerstört worden ist, so bleibt doch die Lebensweise dieser Körnersammler interessant genug. Noch höheres Interesse hat die *Pilzzucht* der Ameisen namentlich seit MÖLLER's Forschungen (1893) erregt. Die Lebensweise der Blattschneiderameisen Amerika's aus der Gruppe der *Attini* ist dadurch in ein neues Licht gerückt, zumal wir seit den Beobachtungen v. IHERING's, E. GÖLDI's und namentlich JACOB HUBER's (1905) auch über die sinnreiche Weise unterichtet sind, wie die Königin von *Atta sexdens* den jungen Pilzgarten anlegt und kultiviert, wenn sie nach dem Paarungsfluge ihre neue Kolonie gründet. Dass es sich bei dieser scheinbar hochintelligenten « Gemüsekultur » der Ameisen um einen erblichen Instinkt handelt, wird übrigens durch die Analogie mit der Pilzzucht der Termiten bestätigt, bei denen diese Sitte namentlich in der Gattung *Termes* noch viel weiter verbreitet ist, obwohl die Termiten psychisch unter den Ameisen stehen. Einen niedlichen wallnussgrossen Pilzgarten aus dem Neste einer kleinen Gasttermite (*Microtermes globicola*), die auf Ceylon in den Hügeln von *Termes Redemanni* lebt, zeigt die beifolgende vergrösserte Abbildung (Fig. 22). Viele andere Ameisen endlich leben von der *Jagd*, besonders auf Insekten. Bei unseren grossen haufenbauenden Wald- und Wiesenameisen (*Formica rufa* und *pratensis*) ist dieser Nahrungserwerb allerdings nur ein sekundärer im Vergleich zum Blatt- und Schildlausbesuch. Immerhin ist er bedeutend genug um diese und andere acervicole *Formica*-Arten der *rufa*- und *exsecta*-Gruppe als eminent « nützlich » unter den Schutz der Forstgesetze zu stellen. Die blutrote Raubameise *Formica sanguinea* beschäftigt sich sogar fast ausschliesslich mit der Jagd und überlässt die Blattlauszucht ihren Sklaven. Weit zahlreicher sind die carnivoren Jagdameisen jedoch unter den tropischen und subtropischen Ameisen, sowohl unter den *Ponerinen* (*Lobopelta*, etc.) als ganz besonders unter den *Dorylinen*.

5. — Die Dorylinen und ihre Gäste.

Die Unterfamilie der *Dorylinen* umfasst *Jagdameisen*, welche teils oberirdisch teils unterirdisch auf Raub ausgehen und deshalb meist unstete *Wanderameisen* sind. Sie sind die Geissel der Klein-tierwelt in den Tropen. Zusammengesetzte Augen (Netzaugen) fehlen ihnen; an ihre Stelle treten einfache Ozellen, die jedoch vielfach rudimentär sind oder ganz verschwinden. Wo sie gut entwickelt sind, wie z. B. bei *Eciton Burchelli*, vermögen sie sogar Farben zu unterscheiden, wie aus dem Vergleich mit der Färbung ihrer Gäste hervorgeht (1). Trotzdem sind die Dorylinen in noch höherem Grade Tast- und Geruchstiere als die übrigen Ameisen. Die meisten *Eciton* des neotropischen Gebietes unternehmen als Wanderameisen (legionary Ants) ihre schon von RENGGER, BELT und BATES erwähnten Jagdzüge oberirdisch in grossen Armeen; desgleichen die schon von SMEATHMAN beobachteten *Anomma* Afrika's, welche in oft riesigen Heereszügen ihre Treibjagden auf alles Kleingetier veranstalten, weshalb sie den Namen « driver Ants » erhielten. *Anomma* ist eine wegen ihres oberirdischen Jagdbetriebes meist dunkelgefärbte Untergattung von *Dorylus*, dessen hellgefärbte Arten ebenso wie jene der Gattung *Ænictus* unterirdisch jagen. Den Trimorphismus der Arbeiterform von *Anomma Wilverthi* ersieht man aus Figur 23. Allerdings ist er durch Uebergänge vermittelt. Die kleinste Arbeiterform könnte man wegen ihrer reduzierten Fühlergliederzahl für eine andere Gattung halten, wenn man sie allein fände. Sowohl bei *Anomma* wie bei *Eciton* sind zweierlei Züge zu unterscheiden, Jagdzüge und Umzüge. Nur bei letzteren werden die riesigen, völlig flügellosen Weibchen und die gleichfalls sehr grossen, häufig entflügelten Männchen mitgeschleppt, und zwar bei *Anomma* über offene Strecken nur in bedeckten Tunnels. Beim Umzug in eine neue Jagdgegend wird ein neues Nest bezogen (VOSSELER in Ostafrika und LUJA am unteren Congo), welches wochenlang als Stützpunkt für die Jagdzüge dient. Dass diese Nester eine Fauna von Gästen (2), namentlich Käfern, enthalten, welche von den Abfällen der Jagdbeute ihrer Wirte leben, kann weniger befremden. Aber dass die Jagdameisen selbst auf ihren Beutezügen von einer ganzen Schaar von Gästen, besonders aus der Käferfamilie der Kurzflügler, begleitet werden, ist um so auffallender, da man denken sollte, die Jäger würden sich an erster Stelle dieses so nahe liegenden Wildes bemächtigen. Und doch besitzen sowohl

(1) Die psychischen Fähigkeiten der Ameisen, 2. Aufl., 1909, Kap. VI.
(2) Die Nestgäste von *Anomma Wilverthi*, welche E. LUJA bei Kondue in mehreren Nestern fand, werden nächstens beschrieben werden.

die *Eciton* Amerika's als die *Anomma-Dorylus* Afrika's die grösste Zahl von Gästen unter allen tropischen Ameisen!

Die Entwicklungstheorie erklärt uns einigermassen dieses Rätsel. Je grösser die Anpassungs*notwendigkeit* gegenüber einem bestimmten Feinde ist, desto grösser wird auch — bei Voraussetzung der Anpassungs*fähigkeit*, die namentlich in der Organisation der Kurzflügler eine sehr weitgehende ist — die Anpassungs*häufigkeit* und die Anpassungs*höhe* sein. Hieraus begreift sich sowohl die grosse Zahl der Dorylinengäste unter den Staphyliniden — bis jetzt sind gegen 40 Gattungen beschrieben — als auch die hohe Anpassungsstufe, welche viele derselben zeigen.

Drei hauptsächliche morphologisch-biologische Anpassungstypen treten uns hier entgegen : der *Symphilentypus*, der *Mimikrytypus* und der *Trutztypus*. Der erste macht die Gäste ihren Wirten durch Exsudate angenehm, der zweite spiegelt sie ihnen als ihresgleichen vor, der dritte macht sie für die Ameisenkiefer unangreifbar. Diese drei Anpassungstypen kommen auch bei anderen Myrmekophilen vor; aber ihre Ausbildung ist bei den Gästen der Wanderameisen eine eigenartige, dem Charakter der Wirte entsprechende; und zwar ist sie bei den neotropischen Ecitongästen eine ganz ähnliche wie bei den afrikanischen Anommagästen, obwohl die Gattungen, welche in beiden Weltteilen jene Typen repräsentieren, untereinander in keiner näheren systematischen Verwandtschaft stehen. Ihre Aehnlichkeit beruht somit auf *Convergenz* infolge ähnlicher Anpassungsbedingungen.

Der Hauptvertreter des *Symphilentypus* der Anommagäste Afrika's ist *Sympolemon anommatis*, der « Kriegskamerad der Treiberameisen » (Fig. 24). Seine schlanke Körperform ist bedingt durch seine Bewegungsweise als Begleiter der raschlaufenden Wirte. Er bewegt sich sogar noch schneller als diese, indem er seinen Hinterleib als Sprungfeder benutzt, um pfeilschnell dahinzuschiessen, wie ein Beobachter (P. Hermann Kohl) sich ausdrückt. Schnittserien des Hinterleibes gaben mir über den Mechanismus jener Sprungfeder Aufschluss. Man sieht auf dem Sagittalschnitt, wie die Chitinspangen in das Hinterleibslumen hineinragen und starken Muskelbündeln als Ansatzfläche dienen. (Siehe Fig. 25.) An den langen Beinen sind die Füsse rudimentär und in pantoffelförmige, dicht behaarte Gebilde umgewandelt, welche teils als Haftorgane dienen, teils aber auch ein Analogon zu den befiederten Füssen des Steppenhuhns (*Syrrhaptes*) darstellen. Unter den brasilianischen Ecitongästen ist *Ecitogaster* dem *Sympolemon* in

Körperform und Fühlerbildung am ähnlichsten. Auch weist die Zungenbildung beider darauf hin, dass diese Gäste aus dem Munde ihrer Wirte gefüttert werden.

Am merkwürdigsten ist der *Mimikrytypus* der Dorylinengäste. Er ist an erster Stelle eine « *Tastmimikry* », welche auf passive und aktive Täuschung des Fühlertastsinnes der Wirte berechnet erscheint.

Die passive Täuschung wird bewirkt durch .die Formähnlichkeit der einzelnen Körperabschnitte des Gastes mit jenen des Wirtes, die aktive Täuschung durch die Aehnlichkeit der Fühlerbildung zwischen Gast und Wirt. Den höchsten Grad erreicht diese Mimikry bei *Mimeciton* (Fig. 26) unter den ecitophilen Staphyliniden und bei *Dorylomimus* (Fig. 27) unter anommatophilen. Bei diesen Gattungen besteht keine gesetzmässige Färbungsähnlichkeit zwischen Gast und Wirt, weil letzterer nur rudimentäre Ozellen besitzt (*Eciton prædator*) oder gar keine Augen hat (*Anomma Wilverthi*). Bei den Gästen solcher *Eciton* dagegen, welche gut entwickelte Ozellen haben, tritt zur Tastmimikry noch eine ganz genaue Aehnlichkeit der Färbung des Gastes mit der gleichgrossen Arbeiterform des Wirtes hinzu, z. B. bei *Ecitophya* (Fig. 28), *Ecitomorpha* und *Ecitonidia*.

Auf den höchsten Stufen des Mimikrytypus, bei *Mimeciton*, *Dorylomimus* und *Ecitophya*, bildet die Mimikry sogar die Grundlage für ein echtes Gastverhältnis; für *Dorylomimus Kohli* steht dies auch durch direkte Beobachtung fest (P. KOHL).

Eine analoge Tastmimikry wie bei den ecitophilen Staphyliniden begegnet uns auch unter den ecitophilen Proctotrypiden (Hymenopteren) in den Gattungen *Mimopria* und *Ecitopria*.

Den *Trutztypus* der neuweltlichen Ecitongäste repräsentiert hauptsächlich die Gattung *Xenocephalus* (Fig. 29), bei welcher der Kopf und die Extremitäten durch ein Schutzdach gedeckt sind. Unter den altweltlichen Dorylinengästen begegnet uns ein analoger, aber minder vollkommener Trutztypus bei *Pygostenus* und verwandten Gattungen. Die höchste Entwicklung erreicht der Trutztypus der dorylophilen Staphyliniden jedoch in der ganz isoliert stehenden Gattung *Trilobitideus* (Fig. 30), die in mehreren Arten bei *Dorylus* und *Anomma* lebt : eine blattförmig flache, an den Boden sich anschmiegende Körpergestalt, die überdies mit kegelförmigen Höckern auf der Oberseite besetzt ist. Greift eine Jagdameise nach diesem Gaste mit ihren Kiefern, so kann sie ihn nur an einen Höcker packen und höchstens fortschleudern, wenn ihre Kiefer nicht sogleich wieder abgleiten. Das Tier, bei dem nicht einmal Flügeldecken vorhanden sind, gleicht eher einer Silphidenlarve als einem erwachsenen Käfer. Leider müssen wir uns hier mit diesen wenigen Beispielen aus dem ausserordentlich reichen Gebiete der Anpassungscharaktere der Dorylinengäste begnügen.

6. — Der Nestbau der Ameisen.

Ein Ameisennest ist ein unregelmässiges System von Gängen und Kammern, welche zum Aufenthalt der Ameisen und ihrer Brut dienen, kein kunstvoller Bau wie die Bienenwabe, dafür aber auch keine starre Schablone, sondern einer fast *unbegrenzten Anpassung* an die verschiedensten Materialien und Oertlichkeiten fähig. Von einer winzigen, kaum einige Millimeter messenden Erdhöhle oder Rindenspalte angefangen bis zu den mächtigen Ameisenhaufen unserer Waldameisen und den noch umfangreicheren Nestern mancher grossen *Atta* Amerika's finden wir alle Uebergänge in der *Grösse* eines Ameisennestes. Ebenso gibt es kaum eine *Oertlichkeit*, wo Ameisen nicht ihr Nest anbringen können, kaum ein *Material*, aus dem es nicht bestehen kann. Durch FOREL und andere Forscher ist die ungeheuere Mannigfaltigkeit der Nestbauarten unter den Ameisen längst bekannt. Man unterscheidet Erdnester, Nester unter Steinen, oberirdische Erdhaufen, Haufen aus trockenem Pflanzenmaterial, mehr oder weniger mit Erde vermischt, Nester unter Rinde, in hohlen Gallen, in hohlen Stengeln und hohlen Bäumen, Nester in morschen Strünken oder ausgemeisselt in festem Holze, Nester aus Carton, die entweder zwischen Wurzeln oder in hohlen Bäumen angebracht sind oder frei an den Zweigen hängen, endlich noch Gespinstnester, die aus zusammengesponnenen Blättern oder aus mit Gespinst austapezierten anderweitigen Höhlungen bestehen können. Es kann ferner irgendwelcher schon vorhandene Hohlraum zu einem Ameisenneste umgestaltet werden, mag es nun um ein Stück Dachpappe, um den Deckel einer Conservenbüchse, um getrockneten Kuhmist oder um einen alten Pferdeschädel sich handeln, in welchem P. SCHUPP in Rio Grande do Sul einmal ein Nest von *Camponotus rufipes* fand. Zahlreich sind auch die gestohlenen Nester, die früher anderen Ameisenarten oder Termiten gehört haben und entweder nach oder schon vor dem Abzug der Erbauer von den Ameisen in Besitz genommen wurden.

Als Beispiel eines typischen Ameisenhaufens diene die Photographie eines Riesennestes von *Formica rufa* bei Luxemburg, das 17 m. im Umfange mass (Fig. 31). Ein Cartonnest von *Cremastogaster Stadelmanni dolichocephala* SANTSCHI von 1,10 m. Länge, das sich im Naturhistorischen Museum von Luxemburg befindet, zeigt die Figur 32, wie es an seinem natürlichen Standort auf einem hohen Baume hängend von E. LUJA zu Kondué photographiert wurde. Ein Gespinstnest von *Polyrhachis laboriosa* aus Kondué (E. LUJA!) zeigt endlich Figur 33. Es besteht aus zusammengesponnenen Blättern, deren Oberflächen mit Gespinst ausgekleidet sind. Die obere äusserste Schicht, in welche auch reichlich Holzmulm eingefügt ist, scheint, wie dies nach P. KOHL's Beobachtungen auch bei *Oecophylla longinoda* am oberen Congo der Fall ist, von den Ameisen mittelst ihrer Oberkiefer und des Sekretes ihrer Kieferdrüsen hergestellt zu sein, während das Gespinst selber aus einer anderen Quelle hervorging, die wir gleich kennen lernen werden.

Diese *Gespinstnester* der Ameisen sind von hohem psychologischem Interesse.

Denn der bei ihnen verwandte Spinnstoff stammt nicht von den Ameisen sondern von ihren *Larven*, die von den Arbeiterinnen ins Maul genommen und als « Webschiffchen » verwendet werden! Sie führen den Mund der Larven, aus welchem der Spinnstoff tritt, von einem Blattrand zum andern und weben so ihr Nest. Als vor 20 Jahren die ersten Nachrichten RIDLEY's hierüber aus Ostindien nach Europa gelangten, klangen sie kaum glaublich. Jetzt sind sie aber durch viele Forscher übereinstimmend bestätigt worden, für *Oecophylla smaragdina* auf Ceylon durch DOFLEIN und BUGNION, für *Oecophylla longinoda* am Congo durch P. KOHL, u. s. w. Bei den *Oecophylla* ist diese Nestbaumethode allgemein, bei der Gattung *Polyrhachis* kommt sie nur einem Teile der Arten zu, während andere Kartonnester verfertigen. In der Gattung *Camponotus* ist sie endlich nur von einer Art, *C. senex* in Brasilien, bekannt, und in der Gattung *Technomyrmex* kennt man ebenfalls nur von einer Art (*T. bicolor textor* FOREL) Gespinstnester.

Dass Ameisen ihre eigenen Larven als Webschiffchen zum Spinnen benutzen und so das Spinnvermögen ihrer Larven in zweckmässiger Weise zum Nestbau verwerten, ist jedenfalls pschychologisch höchst merkwürdig. Es ist ein Gebrauch von « Werkzeugen », die vom eigenen Körper des Tieres verschieden sind und nicht aus demselben stammen, während z. B. das Netz, mit dem die Spinne ihre Beute fängt, ein Produkt ihrer eigenen Körperdrüsen ist. Einen so gut verbürgten und so sinnreichen Gebrauch von Werkzeugen, wie ihn die Ameisen beim Bau ihrer Gespinstnester betätigen, suchen wir selbst bei den höheren Wirbeltieren in freier Natur vergebens. Dennoch dürfen wir ihn psychologisch nicht überschätzen. Es handelt sich hier wie bei den übrigen spezifischen Nestbauarten der Ameisen um *erbliche Instinkte*, über deren phylogenetischen Ursprung jedoch noch tiefes Dunkel ruht. Jedenfalls dürfen wir eine *Polyrhachis*-Art, welche mittelst ihrer Larven ein Gespinstnest verfertigt, deshalb nicht für « intelligenter » halten als eine andere Art derselben Gattung, die das Sekret ihrer Oberkieferdrüsen zum Bau eines Kartonnestes verwendet. Ebenso dürfen wir auch *Camponotus senex* nicht deshalb etwa für die « intelligenteste Art » ihrer Gattung ausgeben, weil nur sie Gespinstnester baut, andere Arten aber Cartonnester, Holznester u. s. w. Alle die verschiedenen Nestbauinstinkte sind objectiv zweckmässig in ihrer Art, beruhen aber nicht auf der intelligenten Ueberlegung des Einzelwesens, weil sie eben erbliche Instinkte sind. Ihre Ausübung ist trotzdem kein blosser Reflexmechanismus, weil sie unter dem Einflusse der Sinneswahrnehmung und Sinneserfahrung des Individuums erfolgt. Auch hier müssen wir uns also in der psychologischen Erklärung des Tierlebens in der richtigen Mitte halten zwischen zwei, gleich verfehlten Extremen. Dann werden auch wir von den Ameisen « *Weisheit lernen* », wenn wir ihre Wege betrachten.

TAFELERKLÄRUNG (1).

PL. XII.

Fig. 1. — *Formicoxenus nitidulus* Nyl. (glänzende Gastameise), Ergatoides Männchen und Arbeiterin (8 : 1).

Fig. 2. — *Formica rufa* L. (rotrückige Waldameise), Königin und kleine Arbeiterin (3 : 1).

Fig. 3. — *Formica truncicola* Nyl., geflügeltes Weibchen (4 : 1).

Fig. 4. — *a*. Arbeiterin von *Formica exsecta* Nyl.;
b. Arbeiterin von *Formica fusca* L. (4 : 1).

Fig. 5. — *a*. *Formica fusca* L., kleine Königin;
b. *Formica exsecta* Nyl., geflügeltes Weibchen (4 : 1).

Fig. 6. — *Formica sanguinea* Ltr., Arbeiterin (3,5 : 1).

PL. XIII.

Fig. 7. — *a*. Kopf von *Formica sanguinea* Ltr., Arbeiterin;
b. Kopf von *Polyergus rufescens* Ltr., Arbeiterin (6 : 1).

Fig. 8. — *Polyergus rufescens* Latr. (Amazonenameise) :
a. Ergatoide Königin;
b. Arbeiterin (4 : 1).

Fig. 9. — *a*. *Harpagoxenus (Tomognathus) sublævis* Nyl., ergatoide Königin;
b. *Leptothorax açervorum* F., Arbeiterin (Sklavin) (5 : 1).

Fig. 10. — *Strongylognathus testaceus* Schenk, Arbeiterin (12 : 1).

Fig. 11. — *Wheeleriella Santschii* For., geflügeltes Weibchen, Ober- und Seitenansicht (5 : 1).

Fig. 12. — *Anergates atratulus* Schenk, puppenähnliches Männchen (12 : 1).

Fig. 16. — *Xenodusa cava* Lec. (Nordamerika) (5 : 1).

(1) Die mikrophotographischen Aufnahmen sind meist mit Zeiss Tessar 1 : 6,3 aufgenommen.

PL. XIV.

FIG. 13. — *Lomechusa strumosa* F. (5 : 1) :
a. Mit aufgerolltem Hinterleib;
b. Mit ausgestrecktem Hinterleib, um die gelben Haarbüschel zu zeigen.

FIG. 14. — Larve von *Lomechusa strumosa* F. (5 : 1).

FIG. 15. — Fütterung von *Atemeles pratensoides* WASM. durch *Formica pratensis* DEG. (6 : 1).

FIG. 17. — *Formica sanguinea* LTR. (4 : 1) :
a. Arbeiterin;
b. Pseudogyne.

FIG. 18. — *Miroclaviger cervicornis* WASM., Madagaskar (12 : 1).

PL. XV.

FIG. 19. — *Pleuropterus Dohrni* RITS. subsp. *Lujæ* WASM., Congo (5 : 1).

FIG. 20. — *Paussus howa* DOHRN, Madagaskar (5 : 1).

FIG. 21. — Schnitt durch das Drüsenzellenlager des Fühlerbechers von *Paussus cucullatus* WESTW. (1000 : 1). [ZEISS, Apochrom. 2.0, 1.30, Compensationsocul. 4.]

FIG. 22. — Pilzgarten von *Microtermes globicola* WASM., Ceylon (1,5 : 1).

FIG. 23. — *Anomma Wilverthi* EM., drei Arbeiterinnen aus derselben Armee (3 : 1), Congo.

FIG. 24. — *Sympolemon anommatis* WASM., Congo (6 : 1).

PL. XVI.

FIG. 25. — Sagittalschnitt durch den Hinterleib von *Sympolemon anommatis*, um die Muskelbündel zu zeigen (20 : 1).

FIG. 26. — *Mimeciton pulex* WASM., Männchen und Weibchen, Brasilien (10 : 1).

FIG. 27. — *Dorylominus Kohli* WASM., Congo (8 : 1).

FIG. 28. — *Ecitophya simulans* WASM., Brasilien (4 : 1).

FIG. 29. — *Xenocephalus gigas* WASM., Brasilien (4 : 1).

FIG. 30. — *Trilobitideus insignis* WASM., Congo (11 : 1).

FIG. 32. — Cartonnest von *Cremastogaster Stadelmanni* subsp. *dolichocephala* SANTSCHI (auf dem Baum hängend), Congo (1 : 44).

FIG. 33. — Gespinstnest von *Polyrhachis laboriosa* SM., Congo (1 : 2).

PL. XVII.

FIG. 31. — Riesennest von *Formica rufa* L., Luxemburg (17 m. Umfang).

FIG. 1.
Formicoxenus nitidulus NYL. (glänzende Gastameise).
Ergatoides Männchen und Arbeiterin (8 : 1).

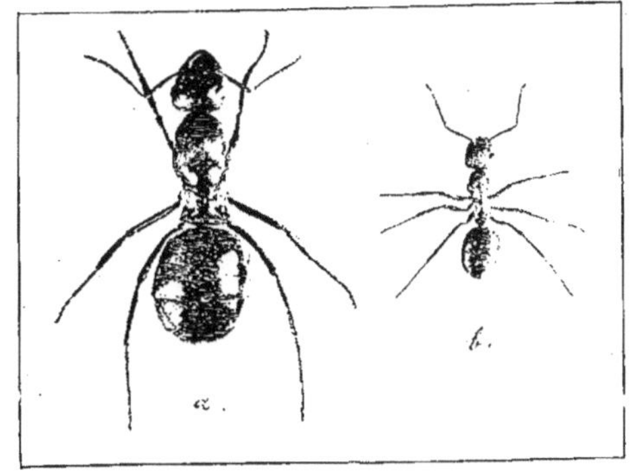

FIG. 2.
Formica rufa L. (rotrückige Waldameise).
a) Königin; *b*) kleine Arbeiterin (3 : 1).

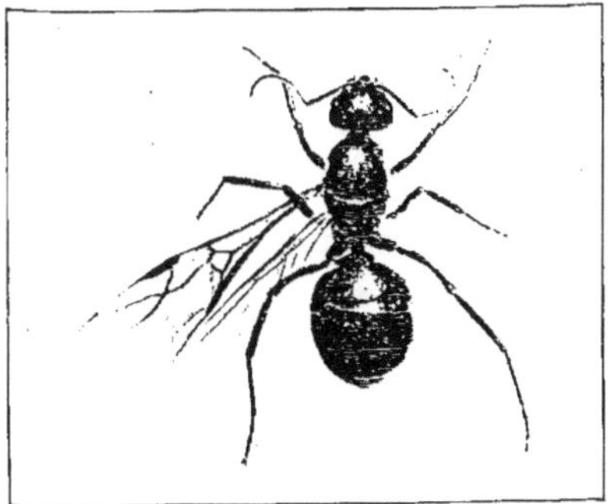

FIG. 3.
Formica truncicola NYL., geflügeltes Weibchen (4 : 1).

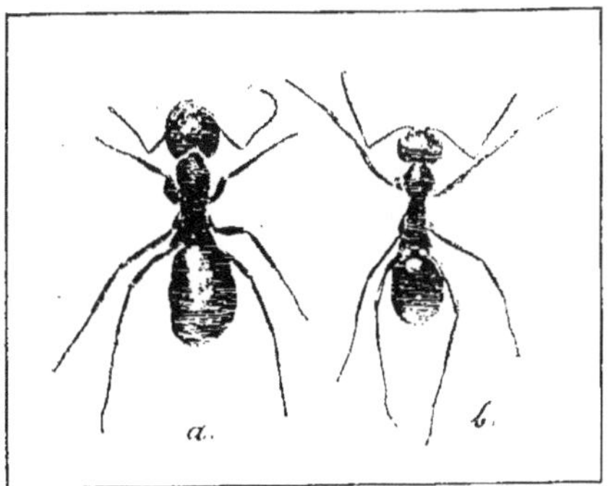

FIG. 4.
a) *Formica exsecta* NYL., Arbeiterin
b) *Formica fusca* L., Arbeiterin.

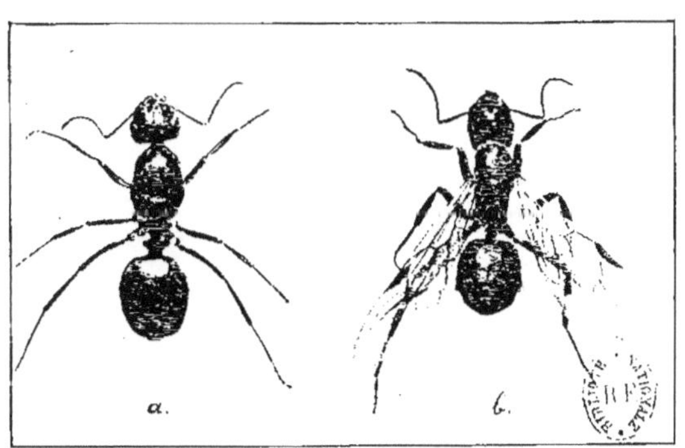

FIG. 5.
a) *Formica fusca* L., kleine Königin;
b) *Formica exsecta* NYL., geflügeltes Weibchen (4 : 1).

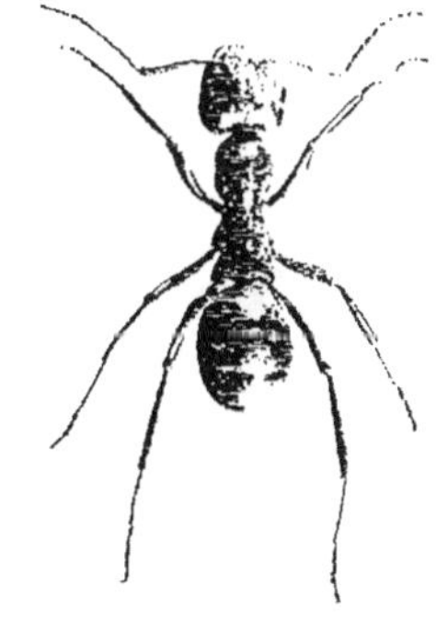

FIG. 6.
Formica sanguinea LATR., Arbeiterin (3,5 : 1).

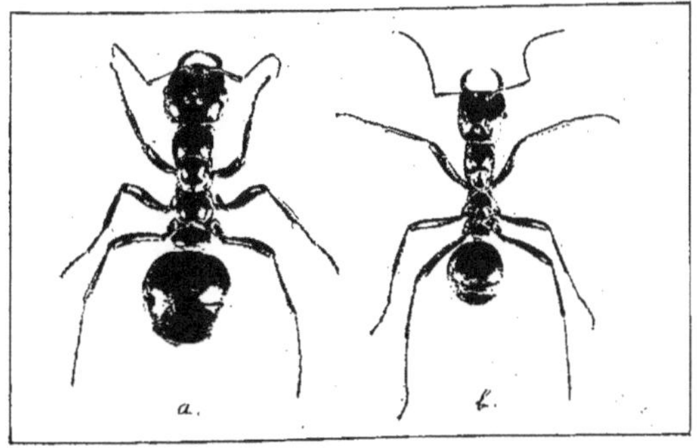

FIG. 8.
Polyergus rufescens LTR., (Amazonenameise);
a) Ergatoide Königin; *b*) Arbeiterin (4 : 1).

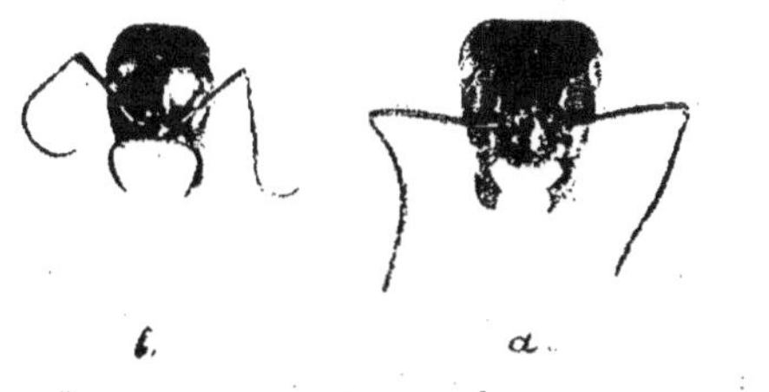

FIG 7
Köpfe der Arbeiterin von { a) *Formica sanguinea* LTR. / b) *Polyergus rufescens* LTR. } (6 : 1)

FIG. 9.
a) *Harpagoxenus (Tomognathus) sublaevis* NYL., ergatoide Königin
b) *Leptothorax acervorum* F., Arbeiterin (Sklavin) (5 : 1).

FIG 10.
Strongylognathus testaceus SCHENK, Arbeiterin (12 : 1).

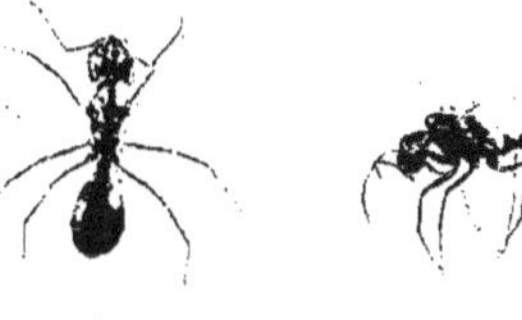

FIG 11.
Wheeleriella Santschii FOR., Weibchen (5 : 1).

FIG 12.
Anergates atratulus SCHENK, Männchen (12 : 1).

FIG. 16
Xenodusa cava LEC., Wisconsin (5 : 1).

FIG. 13.
Lomechusa strumosa F. (5 : 1). { a) mit aufgerolltem Hinterleib. b) mit ausgestrecktem Hinterleib.

FIG. 14.
Larve von *Lomechusa strumosa* F. (5 : 1).

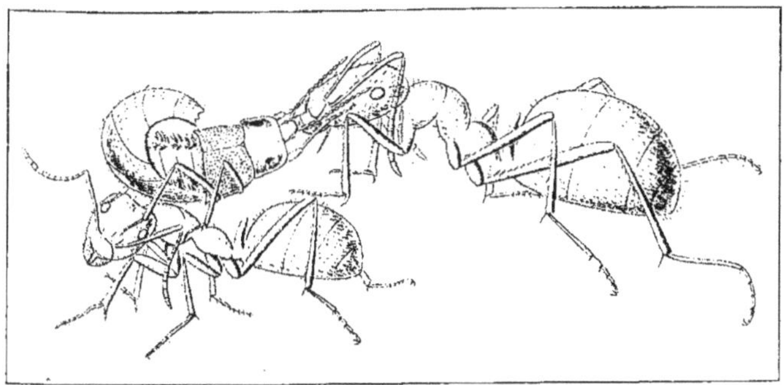

FIG. 15.
Fütterung von *Atemeles pratensoides* WASM. durch *Formica pratensis* DEG. (6 : 1).

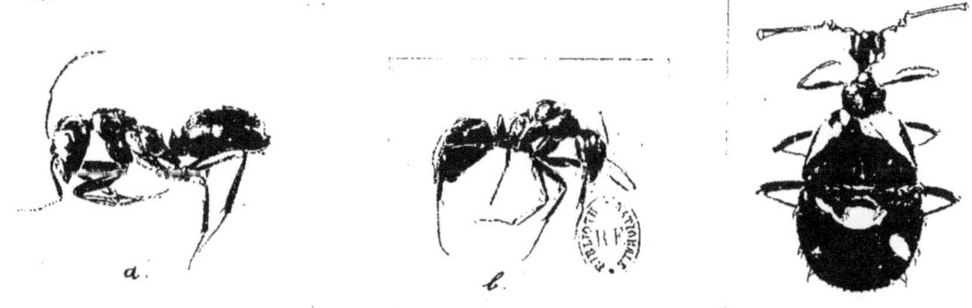

FIG. 17.
Formica sanguinea LTR. (4 : 1) { a) Arbeiterin. b) Pseudogyne.

FIG. 18.
Mirocloriger cervicornis WASM., bei *Camponotus Radamae mixtellus* FOR. (12 : 1).

FIG. 19.
Pleuropterus Dohrni RITS. subsp. *Lujæ* WASM.
(5:1) Kondué (Congo).

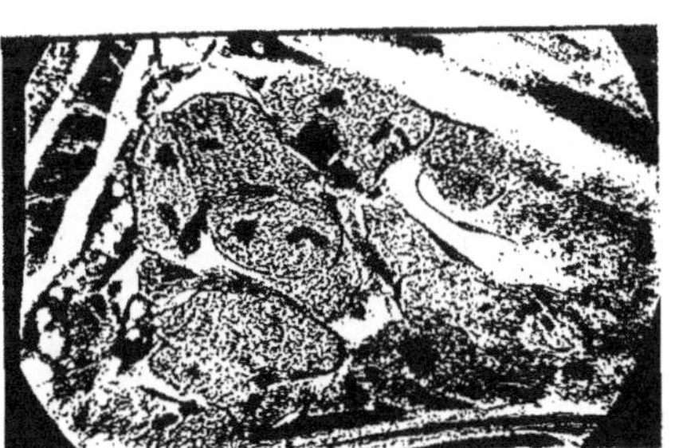

FIG. 21.
Schnitt durch das Drüsenzellenlager des Fühlerbechers von *Paussus cucullatus* WESTW. (ZEISS, Apochrom. 2.0 mm.: 1000:1.)

FIG. 22.
Pilzgarten von *Microtermes globicola* WASM., Ceylon (1,5:1).

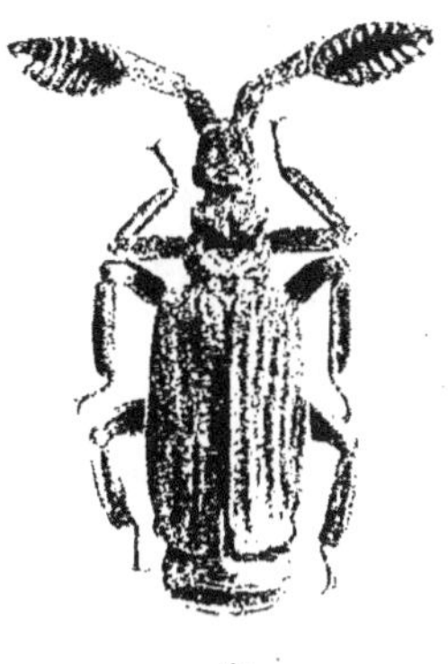

FIG. 20.
Paussus howa DOHRN (5:1).

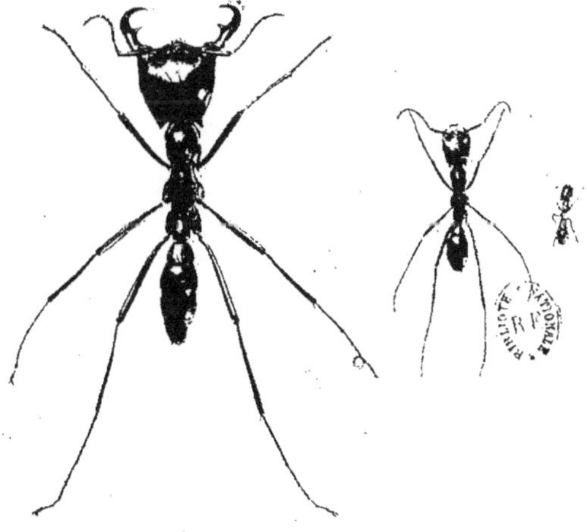

FIG. 23.
Anomma Wilverthi EM., drei Arbeiterinnen aus derselben Armee (3:1), Kondué, Congo.

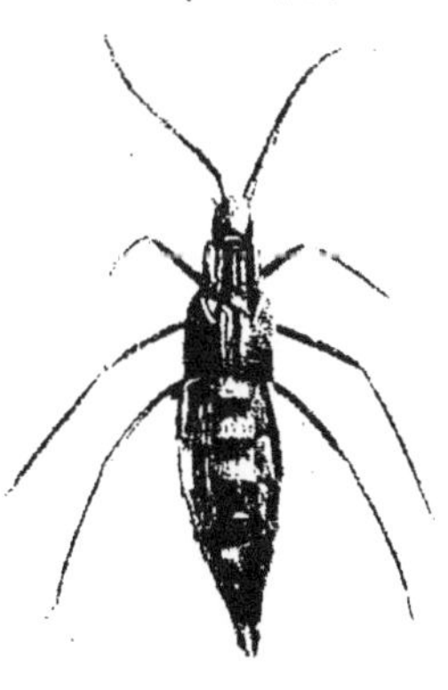

FIG. 24.
Sympolemon anommatis WASM.
bei *Anomma Wilverthi* EM., Congo (6:1).

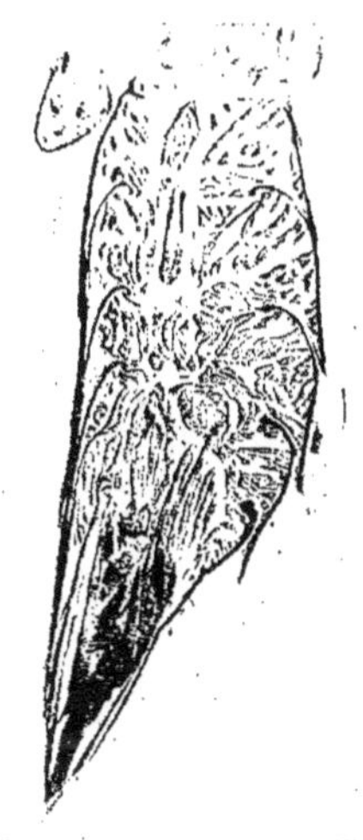

FIG. 25.
Medianer Sagittalschnitt durch den Hinterleib von *Sympolemon anommatis* (20 : 1).

FIG. 26.
Mimeciton pulex WASM., Männchen und Weibchen, bei *Eciton prædator* SM., Brasilien (10 : 1).

FIG. 27.
Dorylomimus Kohli WASM., bei *Anomma Wilverthi* EM., Congo (8 : 1).

FIG. 28.
Ecitophya simulans WASM., Brasilien (4 : 1).

FIG. 29.
Xenocephalus gigas, WASM. bei *Eciton rapax* SM., Amazonas (4 : 1).

FIG. 30.
Trilobitideus insignis WASM. bei *Anomma Wilverthi* EM., Congo (11 : 1).

FIG. 32.
Cartonnest von *Cremastogaster Stadelmanni* subsp. *dolichocephala* SANTSCHI, Kondué, Congo (1 : 44).

FIG. 33.
Gespinstnest von *Polyrhachis laboriosa* SM., Kondué, Congo (1 : 2).

FIG. 31.
Riesennest von *Formica rufa* L., Luxemburg.

Notes on Distribution and Ecology of North American Jassidæ,

by HERBERT OSBORN.

The Insects of this group have been the subject of considerable study, and we are now in position to give something of the limits of distribution for the different species, and a note in this connection may be in place.

Our Jassid fauna in North America is closely related to that of Europe, and while there are not such a very great proportion of the species, there are a number of instances where the species are either identical or so closely related that they must be considered as having had a very recent separation.

In the most part for the Jassidæ proper it appears that the relation is mainly with the Holarctic fauna, and the direction of migration has been either from the northeast, implying an invasion of the Palearctic species from that direction and the dispersal to the south or southwest, or as perhaps more distinctly indicated by certain species, an invasion from the northwest and a dispersal from that quarter. The latter is indicated by the occurrence of certain of our species that are closely related to the Japanese or Siberian fauna. In some of these species there is perhaps equal reason to believe that the direction of dispersal has been from America to the Palearctic region. There are some instances of very peculiar occurrence that need special note. One of the most striking of these is the case of *Cicadula 6-notata*, a species which is widely distributed in the Palearctic region, and is at present distributed throughout the entire Nearctic region. It is a very strange fact that no record of this species appears in the American records up to about the year 1880, and as this species is a conspicuous one and common upon a number of cultivated crops, the failure to record it seems to be of special significance. This is especially true since extended observations by such observers as SAY, HARRIS, FITCH, WALSH and UHLER would seem to have made it almost certain that the species would be recognized if present. While it is hardly safe to conclude that the species is a recent introduction, especially in view of the very wide dispersal of the species as recognized in the last quarter century, it is at least possible that we must consider it as a comparatively recent introduction. Of course the species has been recognized as far north as in Alaska, and it is possible that the point of invasion and dispersal has been from that quarter.

Other cases of interest are to be found in the genera *Scaphoideus* and *Agallia*. In *Scaphoideus* the species are practically all American, and until recently it was supposed to be strictly American. The species occur all the way from the Boreal zone to the Tropics, but the majority are known from the Transition or Austral region. The Exotic species so far as observed occur in Japan, India and Egypt. For this genus it would appear therefore that the center of origin was in the central or southern United States, and that the lines of dispersal have been to the north and south, and especially northwest for the extension to the Palearctic region. For *Agallia* the origin of the group appears rather to have been tropical, possibly an old-world species, and the lines of distribution into south and central America and thence northward The northern limits for the species for the most part are coincident with the Transition or Boreal zone.

Both *Athysanus* and *Deltocephalus* are well represented in North America, and a number of the species show great adaptation for conditions in the central plain and desert region of the United States. Both have evidently been represented in the region for a long period, and it is quite possible, at least for the latter, that the center of dispersal was from this region. For the grasses of the plain and desert there has been a remarkable modification, in some cases the development of distinct dimorphic forms and a definite restriction to certain species of plants.

The genus *Dorycephalus* is represented in the central northern part of the United States, and species here apparently parallel the only other known species which occurs in Russia in their adaptation to particular grasses, and to a dry climate. Our species have a remakable adaptation to the plants on which they live, the larval forms presenting very striking cases of resemblance to certain parts of the plants.

On the whole it may probably be said that the Jassidæ have had more or less of their development in both of the regions, but that considerable number of the genera quite certainly had their evolution in North America, and their dispersal to adjacent countries.

Sur les Fourmis moissonneuses (*Messor barbara*) des environs de Royan,

par le Prof^r E.-L. BOUVIER (Paris).

Les Fourmis moissonneuses sont communes aux environs de Royan, surtout à Saint-Georges-de-Didonne, où leurs colonies se trouvent en grand nombre dans le sol qui recouvre les falaises et, en certains points, à quelque distance dans les terres. Elles appartiennent à l'espèce *Messor barbara* L., mais il s'y trouve aussi des *Messor structor* LATR., quoique en quantité moindre, autant qu'il m'a semblé.

J'ai pu suivre et étudier ces Insectes pendant les mois de juillet, août et septembre, au cours des vacances de 1910. Ils m'ont révélé bien des traits intéressants, mais je me bornerai à mettre en évidence dans la présente note ceux qui sont relatifs au déménagement. Ils me paraissent curieux et propres à retenir l'attention des biologistes.

Mes observations sur le déménagement ont porté sur deux colonies de *Messor barbara*. Je vais les reproduire telles qu'elles se trouvent consignées dans des notes qui furent prises sur l'heure même et photographièrent pour ainsi dire les phénomènes observés. Ce relevé sera suivi des conclusions qu'il comporte.

Histoire de la première colonie.

Le premier nid où j'observai le déménagement se trouvait installé à une faible distance de Saint-Georges, presque sur le bord de la falaise verticale et tout près du chemin battu que suivent les baigneurs quand ils font leur promenade. A une dizaine de mètres, de l'autre côté du chemin, se trouvait un autre nid qui m'avait peu frappé jusque-là, parce qu'il n'était guère actif. C'est entre ces deux nids que s'effectua le déménagement. J'appellerai P le premier nid et R le second. Comme j'ignorais au début la portée du manège auquel se livraient les Fourmis, on verra dans les notes suivantes la trace des hypothèses que je dus ébaucher avant d'atteindre la solution définitive.

17 août. — Journée grise et sans soleil, tiède, avec un peu de pluie. Tous les nids de moissonneuses sont plus ou moins actifs et ouverts.

A *2 heures*, entre deux petites averses, un long convoi double s'établit entre P et R, traversant en ligne très oblique le chemin battu de la falaise. Les Fourmis du double convoi s'entremêlent et ne forment pas deux trains séparés, mais chacune d'elles poursuit sa route dans une même direction et pénètre dans l'orifice du nid à l'extrémité. Ce n'est qu'une procession, car les Fourmis ne portent rien, à part quelques cas très rares. Mais quelle est la signification de ce manège en deux sens opposés?

Vers *5 heures*, la procession a cessé.

22 août. — Après quelques journées très ensoleillées, il a fortement plu la veille. On a donné un coup de pioche en R pour y capturer des reines, et je vois sortir par les orifices mis à nu quelques-unes de ces dernières, entraînées par des travailleuses. L'activité du nid est grande, car il faut réparer les grands dégâts produits par le coup de pioche.

23 août. — Beau jour très ensoleillé, avec vent médiocre et pas trop de chaleur. Il a plu assez fort une partie de la nuit.

Durant la matinée et l'après-midi, des reines continuent à sortir de R, comme de coutume entraînées par les ouvrières. Une fois libres, les reines deviennent très vite fort actives et, soit après avoir un peu erré, soit de suite, prennent le chemin du nid où elles rentrent. Cette rentrée se fait souvent sans encombre, mais souvent aussi la reine rentrante est de nouveau saisie par les ouvrières qui l'entraînent au loin, d'ailleurs sans traitement mauvais.

Je prends une de ces reines et je la porte à l'orifice d'une autre colonie assez lointaine. Aussitôt une quantité d'ouvrières s'acharnent contre l'intruse, essayent de lui arracher les ailes, la tirent par les antennes et les pattes, et finalement l'entraînent loin du nid. Les reines sont accueillies tout autrement au nid P. La première que j'y porte est palpée, puis entre dans le nid. La deuxième est aussi palpée, mais en outre entraînée à quelque distance; elle revient à l'orifice, en est éloignée de nouveau, puis y entre, cette fois pour tout de bon, semble-t-il.

Cependant de nombreuses ouvrières sortent de P chargées de graines, surtout des petites graines jaune pâle de *Medicago minima* LAM., plante très commune sur la falaise. Ainsi chargées, les Fourmis s'en vont très vite au nid R en suivant la même voie que les processionneuses du 17 août. Il leur faut à peu près un quart d'heure pour suivre cette voie, qui mesure environ 10 mètres. Arrivées au nid R, les ouvrières se comportent de façons diverses : tantôt elles pénètrent dans le nid pour abandonner leur fardeau, tantôt elles déposent celui-ci près de l'orifice où il est saisi par des emmagasineuses, parfois même le fardeau est directement passé à l'une de ces ouvrières. Ceci fait, les Fourmis reviennent à vide et rentrent en P, après avoir suivi la même route en sens inverse. Cela dure toute la journée.

Ces allées et venues, de même que les expériences faites avec les reines, prouvent sans conteste que les nids P et R, quoique fort éloignés, appartiennent à une même colonie. Ce déménagement est sans doute exigé par l'abondante population larvaire de R, où les reines écloses sont nombreuses et où l'on voit quantité d'ouvrières (des jeunes très probablement) sorties et portées au dehors par d'autres. Des galeries souterraines font peut-être communiquer les deux nids P et R, mais le trajet par ces galeries serait certainement plus long que celui effectué au dehors pour le déménagement. Les graines transportées sont parfaitement saines, sans trace aucune de germination.

Vers le soir, sur les débris un peu humides qui entourent

l'orifice de P, je vois courir de nombreux Cloportes blancs et aveugles; ces bestioles sont assez nombreuses, et sans cesse on en voit sortir de nouvelles par l'orifice du nid. En peu de temps je puis en capturer une dizaine. Ce sont les commensaux du nid, et les ouvrières n'y prennent aucune garde.

24 août. — Journée pluvieuse avec de nombreuses éclaircies; vent. Le déménagement continue de P en R.

25 août. — Matinée très pluvieuse avec grand vent; après-midi plus calme et un peu ensoleillée.

Vers *midi*, le déménagement continue, mais les ouvrières qui vont de R à P, pour chercher un fardeau, sont bien plus nombreuses que les autres.

Je reviens à *3 heures* et trouve le nid R éventré à coups de bêche par un chercheur de reines. C'est un désastre que les ouvrières s'occupent à réparer. Celles qui reviennent chargées du nid P se sont blotties près de R sous des mottes rejetées par la bêche; elles y ont déposé leurs graines, attendant sans doute que de nouveaux orifices soient pratiqués par les travailleuses dans le nid mutilé.

Le soir, à *7 heures,* toutes ces graines ont disparu, mais le déménagement est peu actif. A l'orifice du nid P se voient un petit nombre d'ouvrières et quelques Lépismes.

26 août. — Assez beau temps, sans pluie, soleil médiocre, vent assez fort.

Le nid P semble mort, à peine y voit-on deux ou trois ouvrières qui s'éloignent peu de l'orifice. Dans le trou creusé en R par le chercheur de reines, très peu de Fourmis et pas d'orifice visible; on doit sans doute travailler sous le sol pour réparer les dégâts. Mais à 80 centimètres de R, dans un petit trou de bêche donné par le même chercheur, l'activité est grande et des balles d'Avoine sont rejetées en abondance. Ce nid R' appartient-il à la même colonie que R et P?

27 août. — Temps magnifique, vent du nord.

L'orifice et les alentours du nid P sont absolument déserts. En R, faible activité à un orifice où l'on sort de menus débris; activité plus grande en R' où les balles d'Avoine s'accumulent

28 août. — Journée belle et assez chaude.

Le nid P comme la veille, pourtant j'y trouve une ouvrière vers 5 heures; en R et R', l'activité est presque nulle, comme dans la plupart des autres colonies d'ailleurs.

29 août. — Très belle journée, chaude et sans aucun nuage.

Le matin, entre 9 et 10 heures, la plupart des colonies sont actives, au surplus assez médiocrement. Je vois en P une ouvrière, une seule; elle déménage une graine qu'elle emporte vers R.

Inutile de pousser plus loin le relevé de ces notes. Après une absence de quelques jours, je pus constater que le nid P était absolument désert et abandonné par ses habitants, que le nid R, depuis les coups de bêche, avait perdu presque toute activité; tandis que le nid R', au contraire, était devenu un centre de grand travail, occupé par les ouvrières de la même colonie.

Pour quelles raisons nos moissonneuses abandonnèrent-elles le nid P? Il est difficile de le dire d'une façon précise. Mais si l'on observe que ce nid était situé tout au bord de la falaise, en un point où la terre se désagrège et tombe peu à peu au pied du rocher, on peut croire que les Fourmis durent forcément quitter un gîte qui ne leur offrait plus protection suffisante.

Histoire de la seconde colonie.

La seconde colonie dont j'ai suivi le déménagement était située beaucoup plus loin de Saint-Georges, au-dessus des roches de Vallières dont elle était séparée par un enclos qui masque en partie le pittoresque paysage.

Le nid A', dont les Fourmis opérèrent le déménagement, se trouvait sur le bord même de la barrière de clôture, au bas d'un fossé presque horizontal qui suivait la route et servait à l'écoulement des eaux; j'ai pu constater qu'il appartenait à la même fourmilière que d'autres nids voisins A'', situés, comme lui, au fond ou sur les parois du fossé. La voie suivie par les déménageuses traversait un peu obliquement la route et, de l'autre côté de cette dernière, aboutissait à un nid A''' où furent transportées les graines. Le nid A''' se trouvait au niveau même de la route, parmi les herbes qui recouvraient les côtés de celle-ci; il se trouvait dès lors dans un lieu sec, tandis que le nid A', par sa situation au bas d'un fossé d'écoulement, occupait en temps de pluie un milieu fort humide.

Mes observations sur ce nid sont un peu postérieures aux précédentes; elles sont aussi plus complètes parce qu'elles furent suivies plus longtemps et reçurent un bénéfice de ces dernières.

24 août. — De nombreuses graines en voie de germination (les graines jaunes d'une petite légumineuse) sont rejetées en dehors sur les monticules de débris de A' et de A'' situés au fond du fossé d'écoulement. (Il y a eu un orage violent dans la nuit du 15 au 16 août, quelques averses le 16 et le 17, et de fortes pluies le 21; le 22 et le 23 août ont été beaux, mais la journée du 24 fut un peu pluvieuse. Ainsi s'explique la germination de certaines graines dans les nids A', A'' situés au fond du fossé.)

27 août. — La matinée du 25 août a été très pluvieuse, mais il faisait assez beau le 26 et un temps magnifique le 27.

Le terrier A'' est ouvert, mais complètement inactif. En A', par contre, l'activité est très grande. Comme au cours des journées précédentes, les ouvrières traversent la route suivant une direction un peu oblique, mais au lieu de se rendre dans les herbes du bord pour y faire récolte, elles se rendent au terrier A''' sans monticule de débris, et un double convoi d'ouvrières cheminant à vide s'établit entre A' et A'''. C'est une double procession identique à celle observée précédemment entre P et R.

Dans l'un ou l'autre sens, les processionneuses voyagent à vide; quelques-unes pourtant sont chargées de larves qu'elles portent de A' à A'''. Ces porteuses de larves sont très rares; j'en ai trouvé cinq pendant une durée d'observation de vingt minutes; c'est bien peu, car des milliers de Fourmis ont dû passer sous mes yeux, dans le double convoi, durant ce laps de temps. Tantôt les larves emportées sont petites et agglomérées en paquets où j'en ai compté jusqu'à quinze, le plus souvent elles sont solitaires et plus grosses. Au cours des mêmes vingt minutes, j'ai vu deux Fourmis chargées de larves, qui allaient de A''' en A'. Serait-ce une erreur de la part des porteuses? En tout cas, le transport des larves s'effectue principalement en sens contraire, de A' à A'''.

Le terrier de A' est le siège d'une activité intense; j'en vois sortir une reine tirée, comme de coutume, par une aile. Entraînée à quelque distance du nid, celle-ci se hâte d'y revenir dès qu'elle est libre; mais bientôt elle rebrousse chemin et se mêle à la procession qui se dirigera vers A'''; d'ailleurs elle est bientôt arrêtée par ses compagnes de voyage et revient vers A'. Je ne l'ai pas suivie plus longtemps.

28 août. — Journée belle et assez chaude encore que beaucoup de colonies soient extérieurement peu actives l'après-midi.

La procession de la veille continue. Les ouvrières qui se dirigent de A' vers A''' sont moins nombreuses, mais l'activité est extraordinaire dans ce dernier nid. Une partie des ouvrières qui en sortent se rendent en A', les autres vont récolter parmi les herbes du voisinage.

29 août. — Très belle journée.

Le matin, les ouvrières des nids A' et A''' se comportent comme la veille, mais avec une activité deux fois moindre. Après-midi, la double procession a pris fin, et seules quelques déblayeuses travaillent sur le monticule de A'.

4 septembre (après une absence de quelques jours). — Il fait beau mais frais le matin, très chaud à midi.

Le matin, très faible activité en A' où je vois deux ouvrières se rendre sans charge en A''' où l'activité n'est pas plus grande.

L'après-midi, vers 2 heures, l'activité extérieure est nulle dans les deux nids, qui sont clos.

Le soir, au *coucher du soleil*, le temps se couvre et devient frais, certaines colonies sont en grande activité récoltante, mais les ouvrières de A' et de A''' se bornent à ouvrir leur terrier.

A *9 heures du soir*, un double train serré d'ouvrières s'est établi entre ces deux derniers nids.

5 septembre. — Le matin, pluie fine et pénétrante qui cesse vers midi.

A *2 heures*, je me rends sur les rochers de Vallières. Je retrouve le double train de la veille, mais il paraît moins dense, et les ouvrières qui vont de A' en A''' sont presque toutes chargées de graines, les autres revenant à vide en A'. C'est le déménagement.

Le soir, vers le *coucher du soleil*, le déménagement continue;

mais l'orifice de A' est entouré de Cloportes blancs et aveugles qui sortent par l'orifice ; il y en a des centaines, et il en sort toujours ; cela fait une large tache blanche sur le monticule du terrier. Les Cloportes suivent le train des ouvrières, sans jamais beaucoup s'en écarter ; il sont encore nombreux à 1 mètre du nid, rares plus loin ; j'en trouve pourtant plusieurs à 3 ou 4 mètres de A''' et même une sur le monticule de ce nid.

A *9 heures*, par une nuit noire et tiède, le double train de déménagement est d'une activité extrême. Les Cloportes pullulent, on les compte par milliers, et leur nombre, certainement, n'est pas inférieur à celui des Fourmis, entre lesquelles ils s'agitent sans être le moins du monde inquiétés. Ils sont plus nombreux autour de A', mais on en trouve partout sur la voie suivie par les Fourmis, et jusque dans l'orifice de A''' où ils pénètrent. Bien que divaguant un peu dans leur marche irrégulière, ils se dirigent visiblement de A' en A''' ; c'est bien une émigration. Ils sont repoussés par la lumière, tandis que les Fourmis en sont simplement un peu étonnées.

6 septembre. — A *6 heures du matin*, par un temps couvert, brouillassant et un peu frais, le double train de déménagement continue, mais singulièrement réduit. Les Cloportes sont beaucoup moins nombreux que la veille et localisés plus ou moins autour de A' et de A''' ; on n'en trouve plus sur la partie de la voie qui traverse la route ; suivant la position qu'ils occupent, ils rentrent dans l'un ou l'autre nid.

Le temps se met au beau et devient magnifique. L'après-midi, de *2 à 3 heures*, les colonies sont inactives, notamment en A' et A''' où le déménagement a cessé, de même que l'émigration des Cloportes : A' est ouvert, A''' presque clos. Aux environs de A', les Cloportes ont abandonné leurs mues par centaines.

Vers le *coucher du soleil*, l'activité reprend un peu ; le temps est couvert. En A', A''', quelques rares ouvrières reconstituent le double train de déménagement. Pas de Cloportes au nid A''', qui est très incomplètement ouvert : par contre, on en voit d'assez nombreux à l'orifice de A', dont ils ne s'éloignent guère.

A *8 1/2 heures*, par un temps clair et tiède, sous un ciel semé d'étoiles, le double train de déménagement est fort actif. Les Cloportes sont au dehors, moins nombreux que la veille, mais toutefois encore très abondants. J'en trouve deux sur la voie suivie par les déménageuses, à mi-chemin entre A′ et A‴. Je n'en vois pas au delà, et il n'y en a aucun à l'orifice de A‴, c'est, probablement, le début de l'émigration ce jour-là.

7 septembre. — A *8 heures du matin*, par un temps couvert mais presque doux, le double train de déménagement fonctionne à peine, et c'est tout au plus si l'on compte une dizaine d'ouvrières entre les deux nids. Les Cloportes sont également rentrés, mais pas tous, et j'observe des traînards en divers points sur toute la longueur de la voie suivie par les Fourmis; ces traînards sont quelque peu errants, et néanmoins ne s'écartent guère de la voie ; presque tous reviennent en A′

Inactivité complète tout le reste du jour jusqu'au coucher du soleil. A *7 1/4 heures*, quelques rares Fourmis s'aventurent aux orifices de A′ et de A‴, mais l'obscurité est déjà si grande qu'il m'est impossible d'apercevoir des Cloportes.

8 septembre. — Il a plu un peu la nuit, et le jour commence très nuageux.

A *10 1/2 heures* du matin, un petit nombre de Fourmis établissent encore un double train de déménagement. Il y a également quelques Cloportes sur la voie suivie, en des points fort divers ; tous, sauf les plus voisins de A′, se dirigent vers A‴ sans s'écarter de la voie qu'ils reconnaissent sans doute à l'odeur (car les ouvrières y sont très rares), grâce aux mouvements actifs et continus de leurs antennes. Je suis particulièrement deux de ces Cloportes : l'un était à 30 centimètres de A‴ où il arrive après quelques minutes, en s'insinuant entre les parcelles terreuses de l'orifice qui est en partie bouché. L'autre était à $1^{m}50$ et fut capturé au moment où il arrivait en A‴.

L'après-midi, presque plus de déménageuses et aucun Cloporte sur la voie de déménagement ; quelques-uns s'aventurent à l'orifice de A′.

9 septembre. — La pluie est menaçante ; pourtant, à 10 heures du matin, quelques ouvrières déménagent encore. Pas de Cloportes.

10 et 11 septembre. — Tempête continue et très fortes pluies.

12 septembre. — Temps nuageux avec éclaircies fréquentes ; il fait assez chaud ; quelques averses dans l'après-midi. Partout les ouvrières s'occupent à rétablir les orifices détériorés par les pluies des jours précédents; dans de nombreuses colonies, des orifices nouveaux sont établis (sans doute pour faciliter l'évaporation), et des éclosions d'ailées se produisent.

Les pluies ont fait disparaître le monticule A' situé au fond du fossé d'écoulement; A'' est clos, mais un peu au-dessus, dans la pente même du fossé, un nouvel orifice s'est ouvert, où l'on voit des ouvrières et quelques Cloportes. A''' est ouvert, mais peu actif; à côté, les ouvrières ont établi un nouvel orifice où l'activité est plus grande.

Le soir, au coucher du soleil, l'activité est plus faible encore ; quelques ouvrières seulement se voient aux orifices. Le nouvel orifice de A'' est occupé par de nombreux Cloportes qui s'aventurent quelque peu au dehors, mais pas d'ouvrières. Tout est donc en A'''.

A partir du 12 septembre, le nid A' semble mort, mais un restant d'activité se manifeste dans les orifices voisins, d'où l'on voit sortir parfois un Cloporte; puis l'activité disparaît partout au fond du fossé, tandis qu'un ou deux orifices supplémentaires s'ouvrent au voisinage de A'''.

Le 24 septembre, j'ouvre le nid A'. Les greniers les plus bas, c'est-à-dire situés dans le sol même du bas du fossé, renferment encore de nombreuses graines, toutes en germination, avec de longues tigelles qui se dirigent vers la surface en s'enchevêtrant. On atteint vite le roc en ce point, mais de multiples galeries s'avancent dans le flanc du fossé. Les greniers annexés à ces galeries renferment encore beaucoup de graines; certaines graines sont en germination dans les greniers les plus bas, c'est-à-dire les plus voisins du fond du fossé; dans les autres, la récolte est restée intacte, et je trouve même quelques rares Fourmis et quelques Cloportes dans les greniers les plus élevés.

Ainsi, aucun doute ne saurait subsister. Les Fourmis, suivies des Cloportes, ont abandonné A' parce que ce nid était situé dans des conditions fâcheuses, exposé aux inondations et à une humidité persistante à cause des eaux des pluies qui venaient se réunir dans le fossé d'écoulement.

J'ai ouvert le nid avant le déménagement complet, mais peut-être ses graines restées intactes auraient-elles été, comme les autres, déménagées dans la suite.

Conclusions.

De ces deux séries d'observations concordantes on doit conclure :

1° Que les Fourmis moissonneuses abandonnent l'un des gîtes de leurs colonies lorsque ce gîte se trouve dans des conditions par trop désavantageuses, et qu'elles déménagent les récoltes de ce gîte pour les porter dans un autre, préexistant, situé en lieu plus sûr ;

2° Que le déménagement est précédé par une double *procession de reconnaissance*, qui s'effectue à vide entre les deux gîtes et qui permet aux Fourmis de bien connaître la voie à suivre, les lieux qu'il faut quitter et ceux qui serviront de gîte définitif;

3° Que, cette connaissance des lieux une fois acquise, les Fourmis établissent, suivant la même voie, un double *train de déménagement* dans lequel les ouvrières emportent au gîte définitif les récoltes du gîte désavantageux, puis reviennent à vide dans ce dernier pour y prendre une nouvelle charge ;

4° Que le Cloporte aveugle des fourmilières, *Platyarthrus Hoffmannseggi* BRANDT, pullule parfois dans les gîtes de notre moissonneuse, qu'il les abandonne pour suivre cette dernière une fois l'émigration commencée, et que son émigration propre s'effectue à peu près exclusivement la nuit, grâce aux odeurs laissées sur le sol par les Fourmis et à la sensibilité olfactive, très grande, des antennes de l'animal.

Je crois bien que ces faits n'ont pas encore été observés chez les moissonneuses, du moins sont-ils passés sous silence par EBRARD, MOGGRIDGE, FOREL, PIÉRON et les autres myrmécologues consultés à ce sujet. Le déménagement et l'émigration consécutive des Cloportes ne sont pas des nouveautés dans l'histoire des Fourmis, mais il semble bien qu'on n'avait jamais observé d'aussi près l'exode du *Platyarthrus* et qu'on ne soupçonnait pas ce Crustacé en telle quantité dans certaines fourmilières. Ce qui me paraît plus neuf, c'est le procédé par lequel les ouvrières préludent au déménagement, je veux dire la double procession de reconnaissance. Ainsi que j'ai pu l'observer à Vallières, et cette année même dans mon jardin de Maisons-Laffitte, le *Lasius fuliginosus* LATR. est coutumier du fait, mais s'il établit des processions doubles, où il semble voyager à vide dans les deux sens, c'est pour aller au loin prendre sur les arbres la nourriture qu'il convoite, non pour se renseigner sur les lieux.

Au surplus, les observations biologiques gagnent à être plusieurs fois répétées, car elles sont délicates, souvent accidentelles, et il faut profiter du hasard heureux qui permet de les faire; celles-ci ont été suivies longuement et de très près, c'est pourquoi j'ai jugé utile de les soumettre aux biologistes.

Les Polycténides et leur adaptation à la vie parasitaire,

par le Dr G. HORVATH (Budapest).

Les Polycténides constituent une petite famille d'Hémiptères parasites des Chauves-souris exotiques. On ne connaît actuellement que dix espèces, dont sept habitent les régions tropicales et subtropicales d'Asie et d'Afrique, et trois qui sont propres à l'Amérique centrale et méridionale. Toutes sont d'une petite taille (2 $^1/_2$ à 4 mill. de long) et sans exception très rares; il n'existe encore jusqu'à présent dans les collections qu'à peine deux douzaines d'exemplaires.

Vu la grande rareté et l'organisation singulière de ces Insectes, leur place systématique est restée assez longtemps méconnue. On les avait rangés tantôt parmi les Anoploures, tantôt parmi les Diptères, tantôt parmi les Hémiptères.

H. GIGLIOLI, qui a décrit en 1864, sous le nom de *Polyctenes molossus*, la première espèce trouvée sur une Chauve-souris de Chine, a pris cet Insecte pour un Diptère appartenant à la famille de *Nyctéribiides*.

WESTWOOD, qui a eu l'occasion de soumettre, en 1874, à un nouvel examen ce *Polyctenes* de Chine et de décrire en même temps encore une deuxième espèce (*Polyctenes fumarius*) de la Jamaïque, a placé ces Insectes dans l'ordre des Anoploures. Il a cependant fait remarquer que ces formes curieuses, pour lesquelles il a créé la famille des *Polyctenidæ*, montrent une certaine affinité avec les Hémiptères.

M. C.-O. WATERHOUSE, en publiant en 1879 deux espèces nouvelles (*Polyctenes lyræ* et *spasmæ*) de l'Asie méridionale, a émis l'opinion que les Polycténides appartiennent aux Diptères et qu'ils se rapprochent beaucoup des Hippoboscides. Mais l'année suivante, à propos de la description d'une seconde espèce américaine (*Polyctenes longiceps*), il a abandonné cette hypothèse et, revenant à l'idée de WESTWOOD, il s'est prononcé en faveur de l'affinité des Polycténides avec les Hémiptères.

Néanmoins on n'a pas cessé de considérer les Polycténides comme Diptères, et les catalogues des Diptères orientaux de J.-M.-F. BIGOT (1892) et de F.-M. VAN DER WULP (1906) les énumèrent encore parmi des Nyctéribiides. Ce n'est que depuis les recherches du Dr P. SPEISER que la place systématique de ces Insectes est définitivement arrêtée. Le Dr SPEISER a démontré, en 1904 (« Zoolog. Jahrb. », Suppl. VII, p. 377), qu'ils sont des Insectes amétaboliques avec l'appareil buccal suceur, c'est-à-dire qu'ils sont de vrais Hémiptères; et dans un récent travail (« Record of the Ind. Mus. », III, 1909, p. 272), il s'est prononcé catégoriquement pour ranger ces Insectes auprès de la famille des Cimicides (*Acanthiadæ*), dont le type universellement connu est la fameuse Punaise de lit.

Le même auteur a décrit encore deux espèces nouvelles, une (*Polyctenes talpa*) de l'île de Nias, l'autre (*Polyctenes intermedius*) d'Égypte. Le nombre des espèces connues est ainsi arrivé à sept.

Toutes les sept espèces ont été placées dans le seul genre (*Polyctenes*). Mais G.-W. KIRKALDY, dont nous déplorons la mort aussi inattendue que prématurée, a divisé, en 1906 (« Canad. Entomol. », XXXVIII, p. 375), l'ancien genre *Polyctenes*, en se basant sur les différences dans la structure de la tête, des antennes, du pronotum et des pattes, en quatre genres : *Hesperoctenes, Euroctenes, Polyctenes* et *Eoctenes*. Trois de ces genres sont en effet valables, mais je pense que le genre *Euroctenes* est à réunir aux *Polyctenes*.

KIRKALDY n'a pas connu les deux espèces décrites par le Dr SPEISER; elles constituent, d'après moi, un genre à part (*Hemischizus*). C'est au même genre qu'il faut rapporter encore une troisième espèce, provenant du Soudan et décrite tout récemment sous le nom d'*Eoctenes eknomius* par MM. V.-L. KELLOG et J.-H. PAINE (« Entomolog. News », XXI, 1910, p. 402).

Une nouvelle espèce de l'Afrique centrale, que je viens de décrire (« Ann. Mus. Hung », VIII, 1910, p. 572), avec une nouvelle provenant du Brésil, est aussi le type d'un genre nouveau (*Ctenoplax*).

Les dix espèces actuellement connues se répartissent dans les cinq genres de la manière suivante :

Hesperoctenes KIRK., 1906.

1. *fumarius* WESTW., 1874. — Jamaïque (sur *Molossus rufus* var. *obscurus*.
2. *impressus* HORV., 1910. — Brésil.
3. *longiceps* WATERH., 1880. — Guatémala (sur *Molossus abrasus*).

Polyctenes GIGL., 1864.
(*Euroctenes* KIRK., 1906.)

1. *molossus* GIGL., 1864. — Chine méridionale (sur *Nyctinomus Cestonii*).
1. *lyræ* WATERH., 1879. — Inde orientale (sur *Megaderma lyra*).

Eoctenes KIRK., 1906.

1. *spasmæ* WATERH., 1879. — Java (sur *Megaderma spasma*).

Ctenoplax HORV., 1910.

1. *nycteridis* HORV., 1910. — Afrique centrale (sur *Nycteris hispida*).

Hemischizus HORV., 1911.
(*Syncrotus* HORV., 1910.)

1. *talpa* SPEIS., 1898. — Ile Nias (sur *Megaderma spasma*).
2. *intermedius* SPEIS., 1904. — Égypte (sur *Taphozous perforatus*).
3. *eknomius* KELLOG et PAINE, 1910. — Soudan.

Si nous examinons ces Insectes de plus près, nous trouvons que c'est, en effet, des Cimicides qu'ils se rapprochent le plus, quoique leur aspect général soit bien différent et que plusieurs de leurs caractères aient subi d'importantes modifications. Ces différences et ces modifications sont incontestablement le résultat de leur adaptation à la vie parasitaire constante. Elles s'expliquent facilement par le fait que les Cimicides sont des parasites temporaires, tandis que les Polycténides sont des parasites stationnaires qui ne quittent jamais leurs hôtes.

Les Cimicides, qui ne s'attaquent à leurs victimes que pendant un certain temps, mais qui en restent le plus souvent éloignés, sont bien conformés, avec leur corps aplati, pour se cacher dans les fissures des murs, des bois, des rochers, etc., et s'y sauver rapidement grâce à leurs pattes assez allongées. Mais pour les Polycténides fixés à demeure dans la fourrure des Chauves-souris, il est de la plus grande importance que leur habitat y soit assuré autant que possible. Ce sont leurs propres hôtes qui seuls ont intérêt à les déranger et qui cherchent, en se grattant, à se débarrasser d'eux. Le seul danger dont les Polycténides soient menacés constamment, c'est d'être bousculés ou expulsés par les griffes de leurs hôtes. Il n'y a donc rien de surprenant à ce que la plupart des modifications que l'on peut observer chez ces Insectes se soient produites dans le but de les protéger contre les brusques mouvements de leurs souffre-douleur et d'y assurer leur stabilité.

La forme générale du corps, qui est très élargie chez les Cimicides, est devenue chez les Polycténides beaucoup plus étroite et graduellement rétrécie vers l'avant. La tête a pris une forme déprimée, parabolique, graduellement dilatée en arrière, mais les expansions latérales du pronotum sont disparues. Les bords latéraux de la tête et du pronotum forment par conséquent une ligne continue, de sorte que les griffes des Chauves-souris n'y trouvent aucune saillie où elles pourraient s'accrocher.

Une telle structure du corps et plus particulièrement de la tête est assez répandue parmi les Insectes habitant la fourrure des Mammifères. Ainsi, nous avons en première ligne deux genres de Streblides (*Strebla*, *Euctenodes*) et les Nycteribiides, qui tous sont des Diptères parasites des Chauves-souris. Il en est de même pour le curieux Coléoptère *Platypsyllus castoris*, qui vit dans la fourrure du Castor. Et, enfin, les Siphonaptères sont aussi ordinairement d'une taille plus ou moins allongée, avec les bords latéraux de la tête, du thorax et de l'abdomen formant une seule ligne continue, sans saillies apparentes.

Un des caractères les plus remarquables que présentent les Polycténides réside dans la structure de leur tête, qui paraît divisée en deux parties. La partie apicale forme une lame semi-lunaire, mobile, réunie à la partie basale par une articulation distincte. Une pareille structure de la tête se trouve aussi chez les deux genres *Strebla* et *Euctenodes* (de la famille des Streblides) dont les espèces ont l'aspect général des Polycténides. Comment peut-on expliquer une telle bipartition de la tête ? et qu'est-ce que cette lame apicale mobile chez les Polycténides ? WESTWOOD a pensé, mais avec quelque doute, que c'est peut-être le clypeus, et le Dr SPEISER, qui a fait un excellent travail sur les Polycténides, a aussi émis la même opinion. Je pense que les deux auteurs sont bien dans le vrai (1).

Cette mobilisation du clypeus a évidemment pour but de mieux assurer à ces Insectes leur séjour dans la fourrure des Chauves-souris. La tête des Polycténides est bien aplatie et amincie à son bord antérieur, mais il est évident qu'elle pourra se serrer encore mieux contre la peau de leur hôte et n'y laisser aucune prise aux griffes de celui-ci, si sa partie apicale est mobile et peut s'appliquer ainsi parfaitement à la surface de la peau.

Un autre caractère très important des Polycténides est l'absence totale des yeux. Ce manque d'organes de la vue s'explique par la vie sédentaire et n'est pas rare chez les Insectes parasitaires, par exemple chez beaucoup de Streblides, Nyctéribiides, Siphonaptères, Pédiculides, Mallophages, le *Platypsyllus castoris*, *Braula cæca*, etc.

Le peu de développement des antennes et du rostre que l'on observe chez les Polycténides doit être attribué aussi à la vie parasitaire. Les antennes sont plus courtes ou seulement un peu plus longues que la tête. Le rostre, qui est triarticulé comme chez les Cimicides, est toujours plus court que la tête.

Les trois caractères précédents, c'est-à-dire l'avortement des yeux, le raccourcissement des antennes et du rostre, sont, avec l'absence de l'écusson, les seules modifications morphologiques qui ne paraissent pas contribuer à assurer et augmenter la bonne et solide fixation dans la fourrure, tandis que toutes les autres se sont produites dans cet unique but.

On peut le constater aussi dans la structure des pattes, à l'aide

(1) La partie apicale mobile de la tête des genres *Strebla* et *Euctenodes* est, d'après le Dr SPEISER (« Archiv für Naturg. », 1900, I, p. 35), très probablement issue de la soudure des palpes maxillaires dilatées.

desquelles l'Insecte s'accroche et se fixe dans la fourrure de son hôte. Les pattes antérieures sont toujours très courtes, avec les cuisses fortement renflées, ce qui laisse deviner la grande force avec laquelle ces Insectes se cramponnent à la place qu'ils occupent. Les pattes intermédiaires et postérieures sont moins épaisses et peu raccourcies chez les espèces américaines, dont les cuisses postérieures dépassent l'extrémité de l'abdomen; chez les espèces de l'ancien monde, les deux dernières paires de pattes sont plus courtes, les cuisses plus robustes et les fémurs postérieurs n'atteignent pas l'extrémité de l'abdomen.

Les espèces asiatiques et africaines diffèrent aussi des espèces américaines par la structure des ongles. Les espèces américaines ont les mêmes ongles simples, pourvus à la base d'un petit tubercule obtus, que ceux des Cimicides. Chez les espèces de l'ancien monde, un des ongles a conservé cette forme primitive, mais l'autre est fortement courbé en crochet à la base, avec le tubercule basal prolongé en une longue dent aiguë qui donne à cet ongle l'apparence d'être bifide. Cette curieuse modification des ongles, dont on trouve l'analogie chez certains genres des Hippoboscides (*Ornithomyia*, *Ornithoctona*, *Lynchia*, etc.), sert évidemment à l'Insecte pour mieux se fixer entre les poils de la fourrure.

La structure des tarses n'était pas encore élucidée suffisamment jusqu'à présent chez les Polycténides, et les auteurs ne sont pas tout à fait d'accord sur le nombre des articles dont les tarses intermédiaires et postérieurs sont composés. On les avait indiqués, décrits et dessinés, tantôt triarticulés, tantôt quadriarticulés. Cette contradiction paraît devoir s'expliquer par des observations trop superficielles, car les tarses ainsi que les tibias intermédiaires et postérieurs sont annelés d'un blanc clair, et ces anneaux clairs paraissent répondre à autant d'articulations. En ce qui concerne les tibias, il n'est encore venu à personne l'idée qu'ils pourraient être vraiment articulés; mais quant aux tarses, on est resté moins rigoureux et, en prenant pour une articulation l'anneau clair qui se trouve sur le deuxième article tarsal, on s'est imaginé y voir un tarse à quatre articles. Or, un tarse à quatre articles serait en effet trop extraordinaire dans l'ordre des Hémiptères. Aussi je ne doute guère que tous les tarses des Polycténides ne soient que triarticulés.

Il est impossible de donner en ce moment une explication suffisante de l'origine et de la destination de ces anneaux clairs des tibias et des tarses. Ils ont certainement une relation quelconque avec la vie parasitaire, puisqu'on les retrouve aussi sur les cuisses et parfois (*Cyclopodia, Eucampsipoda*) même sur les tibias des Nycteribiides, parasites également des Chauves-souris.

Les Polycténides présentent encore un autre caractère bien remarquable et unique dans l'ordre des Hémiptères : c'est la présence des « peignes » (*ctenidium*). On sait que les « peignes » sont des épines spécialisées et disposées en ligne transversale ou oblique sur certaines parties du corps. Ils se trouvent chez beaucoup d'Insectes parasitaires et particulièrement chez ceux qui vivent dans la fourrure des Mammifères, comme les Siphonaptères, les Nycteribiides, quelques Streblides (*Strebla*, *Euctenodes*), le *Platypsyllus castoris*, etc.

Le développement successif des peignes chez les Polycténides est très intéressant et peut nous servir en même temps de guide dans la phylogénèse de cette famille. C'est chez les espèces néotropicales (*Hesperoctenes*) qu'ils sont le moins développés, on n'y trouve qu'un seul peigne à la face inférieure de la tête, et même ce peigne est encore assez incomplet et largement interrompu sur la ligne médiane. Chez les espèces asiatiques et africaines, ce peigne inférieur devient plus complet, et il en existe en outre aussi à la face supérieure du corps. Ils y apparaissent au début seulement sur le bord postérieur de la tête (*Eoctenes*), puis aussi sur le bord postérieur du pronotum (*Polyctenes*) et, enfin, même sur le bord apical des élytres rudimentaires (*Ctenoplax*, *Hemischizus*).

Il est à noter que les peignes qui se trouvent sur l'occiput, le pronotum et les élytres des espèces de l'ancien monde sont, au point de vue phylogénétique, de formations assez jeunes, car ils manquent encore chez les nymphes, c'est-à-dire à un état larvaire où le peigne inférieur de la tête est déjà présent et où un des ongles a déjà subi la modification si caractéristique dont je viens de parler (1).

Ces peignes sont assurément destinés à augmenter la bonne fixation de l'Insecte dans la fourrure. Ils jouent à peu près le rôle des crochets d'un harpon entre lesquels les poils de la fourrure se

(1) La première nymphe connue appartenait à *Polyctenes molossus* GIGL. Elle a été décrite et figurée par WESTWOOD (« Thesaur Oxon. », 1874, p. 199. tab. 39, fig. B), mais cet auteur l'avait considérée comme le mâle (?), tandis qu'il avait pris pour la femelle (?) un individu qui avec son pénis saillant est évidemment un mâle (loc. cit., tab. 39, fig. *A*). Une autre nymphe a été décrite par le Dr SPEISER (« Records of the Ind. Mus. », III, 1909, pp. 272-273) ; elle appartient peut-être à *Hemischizus talpa* SPEIS. où à une espèce nouvelle.

serrent et s'opposent de cette façon à ce que l'Insecte soit enlevé facilement par les griffes de son hôte.

Chez deux genres de l'ancien monde (*Ctenoplax*, *Hemischizus*), dont les espèces sont les plus avancées dans l'adaptation à la vie parasitaire, les élytres sont soudées sur la moitié basilaire de leur suture, ce qui augmente encore davantage l'importance biologique, c'est-à-dire la résistance du peigne apical des élytres.

Les espèces asiatiques et africaines ont le premier article des antennes épais et muni, à la face inférieure, de quelques épines semblables à celles des peignes.

Nous voyons donc que tous, ou à peu près tous, les caractères dont les Polycténides se distinguent des autres Hémiptères, et plus particulièrement des Cimicides, doivent leur origine à l'adaptation de ces Insectes à la vie parasitaire. Cette adaptation se manifeste principalement par une tendance à leur assurer une bonne et solide fixation entre les poils de la fourrure, et destinée à les préserver autant que possible des attaques de leurs hôtes. Les espèces américaines, avec leurs pattes intermédiaires et postérieures peu raccourcies et assez grêles, leurs ongles simples, leur peigne unique et incomplet, représentent un type primitif, tandis que les espèces de l'ancien monde prouvent, par la structure des pattes, le développement plus fort des peignes, la taille plus allongée, les antennes généralement plus courtes et aussi par d'autres caractères, qu'elles sont bien plus avancées dans la voie de l'adaptation à la vie parasitaire.

EXPLICATION DE LA PLANCHE XVIII

Fig. 1. — *Hesperoctenes impressus* Horv., Brésil; grossissement 33.

Fig. 2. — *Ctenoplax nycteridis* Horv., Afrique centrale; grossissement 33.

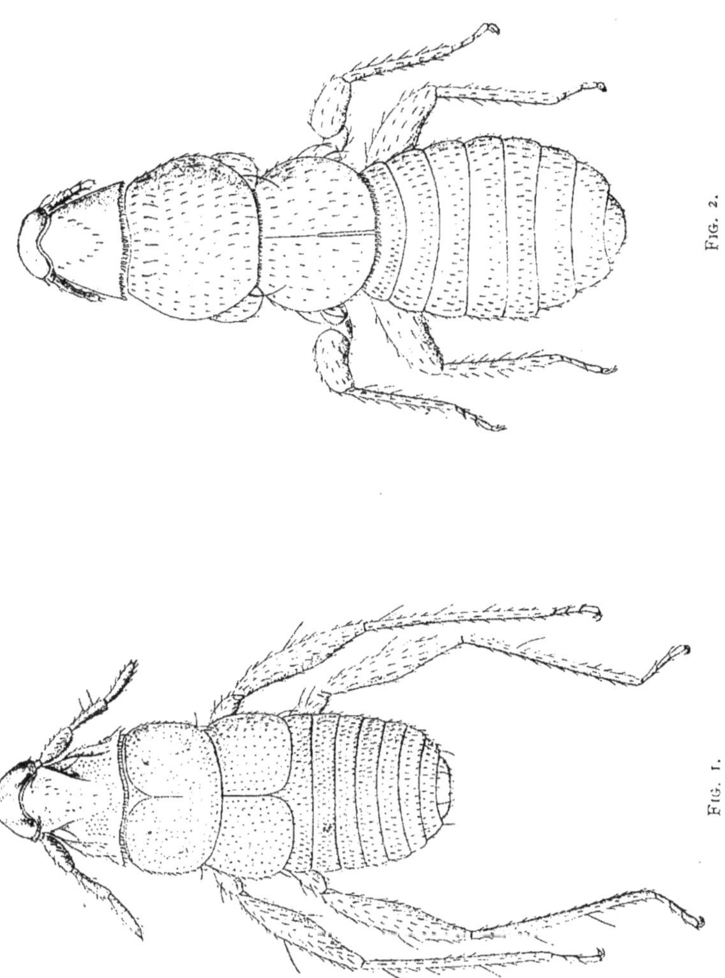

Fig. 2.

Fig. 1.

Dr G. HORVATH. — LES POLYCTÉNIDES.

Kleine Ergebnisse,

von Dr. PHIL. OTHM. EM. IMHOF.

Summarische Uebersicht des *Catalogus Hymenopterorum hucusque descriptorum Auctore Professore Œnipontano* Dr. C. G. DE DALLA TORRE.

Die Cataloge ganzer Classes und Ordines geben die Möglichkeit, eine Vorstellung von der numerischen Repräsentation der bisher entdeckten und beschriebenen Species, Varietäten, Subspecies, Formen, der Subgenera, Genera und der Classificationseinheiten : Familien, Subfamilien, Tribi, Subordines, Ordines zu gewinnen. Nur von der Ordo Hymenoptera besitzen wir einen neueren vollständigen alle Einheiten vereinigenden einheitlichen Catalogus, erschienen von 1893-1902, Bände I bis X.

1894, Bd. I, Bd. IX;	1902, Bd. III;	1892, Bd. VI;	1896, Bd. X.
1892, Bd. II, Bd. VII;	1898, Bd. IV, Bd. V;	1897, Bd. VIII;	

Der Catalog enthält annähernd 43,000 Species mit den Litteraturtiteln und dem geographischen Vorkommen, für einen einzelnen Autor eine unglaubliche Riesenarbeit.

Der Catalog ist eine reiche Basis für eine Reihe wissenschaftlicher Studien wie

Historische Uebersicht der Description der Species und Genera.
Historische Uebersicht der Synonyma.
Geographisches Vorkommen der Genera und Familien.
Uebersicht der Varietäten.

Eine Numerirung der Genera und Species fehlt leider, wohl weil sie den Gebrauch erschwert hätte.

Eine sichere Zählung besonders der Species der vielen Genera mit Hunderten von Species ist ein ziemlich schwieriges Unternehmen und war am ehesten durch Schreibung eines Cataloges ohne die unzählbaren Litteraturtitel möglich und mit diesem Catalog mit Beisetzung des geographischen Vorkommens in besonderen Farben für die Erdhauptteile konnte eine klare Verbreitung der Genera erzielt werden.

Die numerische Vertretung der Familien für die grösseren Familien in Annäherungszahlen ist :

	Genera.	Species.	Zusammenfassungen.
Agriotypidæ	1	1	
Apidæ	136	5,800	14
Braconidæ	239	3,100	27
Chalcididæ	403	5,100	37
Chrysididæ	15	820	8
Crabronidæ	71	3,300	17
Cynipidæ	94	956	7
Evanidæ.	6	284	3
Formicidæ	164	2,500	5
Ichneumonidæ . . .	659	10,600	5 12 10 5 7 6
Lysiognathidæ . . .	1	1	
Megalolyridæ . . .	2	5	
Mutillidæ	5	1,300	
Pelecinidæ . . .	2	13	
Pompilidæ	14	1,800	
Proctotrupidæ . . .	179	1,870	12

	Genera.	Species.	Zusammenfassungen.
	—	—	—
Sapygidæ	1	43	
Scoliidæ	10	673	
Stephanidæ	4	66	
Tenthredinidæ . . .	132	2,670	18
Thynnidæ	3	347	
Trigonaloidæ . .	3	32	
Vespidæ	50	1,700	3

Von den artenreichen Genera zählen

54	Genera mit	101- 200	Species	7,618	Species.
19	—	205- 283	—	4,523	—
10	—	322- 384	—	3,519	—
6	—	414- 482	—	2,715	—
8	—	553-1,675	—	7,039	—

Diese 97 Genera mit 101-1,675 Species zählen zusammen 25,414 Species. Diese Summe ist etwa drei Fünfteile der Gesammtzahl Hymenoptera, die Zahl dieser Genera etwa ein Drieundzwanzigteil aller Genera mit aufgeführten Species.

Vorkommen der Generazahl in den Erdhauptteilen.

	Genera.		Genera.
	—		—
Nord-Amerika	912	Indischaustralischer Archipel .	242
Central-Amerika	323	H.-Ind., China, Japan . . .	204
Centralamerik. Archipel . .	280	Australien	240
Süd-Amerika	456	Sibirien	77
Klein-Asien, Persien, Arabien, Aegypten.	56	Transcaspien	48
Vorder-Indien	309	Caucasus.	79
Grönland	..	N. Pacifische Inseln	49
Europa	1,313	Hinter-Indien	...
Afrika	517	Präliminär	..

Grösste Generazahl der Familien.

Vorder-Indien : *Mutillidæ.*
Afrika : *Chrysididæ*, *Vespidæ.*
Nord-Amerika : *Lysiognathidæ*, *Proctotrupidæ*, *Crabronidæ*,
Europa : *Agriotypidæ*, *Cynipidæ*, *Tenthredinidæ*, *Braconidæ.*
Süd-Amerika : *Trigonaloidæ*, *Pompilidæ*, *Formicidæ*, *Apidæ.*

Varietäten der Familien.

Nur in den Familien : *Agriotypidæ*, *Lysiognathidæ*, *Megalolyridæ*, *Sapygidæ*, *Stephanidæ* und *Thynnidæ* sind keine Varietäten enthalten.

Die Gesammtzahl der variirenden Species beträgt 974 mit 1,655 Varietäten. Besonders bemerkenswert ist die grosse Zahl Varietäten *Vespidæ* 98, *Chrysididæ* 103; die grössten Zahlen der übrigen Familien sind :

Tenthredinidæ . . .	106	Species	190	Varietäten.
Formicidæ	169	—	257	—
Apidæ	170	—	338	—
Ichneumonidæ . . .	209	—	321	—

Ein besonderes Studium wird die geographische Verbreitung der Varietäten bilden.

Diagnose und eigenartige Morphologie der Classificationseinheiten.

Unsere Classificationen geben für die Classificationseinheiten Diagnosen mit den gleichartigen Characteristica, die allen Mitgliedern der Einheit zukommen. Durch diese Diagnosen werden die Einheiten classificatorisch unterschieden.

In die umfassendere Descriptio der Einheiten tritt ein zweites Moment ein : Hervorhebung der besonders manigfaltig ausgebildeten Körperteile.

Wir kennen die Einheit wirklich erst dann, wenn wir wissen, welche Körperteile in der ganzen Mitgliederreihe Manigfaltigkeit der Ausbildung in Form sowie in Macro- und besonders Microskulptur bestehend in Höckern, Dornen, Stacheln, Haaren, grösseren Fortsätzen, flächenartige Verbreiterungen, in der Microskulptur der Skelettoberfläche von zierlichster Netzwerkbildung bis zu ganz glatter Oberfläche der Chitinhaut besitzen.

Ein besonderes Ziel wissenschaftlicher Vergleichung ist die Durcharbeitung der Einheiten, Familien, nicht nur ein und derselben Ordo sondern möglichst aller Familien, ohne Rücksicht auf Zugehörigkeit zu einzelnen Ordines, auf diese Eigenart der Ausbildung einzelner Körperteile, welche Eigenart in Familien ganz verschiedener Ordines gleichartig vorhanden sein kann.

Diese Eigenarten dürften mit der Lebenstätigkeit, mit der Lebensaufgabe der Mitglieder der Einheiten im Haushalte der Natur die hervorragendste Rolle spielen und die Ohnmacht des praktischen Entomologen dürfte gerade in der unzureichenden Kenntnis des zweiten Momentes der umfassenderen Descriptionen der Einheiten als Faunenteile in dem Gesammtleben der Natura Terræ beruhen.

Ausmaasszusammenstellungen.

Wir kennen meist die grösseren Species und die Grössen einzelner Repräsentanten der Einheiten.

Eine besondere Forschungsrichtung repräsentiren Gesammtzusammenstellungen aller Species einer Familie zu einem übersichtlichen Grössenbild.

Als Beispiel, an das sich leicht besondere Fragen knüpfen lassen, bot sich mir die Familie *Tingididæ*.

Ich stelle alle Grössenangaben in der Monographie der *Tingididæ Palæarcticæ* von Horvath in eine geordnete Uebersicht mit der jeweiligen Zahl Species.

In der Monographie sind 2 Subfamilien, 21 Genera, 213 Species, 46 Varietäten. Drei Species sind ohne Grössenangaben.

Alle Species besitzen nur Grössen von 1.6 bis 5 Millimeter.

Das Auffallende dieser *Tingididæ* ist, dass in der Gesammtzusammenstellung, weil die Speciesgrössen variiren, der Grössenangaben mit Bruchteilen von Millimetern sich nicht weniger als 70 Rubriken ergeben.

1.5	Millimeter bis	2	Millimeter	9	Rubriken mit	28	Species.
2	—	3	—	27	—	84	—
3	—	4	—	24	—	80	—
4	—	5	—	10	—	18	—

Die mittlere Grösse ergibt sich zu 2,8 bis 3 Millimeter.

Ergebnisse derartiger Zusammenstellungen sind :

Kleinste Ausmaasse, grösste Ausmaasse, Ausmaasse der Mehrzahl der Species, mittlere Grösse, constante Grössen, variirende Grössen.

Bei Vergleichung vieler Familienzusammenstellungen wird besonders mit Berücksichtigung des geographischen Vorkommens mit Berücksichtigung der orohydrographischen Configuration und mit Gleichstellung der Pflanzen- und Blütengrössen, bei Compositen mit der Grösse der Einzelblüten, die Forschung interessante Ergebnisse erzielen können.

Historische Zusammenstellungen.

Ein höchst interessantes Forschungsgebiet der Entomologie ist die historische Darstellung für die eine besondere Zeitschrift für die gesammte Zoologie ins Leben trat. Die Entomologie ist aber so reich, dass sie fast einer eigenen Zeitschrift bedürfte.

Historische Arbeiten können sehr verschiedene Gesichtspunkte zur Idee haben.

Geschichte der Classification ganzer Ordines und der bisher aufgestellten Ordines in klarer Kürze ist ein wesentliches Desiderat.

Geschichte der Synonymen ist wohl das schwierigste Gebiet.

Geschichte der Descriptionen neuer Species und Genera und ihrer weiteren Teilungen ganzer Familien sind besonders interessant.

Historische Uebersichten der Autoren ganzer Familien geben das Bild der Entwicklung der Forschung.

Eine klare Uebersicht gäbe eine Tabelle mit der letzten Systematik als Basis mit historischer Reihenfolge der Autoren mit Eintragung der Jahreszahlen in den Rubriken der Subgenera und Genera mit historischer Numerirung und Hervorhebung der erstdescripten Species jedes Genus.

Ich gebe die Resultate einer derartigen Zusammenstellung der paläarktischen *Tingididæ*.

Genera.	1803 *Tingis* Fs.	1843	*Cantacader* AMT. SRO.
	1822 *Copium* THNB.	1844	*Campylosteira* FB.
	1825 *Piesma* LPLT. SRV.		*Physatocheila* FB.
	Monanthia LPLT. SRV.	1860	*Monosteira* D. CST.
	1827 *Dictyonota* CTS.	1873	*Stephanitis* STL.
	1833 *Galeatus* CTS.		*Lasiacantha* STL.
	1837 *Derephysia* SPN.		*Oncochila* STL.
	Catoplatus SPN.	1874	*Biskria* PTN.
	Serenthia SPN.		*Elasmotropis* STL.
	1840 *Acalypta* WSTW.	1905	*Aconchus* HRV.

1905 *Hyalochiton*, HRV.

Aelteste Speciesnomina vor 1800 mit Descriptionen.

1758 *cardui* v L. *Tingis.*	1775 *pyri* Fs. *Stephanitis.*
1778 *carthusianus* Gz. *Catoplatus.*	1781 *musci* SCHR. *Acalypta.*
1782 *echi* SCHR. *Monanthia.*	1788 *teucri* HST. *Copium.*

1794 *humuli* Fs. *Monanthia.*

Zahl der Species novæ in 4 Zeitabschnitten von 1800-1906.

1801 bis 1840	. . . .	35	Species.
1843 — 1875	. . . .	48	—
1875 — 1900	. . . .	67	—
1902 — 1906	. . . .	56	—

Jahre mit grösster Zahl novæ Species.

1839.	8 Species.	1844 und 1874. .	11 Species.
1879 und 1902. .	9 —	1905.	19 —
1880.	10 —	1906.	23 —

In den Jahren von 1867 bis 1889 erschienen jedes Jahr neue Species.

Die grössten Anzahlen Species haben

FIEBER. . . .	19 Species.	PUTON. . . .	27 Species.
JAKOWLEW . .	25 —	HORVATH. . .	65 —

Die Anpassungsmerkmale der Atemeles, mit einer Uebersicht über die mitteleuropäischen Verwandten von « Atemeles paradoxus Grav. »

(179. Beitrag zur Kenntnis der Myrmekophilen),

von E. Wasmann, S. J. (Valkenburg).

Unsere systematischen Kategorien sind Abstraktionen, die auf Grund der morphologischen Aehnlichkeit oder Verschiedenheit der betreffenden Individuen gebildet werden. Ihr Zweck ist, die natürlichen Verwandtschaftsverhältnisse der Formen annähernd zum Ausdruck zu bringen. Wo die Variationen und Mutationen, die zur Bildung neuer Arten führen, von den biologischen Anpassungsgesetzen geleitet werden, muss die systematische Betrachtungsweise sich stets mit der biologischen verbinden. Dies gilt insbesonders auch für die Unterscheidung der verschiedenen Formen in der Gattung *Atemeles* Steph. unter den Staphyliniden. Biologisch betrachtet stellt sich uns diese ganze Gattung dar als ein Anpassungsresultat an *die doppelwirtige Lebensweise* jener Käfer einerseits bei *Myrmica rubra* L. (*scabrinodis*, *lævinodis*, etc.), welche deren gemeinschaftlicher Winterwirt (Imagowirt) ist, andererseits an verschiedene Arten und Rassen von *Formica*, welche die Sommerwirte (Larvenwirte) der verschiedenen *Atemeles*-Formen sind. Die Lebensweise der *Atemeles* convergiert daher in den *Myrmica*-Nestern, divergiert in den *Formica* Nestern. Die merkwürdige Erscheinung, dass die *Atemeles* zur Fortpflanzungszeit im Frühjahr die *Myrmica* verlassen um zu den *Formica* überzugehen, bei denen sie ihre Larven erziehen lassen, ist, wie ich schon früher gezeigt habe (1), gleichsam eine stammesgeschichtliche Reminiszenz an die ehemalige einwirtige Lebensweise ihrer Vorfahren bei *Formica*. Diese Erscheinung ist um so merkwürdiger, als die *Atemeles* nach ihrem gemeinschaftlichen Gattungscharakter, der sie von *Lomechusa* unterscheidet, den *Myrmica* allseitiger angepasst sind als den *Formica* und deshalb bei den verschiedenen *Myrmica*-Arten (bezw. Rassen) Aufnahme finden, während jede *Formica*-Art (bezw. Rasse) nur je *eine* bestimmte *Atemeles*-Form pflegt und erzieht.

Der Zweck der vorliegenden Arbeit ist, zu zeigen, wie die verschiedenen *Atemeles*-Formen aus der Verwandtschaft des *At. paradoxus je nach der Verschiedenheit ihrer Formica-Wirte* (Larvenwirte) *sich morphologisch differenzieren*. Derartige Differenzen, mögen sie auch für unser Auge noch so unbedeutend sein, dürfen offenbar nicht zu den auf regelloser individueller Variation beruhenden « Aberrationen » gerechnet werden, weil sie eben auf gesetzmässiger, biologischer Anpassung beruhen. Daher ist es z. B. nicht zutreffend, dass in der 2. Auflage des « *Catalogus Coleopterorum Europæ et Caucasi* » (1906, Spalte 208) die Varietät *Foreli* WASM. des *At. pubicollis* als « Aberration » (a.) aufgeführt wird, da sie eine Anpassungsform an die Lebensweise bei der hellroten *Formica sanguinea* ist. Desgleichen kann die Varietät *nigricollis* KR. des *At. emarginatus* nicht als blosse Aberration bezeichnet werden, weil sie eine Steigerung der Anpassung an den schwarzen Larvenwirt (*Formica fusca*) darstellt.

Die Merkmale, durch welche die verschiedenen *Atemeles*-Formen infolge ihrer Anpassung an verschiedene Arten und Rassen der Gattung *Formica* sich unterscheiden, sind Unterschiede in der Körpergrösse und Körpergestalt, in der Färbung und Skulptur und namentlich auch in der Behaarung. Bei der grösseren Formicaform lebt auch die grössere Atemelesform, bei der dunkleren bezw. helleren Formicaform die dunklere bezw. hellere Atemelesform, und bei der dichter und länger behaarten Formicaform die

---

(1) Zur Lebensweisen von *Atemeles pratensoides* (« Zeitschrift für wissensch. Insektenbiologie », 1906, Hefte 1 und 2, S. 3); Beispiele rezenter Artenbildung bei Ameisengästen und Termitengästen. (« Festschr. für Rosenthal », 1906, und « Biol. Centralbl. », XXVI, Nrn. 17 und 18, S. 47, [569]).

dichter und länger behaarte Atemelesform. So interessant diese Unterschiede in biologischer Beziehung sind, so dornenvoll sind sie für den Systematiker : berechtigen sie zur Aufstellung von Arten oder von Unterarten oder blos von Varietäten? Mögen die betreffenden morphologischen Differenzen für unser Auge noch so unscheinbar sein, so müssen wir ihnen, sobald sie sich als konstant erweisen, doch wegen ihres biologischen Wertes auch einen möglichst prägnanten systematischen Ausdruck verleihen; ob unter dem Namen einer Spezies, einer Subspezies oder einer Varietät, das hängt von der Grösse der Unterschiede in dem betreffenden Falle ab. Wenn die Botaniker sogar auf Grund blosser biologischer Verschiedenheiten sogenannte biologische Arten unterscheiden, so dürfen die Zoologen wohl auf die Kombination geringer morphologischer Unterschiede mit ausgeprägten biologischen Unterschieden ebenfalls neue Arten oder Unterarten gründen.

Ich gehe nun zu der Uebersicht der mitteleuropäischen Verwandten von *Atemeles paradoxus* über. Dieses Beispiel wurde deshalb ausgewählt, weil mir durch fünf und zwanzigjährige Studien ein sehr reiches Sammlungs- und Beobachtungsmaterial für diese Gruppe zu Gebote steht; ferner auch deshalb, weil sich neuerdings ein dringendes Bedürfnis nach morphologischer Sichtung dieser Formen auf biologischer Grundlage ergab. 1903 hatte ich bei der schwarzen, dichtbehaarten *Formica pratensis* zu Luxemburg einen fast schwarzen, borstigen *Atemeles* in grosser Zahl entdeckt, der sich so auffallend von dem bei *F. rufa* lebenden *At. pubicollis* unterschied, dass ich ihn als neue Art, *At. pratensoides*, beschrieb. Als ich ferner im Sommer 1909 bei Lippspringe i. Westfalen in einer Kolonie von *Formica truncicola*, die durch hellrothe Färbung und dichtere Behaarung von *rufa* abweicht, einen *Atemeles* in Menge fand, war ich überrascht darüber, dass derselbe von dem bei *rufa* lebenden *pubicollis* sich nicht zu unterscheiden schien. Ich nahm deshalb irrtümlich an, es handle sich um den echten *pubicollis* (1), und derselbe habe als Larvenwirt bald *F. rufa*, bald *F. truncicola*. Im Sommer 1910 traf ich aber wiederum die Larven desselben *Atemeles* in einer anderen *truncicola*-Kolonie bei

(1) In meiner Arbeit « Wesen und Ursprung der Symphilie » (« Biol. Centralbl. », 1910, S. 100) ist deshalb der Nr. 3-5, *F. truncicola* lebende *Atemeles* als *pubicollis* angeführt. Es handelt sich jedoch um die neue subsp. *truncicoloides*.

Lippspringe. Erst jetzt untersuchte ich das ganze, von meinem Kollegen H. KLENE, S. J., unterdessen präparierte, letztjährige Atemelesmaterial von *F. truncicola*. Da stellte sich nun heraus, dass dieser *Atemeles* trotzdem eine eigene Anpassungsform an *F. truncicola* darstellt, indem er von dem typischen *pubicollis* durch dichtere Beborstung der Unterseite des Hinterleibes, dichtere Punktirung der Oberseite desselben, dichter behaarte Flügeldecken und etwas hellere Färbung abweicht; ich werde ihn deshalb als eigene Rasse (*truncicoloides*) bezeichnen. Dasselbe gilt auch für die bei *F. sanguinea* lebende Varietät *Foreli* des *pubicollis*, die ich ebenfalls jetzt als Rasse (Subspezies) (1) auffasse.

Tabelle der mit « paradoxus » GRAV. verwandten mitteleuropäischen « Atemeles »-Formen.

GEMEINSCHAFTLICHE MERKMALE (gegenüber *emarginatus*) : Drittes Fühlerglied mindestens doppelt so lang als das zweite. Halsschild querrechteckig, mit deutlich vertieften Seitengruben (2). Oberseite des Hinterleibes deutlich punktiert.

DIFFERENZIERUNGSMERKMALE, infolge der Anpassung an verschiedene Larvenwirte aus der Gattung *Formica* :

I. Kleiner und schlanker (4-4.5 mm. l., 1.5 mm. br.), Halsschild ohne gelbe Börstchen, Körper meist heller rotbraun gefärbt (3). Fühler schlanker, den Hinterrand der Flügeldecken erreichend, lose gegliedert (die Verbindungsstiele der Glieder länger), die

---

(1) In der Myrmekologie hat man längst den Begriff der Subpezies neben demjenigen der Varietät eingeführt. In der 2. Auflage des « Catalogus Coleopterorum Europæ, Caucasi et Armeniæ (1906) » fehlt er noch. Vgl. daselbst die Spalte 215, Dinardaformen.

(2) Auf die in der « Deutsch. Ent. Zeitschr. 1887 » von mir erwähnten, wahrscheinlich durch Kreuzung veranlassten Uebergänge in der Halsschildbildung zwischen *paradoxus* und *emarginatus* gehe ich hier nicht ein.

(3) *At. paradoxus* var. *nigricans* WASM. (« Zeitschrift für wissensch. Insektenbiologie », 1906, S. 3) ist wahrscheinlich der dunklen Varietät *fusco-rufibarbis* von *F. rufibarbis* angepasst.

Glieder meist auch in sich selbst länger (Glied 4 und 5 nur selten quer, 8 meist deutlich länger als breit), Hinterleib auf der Oberseite dicht und stark punktiert, auf der Unterseite mit längerer, dichter anliegender gelber Grundbehaarung zwischen den Borsten :

1. « Atemeles paradoxus » GRAV. und seine Varietäten.

[Larvenwirt (Sommerwirt) : *Formica rufibarbis* F. und ihre Varietäten, ferner *Polyergus rufescens* LTR. mit *rufibarbis* als Sklaven (1). Ganz Mittel- (und Nord-) Europa.]

II. Grösser und breiter (5-5.5 mm. l., 2 mm. br.) Halsschild mit feinen, gelben Börstchen, Färbung meist dunkler, Fühler dicker und kürzer, den Hinterrand der Flügeldecken nicht erreichend, enggegliedert (die Glieder sehr kurz gestielt), auch die Glieder selbst meist breiter (Glied 4 und 5 meist quer, 8 nicht länger als breit), Hinterleib auf der Oberseite spärlicher und feiner punktiert, auf der Unterseite mit kurzer spärlicher, gelber Grundbehaarung zwischen den Borsten : *pubicollis*-Gruppe.

II*a*. Halsschild mit deutlicher Punktierung auf chagriniertem Grunde. Färbung dunkler oder heller rotbraun, Kopf und eine Hinterleibsbinde schwarzbraun.

a[1]. Halsschild und Flügeldecken dunkel rotbraun ; Flügeldecken nur mit sehr kurzen, kaum sichtbaren gelben Härchen, Oberseite des Hinterleibes nur äusserst spärlich und fein punktiert, mit sehr kurzen, kaum sichtbaren gelben Härchen in den Punkten. Unterseite des Hinterleibes stark glänzend, nur mit sehr kurzen gelben Härchen als Grundbehaarung, die Borsten spärlich, nur zwei Querreihen auf jedem Segment bildend :

2. « Atemeles pubicollis » BRIS. i. sp.

[Larvenwirt (Sommerwirt) : *Formica rufa* L. i. sp. Ganz Mittel- (und Nord-) Europa.]

a[2]. Ränder des Halsschildes und die Flügeldecken etwas heller rotbraun oder rostrot ; Flügeldecken mit längerer und dichterer

---

(1) Da in den *Polyergus*-Kolonien nur die Sklaven mit der Brut- und Gastpflege sich befassen, sind die daselbst erzogenen *paradoxus* höchstens etwas grösser (reichlichere Ernährung), sonst von der Normalform nicht verschieden. Luxemburg, in mehreren Kolonien.

gelber Behaarung, Oberseite des Hinterleibes dicht punktiert mit deutlich sichtbaren gelben Härchen in den Punkten; Unterseite des Hinterleibes weniger glänzend, mit längeren gelben Härchen als Grundbehaarung, dichter beborstet, die Borsten dicker und länger, drei Querreihen auf jedem Segment bildend :

3. « Atemeles pubicollis » subsp. « truncicoloides » WASM. n. subsp.

[Larvenwirt (Sommerwirt) : *Formica truncicola* NYL. Bisher nur aus Lippspringe in Westfalen bekannt (WASMANN 1909), aber wohl weiter verbreitet.]

a^3. Einfarbig rostrot, nur der Kopf und eine schmale Hinterleibsbinde dunkler; Flügeldecken nur mit sehr kurzer, gelber Behaarung; Oberseite des Hinterleibes fein, etwas dichter als bei *pubicollis* punktiert, mit deutlichen Härchen in den Punkten. Unterseite des Hinterleibes mit kurzer gelber Grundbehaarung und mit spärlichen längeren Borsten (zwei Querreihen auf jedem Segment?) :

4. « Atemeles pubicollis » subsp. « Foreli » WASM. (1).

[Larvenwirt (Sommerwirt) : *Formica sanguinea* LTR. Bisher nur aus den Vogesen bekannt (FOREL !).]

II*b*. Halsschild sehr fein chagriniert, mit völlig verloschener Punktierung. Färbung schwarzbraun, mit braunen Flügeldecken. Flügeldecken ziemlich dicht und lang gelb behaart (wie bei *truncicoloides*), Oberseite des Hinterleibes in den Punkten mit längeren gelben Härchen, und ausserdem nahe den Seiten dichter beborstet, Unterseite schwach seidenglänzend, mit längerer gelber Grundbehaarung (ähnlich wie bei *truncicoloides*) und sehr dicht und lang beborstet, vier Borstenreihen auf jedem Segment; die gelben Haarbüschel an den Hinterleibsseiten stärker als bei allen übrigen Atemelesarten entwickelt :

---

(1) *At. pubicollis* Var. *Foreli*. (« Deutsche Ent. Zeitschrift », 1892, S. 351.)

5. « **Atemeles pratensoides** » WASM. (D. E. Z., 1904, S. 10).
[Larvenwirt (Sommerwirt) : *Formica pratensis* DEG. Bisher nur aus Luxemburg bekannt (WASMANN 1903).]

Ueber den phylogenetischen Zusammenhang der verschiedenen Atemelesformen *untereinander* wissen wir gegenwärtig noch nichts Bestimmtes. Sie stellen auf der gemeinschaftlichen Grundlage der Anpassung an *Myrmica* verschiedene Anpassungen (1) an verschiedene Formicawirte dar. Wie ihre Lebensweise in den Myrmicanestern convergiert, in den Formicanestern divergiert, so convergieren auch ihre Anpassungen gegenüber *Myrmica*, divergieren gegenüber *Formica*. Bezüglich der hypothetischen Stammesgeschichte der *Lomechusini* verweise ich im übrigen auf meine frühere Arbeit « Beispiele rezenter Artenbildung bei Ameisengästen und Termitengästen », 1906, 2. Teil.

Dafür, dass die *Atemeles* zu ihren Formicawirten im Frühling nicht bloss spontan übergehen, sondern manchmal auch von den *Formica* in ihre Nester eingetragen, ja sogar aus den Myrmicanestern gewaltsam geraubt werden, habe ich schon früher einige Beispiele berichtet (2). Einige neue Beobachtungen seien hier noch beigefügt. Am 6. April 1910 sah ich vor einem der *rufibarbis*-Nester in unserem Garten zu Luxemburg eine Arbeiterin mit einem Käfer, den sie im Maule trug, eilig einem der Nesteingänge sich nähern. Ich fing sie in einem Fangglase ab, woselbst sie den Käfer niedersetzte. Es war ein völlig unversehrter *Atemeles paradoxus*, der sich vorher mit eingezogenen Fühlern und Beinen regungslos verhalten hatte — wie gewöhnlich beim Transport der *Atemeles* durch die Ameisen — jetzt aber munter umherlief. Ferner sah ich, wie am 18. Mai 1910 bei Lippspringe in Westfalen eine *Formica fusca* einen *Atemeles emarginatus* von draussen her genau in derselben Weise in ihr Nest trug. Die *Atemeles* drängen sich also nicht blos ihren Formicawirten auf, sondern werden auch von diesen aufgesucht. Die verschiedenen Atemelesformen machen

(1) Diese Anpassungen sind ferner bei verschiedenen Arten und Rassen von *Atemeles* verschieden weit fortgeschritten. Ich werde hierauf in einer anderen Arbeit zurückkommen (« Zur Doppelwirtigkeit der *Atemeles* »).

(2) Wesen und Ursprung der Symphilie. (« Biol. Centralbl. », 1910, S. 101.)

auf die Sinnesvermögen von *Formica* — auf deren Geruchssinn, und, nach den Anpassungscharakteren der Käfer zu schliessen, auch auf deren Gesichts- und Tastsinn — einen hochgradig spezialisierten Eindruck, während sie auf die Sinnesvermögen von *Myrmica* einen allgemeineren angenehmen Eindruck machen. Hieraus erklärt sich, weshalb man normaler Weise stets nur eine bestimmte Atemelesform in den Nestern einer bestimmten *Formica*-Art oder Rasse findet, während in den Myrmicanestern verschiedene Atemelesformen unterschiedslos aufgenommen und gepflegt werden.

Contribution à l'histoire des Lampyrides,

par Ern. Olivier (Moulins).

La faculté lumineuse que possèdent les Coléoptères du groupe des Lampyrides les a fait remarquer depuis longtemps des voyageurs et aussi des physiologistes qui ont publié de nombreux travaux et se sont livrés à de minutieuses recherches sur les causes de ce singulier phénomène et les conditions dans lesquelles il se produit.

Je ne viens pas ajouter un mémoire de plus à ceux qui traitent déjà ce sujet. La liste en est longue (1), et cependant, si je puis m'exprimer ainsi, la lumière n'est pas encore faite : le mode de fonctionnement de l'appareil photogène et la nature de la lueur émise ne me paraissent pas expliqués d'une façon satisfaisante et restent encore obscurs.

Je me bornerai à présenter quelques remarques sur l'organisation de ces Insectes intéressants entre tous.

Les Lampyrides constituent un groupe ou une tribu de la famille des Malacodermes établie par Latreille et dont la place naturelle est entre les Lycides et les Téléphorides.

Ils se rapprochent des premiers par les genres *Pyractonema, Lucidina, Lychnuris*, et les *Photurini* forment la transition avec les seconds. Mais ils se distinguent aisément de leurs voisins par

---

(1) Voir la liste que j'en ai donnée (« Coleopterorum catalogus, editus a. Schenkling. *Lampyridae* », p. 3).

un facies spécial et, en outre de leur appareil lumineux, par plusieurs autres caractères faciles à constater, la forme du prothorax, la position des antennes et des hanches, l'articulation des fémurs, etc.

Bien que les Lampyrides soient connus de toute antiquité, leur étude systématique a été très négligée, et malgré la quantité considérable d'espèces bien caractérisées qui font partie de cette tribu, il n'en avait été décrit jusqu'en ces derniers temps qu'un très petit nombre.

En 1832, LAPORTE DE CASTELNAU constate que les voyageurs en rapportent incessamment de toutes les parties du monde, et il donne comme extraordinaire le chiffre de 200 espèces conservées dans les collections seules de Paris (1).

Depuis cette époque, ce chiffre s'est accru rapidement, grâce aux travaux de BLANCHARD, MOTSCHULSKY, LE CONTE, HORN, KIRSCH et de plusieurs autres entomologistes, et en 1869, MM. GEMMINGER et DE HAROLD enregistrent 446 espèces (2).

Les envois continuèrent à affluer de toutes les régions où les explorations se firent plus méthodiques et plus minutieuses : en 1907, dans le *Genera Lampyridarum*, j'ai énuméré 1002 espèces(3); seulement trois ans après, en 1910, j'en ai catalogué 1109 (4), et il en existe dans les collections un stock important encore inédit.

Mais si la classification systématique de ces Insectes a fait des progrès importants, il n'en est pas de même de leur biologie, et nous sommes encore dans une ignorance à peu près complète du genre de vie et des mœurs de la plupart.

Sauf pour quelques-uns très répandus dans des régions habitées, et par conséquent facilement observables, nous ne savons rien de précis sur l'époque et la durée de l'apparition du plus grand nombre.

Dans beaucoup de genres, nous ne connaissons que des mâles, et c'est souvent avec incertitude que nous identifions une femelle.

---

(1) Essai d'une division du genre Lampyre. (« Ann. Soc. Ent. France », 1832, p. 122.)

(2) « Catalogus Coleopterorum hucusque descriptorum », p. 1636.

(3) « Genera Insectorum » dirigés par P. WYTSMAN. *Lampyridae*, 63 pages, 3 pl. col.

(4) « Catalogus Coleopterorum, » editus a. S. SCHENKLING. *Lampyridae*, 68 pages.

La difficulté est encore plus compliquée pour les larves, qui offrent souvent des formes bizarres et auxquelles nous ne pouvons assigner qu'un état civil très hypothétique.

Ces lacunes ne peuvent être comblées que par les observations accumulées, précises et consciencieuses de naturalistes habitant les régions fréquentées par ces Insectes, ou de voyageurs y séjournant assez longtemps.

Mais les recherches à faire dans ce but, qui doivent avoir lieu la nuit, souvent dans des forêts ou des localités marécageuses et éloignées de toute habitation, ne sont pas toujours faciles, et on y est exposé à des dangers de toute sorte.

C'est ainsi que M. Weyers, qui habitait alors à Païnan (Sumatra), auquel j'avais demandé d'essayer de surprendre des Lampyrides *in copula*, m'écrivait : « Les chasses nocturnes sont assez dangereuses ici, à cause des tigres qui sont nombreux dans nos environs; bien souvent, la nuit, j'entends leur cri court et mélancolique à peu de distance de mon habitation, et il m'est arrivé de trouver leurs traces toutes fraîches, le matin, à une quinzaine de mètres de mon logis ». On comprend facilement que les inconvénients qui peuvent résulter d'une semblable rencontre sont de nature à faire renoncer à une promenade entomologique.

Dans un groupe où les deux sexes sont bien souvent très différents, les caractères génériques établis sur les mâles seuls sont forcément incomplets, et comme, pour le moment, beaucoup de femelles sont inconnues, une classification ne peut être définitive; car leur connaissance entraînera très probablement des modifications importantes dans l'ordre adopté à ce jour et qui ne peut être considéré que comme provisoire.

Actuellement, la conformation des antennes et celle du dernier segment ventral fournissent les meilleurs caractères séparatifs. Sauf chez quelques genres que j'appellerai *aberrants*, les antennes sont composées de onze articles, mais leur dimension et leur structure présentent une grande diversité. Elles peuvent être aussi longues que le corps ou ne pas dépasser les dimensions du prothorax et même de la tête; les deux premiers articles et le onzième varient peu, mais les autres offrent de notables différences : ils peuvent être flabellés ou biflabellés, ou dentés plus ou moins longuement, ou plus ou moins comprimés et élargis, ou bien très allongés, minces et filiformes; quelques-uns affectent des formes toutes particulières.

Le tableau ci-dessous contient les principaux types.

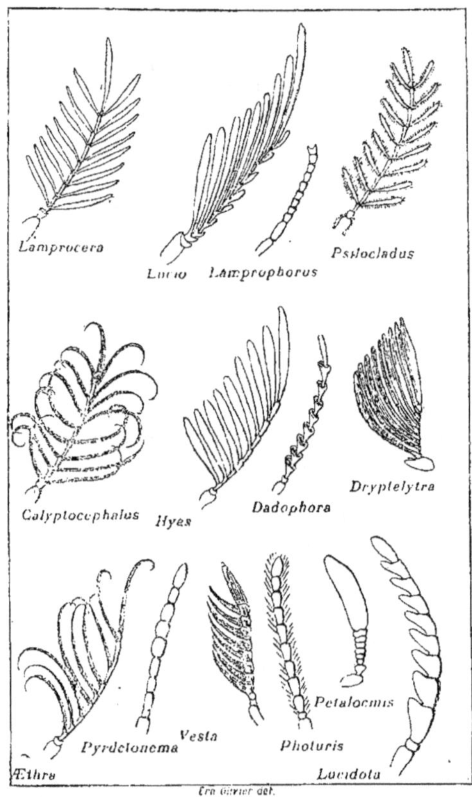

La tête est petite, excavée entre les yeux qui sont très gros, sourtout chez les mâles : dans quelques espèces (*Lucernuta*, *Lampyris*, *Lamprophorus*) ils atteignent une dimension énorme, se touchent presque en dessus, sont tout à fait contigus en dessous et constituent, à eux seuls, à peu près tout le volume de l'organe.

Ces yeux, à cause de leur forte saillie, sont exposés à être continuellement blessés quand l'Insecte, après son vol nocturne, regagne sa demeure sous des écorces, des débris, des détritus végétaux. La nature a pourvu à cet inconvénient en développant le prothorax de façon à les recouvrir complètement et à leur constituer une armure protectrice; mais, en même temps, comme il est absolument opaque, il fait l'effet d'un écran qui intercepte la lumière, et l'Insecte ne pourrait rien apercevoir, si ce prothorax n'avait pas au-dessus de chaque œil une place transparente qui sert comme de petites fenêtres à travers lesquelles les objets environnants sont perceptibles, les yeux restant convenablement abrités.

Cette disposition du prothorax du mâle, en forme de capuchon avec des plaques sus-oculaires diaphanes, préservatrices des yeux, s'observe surtout chez les espèces dont les femelles sont aptères et dont la recherche exige de la part du mâle une circulation pénible à travers les amas de feuilles et les brins de gazons où elles ont élu domicile.

Chez les espèces où les deux sexes, ailés et également agiles, peuvent se rencontrer, sans obstacles, sur les feuilles ou les tiges des plantes, la tête est complètement dégagée du prothorax et même souvent attachée à un cou assez visible (*Photurini*).

Dans toute une série de ces Insectes, les mâles seuls sont pourvus d'ailes et d'élytres; les femelles n'ont point d'ailes et seulement des élytres très rudimentaires, simplement représentés par une sorte de moignon très court, en forme d'écaille. Les antennes sont alors filiformes et très courtes; les mâles ne sont pas ou à peine lumineux; les femelles, au contraire, émettent de presque toutes les parties de leur corps une lueur phosphorescente intense qui n'est pas intermittente, comme chez les espèces ailées, mais persiste d'une façon continue et régulière pendant toute la durée de son émission, qui est de plusieurs heures.

L'abdomen est composé de sept segments et de six seulement chez les Lucioles mâles.

En dessus, si on n'écarte pas les élytres, il n'y a généralement que le dernier segment ou pygidium qui soit visible et souvent seulement en partie. Il affecte chez les différents genres des formes assez variées : arrondi ou ogival, ou tronqué, ou échancré, ou sinué, ou trilobé avec les angles latéraux saillants et une pointe médiane triangulaire.

Le premier segment du ventre est ordinairement caché, en grande partie, par les hanches postérieures et le prolongement du métasternum, et on n'en peut apercevoir que les côtés; les cinq suivants sont à peu près d'égales dimensions; ils peuvent être impressionnés latéralement, et leurs angles postérieurs sont plus ou moins saillants en arrière ; l'abdomen est alors, parfois, denté, lobé ou lacinié; le septième est le plus court, le plus étroit et subit, chez les mâles, de nombreuses modifications : arrondi, tronqué, échancré, sinué, mucroné, triangulaire; chez certains *Luciolini*, il offre, ainsi que le pygidium, une conformation bizarre et bien caractéristique.

Les dessins ci-dessous présentent les principales variations du dernier segment ventral des mâles. Ce segment est chez les femelles d'une forme plus constante, généralement en triangle, à base plus ou moins large, incisé ou échancré au sommet.

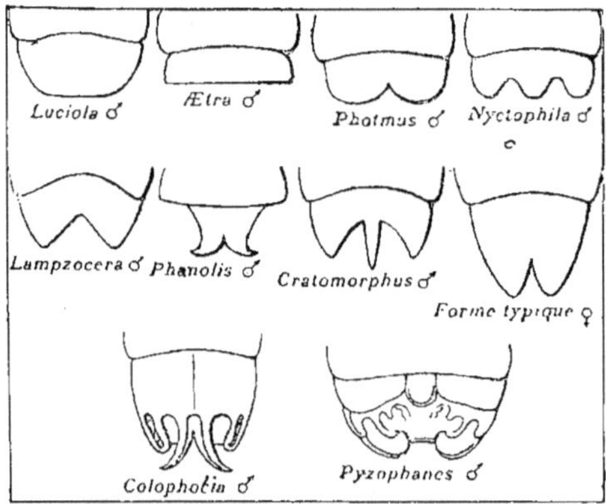

En plus du septième, on observe souvent, chez les mâles, un huitième segment supplémentaire, linéaire, tout petit, qui n'est pas toujours visible et fait partie de l'armure génitale.

Les derniers segments du ventre sont le principal siège de l'appareil lumineux persistant après la mort de l'Insecte sous forme de plaques lisses, luisantes, blanches, jaunâtres ou rosées qui recouvrent entièrement les segments ou seulement leur partie médiane, ou sont réduites à de simples points arrondis situés sur les côtés de l'avant-dernier ou dans son milieu.

Les Lampyrides les plus répandus en France et en Europe sont le *Lampyris noctiluca* et les espèces voisines dont les mâles sont ailés, mais point du tout ou à peine lumineux, tandis que les femelles, aptères, munies de pattes très courtes et chargées d'un lourd abdomen, circulent difficilement, mais, en revanche, sont extrêmement phosphorescentes.

Alors s'est établie cette croyance que la femelle immobilisée sur le sol avait été douée du pouvoir lumineux pour s'en servir, à la tombée de la nuit, comme d'un fanal indicateur de sa présence au mâle ailé.

Cette explication ne peut être acceptée, car chez la grande majorité des espèces, les mâles sont beaucoup plus lumineux que les femelles. Les larves aussi, qui n'ont pas besoin de solliciter la présence des mâles, sont toutes plus ou moins lumineuses, et les plus grosses émettent même une lueur assez vive pour permettre, dans la nuit la plus obscure, de lire avec leur seule clarté.

Et tous les autres Insectes crépusculaires ou noctures, de nombreux Lépidoptères notamment, savent bien se retrouver sans que les femelles aient besoin d'un organe spécial brillant.

Je ne comprends pas bien le but de cette faculté photogène et je crois qu'il faut n'y voir qu'un ornement analogue à la parure de noce des Oiseaux dont le plumage prend à l'époque des amours un éclat inaccoutumé.

Quoi qu'il en soit, il résulte de l'observation que ce pouvoir phosphorescent est soumis aux règles suivantes :

Il est en raison inverse du développement des antennes : plus ces organes sont longs et compliqués, plus il est faible.

Quand les femelles sont aptères, elles sont beaucoup plus lumineuses que les mâles.

Quand les deux sexes sont ailés, les mâles sont plus lumineux que les femelles.

Quand les deux sexes sont aptères (*Phosphœnus*), la faculté lumineuse chez tous les deux est nulle ou insignifiante.

Comme je l'ai déjà dit, la lueur émise par les femelles aptères est continue et régulière pendant tout le temps de son émission; quand l'Insecte éteint son appareil, il ne le rallume plus que la nuit suivante.

Mais la lumière émise par les espèces ailées est intermittente, c'est-à-dire qu'elle disparaît pendant un instant pour reparaître vite à intervalles rapprochés et égaux.

Le mécanisme de ce mouvement rythmique ne s'explique pas d'une façon bien satisfaisante : son but est évidemment de dérober l'Insecte aux poursuites des Chauves-souris et des Oiseaux crépusculaires qui auraient trop de facilités pour s'en emparer si la lumière émise restait constamment visible.

Les espèces aptères, toujours plus ou moins cachées, n'ont pas besoin de ce moyen de défense : suffisamment protégées par les gazons et les plantes sous lesquels elles se trouvent, elles peuvent éclairer constamment.

Il y a peu de familles de Coléoptères où on remarque un aussi grand nombre de cas de mimétisme entre espèces appartenant à des genres bien différents. La coloration et la forme sont identiques ; les antennes seules constituent la différence : aussi la détermination d'un Lampyride n'est certaine que s'il est muni de ces organes.

Voici quelques cas de ressemblance que je soumets à la réunion :

Phænolis ustulatus GORH. Antennes biflabellées.
Vesta xanthura ERN. OLIV. Antennes à articles comprimés, unidentés.

Lucidota ingloria ERN. OLIV. Antennes à articles comprimés, élargis, unidentés.
Photinus simulans ERN. OLIV. Antennes filiformes.

Pyrocœlia rufa ERN. OLIV. Antennes à articles comprimés, élargis, dentés.
Photinus secernatus ERN. OLIV. Antennes à articles cylindriques, linéaires

Luciola ovalis HOPE. Antennes plus courtes que la moitié du corps.
Photinus longicornis ERN. OLIV. Antennes plus longues que le corps.

Luciola insignis ERN. OLIV. Antennes simples, filiformes de 11 articles.
Amydetes lucioloides ERN. OLIV. Antennes longuement flabellées de 37 articles.

La similitude est parfaite sous les rapports de la forme et de la coloration, et si les antennes manquent, ce ne sera qu'avec doute que ces espèces pourront être identifiées.

Les Lampyrides ont des représentants dans toutes les régions du globe et sont particulièrement abondants dans l'Amérique du Sud, dans l'Asie orientale et dans l'Afrique méridionale. Mais chaque espèce semble avoir un habitat spécial, une époque d'apparition différente et être localisée dans un espace relativement restreint.

Malheureusement, comme je l'ai déjà dit, nous manquons de documents précis sur le genre de vie de ces Insectes, et, actuellement, leur répartition géographique ne peut être établie que pour les groupes principaux considérés dans leur ensemble.

Un rapide examen permet de séparer aussitôt les Lampyrides en deux groupes bien distincts : espèces à antennes rameuses ou composées; espèces à antennes simples, linéaires.

Les espèces à antennes rameuses sont à peu près toutes confinées dans le centre de l'Amérique méridionale. Une remonte au Mexique (*Phœnolis nigricollis* Gorh.), une autre aux États-Unis (*Calyptocephalus bifarius* Say); le genre *Vesta* est représenté en Asie orientale, mais déjà il ne porte plus les élégants panaches des grandes espèces américaines et ses antennes ne sont plus guère chargées que d'une dentelure plus ou moins longue.

Les espèces à antennes linéaires, simples ou à peine dentées, sont comprises dans les groupes des *Lampyrini*, des *Photinini*, des *Photurini* et des *Luciolini*.

Les *Lampyrini* sont essentiellement palæarctiques. Sauf les régions boréales, ils habitent toute l'Europe et toute l'Asie, où ils s'étendent, au nord, jusqu'à l'Amour, à la Dzoungarie et à la Finlande; au sud, jusqu'à Bornéo, Sumatra, Java. En Afrique, ils occupent une étroite zone tout le long du littoral méditerranéen du Maroc à Tripoli; on ne les retrouve plus qu'à l'est, en Abyssinie, et ils reparaissent en nombre de l'équateur au cap de Bonne-Espérance. Il n'en existe pas à Madagascar ni dans l'Amérique méridionale, mais deux genres, *Microphotus* et *Phausis*, sont représentés chacun par deux espèces au centre des États-Unis, dont l'une, *Phausis splendidula*, commune dans une grande partie de l'Europe, et qui, d'après Le Conte, ne serait que naturalisée dans quelques localités du Maryland et de l'Illinois où elle aurait été introduite.

Les *Photinini*, très nombreux et d'une étude difficile, habitent exclusivement l'Amérique où on les rencontre partout, du Canada au sud de l'Argentine et dans l'archipel des Antilles ; c'est dans ce groupe que rentre *Photinus frigidus*, habitant de Terre-Neuve, le Lampyride qui, avec *Lampyris noctiluca*, remonte le plus vers le nord.

Les *Photurini* se rencontrent dans les mêmes régions que les *Photinini* et sont, comme ces derniers, exclusivement américains. Toutefois, ils remontent un peu moins au nord des États-Unis et descendent un peu moins bas dans l'Argentine. Ils existent aussi aux Antilles.

Les *Luciolini* n'existent pas dans le Nouveau Monde, mais ils comptent de nombreux représentants en Europe, en Asie, en Afrique, en Australie et dans tous les archipels océaniens. En

Europe, les Lucioles habitent en Portugal, en Espagne, dans l'extrême midi de la France, en Italie, en Grèce, en Turquie et jusqu'en Hongrie. On n'en trouve pas dans le nord de l'Afrique, et c'est un fait remarquable que l'absence sur le littoral sud de la Méditerranée de ces Insectes qui en occupent tous les autres rivages; en revanche, elles abondent à partir du Sénégal et de l'Abyssinie jusqu'au cap de Bonne-Espérance, ainsi qu'à Madagascar et aux Seychelles.

En Asie, leur limite septentrionale est moins avancée que celle des Lampyris, mais elles sont répandues au Japon et dans les îles de la mer de Chine, d'où elles atteignent tous les archipels océaniens. Dans cette partie du monde, les espèces sont très localisées et se partagent en plusieurs formes bien caractérisées. Les *Luciola chinensis* et *ovalis* sont les types dont l'aire de dispersion est la plus considérable, dans tout l'Hindoustan, l'Indo-Chine, la Chine. Les *Ototreta* paraissent spéciaux à Bornéo et à Sumatra, les *Colophotia* aux Philippines, les *Bourgeoisia* à la Nouvelle-Calédonie et îles voisines, les *Atyphella* à la Nouvelle-Guinée et aux régions orientales de l'Australie, Queensland et New South Wales, les *Lampyroidea* à la Syrie et à l'Asie occidentale.

Il est à présumer que chaque archipel ou même chaque île de la Malaisie renferme un type spécial, mais les documents précis sous ce rapport sont complètement insuffisants; beaucoup d'îles n'ont pas encore été explorées, et, d'autre part, les naturalistes qui naviguent au milieu de ces îles sont exposés à faire des erreurs en mêlant des captures provenant de localités différentes.

Ce n'est qu'avec le temps et en multipliant les observations que l'on pourra arriver à établir sur des bases certaines la distribution géographique de ces intéressants Coléoptères.

Preliminary notes on the importance of the new family « Urothripidæ » BAGNALL in the study of the « Thysanoptera »,

by RICHARD SIDDOWAY BAGNALL (Penshaw).

Urothrips paradoxus BAGNALL (1), the type of the family *Urothripidæ*, is a small, recently discovered Ethiopian Insect, which, whilst clearly belonging to the *Thysanoptera*, differs in many important features from all the known forms.

Before its importance can be appreciated, however, we must briefly outline some of the chief characters of the order *Thysanoptera* as recognized before the discovery of the primitive *Urothrips*. Thus in both the sub-orders *Terebrantia* and *Tubulifera*, every known form without exception has the maxillary and labial palpi never less than two-segmented; the intermediate pair of coxæ always more widely separated than either the anterior or posterior pairs, whilst they possess four pairs of stigmata, and in some species of the *Terebrantia* the metathoracic pair is either very small, vestigial or sometimes apparently absent altogether.

Urothrips differs sharply in the following characters : the maxillary and labial palpi are each composed of but a single joint, the apparent division of the maxillary palpus being an oblique series

(1) BAGNALL. « Annales Musei Nationalis Hungarici », 1909, VII, pl. III, pp. 125-136.

of seta-pits; the posterior pair of coxæ is very distinctly more widely separated than either the anterior or intermediate pairs, whilst there are eleven pairs of stigmata present, which are large and well-developed. Other salient characters are seen in the distinctive type of antennæ, which do not appear to possess any type of sense-cone or area known in other forms, and in the elongate form of the ninth abdominal segment, which is distinctly longer than any of the preceding.

It will thus be seen that *Urothrips* is sharply differentiated by the possession of certain characters, which from their constancy throughout the previously known forms might well have been regarded as of ordinal value; and yet in its general form this genus falls into the sub-order *Tubulifera*, despite the fact that it differs from true *Tubulifera* almost more strongly in its structure as above outlined than do members of the other sub-order *Terebrantia*.

Regarding *Urothrips* as Tubuliferous, we find another series of characters which appear to be of more than generic value, but which we do not yet fully understand and appreciate. Firstly there is the presence of at least two minute dorsal papillæ on each of the abdominal segments three to eight, between the lateral spine-like process and the stigma on each side; unfortunately these are very difficult to discern in our two preparations, and no such organs have been described in other *Thysanoptera*. Secondly we notice the absence of the bristles seen in most *Thysanoptera* on the antennæ, head, thorax, legs and abdomen, which reminds us of the vestigial state of the setæ in two Tubuliferous genera, *Lispothrips* REUTER (1), and *Dermothrips* BAGNALL (2), a condition closely approaching that of *Urothrips*. Then, from our preparations, it would appear that *Urothrips* differs in its sexual characters; we have searched for the characteristic organs of the *Tubulifera* without success, but from the nature of the preparations we have been unable to describe the genitalia.

That the two specimens represent the male and female is shown by certain secondary characters, which we may without doubt

(1) REUTER, « Acta Soc. pro Fauna et Flora Fennica ». 1899, XVII, nº 2, 69 pp.

(2) BAGNALL, Thysanoptera in « Fauna Hawaiiensis », 1910, III, pt. VI, pp. 669 701, pl. XVII-XIX.

regard as sexual; thus in the one sex, which we have described as the female, we find a minute tooth or papilla at each hind angle of the prothorax, which does not appear to be present in the other specimen, whilst the long terminal hairs of this latter specimen (which we regard as the male) are whip-like, being uniformly thick for about one-third their length, and then abruptly reduced and continued as a long colorless filament; whereas in the female these long bristles or hairs are gradually and evenly tapered from the base to the extreme apex. These characteristics have not been described in any known Thrips; and in this latter connection it is worth noting that the tube is encircled by six terminal hairs, but that there are no intermediate minor setæ as in the *Tubulifera* known to us, though on the mid-line both dorsally and ventrally there is a specialized and very small seta, which divides the six terminal hairs into two lateral groups of three each.

From the above lines it will be seen that *Urothrips* is characterised very strongly morphologically, and it is difficult to realize the importance of the various features in which this type departs from other *Thysanoptera*, but to us it seems that a study of this Insect, and of other allied forms which we understand have recently been discovered, will cause a change in our present views as to the ancestry or phylogeny of the *Thysanoptera* and their relationship with other groups.

Without venturing any fixed or decided opinion it might be well to review some of these peculiar features.

The structure of the antennæ, of the sternum, and of the eleventh abdominal segment (or end of tube), the presence of the abdominal papillæ, and the diminished or obsolete chætotaxy are difficult features to place in their proper perspective, in viewing the relationship of *Urothrips* with others of the order, and, as we have already said, we can make no deductions from the primary sex characters, as we have been unable to describe the genitalia of either sex, and because of this want of knowledge it is very possible that the remarks and deductions we make under the various headings hereafter may be of less value than we at present think. Though we make the « Stigmata » our last heading, we believe that the number of stigmata is one of the most important features we have been able to bring forward, and that in this feature we may ultimately find the key to the puzzling questions that have arisen since the discovery of *Urothrips*.

Ninth abdominal segment.

Instead of being transverse, and as short as, or shorter than the proceeding segments as is the rule in other *Tubulifera*, the ninth abdominal segment is very distinctly elongated and longer than the preceding. We have had the opportunity of examining the larval forms of several *Tubulifera*, and it is interesting to note that in each case the ninth segment is elongated and much longer than the preceding; it seems quite possible that a study of the Thysanopterous larvæ will prove very helpful.

Secondary sex characters.

Well-marked secondary sex characters appear to be exceptional in the *Thysanoptera*; in the *Tubulifera* the male is generally smaller, and has the abdomen more slender, whilst the thorax is sometimes bigger and more convex; the forelegs are almost always bigger and more or less strongly incrassate, and the fore-tarsus is armed with a stouter tooth.

There are however exceptions, such as the genus *Dinothrips* BAGNALL, wherein the male (but not the female) is armed with a pair of bifurcate mesosternal projections and a strong tubercle on the inner side of each intermediate femur.

Urothrips possesses secondary sex characters of distinct importance.

Maxillary and labial palpi.

These are only single-jointed in *Urothrips*, but in all other species of *Tubulifera* they are distinctly two-jointed.

In the *Terebrantia* the joints in the palpi vary as follows :

Thripidæ : Maxillary palpi 2- or 3-segmented, and the labial palpi always 2-segmented.

Æolothripidæ : Maxillary palpi geniculate, 3- to 7-segmented, and the labial palpi 2-, 4- or 5-segmented.

Tracheal stigmata.

One of the most important characters in *Urothrips* lies in the tracheal system, our species possessing no less than eleven pairs of stigmata; the pro-mesothoracic and meso-metathoracic pairs and a pair in each of the abdominal segments one to nine. These stigmata are all large and very distinct.

In all other forms of *Tubulifera* we find only four moderately large pairs of stigmata, the two first named, and a pair on the first and eighth abdominal segments, whilst in the *Terebrantia* the stigmata are smaller and usually four in number, though the meso-metathoracic pair are often very small, vestigial, and sometimes apparently totally absent. We have only examined a few species of *Æolothripidæ*, in which the stigmata present have been very small, and as yet we have personally been unable to trace the fourth (the meso-metathoracic pair) pair in this family. We should have preferred to have had the time and opportunity to examine more material in this family before making use of this observation, and can only incorporate it with reservation.

These facts and observations give rise to an interesting table :

FAMILIES.	STIGMATA.	PALPI.
Tubulifera.		
Urothripidæ BAGNALL	11 pairs very large.	Both pairs 1-jointed.
Phlæothripidæ HALIDAY . . .	4 pairs usually large	Both pairs always 2-jointed.
Idolothripidæ BAGNALL . . .		
Terebrantia.		
Thripidæ HALIDAY	Usually 4 pairs, small; meso-metathoracic pair often very small, vestigial or altogether absent.	Maxillary pair simple, 2- or 3-jointed. Labial pair always 2-jointed.
Æolothripidæ HALIDAY . . .	3 pairs small ; meso-metathoracic pair apparently absent.	Maxillary pair geniculate, 3- to 7-jointed. Labial pair 2-, 4- or 5-jointed.

The presence of numerous pairs of stigmata in *Urothrips*, in place of the 4 pairs found in all previously known *Tubulifera* and in most, though apparently not in all *Terebrantia*, would seem to suggest a primitive form rather than a more recently evolved type, and it appears to us that the intermediate set of stigmata have been comparatively recently lost, anatomical examination of members of the *Phlœothripidæ* showing that the main tracheal trunk on each side of the abdomen runs out laterally in each of the segments one to eight as though to open into a stigma, whereas the stigmata are only present in the first and eighth segments.

If this is so, then the family *Urothripidæ* would come first in a system of classification, followed by the other Tubuliferous families, then by the *Thripidæ* and lastly the *Æolothripidæ;* and it is interesting to note that where there is a diminution either in the number or size of the stigmata, the external parts appear to be correspondingly more complex and specialized. This is shown in an interesting manner as regards the palpi in the above table.

In closing our memoir on *Urothrips paradoxus* we expressed hopes that, before more fully considering the questions of the systematic grouping and position of the *Thysanoptera* and their possible ancestry, more material might be brought to light to help us to come to some conclusion on those important questions. Such further material as we had hoped for is being discovered. Shortly before our memoir was published, Profr. Pietro Buffa, Rome, briefly described two new genera, *Amphibolothrips* and *Bebelothrips* (1) from Italy, and although he does not point out the chief morphological features on which we have erected the family, yet we believe that these genera belong to the *Urothripidæ;* and we now learn that specimens of *Urothrips* or allied forms collected in Africa have been submitted to our colleague Dr. Trybom, of Stockholm, whom we have asked to make a special study of the genitalia.

---

(1) Buffa, « Boll. del Laboratorio dic Zoologia della R. Scuola Superiore d'Agric. in Portici », janv. 1909, III, pp. 194-196, fig. 1-3.

Notes on the Œstridæ,

by Profr. GEORGE H. CARPENTER, Royal College of Science (Dublin).

I. The Ox Warble-Flies.

Hypoderma bovis (DE GEER) *and H. lineata* (VILL.).

The life-history of the common Warble-Flies of the Ox presents several problems of great interest to the naturalist and of considerable practical importance to the farmer and the cattle-owner. It may be advisable, therefore, to place on record, in the proceedings of the first International Congress of Entomology, a summary of some experiments and observations which have been made on the subject, over a period of six years, under the auspices of the Irish Department of Agriculture (1).

One object of these experiments was to test the value of the washes and smears often recommended for application to Cattle during summer in order to prevent the female Fly from laying her eggs. During three successive summers a number of Animals, Calves, Yearlings and Milch Cows, were dressed in various ways, some on the back, some on the legs and some all over, once a week, or daily through the season, while a number of similar Animals

(1) Three reports of these experiments have been published in the « Journal of the Department of Agriculture and Technical Instruction for Ireland ». The first (by G. H. CARPENTER and J. W. STEEN) in vol. VIII, n° 2, 1908; the second (by G. H. CARPENTER and W. F. PRENDERGAST) in vol. IX, n° 3, 1909, and the third (by G. H. CARPENTER and T. H. CORSON) in vol. X, n° 4, 1910.

remained untreated as a « control ». The detailed records of these experiments may be seen in the first of the three reports already mentioned, and the result is that these dressings are useless for the prevention of egg-laying. Whether the washes were applied to back, legs, or all over made practically no difference. The age of the Cattle, however, is an important factor, for our results show that Yearlings are more subject to attack than Calves, and Calves than Cows. Instead of repeating here the detailed figures already published, a summary of the three years' work is given in the appended table, which shows clearly the general conclusions reached.

GENERAL RESULTS OF WARBLE-FLY EXPERIMENTS DURING THREE YEARS 1904-1905, 1905-1906, 1906-1907.

	Dressed.			Untreated		
	Number of Animals.	Number of warbles.	Average number of warbles per beast.	Number of Animals.	Number of warbles.	Average number of warbles per beast.
Calves	67	677	10	24	259	12
Yearlings	28	680	24.3	11	304	27.6
Cows	8	70	8 7	33	110	3.3

From this table it will be seen that the application of dressings had practically no effect in preventing the Flies from laying their eggs on the Cattle. Accordingly no dressings were applied after the summer of 1906, but in the spring of 1907 and subsequent years all the warble-maggots detected on the Department's farm were squeezed out from the Animals' backs and destroyed. This led to a very marked reduction in the number of maggots in the succeeding year. In 1907, 2,090 maggots had been found in 194 Cattle, an average of nearly 11 per beast, while in the spring of 1908, 132 Cattle harboured only 586 maggots, an average of 4.4 per beast. The reduction in that and the two subsequent years would have been much greater, had it been possible to destroy also the maggots in the Cattle on the neighbouring farms. As it was an instructive contrast was noticed between the liability to warbles of the young Cattle and Calves grazed at the centre of the farm, as compared with those which spent the summers in outlying fields, where they were exposed to the incursion of Flies from the surrounding farms.

GENERAL RESULTS OF SYSTEMATIC MAGGOT-DESTRUCTION DURING THREE SUCCESSIVE YEARS 1907-1908, 1908-1909, 1909-1910.

	In centre of farm.			On outskirts of farm.		
	Number of Animals.	Number of warbles.	Average number of warbles per beast.	Number of Animals.	Number of warbles	Average number of warbles per beast.
Calves	90	811	9	3	55	18.3
Yearlings and Heifers.	43	229	5.32	102	1,275	12.5
Cows.	130	321	2.47	70	241	3.44

The above results show that systematic maggot-destruction on the farm, as carried on during the last three years, is effectual in materially reducing the liability of the Animals — especially Yearlings and Heifers — to attack by the Warble-Flies. It is rather noteworthy that while, before the destruction, yearling Cattle were found to be more liable to attack than Calves, in the subsequent years the figures for Calves have been, on the whole, higher than for Yearlings. That is to say, the practice of maggot-destruction benefits, for some reason, Yearlings and Heifers more than Calves exposed to the same risk of attack. Milch Cows seem to be always freer from warbles than the younger Cattle.

As a practical outcome of the experiments, it is evident that preventive dressings can no longer be recommended for use with any prospect of success. Systematic maggot-destruction can, on the other hand, be recommended with confidence, but its results will be somewhat disappointing unless the practice be carried on over a wide area.

It was hoped that the experiment of dressing various batches of Cattle, some on the back and others on the legs, would have indicated on what part of the body eggs are habitually laid, but, as has been already stated, the dressings were useless wherever applied. During one summer (1905), however, four Calves were kept covered with cotton cloth on the back and sides, and the limbs of four others were protected by leather trousers. There was great difficulty in keeping the latter garments in place, and the protection was consequently imperfect. Nevertheless the former batch had 40 warbles the next spring, an average of 10 per beast, while the latter had 14, an average of only 3.5 per beast. This confirms the modern view that the female Warble-Fly lays her eggs on the limbs of Cattle more frequently than on their backs, and observations in the field supported the result of the experiment.

It is well known that while it was formerly believed that the maggots of *Hypoderma* bore directly through the skin, students of the life-history during the last twenty years have inclined more and more to the conclusion that the eggs or young maggots are licked into the mouth and that the latter, after wandering for several months in the wall of the gullet, muscular tissue, vertebral canal and elsewhere, come finally to their position beneath the skin of the back. I had no difficulty in finding young maggots in the submucous coat of the gullets of slaughtered bullocks from August till December. In order to test the mode of entrance, a number of muzzled Calves were turned out among those grazing in the fields, and at feeding-time these animals had their necks tied between stakes and their fore-limbs protected with leggings or aprons, so that it was believed to be impossible for them to lick themselves. For three years there were as many, or almost as many warbles on the Animals that had been muzzled as on the other Cattle, and it was thought that the experiment supported the older view that the maggots gain entrance through the skin, though they must wander subsequently through the tissues even as far as the gullet-wall. But in 1909 an additional precaution was taken by arranging a wire cage-muzzle around the leather one, so that all chance of eggs or maggots being sucked in through the breathing holes was obviated (pl. XIX, fig. 1). The five Animals treated in this way had on the average only 2 warbles each in the spring of this year (1), while the fifteen « control » Animals, that had been grazed in the same field, had an average of 6.33. From this result one may conclude that the normal mode of entrance is by the mouth.

2. The Reindeer Warble-Fly.
Œdemagena tarandi (Linn.).

The arrival at the Dublin Zoological Gardens in April, 1910 of a young male Reindeer badly infested with warble-maggots, afforded an opportunity, not often granted to a naturalist in the more temperate countries of Europe, of studying some stages in the life-history of *Œdemagena tarandi,* whose attacks on Reindeer in Lapland were long ago described graphically by LINNÉ (2), and whose structure was elucidated and figured in the classical monograph of BRAUER (3).

No less than 104 maggots of *Œdemagena* were taken from the back of the Reindeer above-mentioned, and the Animal was, consequently, in very bad condition. Some three dozen of these maggots came into my possession during the month of May, all being in the final larval stage. One of these, squeezed out from the Animal's back on May 10th, formed a normal puparium a few days later, and on June 22nd a female Fly in beautiful condition emerged. This was an agreeable surprise, as, notwithstanding repeated attempts, I have never been able to rear Flies from squeezed out larvæ of *Hypoderma*.

As regards the external form of the larva, I can add but little to BRAUER's description. A few points of detail respecting the spiny armature, and the mouth hooks with their supporting sclerites have already been published (4), together with some notes on the structure, of the terminal abdominal segments that form the protrusible ovipositor of the adult female. From the abdomen of the female specimen reared, I extracted a number of eggs, and these all

(1) 1910.

(2) Quoted by BRACY CLARK « An Essay of the Bots of Horses and other Animals. London, 1815, pp. 54-59.

(3) F. BRAUER, « Monographie der Œstriden ». Wien, 1863, pp. 133-134, pl. VIII, fig. 4, 8.

(4) *Journ. Economic Biology*, vol. V, 1910.

presented a feature that seems to be new and noteworthy. The general shape of the egg (pl. XIX, fig. 2), with its grooved flange (a) for attachment to a hair of the host-animal, closely resembles that of the *Hypoderma* egg ; it is, however, somewhat more quadrate at the base. But at the apex of the egg, a very distinct, delicate reticulated area (pl. XIX, fig. *2 b*) can be detected; and this readily splits away from the rest of the egg-envelope, so that little doubt can be entertained that it is of the nature of a lid, which opens to allow the escape of the newly-hatched larva. A comparison with the well-known life-history of *Gastrophilus*, whose egg is provided with a very distinct, cap-like lid, suggests that in the case of *Œdemagena* also the young maggot in licked into the host's mouth, and makes its way — like the larva of *Hypoderma* — through the gullet-wall and then through the tissues to the final position beneath the skin of the back. It is remarkable that the egg of *Hypoderma* has no such delicate, reticulated area at the apex ; here the whole envelope is firm and relatively thick (pl. XIX, fig. 3). Perhaps this may be held to support the view lately expressed by JOST (1), that in the life-history of *Hypoderma* not the young maggot, but the egg is swallowed by the host-animal, and that the larva is hatched only in the host's food-canal. Future researches may perhaps show if such a difference in habit between the two nearly-allied genera *Œdemagena* and *Hypoderma* really accompanies the slight but suggestive difference in the egg-envelope described in this communication.

---

(1) *Zeitschr. für wissensch. Zoologie*, vol. XXXVI, 1907, pp. 644-715.

FIG. 1.
Calf with leather muzzle surrounded by wire cage as used in experiments on *Hypoderma*.

FIG. 2.
Egg of *Œdemagena tarandi* LINNÉ
MAGNIFIED.

FIG. 3.
Egg of *Hypoderma bovis* DE GEER
MAGNIFIED.

G. H. CARPENTER. — NOTES ON THE ŒSTRIDÆ.

A quoi sert le mimétisme?

par William Schaus.

Dans la communication que je me permets de vous faire, je tiens à déclarer d'abord que mes observations ont été toutes faites dans la région néotropicale et que mes conclusions sont basées sur l'étude des Lépidoptères de cette partie de l'Amérique où pendant de longues années j'ai eu l'occasion de chasser, surtout au Mexique, dans l'Amérique centrale, dans les Guyanes, au Brésil et aussi dans deux des grandes Antilles, Cuba et la Jamaïque. La faune que j'ai étudiée contient une grande partie des espèces qui, à cause d'une similitude dans leurs couleurs et leurs dessins, ont été citées le plus fréquemment comme exemples du mimétisme.

Les immenses forêts des terres chaudes ont toujours eu pour moi un attrait tout particulier, à cause du mystère silencieux qui les enveloppe et de la merveilleuse richesse de leur faune. Il n'y a que l'homme aimant vraiment la nature qui puisse comprendre à quel degré l'atmosphère calme et accueillante de la forêt, avec ses innombrables sources de pensées et sa prodigalité incessante de nouveautés et de surprises, dépasse les attraits du monde et de ce qu'on appelle la vie civilisée.

Il y a trente ans, lors de mes premiers voyages au Mexique, j'étais sous l'influence de l'idée des couleurs dites protectrices, mais à mesure que je voyageais et que j'approfondissais la nature, je voyais que le coloris était, ou ce qu'on pourrait appeler offensif ou bien produit par l'influence naturelle de l'ambiance. Par offensif je veux dire ce coloris qui permet à un Animal, un Reptile ou un Insecte, de rester inaperçu pendant que sa proie, d'une intelligence moindre, s'approche assez près de lui pour être dévorée. Je savais que les Lépidoptères, auxquels je m'intéressais spécialement, n'avaient nullement besoin de coloris offensif, et cependant je ne pouvais découvrir la raison pour laquelle il aurait été défensif.

Parmi tous les arguments avancés par ceux qui admettent le mimétisme, on attache de l'importance à la destruction des Papillons par les Oiseaux, mais quiconque a vécu quelque temps dans une forêt tropicale saura que cette destruction est un mythe.

Les Papillons rhopalocères, ceux qui volent le jour, sont parmi les aliments les moins attrayants pour les Oiseaux, et, s'il n'en était pas ainsi, un grand nombre d'espèces auraient été exterminées depuis longtemps. C'est un cas très exceptionnel que de voir un Oiseau se lancer sur un Papillon, et s'il le fait, il le manque neuf fois sur dix. N'importe quel chasseur d'Insectes vous dira comment un petit Insecte peut souvent éviter un grand filet, et un Papillon n'a qu'à se poser pour éviter la poursuite des Oiseaux, qui ne l'attaquent qu'au vol. Une seule fois j'ai vu des Oiseaux poursuivre à plusieurs reprises des Papillons : c'était à Sarchi, au Costa-Rica, où des *Eumomota superciliaris* s'élançaient sur des *Heliconius petiveranus*, une des espèces soi-disant protégées par son mauvais goût. Je n'ai jamais vu un Oiseau essayer d'enlever un Papillon au repos; mes remarques s'appliquent aux Papillons diurnes : il faudrait des mouvements bien dissimulés pour réussir, ce dont les Oiseaux sont incapables, les sauts, les mouvements des ailes et de la tête effrayant un Papillon, qui est toujours sur le qui-vive et effarouché par tout objet étranger.

Que de fois, debout dans un chemin de la forêt, j'ai vu un grand Morpho ou une belle Nymphalide s'approcher, et dès qu'il me voyait, même à une distance de 10 mètres, il rentrait sous bois ou, en s'élevant, il passait bien au-dessus de ma tête !

En ce qui concerne le goût désagréable de certaines espèces, nous sommes fort ignorants, et l'on ne peut juger la question par les expériences faites sur les Oiseaux et les Animaux captifs, qui sont nourris avec régularité et n'acceptent de friandises que par avidité. Je trouve fort ridicule les essais faits sur les Singes, qui à l'état sauvage n'ont sûrement jamais attrapé un Papillon, du moins je n'ai jamais entendu parler d'un Singe américain qui se soit montré d'une telle adresse.

Je ne crois pas qu'un Oiseau ayant faim ferait attention au mauvais goût d'un Insecte, pas plus qu'un homme affamé ne serait difficile, et ici nous touchons à la question de l'alimentation. J'ai surtout travaillé dans les forêts humides du versant de l'Atlantique où il n'y a pas de saison fixe, ni sèche ni pluvieuse, et où l'apparition des Papillons dépend plus ou moins de la durée des pluies et du beau temps.

Je vais restreindre, autant que possible, mes remarques à mes expériences au Costa-Rica, où je viens de passer trois ans et où, à cause du peu de largeur du pays, à peu près 200 kilomètres d'une côte à l'autre, il existe deux conditions distinctes qu'on peut rencontrer facilement.

Sur le versant du Pacifique, il y a une saison sèche véritable, et on peut toujours y compter sur l'apparition et l'abondance des Papillons à un moment fixe, tandis que sur le versant opposé une période prolongée de temps triste et pluvieux détruit complètement toute éclosion de Papillons, et une espèce qui apparaît en abondance une année favorable ne semble plus se retrouver pendant plusieurs années suivantes. Pendant la première année de mon séjour au Costa-Rica, j'ai pris communément, pendant les mois de mars et d'avril, certaines espèces ares dans un endroit près de la côte orientale, et cependant pendant les deux années suivantes je n'ai plus rencontré ces mêmes espèces, quoique je sois allé les chercher chaque mois de l'année dans le même endroit. Cette disette de Papillons certaines années a été remarquée au Nicaragua par Belt.

Là où la nature joue un rôle aussi puissant et destructif, pourquoi existerait-il une espèce particulière, protégée contre les Oiseaux, et pourquoi deux espèces chercheraient-elles à se protéger mutuellement par le mimétisme? Il faudrait une raison d'être ou une nécessité absolue pour produire un tel résultat. Permettez-moi d'attirer votre attention sur un des groupes le plus en vue comme exemple des espèces qu'on appelle mimétiques, la faune de l'Amérique centrale, partie nord, dont toutes les espèces se trouvent aussi au Costa-Rica, à l'exception de *Ceratinia dionœa* et *Eresia philyra*, représentés par des espèces voisines. On raconte avec émerveillement qu'un collectionneur a envoyé au Musée d'Oxford plusieurs de ces espèces sous le même nom, ce qui est peu flatteur pour l'intelligence de l'entomologiste en question. Aucun collectionneur expérimenté ne prendrait un *Heliconius telchinia* pour un *Melinœa imitata*, leur vol et leurs mœurs étant dissemblables. Il n'y a pas deux espèces de la faune qui aient au vol une ressemblance autre que superficielle, c'est même beaucoup moins apparent que dans certains cas où les espèces sont tout à fait dissemblables, telles que *Castnia dœdalus* et certains *Caligos* de la Guyane, *Agraulis juno* et *Grapta haroldi* des pentes du Popocatepelt au Mexique, ou *Pieris lypera* femelle et *Eucides olympia*, deux espèces communes dans les forêts de la côte orientale du Costa-

Rica, quoique non encore citées de ce pays. Je ne prétends pas qu'on puisse toujours reconnaître au vol le genre auquel appartient une espèce, mais il n'y a pas moyen de prendre un *Leptalis*, un *Eresia* ou un *Heliconius* pour une espèce de la famille des *Ithomiinæ* : même le *Leptalis fortunata* avec ses ailes transparentes se reconnaît facilement par son vol des *Ithomiinæ* à ailes transparentes. Dans leur vol, les deux espèces qui se ressemblent le plus sont *Melinæa imitata* et *Pericopis angulosa*, mais ce dernier, contrairement à la croyance générale, est nocturne dans ses habitudes. A la lumière électrique, je trouvais de vingt à cinquante exemplaires du *Pericopis* dans une seule nuit, tandis que le jour je ne voyais peut-être qu'un seul exemplaire dans le courant d'un mois. C'est aussi le cas pour les espèces du genre *Centronia*, aux couleurs vives bleues et rouges, quoiqu'on les trouve assez souvent le jour, car ils ont l'habitude de se cacher sous les feuilles et ils sont plus souvent dérangés ; quand cela arrive, ils s'envolent avec la rapidité d'une Noctuelle.

Au Costa-Rica, j'ai passé plusieurs mois chez des amis, à Juan Viñas, où j'ai pu installer une forte lumière électrique sur une grande véranda, au deuxième étage, où je pouvais travailler toute la nuit en dépit de l'inclémence du temps, et l'on sait que les nuits orageuses sont les meilleures pour la récolte des Insectes. Dans les jardins qui entouraient la maison, plusieurs Oiseaux avaient construit leur nid, et j'avais l'occasion d'étudier de près leurs habitudes. Deux Gobe-mouches du genre *Myiozetetes* passaient une grande partie de la journée perchés sur un fil de fer du téléphone, à quelques mètres de la maison ; après que je les eus nourris pendant quelques jours, ils sont devenus tout à fait apprivoisés et ne s'effrayaient pas du mouvement de mon bras quand je leur jetais un Insecte ; il était très amusant de voir leur adresse, ainsi que les erreurs qu'ils commettaient. Ils ne suivaient un Insecte que sur une petite distance et, s'ils le manquaient, ils revenaient à leur perchoir. Jamais ils ne touchaient à un Insecte qui tombait par terre, comme le faisaient les Arctiens des genres *Halisidota*, *Ecpantheria* et autres, qui refusaient de voler quand je les jetais en l'air. Les grands Sphingides étaient trop formidables, mais les espèces de taille moyenne ainsi que les Cossides étaient fort appréciés, mais eux aussi n'étaient attaqués qu'au vol. A plusieurs reprises, je ne leur jetais rien et je les observais pendant des heures pour voir s'ils attaqueraient les nombreuses espèces de Papillons diurnes qui voltigeaient dans le jardin, mais je ne les ai pas vus une seule fois s'occuper d'eux, quoiqu'il y eût plusieurs espèces qui n'appartenaient pas aux soi-disant formes protégées. Ma plus grande surprise était de les voir attraper et manger avec délice des Hémiptères qui avaient une odeur des plus infectes, mais on ne peut les critiquer, puisque au Mexique les Indiens, et même quelques blancs, mangent avec gourmandise une espèce d'Hemiptère qu'on vend vivant dans les marchés et dont on fait un hachis menu, goût qui ne s'explique pas. Il y avait aussi dans le jardin deux jolis petits *Miloulus forficatus* qui ne mangeaient que les Nocturnes de petite taille, non seulement au vol, mais aussi en les poursuivant jusque dans les herbes. Ces grands Oiseaux noirs, *Crotophaga ani*, si abondants sous les tropiques de l'Amérique, ne tardèrent pas à découvrir la bonne aubaine : dès le lever du soleil, ils apparaissaient, et il y avait une lutte entre eux et les Oiseaux de basse-cour pour dévorer les nombreux Hétérocères que je jetais à moitié écrasés, de la véranda, par terre.

Rien n'était négligé, même les espèces aux odeurs répugnantes, comme *Estigmene albida*, les *Ecpantheria* et *Pericopis*. Pendant la nuit, il y avait toujours un rassemblement de grands Crapauds, dont quelques-uns avaient même l'audace de monter les escaliers pour se rapprocher de la lumière. Un Tatou venait toutes les nuits dans le jardin, et un gros Rat venait jusqu'à mes pieds, quand j'étais assis sans mouvement dans un fauteuil, pour saisir un Insecte par terre, puis il décampait lestement en emportant sa proie et sans faire attention à sa qualité.

Le jour, pas mal de petits Oiseaux, surtout des Roitelets, entraient sous la véranda et enlevaient des murs et du plafond tous les petits Nocturnes dont je ne pouvais me servir. Les petites espèces blanches du genre *Acidalia*, à peine perceptibles sur la peinture blanche, furent ramassées aussi promptement que les espèces plus visibles, et ceci démontre le peu de protection qu'ont ces espèces qui se reposent sur l'écorce des arbres et ne sont visibles qu'à l'œil expérimenté. Quand on voit la sagacité des Oiseaux, on se demande comment il se fait que certaines espèces n'ont pas été exterminées, et il faut connaître l'abondance prodigieuse des Insectes pour le comprendre. Une seule nuit, mon compagnon et moi nous avons rempli trois grands seaux de Coléoptères ramassés sur une étendue de quelques mètres carrés, et à Tuis, me tenant debout sous une lumière électrique, je fus, en moins de cinq minutes, couvert de la tête aux pieds de ces mêmes Insectes. A Cuernavaca, au Mexique, j'ai vu une Pyrale blanche si abondante autour des lumières, qu'on ne pouvait les comparer qu'à un tourbillon de neige. A Cordoue, dans le même pays, les pierres d'un grand mur disparaissaient complètement sous des Sphingides qui étaient entassés les uns sur les autres, et cependant le jour on ne retrouvait pas un seul exemplaire.

Les Oiseaux détruisent surtout les chenilles, dont la majorité sous les tropiques ne se nourrissent que la nuit, et dès l'aurore on voit les Oiseaux en train de chercher celles qui ne se sont pas cachées à temps. Les Hétérocères sont beaucoup plus nombreux que les Rhopalocères, et leur variété et leur abondance dans les pays humides sont produits par le fait qu'ils n'ont pas besoin du soleil pour leur développement, d'autant plus que pendant de longues périodes de pluies les ennemis de leurs chenilles sont chassés dans des endroits plus secs. Dans la végétation plus épaisse des tropiques, il est impossible de voir même la plus minime partie des Insectes qui s'y trouvent, et l'on juge d'habitude de leur nombre par les Papillons mis en évidence par leur vol tant que le soleil reluit, mais qu'un nuage l'obscurcisse et immédiatement tout disparaît. La masse des Papillons reste dans le haut des arbres jusqu'à ce que le soleil ait pénétré la végétation, et dans la forêt c'est surtout entre 11 et 2 heures qu'on peut chasser avec un bon résultat; alors bien des espèces se sont fatiguées de voltiger et de se nourrir, et elles descendent se reposer sur une feuille baignée de soleil. Les *Erycinides*, moins rares qu'on ne le suppose, et à part quelques genres comme les *Mesosemia* qui volent toujours près du sol, restent dans les grands arbres, et lorsqu'ils descendent, ils s'écartent dans la forêt, ne cherchant pas les éclaircies ni les chemins, comme les espèces de tant d'autres groupes.

Mais revenons au groupe intéressant le mimétisme. Les *Ithomiinæ* sont restreints à la forêt, exigent moins de soleil que n'importe quelle autre famille de Rhopalocères en Amérique et se rassemblent en grand nombre sur certaines fleurs tant qu'elles sont à l'ombre, mais s'envolent dès que le soleil les touche. *Lycorea atergatis* et *Heliconius telchinia* ont besoin de plus de soleil, le dernier voltigeant de long en large au-dessus des chemins de la forêt ou en cercle dans une clairière pendant des heures entières, se reposant de temps en temps sur un point ensoleillé. *Eueides zorcaon* se trouve surtout dans les champs ou bien au-dessus des arbres. Les *Eresia*, habitant la forêt, descendent se reposer un instant sur le sol ou sur une plante basse, puis remontent. *Dismorphia praxinoe* vole près du sol et se repose fréquemment, et il serait impossible de le prendre pour quelque autre espèce que ce soit. *Pieris malenka* femelle a un vol tout particulier et se distingue de loin. *Protogonius cecrops* a un vol saccadé et se repose par instants sur les feuilles, à 2 mètres du sol. Les soi-disant espèces protégées, en dehors de *Eueides zorcaon*, habitent surtout les taillis, les endroits où l'on trouve le moins d'Oiseaux.

Les Papillons ont des ennemis bien plus importants que les Oiseaux, surtout les Lézards qui en détruisent un grand nombre et qui, avec leurs couleurs offensives, restent immobiles pendant de longues heures, saisissant le Papillon moins intelligent qui se pose à leur portée; quelquefois même ils s'en rapprochent doucement par derrière et, jugeant mal la fragilité de leur proie, ne font que détruire le contour des ailes, ce qui explique le grand nombre de Papillons qu'on trouve avec une grosse échancrure dans les ailes postérieures; ceci est surtout le cas pour les Papillons qui se reposent sur le tronc des arbres, comme les *Anæa*, *Prepona*, *Aganisthos* et autres. J'ai même vu des Lézards immobiles à côté d'un endroit humide, sur la route, prêts à se lancer sur le premier Papillon qui viendrait étancher sa soif.

Les *Odonata* ou Libellules occasionnent aussi de grands dégâts et ne font aucune différence dans le choix de leurs victimes. Je ne m'expliquais l'absence complète de papillons le long de certains sentiers que par l'abondance des Libellules, car au delà du rayon de leur vol les Papillons reparaissaient. Un grand Diptère du genre *Promachus* en détruisait aussi pas mal, et plusieurs fois, quand je m'approchais d'une belle *Thecla*, je l'ai vu saisie et emportée par une de ces Mouches voraces. Sur les hautes et froides pentes des montagnes, les Papillons fréquentent les endroits ouverts et ensoleillés, et dans les fleurs jaunes qui les attirent, on trouve des Araignées d'une couleur semblable, qui s'y cachent. C'est remarquable comme elles saisissent et retiennent des Insectes très supérieurs à elles comme taille : aussi n'en font-elles aucun choix, la victoire restant au plus fort.

J'ai déjà dit que je n'ai jamais vu, durant toutes mes années de chasse, un Rhopalocère au repos saisi par un Oiseau, de même que je n'ai jamais vu un Papillon diurne ou nocturne qui fût protégé ou dissimulé par son coloris; je n'ai jamais vu une *Anæa* échapper à cause de la ressemblance du dessous de ses ailes à une feuille, son salut dépendant de sa prudence et de son agilité.

Que les Papillons soient attirés par les couleurs pareilles aux leurs est un fait bien connu, et dans l'île de Cuba je commençais ma chasse aux belles *Callidryas* qui sont propres à l'île, en plaçant un morceau de peau d'orange sur la route qu'elles fréquentaient; le premier exemplaire qui passait avait toujours la curiosité de s'arrêter, et son hésitation lui coûtait la vie. De même, dans l'isthme de Tehuantepec, mon premier *Morpho thoosa* se laissa attirer par un morceau de papier blanc.

Si vraiment certains Diurnes avaient besoin de couleurs protectrices pour sauver leur vie, pourquoi les Nocturnes, dont les couleurs pendant le vol deviennent confuses, en auraient-ils besoin ?

La plupart des *Syntomides*, qui ressemblent étonnamment aux Hyménoptères, sont essentiellement nocturnes, tandis que les Hyménoptères, qui volent la nuit, n'ont aucune forme mimétique parmi les *Syntomides*.

Pourquoi s'embarrasser de difficultés quand les lois naturelles de l'évolution et les influences du milieu produisent des changements visibles et acceptés dans tous les phénomènes biologiques?

Les groupes synchromatiques qui ont attiré le plus l'attention montrent un changement pareil dans leur distribution à travers la région néotropicale. BATES et WALLACE se sont aperçus que les changements étaient d'autant plus rapides aux Amazones, que l'élévation du terrain augmentait et que l'humidité devenait plus grande.

Au Costa-Rica, près de San-Mateo, sur le versant du Pacifique, *Melinæa scylax* était très commun, tandis que *Heliconius clarescens*, qui lui ressemble, ne s'y trouvait pas, quoique *Heliconius telchinia* fût abondant, et que l'espèce de *Melinæa*, qui lui correspond, *imitata*, fût très rare. *Heliconius clarescens* se trouve beaucoup plus au sud sur la côte du Pacifique, mais sur le versant de l'Atlantique j'ai découvert, dans des endroits très éloignés l'un de l'autre, une forme intermédiaire plus près de *Heliconius clarescens* que de *Heliconius telchinia ; Melinæa scylax* n'y existait pas, ce qui montre que *Heliconius clarescens* est une évolution de l'abondant *Heliconius telchinia*, sans aucune influence du *Melinæa scylax*, et que le changement est dû à des conditions locales.

Heliconius galanthus se trouve au Costa-Rica, avec les ailes inférieures complètement noires et puis avec une bordure de taches blanches qui augmentent en grosseur à mesure qu'on se rapproche de Panama; là il acquiert une bordure blanche et porte le nom de *Heliconius chioneus*. Un fait prédominant au Panama : dans les forêts humides du versant oriental, le jaune se change en blanc, ou bien il y a un développement des taches blanches n'existant pas chez des espèces voisines. C'est là que j'ai trouvé la belle *Pierella ocreata*, remarquable par sa grosse tache blanche sur les ailes inférieures, qui n'existe pas chez la *Pierella incanescens* du Costa-Rica. Sur le versant opposé, j'ai vu des *Morpho peleides* qui paraissaient si purement blancs au vol, que j'étais convaincu que c'étaient des *Morpho polyphemus*, le bleu disparaissant complètement. On a parlé d'une ressemblance entre *Heliconius leuce* et la femelle de la *Pieris noctipennis*. *Heliconius leuce* ne se trouve que sur la côte au-dessus de 300 mètres et vole très lentement, tandis que *Pieris noctipennis* se trouve jusqu'à 2,000 mètres et vole avec une impétuosité qui la fait ressembler bien plus à une grande Hespéride.

Les Papillons de la région néotropicale sont essentiellement des espèces forestières, et les espèces qui se trouvent dans les régions ouvertes augmentent rapidement, à mesure que les forêts disparaissent par la culture, et sont presque toutes des espèces très répandues.

Que des espèces appartenant à des familles différentes se développent en suivant des directions pareilles, me paraît être une loi naturelle de la nature ; de même, par la migration et le changement des conditions climatériques, les espèces qui ont une tendance à changer acquièrent des traits distinctifs, tel qu'on le voit chez les *Eueides* de Cuba et d'Haiti, où il n'y a pas d'espèces correspondantes dans les autres familles, le *Lycorea* qui se trouve à Cuba étant bien différent. Les aberrations et les races locales ne sont que des espèces naissantes, montrant peu à peu des traits plus caractéristiques.

Dans l'évolution des espèces, des formes intermédiaires ont disparu en si grand nombre, que les formes divergentes qu'on trouve aujourd'hui sont facilement expliquées, et il n'y a pas de raisons pour lesquelles les races susceptibles de changement ne suivraient des lignes parallèles sous l'influence de conditions absolument semblables.

Nous ne pouvons pas expliquer pourquoi la femelle est plus sujette à la variabilité que le mâle, à moins que ce ne soit parce qu'elle est plus faible et que par cela ses caractères sont moins arrêtés.

Je ne puis accepter la théorie du Prof[r] POULTON qui, dans les « Transactions de la Société Entomologique de Londres 1908 », prétend que la femelle noire du *Papilio glaucus* a pris comme modèle primaire le *Papilio philenor*. Dans la Floride méridionale, je n'ai pris que des femelles noires, tandis que le *Papilio philenor* était plutôt rare. Dans le nord des États-Unis, où le *Papilio glaucus* est commun, la femelle jaune du *Papilio glaucus* prédomine. Que le *Papilio palamedes* soit très voisin du *Papilio glaucus*, cela est démontré par la chenille, qui est presque pareille à celle de *Papilio pilumnus*, tandis que *Papilio pilumnus* ne ressemble au *Papilio glaucus* qu'à l'état parfait. Le *Papilio palamedes* ne pourrait guère prendre le *Papilio philenor* comme modèle, puisque c'est une espèce forestière, tandis que *Papilio philenor* fréquente les champs.

Que des espèces soient influencées par leur ambiance, c'est, à mon avis, la vérité, et il y a dans la nature une puissance réflectrice qui produit ses impressions sur beaucoup d'objets dans des conditions favorables; mais comme nous ne connaissons pas la chimie de la nature, ni les influences premières qui ont été l'origine d'espèces nouvelles, c'est dans le changement continuel de l'évolution qu'il faudra chercher les explications du mystère.

Il en est des genres comme des espèces, et j'ai trouvé chez les *Rosema* et *Datana* la nervulation variant non seulement dans une même espèce, mais aussi dans les ailes opposées d'un même exemplaire. Examinant tout le groupe des Lépidoptères avec ses formes génériques et spécifiques infinies, il me paraît tout naturel que certaines couleurs d'espèces ancestrales se soient maintenues dans des espèces maintenant bien éloignées les unes des autres. La plus grande variabilité se trouve dans les espèces qui fréquentent les forêts des terres d'alluvion, ces terrains plutôt modernes des Guyanes et des Amazones. Sur le versant occidental du Mexique, il y a une tendance au mélanisme aussi prononcée dans les Hétérocères que dans les Rhopalocères. Je ne vois pas comment ceux qui admettent le mimétisme peuvent tirer des conclusions du nombre d'exemplaires de leurs groupes mimétiques qui se trouvent au même moment, car au Costa-Rica je n'ai trouvé aucune règle fixe. Les *Mechanitis* volaient toute l'année, montrant la même variabilité dans les endroits secs que dans les endroits humides. *Melinœa imitata* abondait, alors que l'on ne voyait pas un seul *Heliconius telchinia*. *Eueides zorcaon* était commun dans des endroits où l'on ne voyait jamais d'*Heliconius*, de *Melinœa* ou de *Mechanitis*.

Je cite des faits et j'espère que mes observations auront un intérêt pour ceux qui étudient la faune américaine.

J'ai causé avec beaucoup de chasseurs, collectionneurs d'Oiseaux et d'Insectes, et je n'ai jamais pu apprendre que les Oiseaux attrapent des Papillons si ce n'est dans des cas exceptionels, et c'est pour cela que je demande : A quoi sert le mimétisme?

De l'utilisation des Insectes et particulièrement des Coléoptères dans les questions de zoogéographie,

par JEAN SAINTE-CLAIRE DEVILLE (Épinal).

Il semble que jusqu'à présent, dans les questions de zoogéographie, les matériaux et les arguments aient été empruntés de préférence aux Animaux supérieurs, tels que les Mammifères et les Reptiles, et aux Mollusques terrestres. Je me propose de montrer, dans cette note, quels services peut rendre l'étude des aires de dispersion actuelles des Insectes et notamment des Coléoptères, à condition que les résultats en soient judicieusement utilisés.

Comparés aux Animaux supérieurs et aux Mollusques terrestres, les Coléoptères ont un avantage incontestable, qui est le grand nombre de leurs espèces. Sous le climat de l'Europe moyenne, une petite circonscription, telle qu'un département français, abrite une faune coléoptérique qui atteint presque toujours et dépasse fréquemment le chiffre de 3,000 espèces; une province d'une étendue dix fois plus grande, telle que le bassin de la Seine ou la Silésie, compte facilement 4,500 espèces recensées, ce qui correspond, en réalité, à près de 5,000; enfin, un grand État comme la France ou l'Empire d'Allemagne ne contient guère moins de 8,000 espèces.

Parmi cette multitude d'Insectes, il en est un grand nombre, comme je le montrerai plus loin, qui n'offrent pas d'intérêt et doivent être laissés de côté pour les recherches de zoogéographie. Toutefois, on peut compter que sur le chiffre de 8,000 espèces, que je cite plus haut, 2,000 environ, soit à peu près le quart, méritent qu'on étudie avec soin leur dispersion.

En marquant sur une carte les localités où une espèce a été authentiquement capturée, on obtient une figure constituée par un ensemble de points qu'on peut, lorsqu'ils sont assez nombreux, circonscrire par une courbe enveloppe délimitant la surface actuellement occupée par l'espèce. D'excellents modèles de ces cartes de dispersion ont été donnés par le Dr R. F. SCHARFF dans son beau livre : « European Animals, their geological history and geographical distribution » (Londres, 1907).

En rapprochant les unes des autres celles de ces figures dont les tracés coïncident ou offrent une évidente affinité, on arrive assez aisément à grouper les éléments d'une faune donnée en séries correspondant chacune à une même direction d'origine et à une même époque d'invasion. Ce qu'il importe de faire ressortir, c'est que sur plusieurs milliers de cartes de dispersion fournies par les Coléoptères européens, il sera très facile de trouver les éléments d'une statistique sérieuse et féconde. Avec les grands Animaux et les Mollusques terrestres, le nombre des espèces est réellement un peu trop restreint pour permettre d'aboutir toujours à des résultats concluants.

Par rapport aux autres ordres d'Insectes, un avantage sérieux des Coléoptères est que leur étude, au moins en ce qui concerne la région paléarctique, est déjà fort avancée et que leur détermination est, en général, assez facile et assez sûre. Les renseignements sur leur dispersion sont nombreux et peuvent être, la plupart du temps, bien contrôlés. De plus, dans cet ordres d'Insectes, la proportion des espèces aptères ou peu capables d'un vol soutenu est importante; ce fait diminue les chances de dispersion accidentelle et rend les considérations tirées de l'étude des Coléoptères plus probantes que celles fournies par les Diptères ou les Hyménoptères par exemple.

Ces avantages sont contrebalancés par un inconvénient assez sérieux. Qu'il s'agisse des Vertébrés ou des Mollusques, les fossiles sont très nombreux et très bien connus. Leur étude fournit, au sujet de la dispersion des espèces actuelles aux époques antérieures des renseignements inappréciables. C'est ainsi que les recherches des paléontologistes nous ont appris que le Lion, à l'époque pliocène, a existé jusqu'en Angleterre ; que l'Antilope Saïga, la Marmotte des steppes (*Arctomys bobac*) et divers autres représentants d'une faune semi-désertique ont pénétré autrefois jusque dans l'Europe occidentale. On a pu de même, sur certains points, comparer d'une manière très complète la faune malacologique d'une époque donnée avec celle de l'époque actuelle et évaluer le pour-cent des espèces éteintes. Avec les Coléoptères il n'y a malheureusement rien d'analogue. Les fossiles appartenant à des espèces encore vivantes sont très peu nombreux et leurs gisements très clairsemés. Nous savons, par exemple, par les recherches du Dr FLACH (Aschaffenburg), que le *Carabus Menetriesi* FALD., espèce dont la limite est actuellement jalonnée par la Finlande, les provinces baltiques de la Russie et la Prusse occidentale, a existé en Bavière à la fin de la période tertiaire. Mais le nombre des espèces sur lesquelles nous avons des renseignements de ce genre est infiniment restreint et ne paraît pas destiné à s'accroître beaucoup.

Est-ce à dire que l'étude des Coléoptères, en nous refusant les jalons positifs témoins des anciennes migrations, ne nous donnera aucun indice sur le passé et l'avenir d'une espèce? Il n'en est rien, à condition que nous sachions interpréter certains détails de géographie zoologique. La dispersion d'une espèce n'est fixe, ni dans le temps, ni dans l'espace ; nous n'en pouvons fixer qu'un moment très fugitif, quelque chose comme une photographie instantanée. Mais elle se traduit pour nous par une figure dont le tracé, si nous savons l'examiner, peut nous mettre sur la voie des hypothèses les plus ingénieuses.

L'un des indices les plus importants est celui que l'on peut tirer de la répartition « sporadique ». On sait ce que signifie ce terme. J'en donnerai un exemple ou deux pour en préciser encore l'acception.

Une des espèces caractéristiques du massif central de la France est un *Trechus* découvert au Mont-Dore et décrit par FAIRMAIRE, en 1859, sous le nom de *Trechus amplicollis*. Il a été retrouvé depuis en plusieurs points de la même région, et je lui connais six stations principales, comprises assez exactement dans un triangle ayant ses sommets à Montluçon, Limoges et Aurillac. Or, vingt ans environ après sa découverte, le *Trechus amplicollis* a été reconnu identique au *Trechus sculptus* SCHAUM, Insecte principalement répandu dans le nord de la chaîne des Karpathes, les Sudètes, le Riesengebirge, et qui, d'après SCHILSKY, se trouve jusque dans le Harz. Dans son ensemble, l'espèce occupe donc aujourd'hui deux centres de dispersion assez étendus, complètement isolés l'un de l'autre et situés aux deux extrémités des anciennes chaînes varisques. Elle a dû s'étendre autrefois sur une surface beaucoup plus considérable. On peut supposer que ce *Trechus*, d'un type assez différent des *Trechus* alpins en état d'évolution contemporaine, est un des derniers débris d'une ancienne faune de montagne antérieure au soulèvement des Alpes.

Si je porte sur une carte les localités où se trouve actuellement le *Carabus glabratus* L., j'observe que cette espèce, répandue abondamment et d'une manière à peu près continue dans tout le nord de l'Europe, devient de plus en plus rare à mesure que l'on s'avance vers l'ouest et le midi. Encore commun dans les plaines de l'Allemagne du nord, il est déjà peu fréquent et localisé en Westphalie, dans les Pays-Bas et dans la vallée du Rhin ; il occupe, comme autant d'îlots, les massifs élevés du Taunus, du Schwarzwald et des Vosges. A l'ouest de la Meuse et de la Moselle, il disparaît presque complètement et ne se retrouve plus que sous la forme de colonies isolées dont les plus méridionales atteignent le Dauphiné. De ces colonies, trois sont particulièrement remarquables : celle de la forêt de Compiègne, encore très nombreuse et très florissante, celle de la forêt de Bellesme, dans les collines de Normandie, et celle de Pierre-Perthuis près Avallon (Yonne). Portées sur la carte, elles semblent jetées en arrière de la zone de dispersion continue comme les postes d'arrière-garde d'une armée en retraite. Le cordon sporadique adjacent à une zone de dispersion continue nous apparaît donc comme le témoin d'une ancienne répartition plus étendue, qui devait largement englober tous les îlots. Pour en revenir au *Carabus glabratus*, l'hypothèse la plus probable est que, pendant la période glaciaire, il devait habiter les plaines de l'Europe occidentale ; au fur et à mesure du retrait des glaces et du retour d'une température plus clémente, il a dû coloniser les plaines de l'Europe septentrionale et abandonner la zone climatérique de Paris, où il n'est plus représenté que par des îlots. Sa présence en Irlande semble indiquer que, à l'encontre de beaucoup d'espèces de la faune fenno-scandinave actuelle, il n'est pas originaire de Sibérie.

L'étude des faunes insulaires est en effet très féconde et donne un autre moyen de déceler les anciennes extensions. Considérons une espèce en voie de progression graduelle dans une direction donnée et ayant envahi plus ou moins complètement une presqu'île. A une époque géologique postérieure, sous l'influence d'une cause dont la nature importe peu, un mouvement en sens inverse se dessine. Suivant une loi admise par le Dr SCHARFF, la retraite s'effectue dans une direction opposée à celle de l'invasion. Mais entre-temps la péninsule est devenue une île. Plus d'une fois tout se passe comme si les émigrants, cherchant à suivre le mouvement de leurs congénères continentaux, en étaient empêchés par la mer et se trouvaient prisonniers dans leur île comme des poissons dans une nasse. Inutile de dire que cette comparaison familière ne représente que l'apparence du phénomène. Du fait de l'insularité, les faunes des îles acquièrent simplement une évolution indépendante et en général retardée, qui favorise le maintien d'une certaine proportion de « relicts ».

Peu importe d'ailleurs le mécanisme; c'est un fait incontestable, qu'il y a très fréquemment discontinuité entre la limite de dispersion d'une espèce dans une île et la même limite sur le continent le plus voisin. On constate, par exemple, que la faune de la côte sud de la Grande-Bretagne, et spécialement la faune littorale, a conservé une série d'espèces méridionales qui font absolument défaut sur la rive opposée de la Manche; la plupart réapparaissent sur la côte atlantique de la France : tels sont *Nebria complanata* L., *Myrmecopora uvida* Er., *Psammobius porcicollis* F., *Helops cœruleus* L.; d'autres, par exemple *Medon pocofer* Aubé et *Ceuthorrhynchus verrucatus* Chevr., ne se retrouvent que sur nos côtes méditerranéennes. De l'existence de ces Coléoptères en Angleterre nous pouvons inférer qu'ils ont occupé autrefois, sur l'emplacement de notre pays, des stations beaucoup plus septentrionales qu'aujourd'hui.

De tout ce qui précède il résulte que nous ne sommes pas absolument désarmés pour retrouver, même sans le secours des fossiles, les traces des fluctuations de nos espèces actuelles dans le passé.

J'en arrive maintenant aux précautions qu'il convient de prendre dans le choix des Coléoptères dont la dispersion mérite d'être étudiée.

Il y a lieu, tout d'abord, de n'employer qu'avec une extrême prudence les espèces cosmopolites ou très bien adaptées au voisinage de l'homme; l'extension de certaines d'entre elles, favorisée par le commerce, est un fait d'observation journalière. De ce nombre sont la plupart de nos *Lathridiidæ*, les *Palpicornia* des genres *Sphæridium* et *Cercyon*, beaucoup de Staphylins (*Medon*, *Oxytelus*, *Philonthus*, etc.) et bien d'autres dont l'énumération n'est pas à sa place ici.

Il y a lieu, en revanche, d'accorder une valeur particulière aux espèces aptères, dont les migrations sont forcément lentes et la dispersion accidentelle difficile. Les Curculionides épigés des hautes montagnes, *Otiorrhynchus*, *Dichotrachelus*, etc., incapables de franchir les vallées profondes et les hautes chaînes neigeuses, sont à citer parmi les plus précieux. Il en est de même de toutes les petites espèces des faunes hypogée et frondicole de l'Europe méridionale, ainsi que de la faune des cavernes.

La connaissance de l'éthologie des espèces considérées est de la plus haute importance. Beaucoup de nos Coléoptères sont liés d'une manière exclusive à une espèce végétale déterminée, ou encore à un autre Animal, par exemple à telle ou telle espèce de petits Mammifères ou de Fourmis. En principe, jamais la dispersion d'un Coléoptère phytophage ou parasite ne devra être envisagée indépendamment de celle de son hôte ou de son végétal nourricier. La comparaison des deux aires de dispersion, dont l'une est nécessairement incluse dans l'autre, est seule instructive. Quelques exemples vont éclairer cette notion.

Les contrées où croissent spontanément nos grands Résineux sont aujourd'hui très exactement connues. Les aires de dispersion de ces arbres ne coïncident d'ailleurs qu'imparfaitement avec les zones climatériques. C'est ainsi que l'Epicéa (*Picea excelsa* LINK.), qui s'est étendu jusqu'aux Alpes-Maritimes, n'a pu atteindre certains massifs montagneux dont le climat lui aurait parfaitement convenu, tels que l'Écosse, l'Auvergne et les Pyrénées. Si on porte sur une carte les zones habitées par certains Longicornes bien connus, tels que *Pachyta Lamed* L., *P. quadrimaculata* L., *Gaurotes virginea* L., *Leptura virens* L., *Callidium coriaceum* PAYK., on constate qu'elle coïncide d'une manière très satisfaisante avec l'aire de croissance spontanée de l'Epicéa. Il en est à peu près de même pour un Curculionide, *Pissodes harcyniae* GYLLH., sauf que ce dernier n'accompagne pas son arbre nourricier jusqu'à sa limite vers le sud-ouest et s'arrête aux Vosges. La question se complique lorsqu'il s'agit d'un parasite d'ordre secondaire. Le plus gros de nos Scolytides, *Dendroctonus micans* F., commun dans tout le nord de l'Europe, se trouve aussi, bien que plus rarement, dans nos montagnes françaises et suit son arbre nourricier (toujours l'Epicéa) jusque dans le comté de Nice; mais un parasite du second degré, *Rhizophagus grandis* GYLLH., qui poursuit dans ses galeries la larve du *Dendroctonus*, n'atteint ni le Jura, ni les Alpes. En revanche d'autres espèces, douées d'une grande faculté d'expansion, sont sorties de leurs forêts natales et ont envahi petit à petit les grandes plantations : tels sont, parmi les parasites directs, *Hylastes cunicularius*, *Ernobius abietis*, *Magdalis nitida*, etc. ; parmi les parasites certainement indirects, certains Cryptophagides et Lathridiides dont l'éthologie est encore obscure (*Micrambe abietis*, *Cryptophagus subdepressus*, *Corticaria foveola*).

Il peut arriver qu'un Insecte, moins exclusif, s'accommode de deux ou plusieurs espèces végétales. Un beau Longicorne, remarquable par son dimorphisme sexuel, *Oxymirus cursor* L., se développe également bien dans les souches et les troncs morts de nos deux grands Sapins (*Picea excelsa* LINK. et *Abies pectinata* DC.); il nous faudra donc comparer son aire de dispersion à l'ensemble de celles des deux arbres; or elles se recouvrent partiellement, mais sont loin de coïncider. Nous aurons ainsi l'explication de la présence simultanée de l'*Oxymirus* dans les Pyrénées et en Scandinavie.

Au premier abord, l'aire de dispersion du plus grand de nos Buprestes indigènes, *Chalcophora mariana* L., peut paraître paradoxale, surtout pour ceux qui seraient tentés d'y chercher un rapport quelconque avec les zones climatériques. Superposons lui un calque représentant la surface occupée dans la région paléarctique par l'ensemble des espèces du genre *Pinus* croissant *spontanément* : le mystère s'éclaire subitement et les singularités s'expliquent. Les discordances entre les deux aires (car il en existe) n'en acquièrent que plus d'intérêt.

Il serait facile de multiplier ces exemples, mais le peu qui vient d'être dit suffira à faire comprendre la complexité du sujet et l'extrême variété des cas qui peuvent se présenter. Ce qu'il importe de retenir, c'est que la dispersion d'un Insecte phytophage ou parasite n'a pas grand intérêt en elle-même et qu'elle ne doit être envisagée que par comparaison avec celle de l'hôte ou du végétal nourricier.

Il y a lieu de faire une réserve analogue pour les Coléoptères dont la présence est liée à la nature physique ou chimique du sol. Presque toujours la liaison a lieu par l'intermédiaire de la végétation, et alors nous retombons dans le cas précédent; mais il n'en est pas toujours ainsi, puisque certains Insectes, indifférents à la végétation, comme *Cicindela silvatica*, *Carabus nitens*, plusieurs *Nebria*, *Hydræna*, *Ochthobius*, etc., sont, les uns, exclusivement silicoles, les autres, exclusivement calcicoles. Si l'on veut que leur dispersion soit explicable et intéressante, il est nécessaire de l'étudier avec une carte sur laquelle on aura teinté de couleurs différentes les affleurements purement calcaires, les affleurements purement siliceux et ceux qui participent des deux natures de sol.

Il va sans dire, enfin, qu'il faut éliminer des statistiques les espèces incertaines ou trop rares, sur lesquelles les renseignements font à peu près complètement défaut.

Ainsi s'explique la faible proportion ($^1/_4$ environ) des espèces qui pourront réellement servir à la solution des problèmes de zoogéographie. Ce choix judicieux à exercer parmi les espèces de nos catalogues nécessite une juste appréciation de leur valeur et une connaissance aussi approfondie que possible de leurs conditions d'existence : il ne peut être fait utilement que par un spécialiste. Il en est de même de l'établissement des cartes de dispersion. Nos catalogues et nos faunes locales, malheureusement encore trop peu nombreux, sont de valeur très inégale ; à côté d'œuvres très consciencieuses et vraiment scientifiques, il en est qui fourmillent d'erreurs et dont les indications doivent être passées au crible d'une critique toujours en éveil. Supposons qu'un savant de haut mérite, mais étranger à l'entomologie, entreprenne un travail de cette nature : il sera tenté d'accorder la même valeur à tous les document écrits qui lui tomberont sous la main, et de mettre sur le même pied des renseignements indiscutables et d'autres fort douteux. Finalement, après un travail des plus pénibles, il arrivera à des résultats confus et fort inexacts.

C'est donc à nous, « coléoptérologistes », qu'il appartient de présenter aux savants moins spécialisés, mais plus familiarisés avec l'histoire naturelle générale, des matériaux de choix à l'aide desquels ils puissent édifier des théories solidement assises et reconstituer l'histoire des continents disparus.

Libre à nous d'ailleurs, si nos connaissances nous le permettent, d'essayer d'interpréter nous-mêmes les résultats de nos travaux et de porter un peu de lumière sur ces questions encore si obscures.

Die « Weddabrücke »,

von WALTHER HORN (Berlin).

Schon R. WALLACE hat (« Geographical Distribution of Animals », I, 1876, S. 328 und 359-362) die Vermutung ausgesprochen, dass einst im Tertiär eine direkte Landverbindung zwischen den Malediven und Ceylon, einerseits, und Hinterindien bis zu den Philippinen, andererseits, existiert habe. Dr. BLANFORD erwähnt gleichfalls (« Manual of the Geology of India », I, 1879, S. LIII und LXVIII) die Möglichkeit eines Zusammenhanges zwischen Ceylon und den Malayischen Inseln für die I. Hälfte des tertiären Zeitalters. Neuerdings ist diese Frage etwas in den Hintergrund gerückt, da andere wichtigere Landbrücken das Interesse der Zoogeographen mehr auf andere Erdteile lenkten, vor allem auf den früheren Konnex von Afrika mit Amerika, einerseits, und Asien, andererseits, auf die früheren Landmassen beziehungsweise Landverbindungen am Südpol, etc.

Bei dem Studium der Faunistik der asiatisch-orientalischen Cicindelinen fiel mir auf, dass einige Arten eine merkwürdig diskontinuierliche Verbreitung aufweisen, welche nur zum Teil auf unserer lückenhaften Kenntnis mancher Zwischengebiete beruhen konnte. In der « Deutschen Entomolog. Zeitschrift », 1909, S. 461/2 habe ich diese Species kurz angeführt, es waren :

1. *Cicindela angulata* subsp. *plumigera* W. HORN, bekannt aus Vorderindien südlich Mysore und Sumatra ;

2. *Cicindela aurulenta* F., bekannt aus Ceylon, einerseits, und, andererseits, dem ganzen Gebiet zwischen Bengalen und Sickim, Java und Borneo, Südchina und Formosa ;

3. *Cicindela aurovittata* AUD. und BRL., bekannt von Ceylon und Pondicheri, einerseits, Andamanen, Nikobaren, Rangoon und den Philippinen, andererseits (angeblich auch auf Japan !) ;

4. *Cicindela fuliginosa* DEJ., bekannt von Ceylon, einerseits, und, andererseits, dem ganzen Gebiet zwischen Shanghai, Birma, Java und Borneo ;

5. *Cicindela limosa* SAUND., bekannt von Ceylon, einerseits, und, andererseits, Sickim bis Süd-Bengalen, Birma, Andamanen, Nikobaren und schliesslich von Tschusan (Shanghai).

Bereits damals wies ich auf das übereinstimmende Characteristikum dieser fünf zerrissenen Verbreitungsgebiete hin : das Fehlen der Arten im Gebiet zwischen Süd-Vorderindien und Birma beziehungsweise Bengalen und Sumatra. Da die Cicindelinen von jeher mit besonderem Eifer gesammelt und die oben angeführten Arten nicht schwer zu finden sind, und da es sich obendrein bei ihnen um mehr oder weniger häufige Tiere (nur *Cic. aurovittata* ist etwas seltener) handelt, so kann nicht bei all den angeführten Spezies ein Spiel des Zufalls vorliegen, wenn überall dieselbe Lücke in ihrer Verbreitung resultiert, während alle anderen nach einigen Hunderten zählenden Cicindelinen des südlichen Asiens ein einfach geschlossenes Gebiet bewohnen. Die fünf angegebenen Arten sind hochentwickelte Formen des Genus *Cicindela*, welche wir uns etwa als im mittleren Tertiär entstanden zu denken haben.

Zur Erklärung benutzte ich die WALLACE-BLANFORD-Theorie und stellte für die mutmassliche Landbrücke zwischen Ceylon und Süd-Vorderindien, einerseits, den Andamen, Nikobaren, Birma bis Sumatra, andererseits, die Bezeichnung *Weddaland* oder *Weddabrücke* auf.

Seitdem habe ich einige weitere unterstützende Momente aufge funden und zwar zunächst drei neue charakteristische Verbreitungsgebiete :

6. *Collyris punctatella* CHAUD., Ceylon und Nias (Gunong Sitoli, R. MITSCHKE !);

7. *Cicindela discrepans* subsp. *lacrymans* SCHAUM, Ceylon und Nias (Ombolata);

8. *Cicindela foveolata* SCHAUM, bekannt von Süd-Kanara, Mysore und Malabar, einerseits, und, andererseits, Birma, Tonkin, Philippinen, Sumatra, Celebes, ? Bengalen.

Als ich die betreffenden Exemplare von *Cic. discrepans* aus Nias erhielt, glaubte ich (da man diese Art bisher als ein für Ceylon charakteristisches Tier anzusehen gewohnt war) naturgemäss an einen Irrtum im Fundort. Ausdrücklich wurde mir jedoch gesagt, dass die Exemplare direkt von einem sonst stets als zuverlässig bekannten Missionar auf Nias stammten. Das Auffinden von *Collyris punctatella* CHAUD. in Nias lässt meine Zweifel verstummen, da beide Arten zu ganz verschiedenen Zeiten von verschiedenen Sammlern erbeutet sind.

Nicht besonders betont habe ich früher zwei weitere analoge Züge in der Verbreitung zweier der zuerst angeführten Spezies : *Cicindela aurovittata* und *C. limosa*, von denen die erstere in den Zwischengebieten zwischen Birma und den Philippinen, die letztere in denen zwischen Birma und Shanghai bisher unbekannt geblieben ist. Jetzt bin ich in der Lage, von einer dritten Art (*Cicindela despectata* W. HORN) eine ähnliche diskontinuierliche Verbreitung zu konstatieren : Perak und Philippinen. Ausdrücklich wurde mir versichert, dass der betreffende Zoologe, der dies bisher für « typisch » philippinensisch gehaltene Tier in Perak gesammelt hat, niemals östlich über Perak hinausgereist sei.

Es ergeben sich also ausser den acht im selben Sinne wirkenden Fällen einer durch die hypothetische *Weddabrücke* im *Westen* (zwischen Ceylon beziehungsweise Süd-Vorderindien, einerseits, und den Andamanen, Birma und dem west-malayischen Gebiet, andererseits) erklärlich werdenden Verbreitung noch drei analoge Fälle für die *östliche* Verlängerung dieses gesunkenen Landes (zwischen Birma und der malayischen Halbinsel, einerseits, und den Philippinen beziehungsweise noch nördlicheren Gebieten, andererseits). Auch das angebliche Vorkommen von *Cicindela aurovittata* in Central-Japan erscheint unter diesen Gesichtspunkten jetzt weniger unwahrscheinlich.

Zum Schluss weise ich darauf hin, dass die Coleopteren-Gruppe der Cicindeliden für faunistische und zoogeographische Fragen ein besonders günstiges Arbeitsfeld abgibt. Sie ist nicht nur verhältnismässig sehr gut bekannt (siehe oben), sondern die Arten haben sich zu einem ganz besonderen Grade von Variationsfähigkeit (sowohl bezüglich lokalisierter wie nicht lokalisierter Formen) entwickelt. Ihre Flugfähigkeit und Wanderlust hat in vielen Fällen zu riesigen Verbreitungsgebieten geführt, wohingegen die Lebensweise ihrer Larven und Imagines die Möglichkeit künstlicher Transporte und Transportmittel ausschliesst.

Bemerkungen über die den Baumwollpflanzen in Egypten schädlichen Schmetterlinge und über die Methoden sie zu vertilgen,

von Adolf Andres, Bacos-Ramleh (Egypten).

Egypten ist durch sein vom Nil befruchtetes Delta ein ausschliesslich Ackerbau treibendes Land und ganz und gar abhängig von den Produkten, die sein Boden hervorbringt. Und fast ausschliesslich wird Baumwolle angepflanzt, die mit Ausnahme einiger kleiner Quantitäten, welche auf Sea-Island produciert werden, alle anderen Sorten an Qualität übertrifft.

Leider ist in den letzten Jahren der Ertrag der Ernte trotz grösserer Anpflanzung bedeutend zurückgegangen und während man früher bis zu 5 Centner Baumwolle per Feddan (ein Feddan ist etwas weniger als ein englische Acre) erzielte, ist der Ertrag jetzt ungefähr 3 Centner per Feddan. So ergab die diesjährige Baumwöllernte kaum 5 Millionen Centner, was für das Land einen Verlust von einigen Millionen Pfund Sterling zur Folge hatte. Die Ursache dieser Abnahme in dem Rendement der Ernte ist auf verschiedene Faktoren zurückzuführen. Es kommt sowohl die durch schlechte Drainage hervorgerufene Verschlechterung des Bodens in Betracht als auch hauptsächlich der durch Raupen hervorgerufene Schaden an den Baumwollpflanzen. Ich werde mir gestatten auf diesen letzteren Punkt hier kurz etwas näher einzugehen. Vorher möchte ich nur noch erwähnen, dass der hier in Brüssel vor einigen Wochen tagende Congress der Cotton Growing Association den Wunsch aussprach, dass in Egypten in Bezug auf Erforschung der Ursachen der Abnahme der Baumwollernten von der egyptischen Regierung mehr getan werden möge. Diesem

Wunsche kann man nur beistimmen, besonders was die tierischen Schädlinge anbelangt, deren Biologie noch durchaus nicht in allen Punkten genügend bekannt ist.

Ich habe lange Jahre hindurch die Feinde der Baumwollpflanze studiert und gefunden, dass es hauptsächlich die Raupen von verschiedenen Schmetterlingen sind, die den grössten Schaden tun. Vertreter anderer Klassen, wie z. B. eine kleine Hemiptere (*Oxycarenus hyalinipennis*) kommen weniger in Betracht, da der Schaden, den sie anrichten, verhältnissmässig unbedeutend ist.

Wir können die der Baumwolle schädlichen Raupen in folgende drei Gruppen einteilen :

1° *Agrotis*-Gruppe mit *ypsilon*, *pronuba segetum*, etc.;
2° *Prodenia littoralis*;
3° *Earias insulana*.

Alle drei Gruppen sind je nach ihrem Auftreten der Baumwollpflanze auf verschiedene Weise schädlich, sei es dass sie die junge Pflanzen angreifen oder sei es dass sie die Kapseln zerstören.

Unter den Agrotis ist die Raupe von *Agrotis ypsilon* die bei weitem häufigste und gefährlichste. Sie erscheint in grösserer Anzahl anfangs September, wo sie in den mit Bersim (*Trifolium alexandrinum*) bebauten Feldern auftritt und wo der Schaden, den sie dieser Kultur tut, manchmal ganz bedeutend sein kann. Im März erreicht sie ihren Höhepunkt und wird von diesem Datum ab den frisch gesäten jungen Baumwollpflanzen schädlich, indem sie die jungen Blätter kurz oberhalb der Wurzel abbeisst und verzehrt. Die anderen *Agrotis*-Arten, die ich bereits oben aufführte, und noch einige andere seltener vorkommende Arten tun in gleicher Weise Schaden; doch treten dieselben nie so häufig auf wie der *Agrotis ypsilon*. Zu dieser Gruppe können wir auch die *Caradrina*-Arten rechnen, die hier mit drei verschiedenen Specien vertreten sind, nämlich : *Caradrina exigua*, *latebrosa* und *ambigua*.

Der Schaden, den sie tun, ist öfters sehr beträchtlich : so hatte ich im Mai dieses Jahres Gelegenheit mit junger Baumwolle bepflanzte Felder zu inspicieren, die fast total von Raupen zerstört waren und handelte es sich jedesmal um Raupen von *Caradrina exigua*. Aber der schlimmste Feind der Baumwollpflanze ist ohne Zweifel die Raupe von *Prodenia littoralis*. Schon allein die Menge, in der diese gefrässigen Tiere auftreten, machen sie zu einem gefährlichen Feinde. Ihr Erscheinen beginnt im April, ihren Höhepunkt erreichen sie im Juli, um dann allmählich wieder zu verschwinden. Sie sind der Baumwolle umso gefährlicher, als sie während der Sommermonate ihres Auftretens auf den Feldern Egypten's keine andere Nährpflanze finden und daher ganz auf die Baumwolle angewiesen sind. Im Herbste, wenn die Baumwolle zu gross geworden ist, und im Winter finden wir diese polyphage Raupe auf jungem Mais, süssen Kartoffeln und anderen Kulturpflanzen übergehen.

Wir kommen nun zu dem sogenannten Kapselwurm oder *Earias insulana*. Es ist dies ein kleiner zu den Cymbiden gehöriger Schmetterling von meist grüner Färbung, ich sage meist, denn er ist sehr variabel und kommt häufig auch in gelber Färbung oder mit braunen Flecken auf den Vorderflügel (*Ab. dorsivitta*) vor. Die Raupe von *Earias* bohrt sich in die Blüten oder kleine Kapseln ein, deren Inhalt sie zerstört. Zur Verpuppung schreitet sie, indem sie sich ein kahnförmiges Gespinnst ausserhalb der Kapseln entweder an diesen selbst oder am Stamme des Baumwollstrauches anfertigt. Der Schaden, den diese Raupe besonders letztes Jahr der Baumwollernte zugefügt hat, war ganz ungeheuer. Man fand auf den Feldern fast keine Kapsel, in der man nicht eine Raupe antraf oder die durch eine solche mehr oder weniger zerstört war.

Es ist einleuchtend, dass je nach der Lebensweise dieser Raupen die Art und Weise ihrer Verrichtung eine andere sein muss : bei *Agrotis*-Arten und *Caradrina*-Arten, sowie bei *Earias insulana* kommt die Bekämpfung im Eizustande nicht in Betracht, denn erstere legen ihre Eier einzeln meist am Boden in der Nähe der Futterpflanze ab und die letztere dieselben an die Blüten oder Kapseln der Baumwollpflanze, sodass ein Aufsuchen und Ablesen derselben unmöglich sein dürfte. Dies ist jedoch bei den Eiern von *Prodenia littoralis* nicht der Fall. Dieselben werden auf der Unterseite der Blätter abgelegt und zwar in Paketen von 300 bis 600 Stück und manchmal sogar mehr, die das Weibchen mit seiner Afterwolle bedeckt. Auf diese Weise sind sie leicht sichtbar und können die mit Eiern behafteten Blätter leicht abgepflückt und zerstört werden. In der Tat hat die Regierung auch als einziges Mittel diese Methode, die man « Effeuillage » nennt, eingeführt und bringt dieselbe sogar zwangsweise zur Ausführung. Um aber diese Methode wirklich mit Erfolg durchzuführen, bedarf es nicht allein einer minitieusen Beaufsichtigung, sondern auch finanzieller Opfer, denen viele Landeigentümer nicht gewachsen sind.

Die Vernichtung der Raupen selbst hat bis jetzt keine nennenswerte Resultate gegeben. Mit Arsenik vergiftete Lockspeise um die Pflanzen zu streuen geht in Egypten nicht an, da es unmöglich ist den Arabern Gift anzuvertrauen, und auch die Besprengung der Blätter mit verschiedenen Chemikalien ist nicht angebracht, einesteils wegen der eng gepflanzten Baumwolle, die ein Eintreten in die Felder nicht gestattet, und andererseits wegen der Eiablage des Prodenia, der, wie gesagt, seine Eier unterhalb der Blattoberfläche ablegt. Den Puppen, die mit Ausnahme von *Earias* sämmtlich in der Erde ruhen, beizukommen, ist ebenfalls sehr schwer, da sie ein Ueberschwemmen des Bodens jedenfalls länger ohne Schaden aushalten können, als die darauf stehende Baumwolle.

Es bleibt daher nur noch der Schmetterling selbst zur Bekämpfung übrig. Hier in Egypten hat man viele Versuche mit Lichtfallen gemacht ; es hat sich aber meistens herausgestellt, dass die Wirkung derselben den Hoffnungen nicht entsprach. Dies hat seinen Grund darin, dass man mit denselben meistens nur Männchen zerstört.

Es war daher angebracht, ein anderes Mittel ausfindig zu machen, das den Vorteil hätte, nicht allein billig, als auch im Praktischen verwendbar zu sein.

Ich glaube mit meinem Partner, Herrn G. Maire, dieses Mittel in den sogenannten Kioskfallen gefunden zu haben. In den von uns publicierten Broschüren ist dasselbe genau beschrieben. Es beruht auf das Anlocken der Schmetterlinge durch einen Köder, ein Verfahren, das jedem Schmetterlingssammler bekannt ist. Die angelockten Schmetterlinge dringen in die Falle ein, aus welcher sie den Ausweg nicht mehr finden und in einem in derselben angebrachten Kasten mit Wasser umkommen. Dieser Apparat und die zum Anlocken verwendete Flüssigkeit, das sogenannte Prodenine, ist in allen Kulturstaaten patentiert oder zum Patent angemeldet.

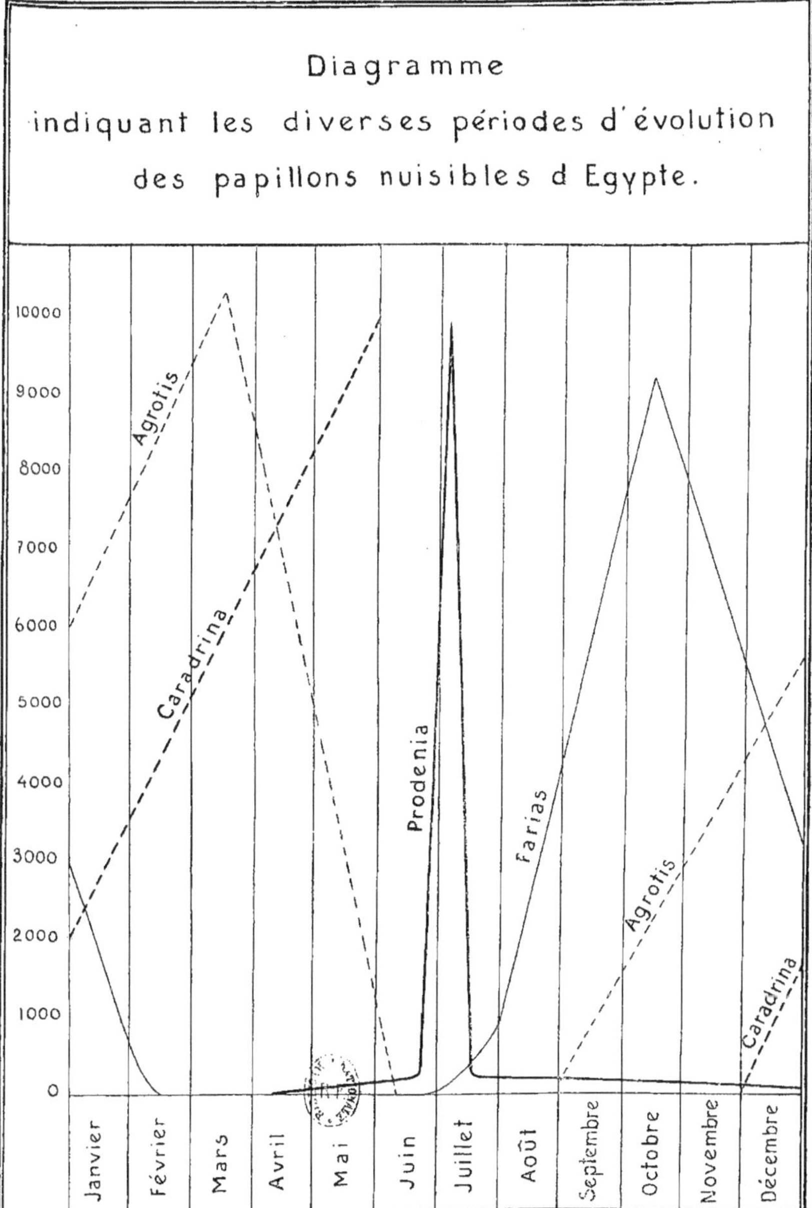

AD. ANDRES. — SCHAEDLICHE INSEKTEN IN EGYPTEN.

Ueber die Abhängigkeit der Fauna vom Gestein,

von KARL HOLDHAUS (1).

Die aus den Gesteinen, welche die feste Erdrinde zusammensetzen, bei der Verwitterung hervorgehenden Nährsalze sind eine Quelle und Grundbedingung alles Lebens. Kein Tier und keine Pflanze vermag diese Nährsalze zu entbehren. Die Untersuchung der Abhängigkeit der Fauna vom Gestein berührt daher eines der Grundprobleme des Lebens.

Während von Seite der Botaniker und Pedologen der Einfluss des Gesteins auf die Pflanzenwelt seit Jahrzehnten den Gegenstand erfolgreicher Forschungen bildet, wurde die Frage der Abhängigkeit der Fauna vom Gestein bisher gänzlich vernachlässigt. Als ich mich vor nunmehr sechs Jahren dem Studium dieses Problems zuwandte, betrat ich durchaus unbebautes Land. Wenn ich mir heute erlaube, Sie mit den bisherigen Ergebnissen meiner Untersuchungen bekannt zu machen, so kann es sich naturgemäss nur um eine kurze vorläufige Mitteilung handeln, welcher ich in einigen Jahren eine ausführliche Darstellung des Gegenstandes folgen zu lassen gedenke (2).

(1) Der Vortrag wurde in der Sitzung von 5. August in sehr verkürzter Form gehalten.

(2) Eine teilweise gleichlautende, aber auf zahlreiche andere Tiergruppen ausgedehnte und die Insekten in kürzerer Weise behandelnde Darstellung des Gegenstandes wird in den « Verhandlungen des VIII. internationalen Zoologenkongresses, Graz 1910 » erscheinen.

I. — Die Landfauna.

Ich bespreche zunächst den Einfluss des Gesteins auf die Landfauna. Zur Landfauna zähle ich hier alle jene Insekten, welche ausserhalb des Wassers und ausserhalb der durchfeuchteten Uferzone der Gewässer leben.

Wenn wir in Mittel- oder Südeuropa beim Sammeln von Landtieren auf die Gesteinsbeschaffenheit achten, so können wir sehr bald die Erfahrung machen, dass die verschiedenen Tierformen sich den wechselnden Gesteinen gegenüber sehr verschieden verhalten. Es gibt sehr viele Tiere, welche wir auf jedem beliebigen Gestein antreffen, es gibt zahlreiche andere Tiere, welche nur auf ganz bestimmten Bodenarten gefunden werden. Nach ihrem Verhalten zu den verschiedenen Gesteinen kann man unter den Landtieren der mittel- und südeuropäischen Fauna folgende Gruppen unterscheiden :

1. Gesteinsindifferente Arten. — Es sind dies Arten, die auf jedem beliebigen Gestein vorkommen können. Hieher gehört die überwiegende Mehrzahl aller Tierformen. Die meisten gesteinsindifferenten Arten besitzen eine ausgedehnte geographische Verbreitung, sie finden sich sowohl im Gebirge als auch in den grossen, aus lockerem Sediment bestehenden Ebenen von Europa. Viele dieser Arten leben auch in Nordeuropa.

Der Terminus gesteinsindifferent, den ich aus der botanischen Literatur übernehme, darf übrigens nicht missverstanden werden. Die gesteinsindifferenten Arten sind nur bis zu dem Grade gegen Differenzen der Bodenbeschaffenheit gleichgiltig, als sie, soweit wir wissen, kein Gestein principiell meiden. Im übrigen wird auch die gesteinsindifferente Fauna durch den Boden beeinflusst, aber dieser Einfluss äussert sich im wesentlichen in grösserem oder geringerem Individuenreichtum, nicht in dem Fehlen oder Vorhandensein ganzer Artencomplexe. In der Ebene finden wir unter sonst gleichen Umständen auf nährstoffreichem Lehmboden eine viel individuenreichere Fauna als auf armem Sand- oder Schotterboden (1). Auch im Gebirge zeigt die gesteinsindifferente Fauna

(1) Es gibt aber auch einzelne Tierformen, die gerade auf magerem Boden einen besonderen Individuenreichtum entfalten.

ähnlich wie die echte Gebirgsfauna eine deutliche Beeinflussung durch den Boden.

2. PSAMMOPHILE ARTEN. — Es sind dies solche Arten, die nur auf tiefgründigem Sandboden leben. Es ist dies eine überaus interessante Biocönose, die aber noch sehr des näheren Studiums bedarf. Zu den exclusiv psammophilen Arten gehören zahlreiche Koleopteren, eine geringere Zahl von Orthopteren, Dipteren, Lepidopteren und wohl noch Vertreter einzelner anderer Insektenordnungen. In Mitteleuropa besitzen wir nur sehr verarmte Reste einer Psammophilfauna in den Dünengebieten von Norddeutschland und in den Sandgegenden der ungarischen Ebene, besonders im Deliblat. Eine viel artenreichere Psammophilfauna lebt in den südlichsten Teilen von Europa und namentlich im Bereiche des Wüstengürtels, der sich von Nordafrika über Südwestasien und Turkestan bis in die Mongolei erstreckt. Im Zusammenhang mit der Lokalisierung vieler Sandgebiete besitzen zahlreiche psammophile Arten eine beschränkte geographische Verbreitung.

3. HALOPHILE ARTEN. — Es sind dies Arten, die nur auf Salzboden leben. Es gibt ziemlich zahlreiche Insekten, welche bis jetzt ausschliesslich auf Salzboden gefunden wurden und aus diesem Grunde allgemein als halophil bezeichnet werden (1). Trotz dieser Einhelligkeit der Meinungen und der leichten Zugänglichkeit der Halophilfauna auch in Mitteleuropa ist die Oekologie der halophilen Tiere doch noch in allen wesentlichen Punkten ungeklärt. Auch ich habe mich mit diesem Gegenstand bisher nicht beschäftigt. Die meisten halophilen Arten besitzen eine weite Verbreitung, die Halophilfauna von Mitteleuropa (z. B. Salzsee von Eisleben, Salzgegenden in der pannonischen Niederung und in Centralsiebenbürgen) streicht mit fast allen ihren Componenten bis weit nach Centralasien hinein und wiederholt so die Verbreitung der ehemaligen sarmatischen und pontischen Wasserbedeckung.

4. PETROPHILE ARTEN. — Es sind dies solche Arten, die nur auf festem Gestein leben, d. h. auf solchen Böden, die an Ort und Stelle aus festem Fels hervorgegangen sind. Die petrophilen Arten

---

(1) Die meisten halophilen Tiere sind übrigens Ufertiere und daher in ökologischer Hinsicht eigentlich mehr vom Wasser als vom Boden abhängig.

meiden alle lockeren Gesteine, sie fehlen daher auf tiefgründigen Sanden und Schottern, auf tiefgründigen lehmigen Sedimenten, mit sehr wenigen Ausnahmen auch auf Löss und Tegeluntergrund (1).

An der Zusammensetzung der einheimischen Petrophilfauna beteiligen sich, neben einer geringen Anzahl von Wirbeltieren, zahlreichen Landschnecken, Myriopoden, Arachniden, manchen Isopoden und terricolen Oligochäten eine sehr grosse Zahl von Insekten aus verschiedenen Ordnungen. Besonders zahlreiche petrophile Arten finden wir in solchen Insektengruppen, deren Vertreter eine terricole Lebensweise führen. Ich gebe im folgenden einen kurzen, naturgemäss sehr lückenhaften Ueberblick über die ausserhalb des Wassers lebenden petrophilen Insekten von Europa (2).

ORTHOPTERA. — Petrophil sind die meisten Arten der Gattungen *Pœcilimon*, *Barbitistes* und *Isophya*, vermutlich *Orphania*, *Analota*, ferner die Mehrzahl der Arten der Gattungen *Rhacocleis*, *Antaxius*, *Thamnotrizon*, *Ephippigera* und *Saga*, alle *Troglophilus* und *Dolichopoda*, *Cuculligera*, die meisten (vielleicht alle)

---

(1) Welche äusseren Faktoren es den petrophilen Arten verwehren, auf lockeres Gestein hinauszutreten, ist noch nicht festgestellt. Jedenfalls sind sowohl physikalische als chemische Eigenschaften dieser Böden (in erster Linie wohl die Nährstoffarmut vieler lockerer Gesteine und die intensive Auslaugung der darauf ruhenden Böden) hiebei von Belang. Eine intermediäre Stellung zwischen lockerem und festem Gestein nehmen gewisse mangelhaft verfestigte Sandsteine und die tertiären und diluvialen Tone ein. Diese Gesteine tragen eine extrem verarmte Petrophilfauna, wobei es stets nur einige wenige, ganz bestimmte petrophile Tiere sind, die auf diese Böden übergreifen. Auch auf den osteuropäischen Löss (in der Walachei und in Südrussland) transgrediren anscheinend einzelne (sehr wenige) petrophile Tierformen. Dieser Löss hat die Fazies eines halbverfestigten, sehr kalkreichen, lehmig-sandigen Sedimentes, dem oft einzelne Tegellagen oder auch Bänke vollkommen verfestigten Gesteins eingeschaltet sind. Auf dem Löss der Wiener Gegend traf ich noch niemals petrophile Faunenelemente. Jedenfalls bedarf die Lössfauna noch eingehenden Studiums. [Vergl. auch K. HOLDHAUS und F. DEUBEL, Untersuchungen über die Zoogeographie der Karpathen. (« Abhandl. zool.-botan. Ges. Wien », Bd. VI, Heft 1, Iena 1910, p. 18.)]

(2) Die meisten der im folgenden genannten Gattungen enthalten neben petrophilen Arten auch gesteinsindifferente Elemente. Das Verhalten zum Gestein muss daher für jede einzelne Art gesondert festgestellt werden.

Arten der Gattungen *Nocarodes*, *Ocnerodes*, *Pamphagus* und *Eunapius*, *Paracaloptenus*, ein Teil der Arten von *Podisma* (*Pezotettix*).

Phasmoidea. — Anscheinend sind alle Phasmiden-Arten des europäischen Faunengebietes petrophil.

Dermaptera. — Petrophil sind einige ungeflügelte Arten von *Forficula*, *Anechura bipunctata*, vermutlich auch einige Arten der Gattung *Chelidura*.

Blattoidea. — Petrophil sind die Mehrzahl der Arten der Gattung *Aphlebia*.

Koleoptera. — Die petrophilen Koleopteren der einheimischen Fauna sind zum grösseren Teil terricol, zum kleineren Teil planticol, nur sehr wenige petrophile Arten (abgesehen von der artenreichen Torrenticolfauna) gehören anderen Biocönosen an. Im Gegensatz zu den gesteinsindifferenten Arten, welche zum grössten Teil geflügelt sind, finden wir unter den petrophilen Koleopteren überaus zahlreiche ungeflügelte Arten (1). In nicht wenigen Gattungen sind die ungeflügelten Arten petrophil, die geflügelten ausnahmslos oder grossenteils gesteinsindifferent. Die blinden Koleopteren gehören mit äusserst wenigen Ausnahmen (2) der Petrophilfauna an. Innerhalb der europäischen Koleopterenfauna sind petrophil : *Procerus gigas*, zahlreiche ungeflügelte Arten der Gattung *Carabus* (3), fast alle *Cychrus* (Ausnahme *C. rostratus*),

(1) Unter den petrophilen Koleopterenarten von Mitteleuropa dürften wohl gegen 70 Proc. der Flügel entbehren, unter den gesteinsindifferenten Koleopteren Mitteleuropa's sind höchstens 5 Proc. aller Arten flugunfähig.

(2) Gesteinsindifferente Blindkäfer der mitteleuropäischen Fauna sind *Claviger testaceus* Preyssl., *Leptinus testaceus* Müll., *Langelandia anophthalma* Aub., *Aglenus brunneus* Gyllh., *Anommatus duodecimstriatus* Müll. Diese Arten besitzen weite Verbreitung in der Ebene und im Gebirge.

(3) Von mitteleuropäischen Caraben sind beispielsweise petrophil *C. irregularis*, *caelatus*, *croaticus*, *Creutzeri*, *depressus*, *Fabricii*, *auronitens*, *obsoletus*, *catenatus*, *Linnei*, *concolor*, *carinthiacus* ; hingegen sind gesteinsindifferent *Carabus coriaceus*, *violaceus*, *intricatus*, *catenulatus*, *variolosus*, *nitens*, *auratus*, *cancellatus*, *granulatus*, *arvensis*, *clathratus*, *Ullrichi*, *Scheidleri*, *hortensis*, *convexus*, *nemoralis*, *glabratus*.

ein Teil der Arten der Gattung *Leistus* (darunter anscheinend alle *Oreobius* und *Leistidius*), sehr zahlreiche Arten der Gattung *Nebria*, anscheinend alle Arten der Gattung *Reicheia, Spelæodytes*, einige *Dyschirius* (subgen. *Reicheiodes*), eine geringe Zahl von *Bembidium*-Arten, vermutlich alle Arten der Gattungen *Scotodipnus, Anillus, Typhlocharis*, fast alle constant ungeflügelten *Trechus*-Arten, alle *Anophthalmus*, einige *Licinus* (subgen. *Neorescius*) zahlreiche *Fabrus*, eine Anzahl ungeflügelter *Amara*-Arten (namentlich die Subgen. *Leiromorpha* und *Leirides*), einige *Abax*, vielleicht alle *Percus*, zahlreiche *Molops*, ein grosser Teil der ungeflügelten *Pterostichus*-Arten, einige *Læmostenus* (z. B. *janthinus, elegans, Schreibersi, cavicola*), einige *Calathus* (z. B. *metallicus*), alle Arten des Subgenus *Platynus*, anscheinend alle *Aptinus*, ein Teil der Arten der Gattungen *Anthobium, Omalium, Geodromicus, Anthophagus, Boreaphilus*, alle *Niphetodes*, vermutlich alle *Leptotyphlini*, eine relativ geringe Anzahl von *Stenus*-Arten (namentlich aus dem subgen. *Parastenus*), manche Arten der Gattung *Lathrobium* (darunter alle *Glyptomerus*), einige *Xantholinus* (subgen. *Typhlodes* und *Vulda*), einige *Othius*, wenige *Philonthus*, wenige *Staphylinus* (z. B. *tenebricosus?, macrocephalus, megacephalus*), ziemlich zahlreiche *Quedius* (namentlich aus den Untergattungen *Sauridus* und *Raphirus*), wenige *Mycetoporus*, anscheinend alle ungeflügelten Arten der Gattung *Leptusa*, relativ wenige *Atheta*-Arten, die Mehrzahl der Arten der Gattung *Sipalia*, einige *Ilyobates*, eine Anzahl ungeflügelter Arten der Gattung *Oxypoda*, einige *Faronus*, alle *Pygoxyon*, die meisten *Trimium* (Ausnahme *Tr. brevicorne*), einige *Euplectus, Scotoplectus, Trogaster*, wohl alle *Amaurops*, sehr wenige *Brachygluta*, sehr zahlreiche Arten der Gattung *Bythinus*, einige *Tychus*, alle ungeflügelten *Pselaphus*, fast alle Arten der Gattung *Cephennium* (Ausnahme vermutlich *C. turgidum*), viele *Neuraphes*, einige ungeflügelte *Stenichnus*, zahlreiche *Euconnus* (darunter alle ungeflügelten Arten), wenigstens ein Teil der *Leptomastax*-Arten, *Ablepton*, vielleicht alle *Mastigus*, anscheinend alle blinden *Silphiden* (*Leptoderinæ*), *Necrophilus*, einige *Choleva*-Arten, einige *Liodes* (subgen. *Oreosphærula*) und *Agathidium*, ein Teil der Arten der Gattungen *Eros, Cantharis, Absidia, Rhagonycha, Pygidia, Malthinus, Malthodes, Dasytes*, einige wenige *Meligethes*, eine Anzahl *Cryptophagus* (anscheinend alle *Mnionomus*), einige *Atomaria* (*carpathica, grandicollis*, etc.), vielleicht alle

Sternodea, die meisten *Anommatus* und *Sphærosoma*, vielleicht die *Hylaia*-Arten, *Montandonia*, ein Teil der Arten der Gattungen *Simplocaria*, *Pedilophorus*, *Byrrhus* (alle *Seminolus*) und *Syncalypta*, vielleicht alle *Curimus*, *Dima*, einige wenige *Corymbites*, zahlreiche *Athous*, eine Anzahl Arten der Gattungen *Niptus* und *Plinus*, einige *Anoncodes* und *Oedemera*, vielleicht alle *Tomoderus*, mehrere *Orchesia*-Arten (*blandula*, *grandicollis*), vermutlich alle *Læna*-Arten und überhaupt sehr zahlreiche südeuropäische Tenebrioniden, wohl die meisten *Dorcadion*, wenige *Cryptocephalus*, zahlreiche *Timarcha*, viele ungeflügelte *Chrysomela*-Arten, vermutlich alle *Orina*-Arten mit Ausnahme von *Or. rugulosa*, ein Teil der Arten der Gattungen *Sclerophædon*, *Phædon*, *Luperus*, *Galeruca*, *Crepidodera*, wohl alle *Orestia*, eine Anzahl ungeflügelter *Psylliodes*-Arten, einige *Longitarsus* (1), anscheinend die *Phænotherion*-Arten, überaus zahlreiche *Otiorrhynchus*, viele *Peritelus*, manche *Argoptochus*, *Phyllobius*, *Polydrusus* und *Brachysomus*, die Mehrzahl der Arten von *Barypithes* und *Omias*, manche *Trachyphlœus*, einige *Tropiphorus*, vielleicht alle *Dichotrachelus*, die Mehrzahl der *Liparus*-Arten, die meisten *Plinthus* und *Liosoma*, *Adexius?*, zahlreiche ungeflügelte *Hypera*-Arten, wohl alle *Alaocyba* und verwandten Blindrüssler, ein Teil der *Acalles*-Arten, wohl alle *Torneuma*, einzelne *Ceuthorrhynchinen*, manche *Aphodius* (z. B. alle *Agolius*), *Chætonyx?*, einige *Geotrupes* (z. B. *alpinus*), manche *Rhizotrogus?*, *Geotrogus?*, vielleicht einige *Anisoplia*. Insgesamt dürfte die petrophile Koleopterenfauna von Europa etwa 3000-4000 Arten umfassen (2).

LEPIDOPTERA. — Petrophil sind Arten der Gattungen *Papilio* (wenige Arten), *Melanargia*, *Erebia* (anscheinend die grosse Mehrzahl aller Arten), *Hesperia*, *Perisonema*, *Agrotis* (ein kleiner Teil der Arten), *Polia*, *Phlogophora*, *Hiptelia*, *Omia*, *Catocala* (einige Arten), *Apopestes*, *Orectis*, *Acidalia*, *Larentia*, *Tephroclystia*,

---

(1) Vielleicht auch einige Arten der Gattungen *Dibolia* und *Apteropoda*.

(2) Vergl. auch K. HOLDHAUS und F. DEUBEL, Untersuchungen über die Zoogeographie der Karpathen (« Abhandl. zool. bot. Ges. Wien », Bd. VI, Heft 1, Iena, 1910, p. 42); daselbst eine Zusammenstellung der petrophilen Koleopteren der Karpathen.

Gnophos, Dasydia, Ocnogyna (anscheinend alle Arten), *Aglaope, Oreopsyche* (anscheinend fast alle Arten), *Scioptera* (1).

Zweifellos finden sich petrophile Landinsekten auch in den Ordnungen der Dipteren, Hymenopteren, Hemipteren und Neuropteren, die faunistische Erforschung dieser Insektengruppen ist indess selbst für Mitteleuropa noch so fragmentarisch, dass sich ohne langjährige Eigenuntersuchungen keine verlässliche Zusammenstellung der petrophilen Vertreter dieser Ordnungen geben lässt. Leider gilt dasselbe in noch höherem Maasse auch von allen apterygoten Insekten, welche zweifellos eine sehr grosse Zahl petrophiler Faunenelemente enthalten.

Sie sehen schon aus dieser kurzen und ganz provisorischen Zusammenstellung, dass die Petrophilfauna eine sehr formenreiche Tiergesellschaft darstellt. Alle hieher gehörenden Tiere sind also dadurch ausgezeichnet, dass sie ausschliesslich auf Böden leben, die an Ort und Stelle aus festem Fels hervorgegangen sind. Damit ist aber der Einfluss des Gesteins auf die Petrophilfauna noch nicht erschöpft. Wenn wir im Gebirge auf verschiedenen festen Gesteinen sammeln, so können wir die Erfahrung machen, dass nicht alle festen Gesteine für die Petrophilfauna gleich günstig sind. Es gibt Felsarten, auf denen wir eine extrem reiche Petrophilfauna antreffen, es gibt andere feste Gesteine, die stets eine um vieles ärmere Petrophilfauna tragen. Ich habe mich bemüht, im Laufe der Jahre möglichst viele Gesteine wiederholt in verschiedenen Teilen von Mittel- und Südeuropa auf ihr faunistisches Verhalten zu prüfen (2). Aus meinen diesbezüglichen überaus zahlreichen

---

(1) Ich besitze bei Lepidopteren leider keine eigenen Sammelerfahrungen. Die obige fragmentarische Liste habe ich durch Literaturstudium gewonnen. Herrn Prof. REBEL bin ich für einige Auskünfte zu Dank verpflichtet.

(2) Meine eigenen diesbezüglichen Untersuchungen wurden in folgenden Gebirgsgegenden auf den jeweils in Klammern genannten Gesteinen angestellt : *Niederösterreichische Alpen* (kristalline Schiefer, verschiedene Kalke und Sandsteine, Triasdolomit, Flysch), *Gurhofgraben* (Serpentin), *Ostkarpathen* (kristalline Schiefer, Dolomit und Quarzit der Permformation, mesozoische Kalke, Hornsteine der Trias, Quarzsandsteine und Conglomerate der Oberkreide, Andesit), *Südkarpathen* (mesozoische Kalke bei Kronstadt, kalkreiches Oberkreideconglomerat am Bucsecs, Granitgneisse der Präsbe, Grünschiefer bei Vöröstorony), *Alpen von Kärnten :* Koralpe, Karawanken, Gailtaler Alpen, Königstuhlgebiet, Hohe Tauern (verschiedene kristalline Schiefer, verschiedene Kalke, Dolomit, Sandsteine und Conglomerate), *Cadorische Alpen* (Kalk, geschieferte Sandsteine,

Einzelbeobachtungen lässt sich folgende allgemeine Regel ableiten : Jene Gesteine, welche bei der Verwitterung einen nährstoffreichen Boden von hoher Wassercapacität geben, tragen eine um vieles reichere Fauna als solche Gesteine, deren Verwitterungsrinde geringen Nährstoffgehalt oder geringe Wassercapacität besitzt. Faunistisch sehr reiche Böden geben daher die meisten Kalke und basischen (d. i. kieselsäurearmen) Eruptivgesteine, quarzarme Sandsteine und Conglomerate, kalkreiche Tonschiefer, sowie die meisten basischen kristallinen Schiefer. Faunistisch sehr arme Böden geben Dolomit, Quarzit und quarzreiche Sandsteine und Conglomerate, alle sehr saueren (d. i. kieselsäurereichen) Eruptivgesteine und kristallinen Schiefer, viele schwer verwitternde Tonschiefer, sowie die plastischen Tone (1). Eine intermediäre Stellung nehmen gewisse Sandsteine und Conglomerate, manche Eruptivgesteine und kristalline Schiefer von mittlerem Kieselsäuregehalt und kalkige Dolomite ein. Die faunistischen Unterschiede zwischen den extremen Gesteinen sind dabei so beträcht-

---

Porphyr), *Monte Baldo, Monti Lessini, Monte Cavallo in den Venezianer Alpen* (Kalk, Mergel), *Euganeen* (Trachyt, basische Tuffe, Mergel), *Montagnola Senese* (Kalk), *Monte Argentaro* (Kalk, paläozoische Schiefer), *Insel Elba* (Quarzporphyr, Granit, Diabas, Serpentin eozäne Sandsteine), *Albaner Berge* (Basalt). *Monte Gargano* (Kalk), *Sizilien :* Gegend von Palermo, Ficuzza, Peloritanisches Gebirge (mesozoische Kalke, Hornsteine, eozäne Sandsteine, kristalline Schiefer), *Süddalmatien* (Kalk). Ausserdem untersuchte ich quaternäre Conglomerate und Sandsteine in den Küstenebenen der *Toskanischen Maremmen* und bei *Palermo,* tertiäre Tone in *Niederösterreich* und im *Becken von Siena*, Löss in *Niederösterreich, Centralungarn*, der *Walachei,* lockere Sande, Schotter und Lehme in *Niederösterreich*, *Ungarn*, *Kärnten*, in der *Poebene bei Mantua (Bosco Fontana)*, auf *Elba*, *bei Messina*, *am Monte Gargano*, Moränen am *Bucsecs in den Südkarpathen*, in der *Tatra im Glocknergebiet*, in den *Cadorischen Alpen*, *beim Wolaya-See*. Ergänzt wurden meine eigenen Sammlerfahrungen durch das Studium der faunistischen Literatur und durch wertvolle Auskünfte seites zahlreicher befreundeter Sammler. Namentlich den Herrn F. DEUBEL in Kronstadt, A.L. MONTANDON in Bukarest, G. PAGANETTI-HUMMLER in Vöslau, I. SAINTE-CLAIRE-DEVILLE in Epinal, HANS WAGNER in Zürich bin ich in dieser Hinsicht zu Dank verpflichtet. Die Sammelreisen nach Elba, Sizilien und auf den Monte Gargano wurden mir ermöglicht durch Reisesubventionen seitens der Hohen kaiserl. Akademie der Wissenschaften in Wien, für deren Gewährung ich auch an dieser Stelle den geziemenden Dank ausspreche.

(1) Auch Serpentin gibt einen sehr trockenen, erdarmen Boden mit sehr merkwürdiger, artenarmer Fauna und Flora. Eingehendere Untersuchungen über die Serpentinfauna wären von theoretischem Interesse.

lich, dass man auf einem sehr günstigen Gestein, wie etwa Kalk oder Diabas in derselben Zeit im Durchschnitt wohl eine 40-50mal so reiche Sammelausbeute zustandebringen kann wie auf einem extrem ungünstigen Gestein wie etwa Quarzit.

Es entsteht nun die Frage, ob innerhalb der Petrophilfauna eine Spezialisirung einzelner Arten auf ganz bestimmte Gesteine zu beobachten ist, d. h. ob es petrophile Arten gibt, die nicht auf allen festen Gesteinen, sondern nur auf ganz bestimmten Felsarten zu leben vermögen. Für eine Tiergruppe ist eine solche Beschränkung auf bestimmte Felsarten tatsächlich seit langem nachgewiesen. Es sind dies die Landschnecken. Die Landschnecken benötigen zum Bau ihres Gehäuses sehr viel Kalk und viele Arten finden sich daher nur auf kalkreichen Gesteinen, also in erster Linie auf dem gewöhnlichen Kalkgestein, aber auch auf einigermassen kalkreichen Sandsteinen und Conglomeraten, auf Mergeln, auf kalkreichen kristallinen Gesteinen. Man kann daher diese hieher gehörenden Landschnecken mit Recht als chalikophil bezeichnen. Abgesehen von dieser Chalikophilie der Landschnecken konnte ich aber in keiner anderen Tiergruppe mit Sicherheit eine durch physiologische Faktoren bedingte Beschränkung einzelner petrophiler Arten auf ganz bestimmte Felsgesteine nachweisen und dieser Befund deckt sich mit den analogen Ergebnissen der ökologischen Pflanzengeographie. Der Einfluss der verschiedenen Felsarten auf die petrophile Insektenfauna äussert sich also im wesentlichen nur in grösserem oder geringerem Individuenreichtum der Fauna, nicht aber in der Lokalisirung ganzer Artencomplexe auf bestimmte Gesteine (1).

Ein ganz besonderes Interesse bietet *die geographische Verbreitung* der petrophilen Tierformen. Die Empfindlichkeit der petrophilen Tiere gegen Differenzen der Gesteinsbeschaffenheit, insbesondere ihre Abneigung gegen lockeres Gestein, bringt es mit sich, dass die petrophilen Tiere ganz andere Phänomene der geographischen Verbreitung zeigen als die durch keinerlei Schranken der Bodenbeschaffenheit gehemmten gesteinsindifferenten Arten. Die geographische Verbreitung der petrophilen Fauna zeigt folgende Eigentümlichkeiten :

1. Petrophile Tiere finden sich hauptsächlich im Gebirge, da die

(1) Eine Ausnahme macht hier natürlich die Höhlenfauna, welche nur in Kalkgebieten lebt, da eben Höhlenbildungen normal nur im Kalk auftreten

Böden im Gebirge ja zum allergrössten Teil unmittelbar aus festen Muttergestein hervorgegangen sind. Hingegen fehlt die petrophile Fauna in den grossen Ebenen von Europa, soweit deren Untergrund aus lockerem Gestein besteht. Wir finden petrophile Tierformen nur in jenen Teilen unserer europäischen Tiefebenen, deren Grund aus festem Gestein (einschliesslich Tegel und gewisser Lössarten) besteht. Solche Tiefebenen mit festem Gestein, die infolgedessen eine Petrophilfauna besitzen, sind beispielsweise das Flachland der Toscanischen Maremmen bei Cecina und Castagneto, die Ebene von Palermo, die Podolische Platte, Teile der walachischen Ebene (sehr kalkreicher Löss), etc. Andererseits fehlt Petrophilfauna im Gebirge an solchen Stellen, wo der Untergrund auf grössere Erstreckung aus lockerem Gestein besteht, also auf lockeren tertiären Auflagerungen, auf breiten Alluvialböden, auf ausgedehnteren lockeren Moränen (1), auf den Lössdecken der niederösterreichischen und zentral-ungarischen Gebirge. Solche lockere Sedimente finden sich im Gebirge stellenweise in beträchtlicher Ausdehnung und bis ins hochalpine Areal (Moränen). Abgesehen von diesen lockeren Auflagerungen treffen wir aber überall in unseren Gebirgen typische Petrophilfauna an, vom Fuss des Gebirges bis zu den höchsten Grenzen tierischen Lebens.

2. Die petrophilen Tiere besitzen im Durchschnitt eine um vieles geringere geographische Verbreitung als die gesteinsindifferenten Arten. Es hängt dies mit der durch die Petrophilie bedingten Einschränkung der Migrationsfähigkeit zusammen. Viele petrophile Arten sind extrem lokalisirt und finden sich nur auf einem einzelnen kleinen Gebirgsstock, mitunter nur auf einem einzelnen Berge.

---

(1) Die Fauna der Moränen in unseren Hochgebirgen bedarf noch eines eingehenden Studiums. Die diesbezüglichen Verhältnisse sind in mannigfacher Weise complicirt. Verfestigte oder sehr tonreiche Moränen tragen Petrophilfauna. Dasselbe gilt von Moränen, welche zahlreiche sehr grosse Felsblöcke enthalten. Auf lockeren Moränen von grösserer Ausdehnung in den höheren Lagen der Alpen und Karpathen traf ich stets nur äusserst wenige petrophile Tiere, deren Vorkommen durch aktives oder passives Ueberwandern von den benachbarten Felsböden wohl genügend erklärt ist. Namentlich im Glocknergebiet (siehe hierüber « Verh. zool. bot. Ges. Wien », 1909. Sitz. ber. p. 365-368) und in der oberen Waldzone der Malajeschter Schlucht am Bucsecs konnte ich den tiefgreifenden faunistischen Gegensatz zwischen lockeren Moränen und Felsböden klar beobachten.

3. In sehr eigenartiger Weise wurde die geographische Verbreitung der petrophilen Fauna durch die Eiszeit modificirt. Wir finden petrophile Tierformen in Europa in grosser Anzahl in den Gebirgen der Mittelmeerländer, wir finden typische Petrophilfauna in den Gebirgen von Mitteleuropa einschliesslich Frankreich's (einzelne petrophile Arten auch in Grossbritannien), aber wir kennen keine einzige petrophile Tierart aus den Gebirgen von Skandinavien und Finnland. Dieses Fehlen der Petrophilfauna in den Gebirgen von Nordeuropa (Fennoskandia) erklärt sich daraus, dass durch die Eiszeit die ganze Tierwelt daselbst zum Aussterben gebracht wurde. In postglazialer Zeit war aber eine Neubesiedelung Fennoskandia's mit petrophilen Arten von Süden her nicht möglich, da die norddeutsche Ebene und das russische Flachland, auf weite Erstreckung aus lockeren Sedimenten bestehend, für diese Tiere eine unüberschreitbare Barriere bildeten (1). Auch die

(1) Das Fehlen petrophiler Arten in Fennoskandia wurde durch Michaelsen (« Die geographische Verbreitung der Oligochäten », Berlin. 1903) für die Oligochäten, durch Kobelt (« Studien zur Zoogeographie » 2 Bde ,Wiesbaden. 1908; « Die geographische Verbreitung der Mollusken in dem paläarktischen Gebiet ». Wiesbaden, 1904) für die Mollusken erwiesen. Meine eigenen diesbezüglichen Studien erstreckten sich auf die Wirbeltiere, Mollusken, Koleopteren, Orthopteren s. l., innerhalb der petrophilen Wasserfauna noch auf die Dipteren und Phryganiden (an der Hand der trefflichen Arbeiten von Ulmer) Für die übrigen Tiergruppen sind ähnliche Feststellungen noch ausständig. Diese Untersuchungen sind sehr mühsam und zeitraubend infolge des grossen Umfanges der zu bewältigenden faunistischen Literatur. Manche Tiergruppen, die zweifellos sehr plastische Resultate ergeben würden, wie die Arachniden (namentlich die torrenticolen Milben), Myriopoden, apterygoten Insekten, etc. sind leider derzeit noch zu mangelhaft erforscht, um in die Untersuchung einbezogen werden zu können.

Ich habe die Theorie hier absichtlich in voller Schärfe hingestellt. Es wäre aber möglich, dass doch ganz vereinzelte petrophile Tierformen auch in Fennoskandia gefunden werden, und zwar entweder Arten, welche dort die Eiszeit überdauerten oder solche Formen, die auf dem Wege der Verschleppung oder aktiver Wanderung (sehr gut fliegende Insekten) in postglazialer Zeit die norddeutsche Ebene zu überschreiten vermochten. Falls sich solche Tierformen nachweisen liessen, ist ihre Zahl jedenfalls äusserst gering. Eine Immigration petrophiler Arten nach Fennoskandia von Osten her ist kaum in Rechnung zu ziehen. Denn erstens versperrt auch hier der Gürtel glazialer Ablagerungen und lockerer Alluvionen den Weg und zweitens scheint die Petrophilfauna des Ural selbst überaus verarmt. Westsibirien, im Gegensatz zu dem gebirgigen und faunistisch extrem reichen Ostsibirien grösstenteils aus quaternärem und rezentem Schwemmland bestehend, besitzt eine sehr arme und monotone Fauna.

während der Eiszeit intensiver vergletscherten oder dem nordischen Inlandeis sehr genäherten Gebirge von Mitteleuropa (Deutschland, Böhmische Masse, Alpen mit Ausnahme der während der Eiszeit unvergletscherten Randzone von massifs de refuge im Süden und Südosten, Nordkarpathen) zeigen noch in der Gegenwart eine wesentlich ärmere Petrophilfauna als die niemals in grösserem Ausmasse vergletscherten Gebirge (z. B. Ost- und Südkarpathen, Gebirge von Südeuropa, Südrand der Alpen, etc.). Es erklärt sich das wohl daraus, dass zahlreiche in ökologischer Hinsicht sehr anspruchsvolle und wenig mobile petrophile Arten sich an der Reimmigration in das durch die Eiszeit devastirte Gebiet nicht beteiligten.

II. — Die Wasserfauna.

Ich habe bisher nur den Einfluss des Gesteins auf die Landfauna besprochen und möchte nunmehr zur Erörterung der diesbezüglich für die Wasserfauna bestehenden Verhältnisse übergehen. Meine folgenden Ausführungen werden Ihnen zeigen, dass zwischen Land- und Wasserfauna hinsichtlich des Verhaltens den einzelnen Gesteinen gegenüber weitgehende Analogien bestehen (1).

Nach dem Grade ihrer Abhängigkeit vom Gestein kann man unter den in unseren einheimischen Binnengewässern lebenden Insektenarten folgende ökologische Einheiten unterscheiden.

1. Gesteinsindifferente Arten. — Hieher gehört die grosse Mehrzahl aller im stehenden Wasser lebenden Insekten

---

(1) Eine Mittelstellung zwischen Landfauna und Wasserfauna nimmt in ökologischer Hinsicht die Uferfauna ein. Ich habe über den Einfluss des Gesteins auf die Uferfauna bisher noch keine zureichenden Erfahrungen gesammelt und muss daher davon absehen, diese Biocönose in die Darstellung einzubeziehen. Die meisten Ufertiere scheinen gesteinsindifferent, manche halophil, eine geringe Anzahl von Arten lebt ausschliesslich am Ufer von Gebirgsbächen und ist daher petrophil. In jedem Fall muss die Uferfauna bei ökologischen Untersuchungen von der eigentlichen Landfauna streng gesondert werden, da sie in vieler Hinsicht abweichende Lebensverhältnisse zeigt und in verschiedenen ökologischen Eigentümlichkeiten der Wasserfauna viel näher steht als der Landfauna. Namentlich die Tierwelt am Ufer der Gebirgsbäche ist in ökologischer Hinsicht vom Wasser abhängig, nicht vom Boden.

sowie alle in den Bächen und Flüssen der aus lockerem Gestein bestehenden Ebenen vorkommenden Tierformen. Die gesteinsindifferenten Wassertiere besitzen grossenteils eine überaus weite Verbreitung. Gesteinsindifferente Faunenelemente bilden die Wasserfauna der grossen Ebenen und des hohen Nordens von Europa.

Die gesteinsindifferenten Wassertiere sind ebenso wie die gesteinsindifferenten Landtiere gegen Differenzen der Gesteinsbeschaffenheit nur insofern gleichgiltig, als sie kein Gestein principiell meiden. Wir finden aber unter sonst gleichen Verhältnissen auf einem nährsalzreichen Gestein, also auf Lehm oder Mergeluntergrund oder auf nährstoffreichen festen Gesteinen eine viel reichere Fauna als auf armem Quarzsand, auf plastischem Ton, auf nährstoffarmen festen Gesteinen oder in Moorwässern. Es ist dies eine Tatsache, die den Fischzüchtern seit langem bekannt und in jedem Handbuch des Fischereibetriebes erörtert ist. Auf kurzen Sammelreisen wird man allerdings über den Einfluss des Gesteins auf die Fauna des stehenden Wassers nur schwer befriedigende Beobachtungen sammeln können, denn die Fauna des stehenden Wassers ist ja nicht nur vom Gestein abhängig, sondern noch von einer Menge anderer Faktoren, von der Temperatur des Wassers, der Tiefe des Wasserbeckens, der Zufuhr organischer Substanzen von aussen, der Art der Zersetzung der im Wasser befindlichen abgestorbenen organischen Reste. Zu einwandfreien Resultaten gelangt man nur, wenn man imstande ist, diese Faktoren aus dem Vergleiche auszuschalten oder in entsprechender Weise zu reguliren und hiezu bietet besonders die Teichwirtschaft günstige Gelegenheit.

2. Halophile Arten. — Wir treffen unter den Wasserinsekten manche Formen an, welche bis jetzt nur in salzhältigen Binnengewässern gefunden wurden. Die meisten dieser halophilen Arten besitzen gleichwohl eine weite Verbreitung. Ein genaueres Studium der Oekologie und geographischen Verbreitung der halophilen Wasserfauna ist noch ausständig.

3. Petrophile Arten. — Hieher gehören alle Tierformen, die nur in solchem Wasser vorkommen, das mit festem Felsgestein in Berührung steht. Die Fauna des stehenden Wassers scheint nur sehr wenige petrophile Tierformen zu enthalten (hieher gehören beispielsweise die Amphibien *Bombinator pachypus*, *Molge Mon-*

tandoni und anscheinend auch *alpestris*), unter den im stehenden Wasser lebenden Insekten kenne ich bisher keine Art, welche ich mit voller Sicherheit als petrophil ansprechen möchte(1). Sehr gross hingegen ist die Zahl der petrophilen Arten in der Fauna des fliessenden Wassers, denn zur Petrophilfauna gehören alle exclusiv torrenticolen Tierformen, d. h. alle jene Tierformen, welche ausschliesslich im Gebirgsbach leben. Der Gebirgsbach unterscheidet sich aber in Hinblick auf das Gestein von dem Bach und Fluss der Ebene (die oft ein ebenso starkes Gefälle haben wie manche Gebirgsbäche mit normaler Torrenticolfauna) dadurch, dass das Wasser des Baches der Ebene nur mit lockeren Sedimenten in Berührung steht, während das Bett des Gebirgsbaches in festes Felsgestein eingegraben ist (1).

Die petrophile Torrenticolfauna ist eine in ökologischer und zoogeographischer Hinsicht überaus interessante Biocönose. An der Zusammensetzung der exclusiv torrenticolen Insektenfauna von Mitteleuropa beteiligen sich folgende Ordnungen :

Ephemeriden. — Ein Teil der Arten der Gattungen *Baetis Epeorus*, *Rhithrogena* und *Ecdyurus*.

Perliden. — Nach freundlicher Mitteilung von Seite des Herrn Prof. Klapalek finden sich exclusiv torrenticole Arten in den Gattungen *Perlodes*, *Dictyogenus*, *Perla*, *Chloroperla* und *Nemura*.

Koleopteren. — Zahlreiche Arten der Gattungen *Hydræna* und *Ochthebius*, ferner Arten der Gattungen *Elmis*, *Esolus*, *Lareynia*, *Riolus*, *Latelmis*, *Stenelmis*, einige Arten der Gattung *Helodes*, möglicherweise auch vereinzelte *Dytisciden*.

Dipteren. — Anscheinend alle Arten der Familie der *Blepharoceriden* und zweifellos noch eine Anzahl anderer Dipteren, deren Larven ausschliesslich im Gebirgsbach leben.

---

(1) Doch werden sich solche Arten wohl zweifellos auffinden lassen.

(1) In Ebenen, deren Untergrund aus festem Fels besteht, müssten die Bäche wenigstens teilweise exclusiv torrenticole Faunenelemente beherbergen. Untersuchungen hierüber liegen nicht vor.

TRICHOPTEREN. — Arten der Gattungen *Rhyacophila* (die grosse Mehrzahl der zahlreichen Arten), *Glossosoma*, *Ptilocolepus*, *Stactobia*, *Philopotamus*, *Dolophilus*, *Tinodes*, *Diplectrona*, *Halesus*, *Psilopteryx*, *Drusus*, *Anomalopteryx*, *Potamorites*, *Micrasema*, *Sericostoma*.

Ebenso wie bei den petrophilen Landtieren lässt sich auch bei der Torrenticolfauna eine weitgehende Beeinflussung durch das Gestein beobachten. Es lässt sich die allgemeine Regel aufstellen, dass solche Gebirgsbäche, deren Wasser durch nährstoffreiches Gestein fliesst und daher reich ist an gelösten Nährsalzen, ceteris paribus eine viel reichere Fauna besitzen als Bäche, deren Wasser durch nährstoffarmes Gestein fliesst und daher selbst arm an Nährsalzen ist. Ungünstige Gesteine für die Bachfauna sind daher in erster Linie alle sehr sauren Eruptivgesteine und kristallinen Schiefer, ferner Quarzit und nährsalzarme Tonschiefer. Eine sehr arme Torrenticolfauna besitzen auch Gletscherbäche und Abflüsse aus Hochmooren, da solche Bäche sehr arm an Nährstoffen zu sein pflegen. Günstige Gesteine sind basische kristalline Gesteine, ferner Kalke und kalkreiche Schiefer und Sandsteine. Ueber das Verhalten von Dolomit konnte ich bisher keine einwandfreien Beobachtungen sammeln.

Die geographische Verbreitung der exclusiv torrenticolen Tiere zeigt jene karakteristischen Eigentümlichkeiten, welche wir auch bei den petrophilen Landtieren feststellen konnten :

1. Exclusiv torrenticole Tiere kennt man einstweilen nur aus dem Gebirge, hier aber findet man typische Torrenticolfauna in jeder Höhenlage, vom Fuss des Gebirges bis zu der oberen Grenze tierischen Lebens. In den grossen, aus lockerem Gestein bestehenden Tiefebenen fehlt die echte Torrenticolfauna vollständig.

2. Die exclusiv torrenticolen Tierformen besitzen im Durchschnitt eine um vieles geringere Verbreitung als die übrigen Wassertiere. Viele Arten sind extrem lokalisirt. Es hängt dies mit der durch die exclusiv torrenticole Lebensweise bedingten Einschränkung der Migrationsfähigkeit zusammen.

3. Während wir in den Gebirgen der Tropen ebenso wie in den Gebirgen der Mittelmeerländer und Mitteleuropa's eine überaus reiche petrophile Torrenticolfauna antreffen, scheinen den Gebirgen von Fennoskandia exclusiv torrenticole Tierformen zu fehlen. Es

ist dies dasselbe Phänomen wie bei der petrophilen Landfauna und in derselben Weise zu erklären.

III. — Allgemeine Gesichtspunkte.

Ich habe mich bisher darauf beschränkt, eine Darstellung des in rein empirischer Weise gewonnenen Tatsachenmaterials zu geben und möchte mir nun erlauben, daran einige Erörterungen allgemeinerer Natur anzuschliessen, welche geeignet sind, das bisher Gesagte dem Verständnis näher zu bringen. Der Versuch einer ursächlichen Erklärung der Phänomene der Abhängigkeit der Fauna vom Gestein begegnet derzeit grossen Schwierigkeiten, da nicht nur das vorliegende faunistische Beobachtungsmaterial noch grosse Lücken aufweist, sondern auch die zur Lösung der verschiedenen Probleme heranzuziehenden Disciplinen der Bodenkunde, Gewässerkunde, chemischen Physiologie der Tiere und Pflanzen grossenteils erst im Beginne ihrer Entwicklung stehen und in grundlegenden Fragen ungeklärt sind (1).

Wenn wir uns auf unseren Sammelexcursionen überzeugen, in

(1) Im folgenden eine Auswahl der wichtigsten Literatur : ROSENBUSCH, Elemente der Gesteinslehre, 3. Aufl., Stuttgart, 1910; RAMANN, Bodenkunde, 2. Aufl., Berlin, 1905; F. CORNU, Die heutige Verwitterungslehre im Lichte der Kolloidchemie, C. R. de la première Conférence internat. agrogéol., Budapest, 1909, pp. 123-130, und mehrfache kurze Ausführungen desselben Autors ebenda pp. 23, 26, 29, 33, 36, 37; RAMANN, Ueber Beziehungen der Gele in der anorganischen Natur zu den Gelen der lebendigen Substanz, Ann. der Naturphilosophie, VIII, 1909, pp. 329-333; JUSTUS v. LIEBIG, Die Chemie in ihrer Anwendung auf Agrikultur und Physiologie, 9. Aufl., Braunschweiz, 1876; ADOLF MAYER, Lehrbuch der Agrikulturchemie, Bde. I, II und IV, 6. Aufl., Heidelberg, 1905-1908; SCHIMPER, Pflanzengeographie auf physiologischer Grundlage, Iena, 1898; WARMING, Lehrbuch der ökologischen Pflanzengeographie, 2. Aufl., Berlin, 1902; WIESNER, Anatomie und Physiologie der Pflanzen, 5. Aufl , Wien, 1906; G. v. BUNGE, Lehrbuch der Physiologie des Menschen, II. Bd., 2. Aufl., Leipzig, 1905; PAUL VOGEL, Ausführliches Lehrbuch der Teichwirtschaft, 3 Bde, Bautzen, 1898-1904; KNAUTHE, Das Süsswasser, Neudamm, 1907; STEUER, Planktonkunde, Leipzig, 1910; PÜTTER, Die Ernährung der Wassertiere und der Stoffhaushalt der Gewässer, Iena, 1909; WALTER, Das Gesetz vom Minimum und das Gleichgewicht im Wasser, Archiv für Hydrobiol. und Planktonkunde, IV (1909), pp. 339-366.

Aus den genannten Handbüchern ist die weitere Literatur zu entnehmen.

wie tiefgreifender Weise die Fauna und Flora von der Gesteinsbeschaffenheit beeinflusst wird, so stellt sich naturgemäss die Frage ein, mittelst welcher Faktoren das wechselnde Gestein diesen Einfluss auszuüben vermag. Zur Klärung dieser Frage dürften die folgenden Erwägungen beitragen :

Betrachten wir zunächst die Landfauna. Die Landfauna und ebenso die Landflora werden naturgemäss nicht direkt durch das Gestein beeinflusst, sondern beide sind abhängig von der Beschaffenheit des Bodens, der durch die Verwitterung der Gesteine entstanden ist. Wenn wir nun beispielsweise auf Kalk eine ganz andere Fauna antreffen als auf Quarzsandstein und nun festzustellen suchen, wodurch der Kalkboden sich von dem Quarzsandsteinboden unterscheidet, so finden wir, dass diese beiden Böden in zahlreichen Eigenschaften wesentlich von einander abweichen. Wir finden 1° chemische Differenzen, d. h. der Kalkboden hat eine ganz andere chemische Zusammensetzung als der Quarzsandsteinboden, wir finden 2° physikalische Differenzen, d. h. der Kalkboden unterscheidet sich vom Quarzsandsteinboden hinsichtlich der Wassercapacität, hinsichtlich des thermischen Verhaltens, hinsichtlich der Struktur, der Tiefe der Verwitterungsrinde, etc.

Es entsteht nun die Frage, sind es die chemischen oder die physikalischen Differenzen in der Bodenbeschaffenheit, welche die Verschiedenheit der Lebewelt auf den einzelnen Böden bedingen. In der botanischen Literatur ist diese Frage seit Jahrzehnten viel erörtert und es bildeten sich zwei Lehrmeinungen, von denen die eine den chemischen Faktoren, die andere den physikalischen Faktoren, in erster Linie der Wassercapacität und dem thermischen Verhalten den entscheidenden Einfluss auf die Vegetation beimass. Eine befriedigende Klärung dieser Frage ist auf botanischem Gebiete bis heute nicht erfolgt.

Für die Landfauna besteht naturgemäss dieselbe Controverse. Eine Beurteilung des ganzen Phänomens wird jedoch hier dadurch erleichtert, dass von den physikalischen Faktoren nur ein einziger eine grössere Bedeutung für die Fauna zu besitzen scheint. Es ist dies die Wassercapacität, d. h. der Grad der Fähigkeit des Bodens, Wasser in sich aufzunehmen und längere oder kürzere Zeit festzuhalten. Die Bedeutung der Wassercapacität ergibt sich aus der Erfahrung, dass überaus zahlreiche Tiere, namentlich sehr viele im Erdboden lebende Arten, ein sehr hohes Feuchtigkeitsbedürfnis besitzen. Böden mit hoher Wassercapacität sind daher für die Fauna günstiger als Böden mit geringer Wassercapacität, ein Satz, der durch die tägliche Sammelerfahrung durchaus bestätigt wird. Hingegen scheinen alle übrigen physikalischen Eigenschaften des Bodens nur untergeordneten Einfluss auf die Fauna auszuüben und können daher einstweilen vernachlässigt werden. Wir können uns somit auf die Untersuchung beschränken, ob der chemischen Beschaffenheit des Bodens oder der Wassercapacität der grössere Einfluss auf die Fauna zukommt.

Seinem inneren Aufbau nach enthält jeder Boden folgende Elemente :

1. Das Skelett, es sind dies die Rückstände der Verwitterung, welche in Form von grösseren oder kleineren Gesteinstrümmern und Gesteinskörnchen in jedem Boden enthalten sind;
2. Die Feinerde;
3. Lösliche Mineralsalze;
4. Die Zersetzungsprodukte organischer Substanzen.

Von diesen Elementen kommt der Feinerde eine besondere Bedeutung zu. Diese Feinerde hat kolloidale Beschaffenheit, d. h. sie besteht aus verschiedenen durch die Verwitterung entstandenen Gelen. Diese Gele sind nicht selbst Pflanzennährstoffe, aber sie haben für die Ernährung der Pflanze (und damit indirekt für jene des Tieres) doch eine grosse Bedeutung, denn sie besitzen infolge ihrer kolloidalen Beschaffenheit die Fähigkeit, die aus den Gesteinen bei der Verwitterung hervorgehenden löslichen Mineralsalze zu absorbiren und dadurch im Boden festzuhalten. Solche Böden, welche viel Feinerde enthalten, vermögen daher ceteris paribus viel mehr Nährsalze in sich aufzuspeichern als Böden mit wenig Feinerde. Der Gehalt an Feinerde bestimmt aber auch den Grad der Wassercapacität eines Bodens. Böden mit viel Feinerde vermögen ceteris paribus viel mehr Wasser in sich aufzunehmen als Böden mit wenig Feinerde. Hohe Wassercapacität und Nährstoffreichtum sind daher in vielen Böden vereinigt (z. B. bei den meisten Kalken und kalkreichen Sandsteinen und Conglomeraten, bei den meisten basischen Eruptivgesteinen und kristallinen Schiefern), ebenso wie auch geringe Wassercapacität und Nährstoffarmuth des Bodens häufig zusammenauftreten (z. B. bei Quarzit und vielen saueren Eruptivgesteinen und kristallinen Schiefern). Es lässt sich also in allen diesen Fällen nicht entscheiden, ob die Wassercapacität oder der Nährstoffgehalt für die Beschaffenheit der Fauna ausschlaggebend ist, da beide Faktoren in diesen Gesteinen nach derselben Richtung wirken. Es gibt jedoch Gesteine, welche zwar einen Boden mit hoher Wassercapacität, aber nur mit geringem Nährstoffgehalt liefern und das sind gewisse nährsalzarme Tone und Tonschiefer. Die aus diesen nährstoffarmen Gesteinen hervorgegangenen Böden tragen trotz hoher Wassercapacität nur eine arme und monotone Fauna und wir sehen daraus, dass günstiges physikalisches Verhalten eines Bodens allein nicht genügt, sondern dass Reichtum des Bodens an Nährsalzen in jedem Falle eine Grundbedingung der Entwicklung einer reichen Fauna ist. Es scheint mir wahrscheinlich, dass die chemische Beschaffenheit eines Bodens einen viel grösseren Einfluss auf die Fauna ausübt als die physikalischen Faktoren (1).

In viel klarerer Weise als bei der Landfauna lässt sich die grosse Bedeutung der chemischen Factoren bei der Wasserfauna nachweisen. Wir haben gehört, dass die Landfauna nicht direkt durch das Gestein beeinflusst wird, sondern indirekt, durch die Beschaf-

---

(1) Die Beschaffenheit der Verwitterungsrinde ist jedoch nicht nur von der Natur des Muttergesteins, sondern auch von den klimatischen Verhältnissen (u. z. sowohl Temperatur als Niederschlagsmenge) abhängig, unter denen die Verwitterung vor sich geht. Ein und dasselbe Gestein ergibt beispielsweise im arktischen Klima einen erdarmen Steinboden, in Mitteleuropa Braunerde, in Südeuropa Terra rossa, in den Tropen Laterit. Man kann daher, entsprechend den Klimaprovinzen, die Erde in klimatische Bodenzonen einteilen, welche sich hinsichtlich der Bodenbeschaffenheit teilweise in überaus tiefgreifender Weise voneinander unterscheiden (Gelgeographie Dr. Cornu's). Wie jedoch unsere Kenntnis der klimatischen Bodenzonen der Erde noch überaus lückenhaft ist, so lässt sich auch derzeit nicht sagen, bis zu welchem Grade der Karakter der Fauna und Flora durch diesen vom Klima bedingten Wechsel der Bodenbeschaffenheit beeinflusst wird. Es ist jedoch sehr wahrscheinlich, dass in dieser Hinsicht eine sehr weitgehende Abhängigkeit besteht. Das Klima wirkt sonach auf die Fauna und Flora in zweifacher Weise ein, 1° direkt, 2° durch die Differenzen der Bodenbeschaffenheit (Unterschiede in der Wassercapacität, in der Struktur und namentlich auch in der chemischen Beschaffenheit) in den einzelnen klimatischen Bodenzonen. Der ganze Habitus der Fauna scheint durch die klimatischen Bodenzonen weitgehend beeinflusst zu werden. Wenn wir an den Insekten des arktischen Klimas düstere Färbung, im tropischen Regenwald üppigste Farbenpracht, im tropischen Steppenklima viel schmucklosere Gestalten antreffen, so ist dies gewiss nicht ausschliesslich eine Folge von unmittelbarer Klimawirkung, Mimikry, etc., sondern vielfach auch eine Begleiterscheinung der durch die Unterschiede der Bodenbeschaffenheit bedingten Ernährungsdifferenzen. Der Zusammenhang zwischen Gelgeographie und Biogeographie wurde zuerst von F. Cornu in seiner vollen Bedeutung erkannt (vergl. « C. R. de la première Conférence internat. agrogéol. », Budapest, 1909, p. 130).

fenheit des Bodens, der aus dem Gestein hervorgegangen ist. Ebenso ist auch die Wasserfauna, von einzelnen Spezialfällen abgesehen, nicht unmittelbar vom Gestein abhängig, sondern vielmehr von den chemischen und physikalischen Eigenschaften des *Wassers,* die aber ihrerseits wieder zum grossen Teil durch die Natur des mit dem Wasser in Berührung stehenden Gesteins bedingt werden. Es ist nun aber bei der Wasserfauna sehr leicht möglich, Verhältnisse aufzusuchen, bei denen die physikalischen Faktoren aus dem Vergleich ganz ausgeschaltet sind und wir also die Wirkung der chemischen Faktoren für sich gesondert studiren können. Es macht im Gebirge keine besondere Schwierigkeit, Gebirgsbäche aufzusuchen, welche hinsichtlich Wassermenge, Gefälle, Temperatur, etc., weitgehend übereinstimmen und sich nur dadurch unterscheiden, dass ihr Wasser verschiedenes Gestein durchfliesst und daher abweichende chemische Beschaffenheit besitzt. Wenn wir bei Gleichheit dieser physikalischen Faktoren in einem Bach, der aus Kalk kommt, eine ganz andere Fauna antreffen als in einem Bach, der aus Quarzsandstein kommt, so können diese faunistischen Differenzen eben nur durch die differente chemische Beschaffenheit der beiden Gewässer hervorgerufen worden sein.

Auf diesem Wege kann man sich also durch vergleichendes Studium der Fauna zahlreicher Gebirgsbäche leicht überzeugen, dass die chemische Beschaffenheit des Wassers einen überaus weitgehenden Einfluss auf die Fauna ausübt.

Ebenso wie beim Studium der Torrenticolfauna lassen sich auch in der Teichwirtschaft zahllose Erfahrungen sammeln, welche die ausschlaggebende Bedeutung der Chemie des Wassers für die Beschaffenheit der Fauna zeigen. Besonders instruktive und plastische Resultate ergeben in dieser Hinsicht die Versuche der Teichdüngung mit anorganischen Dungmitteln (1).

---

(1) Dass neben der chemischen Zusammensetzung des Wassers auch verschiedene physikalische Faktoren — die teilweise gleichfalls vom Gestein beeinflusst werden, teilweise aber von der Beschaffenheit des Gesteins ganz unabhängig sind — in der Faunistik unserer Binnengewässer eine grosse Bedeutung besitzen, ist eine längstbekannte und vielfach in der Literatur behandelte Tatsache. Es scheint mir aber, dass viele Hydrobiologen auf diese physikalischen Faktoren ein viel zu grosses Gewicht legen. Namentlich die jetzt so beliebte Thermometerbiologie wird wohl in vielen Fällen anderen Erklärungsmitteln weichen müssen.

Eine Zusammenfassung des bisher Gesagten führt zu folgenden Resultaten :

1. *Die Landfauna steht in weitgehender Abhängigkeit von der Bodenbeschaffenheit, die Wasserfauna wird beeinflusst durch die Beschaffenheit des Wassers, in dem die Tiere leben. Boden und Wasser sind ihrerseits hinsichtlich ihrer chemischen und physikalischen Eigenschaften in grossem Umfange abhängig vom Gestein.*

2. *Sowohl die chemischen als die physikalischen Eigenschaften des Bodens, beziehungsweise des Wassers, üben einen Einfluss auf die Fauna aus. Es scheint indess, dass den chemischen Faktoren hiebei die grössere Bedeutung zukommt.*

Es entsteht nun die weitere Frage, in welcher Weise die chemischen Eigenschaften des Bodens, beziehungsweise des Wassers, auf die Fauna einzuwirken vermögen. Es ist dies in erster Linie ein Problem der chemischen Physiologie, zu dessen Lösung bisher nur ganz unzureichende Vorarbeiten vorliegen. Im allgemeinen lässt sich etwa folgendes sagen.

Die unmittelbare Aufnahme anorganischer mineralischer Nährstoffe durch den Tierkörper spielt anscheinend eine geringe Rolle. Das Tier entnimmt die zu seinem Aufbau nötigen Nährsalze zum allergrössten Teil seiner pflanzlichen oder animalischen Nahrung. Der Chemismus des Bodens oder Wassers wirkt daher im wesentlichen nicht direkt, sondern auf dem Umwege über das Pflanzenreich auf die Fauna ein. Der grosse Zusammenhang ist daher folgender :

Die carnivoren Tiere sind hinsichtlich ihrer Nahrung abhängig von den phytophagen Tieren.

Die phytophagen Tiere sind hinsichtlich ihrer Nahrung abhängig von den Pflanzen.

Die Pflanzen sind hinsichtlich ihrer Nahrung abhängig vom Boden oder Wasser.

Wir haben daher das Abhängigkeitsverhältnis : carnivores Tier; phytophages Tier; Pflanze ; Boden (Wasser); Gestein (1).

---

(1) Dieser Gedankengang gilt in voller Schärfe für die Landfauna. Inwieweit die Wassertiere in der Lage sind, die zu ihrem Gedeihen nötigen Nährstoffe dem Wasser direkt zu entnehmen, bedarf noch eingehender experimenteller Untersuchung. (Vergl. hierüber die bei STEUER, « Planktonkunde », citirte Literatur namentlich HERBST, PÜTTER).

Die Ausnützung der in einem Boden oder Wasser vorhandenen Nährstoffe seitens der Fauna und Flora erfolgt im Sinne des *Gesetzes vom Minimum*. Dieses von dem grossen Chemiker LIEBIG entdeckte Gesetz lässt sich in folgender Weise formuliren: Die Beschaffenheit der Fauna und Flora auf einem bestimmten Boden oder in einem bestimmten Wasser richtet sich nach demjenigen für die Ernährung der Organismen unentbehrlichen chemischen Bestandteil, welcher in der relativ geringsten Menge vorhanden ist. Wenn daher zum Beispiele in einem Boden alle übrigen für die Organismen nötigen Nährstoffe im Ueberfluss enthalten sind, hingegen nur sehr wenig Phosphor vorhanden ist, so wird die Entwicklung der Fauna und Flora begrenzt durch das zur Verfügung stehende Phosphorquantum und die übrigen reichlich vorhandenen Nährstoffe können nur bis zu dem Grade ausgenützt werden, als der im Minimum vorhandene Phosphor dies zulässt. Die grosse biologische Bedeutung des Gesetzes vom Minimum ist leider noch viel zu wenig erkannt, nur in der Ackerbaulehre und neuerdings auch in der Teichwirtschaft wird die Wichtigkeit dieses Gesetzes in theoretischer und praktischer Hinsicht entsprechend gewürdigt.

Ich bin am Schlusse meiner Ausführungen. Ich habe mich bemüht, unter Beiseitelassung alles Unwesentlichen eine kurze zusammenfassende Darstellung des Problems der Abhängigkeit der Fauna vom Gestein zu geben, soweit ich dasselbe auf Grund meiner bisherigen Studien zu überblicken vermag. Ich bin mir wohl bewusst, dass meine Ausführungen grosse Lücken und manches Hypothetische enthalten. Es wird die Aufgabe von Jahren sein, diese Lücken allmählich auszufüllen, die Hypothesen zu überprüfen. Es wird nötig sein, die Untersuchung auf verschiedene bisher nicht berücksichtigte Tiergruppen auszudehnen, es wird nötig sein, eine Anzahl minder verbreiteter und daher von mir bisher nicht oder nur ungenügend beobachteter Eruptivgesteine auf ihr faunistisches Verhalten zu prüfen, es wird nötig sein, den Einfluss des Gesteins auf die Fauna auch im arktischen Klima, in den Steppen der gemässigten Zone, in den verschiedenen Klimagebieten der Tropen zu untersuchen, es wird endlich nötig sein, durch umfangreiche physiologische Untersuchungen (besonders auch Experimente über Ernährungsphysiologie) eine gesicherte Erklärung der in der freien Natur zu beobachtenden Phänomene anzubahnen. Alle diese Arbeiten übersteigen aber um ein Vielfaches die Kraft eines Einzelnen. Mögen sich im Laufe der Jahre Mitarbeiter finden, möge mein heutiger Vortrag als Basis der Verständigung einige Dienste leisten.

Les Pycnogonides décapodes et la classification des Pycnogonides,

par M. E.-L. BOUVIER (Paris).

Dans une communication présentée récemment à l'Académie des Sciences, de Paris, j'ai fait connaître (1910) et désigné sous le nom de *Pentapycnon Charcoti* un nouveau Pycnogonide décapode recueilli dans les mers antarctiques par l'expédition de M. JEAN CHARCOT, à bord du « *Pourquoi pas?* » Cette curieuse espèce est représentée par un grand mâle ovigère et par deux immatures bien plus petits que j'avais pris d'abord pour des femelles; les trois exemplaires furent capturés aux Shetland du Sud, île du Roi Georges, dans la baie de l'Amirauté, sur un fond de 420 mètres où la température dépassait à peine 0° (+ 0.3). Les zoologistes du « *Pourquoi pas?* » ont également rapporté deux autres Pycnogonides décapodes : le *Pentanymphon antarcticum* HODGSON, espèce commune sur tout le pourtour du continent antarctique, et la *Decolopoda australis* EIGHTS (1834), qui paraît localisée dans les îles situées un peu au nord de ce continent (1).

(1) La *Decolopoda australis*, retrouvée par l'Expédition antarctique écossaise, a été complètement décrite par M. HODGSON (1908); le même zoologiste a fait très exactement connaître (1907) le *Pentanymphon antarcticum* découvert par les naturalistes de la « *Discovery* ».

La collection contient des exemplaires nombreux et variés de la première espèce, mais elle n'en renferme qu'un seul de la seconde; ce dernier provient des Shetland du Sud où l'espèce avait été signalée pour la première fois par EIGHTS. Quant à la *Decolopoda antarctica* BOUVIER (1906), qui fut prise par les naturalistes du « *Français* », à Port-Charcot, elle n'a pas été retrouvée au cours de la campagne du « *Pourquoi pas?* ».

Ces Pycnogonides décapodes sont les seuls actuellement connus; comme on vient de le voir, ils habitent tous la région antarctique et appartiennent aux trois genres *Decolopoda*, *Pentanymphon* et *Pentapycnon*. Quelle que soit l'opinion que l'on professe au sujet des affinités zoologiques des Pycnogonides, il n'est pas douteux que ces Articulés présentent une segmentation réduite et que, par le nombre des segments et des appendices, leurs formes décapodes se rapprochent des formes ancestrales d'où sont issus les Pycnogonides octopodes qui constituent la presque totalité du groupe. Ainsi, les Pycnogonides décapodes ont été les ancêtres, mais ne peuvent être les descendants des Pycnogonides octopodes.

Or, quand on étudie les divers Pycnogonides décapodes, on trouve que les *Decolopoda* présentent tous les caractères essentiels des Colossendéidés, les *Pentanymphon* des Nymphonidés et les *Pentapycnon* des Pycnogonidés. Les affinités des deux premiers genres ne sont pas douteuses, elles frappent au premier abord, et je les ai mises en évidence dans un travail antérieur; quant à celles des *Pentapycnon*, elles sont plus évidentes encore, et l'on peut dire sans trop exagérer que le *Pentapycnon Charcoti* ne diffère des *Pycnogonum* que par sa paire de pattes supplémentaire (1).

Dans le travail auquel j'ai fait allusion plus haut (1906), je divisais les Pycnogonides en deux ordres ou séries : les *Colossendéomorphes*, qui ont pour point de départ les *Decolopoda* et les *Pycnogonomorphes*, dont les *Pentanymphon* représentent l'état ancestral; dans la première série se rangeaient les Décolopodidés et les Colossendéidés, dans la seconde tous les autres Pycnogonides y compris les Phoxichilidés et les Pycnogonidés que je réunissais, à l'exemple de M. G.-O. SARS (1891), dans le sous-ordre des Achélates, c'est-à-dire parmi les formes dépourvues de chélicères et de palpes.

---

(1) C'est à tort que j'avais cru y reconnaître des orifices sexuels sur toutes les pattes; ces orifices sont de simples pores coxaux.

L'importante découverte des *Pentapycnon* nous oblige à modifier ce système. Il n'est plus possible, en effet, de placer la famille des Pycnogonidés, dans laquelle il se range sans conteste, à la fin de la série des Pycnogonomorphes, encore moins de la réunir aux Phoxichilidés dans le groupe des Achélates établi par M. Sars. Car la famille des Pycnogonidés a eu pour point de départ les *Pentapycnon* ou une forme décapode très voisine, et les Phoxichilidés sont octopodes comme d'ailleurs tous les Pycnogonomorphes, excepté les *Pentanymphon*.

Dès lors une solution s'impose, c'est l'établissement d'une série nouvelle pour les *Pentapycnon* et les Pycnogonidés octopodes qui en dérivent. Il conviendra de laisser le nom de *Pycnogonomorphes* aux représentants de cette série et d'attribuer celui de *Nymphonomorphes* à tous les autres Pycnogonides, abstraction faite des Colossendéomorphes.

Ainsi les Pycnogonides nous apparaissent comme devant se diviser en trois séries qui débutent chacune par une forme décapode : les *Colossendéomorphes*, les *Nymphonomorphes* et les *Pycnogonomorphes*.

La série des Colossendéomorphes est essentiellement caractérisée par la position des palpes et des ovigères dont les bases sont plus ou moins rapprochées et, ordinairement même, presque contiguës du côté ventral ; elle se distingue également par la longueur du tarse des pattes, d'ordinaire, sinon toujours, par la brièveté de la partie coxale, qui est beaucoup plus courte que le fémur et dont les articles sont subégaux. Les représentants décapodes de la série sont les Décolopodides, dont les puissantes chélicères ont un scape de deux articles et les palpes huit ou neuf segments bien distincts; elle se continue et se termine par les Colossendéides, qui sont octopodes, dépourvus de chélicères et dont les palpes sont d'ailleurs munis de neuf articles. Dans cette série, la taille est généralement grande ou très grande, et les segments du tronc sont presque toujours confondus, c'est-à-dire sans articulation distincte.

La série des Nymphonomorphes se distingue de la précédente par la position des palpes qui, lorsqu'ils existent, sont insérés frontalement à une distance assez grande des ovigères, par l'écartement de ces derniers, par le développement de la partie coxale des pattes dont le deuxième article est toujours beaucoup plus long que les articles contigus, enfin, dans la majorité des cas, par la grande brièveté des tarses des mêmes appendices. Les segments du tronc présentent dans la règle une articulation distincte, et le plus souvent aussi le céphalon est bien développé ; la taille n'est jamais très grande et peut devenir fort petite. La série commence par la famille des NYMPHONIDÉS et, dans cette famille, par les formes décapodes du genre *Pentanymphon* ; les chélicères et les palpes sont parfaitement constitués dans cette famille, mais ces derniers ne présentent que cinq ou sept articles. Dans les PALLÉNIDÉS, les palpes sont réduits à l'état de court bourgeon chez les *Pallenopsis*, où ils existent dans les deux sexes, et dans les *Neopallene*, où ils disparaissent chez les femelles adultes ; ils font totalement défaut dans les autres genres. Ils disparaissent également chez les PHOXICHILIDIIDÉS où les ovigères n'existent que chez le mâle, réduits d'ailleurs à cinq ou six articles, tandis qu'ils appartiennent aux deux sexes et sont divisés en huit ou dix articles dans les deux familles précédentes. Nous arrivons ainsi par degrés aux formes terminales de la série, les PHOXICHILIDÉS qui ressemblent tout à fait aux Phoxichilidiidés, mais sont dépourvus de chélicères, ce qui les avait fait ranger, par M. SARS (1891), dans le groupe des Achélates, tout à côté des Pycnogonidés.

Ainsi constituée, la série des Nymphonomorphes me paraît être le type d'un groupe par enchaînements qui commence par les *Pentanymphon* (décapodes) et les Nymphonidés (octopodes), pour se continuer par les Pallénidés (disparition des palpes), puis par les Phoxichilidiidés (disparition des palpes et, chez les ♀, des ovigères), et aboutir aux Phoxichilidés (où disparaissent en outre les chélicères). Mais la série me paraît comprendre deux autres groupes ou rameaux, issus, l'un et l'autre, des Nymphonidés. L'un de ces rameaux est constitué par la famille des AMMOTHÉIDÉS qui semble se rattacher aux Nymphonidés par l'intermédiaire des *Leiconymphon* où les chélicères sont bien souvent encore très développées chez les jeunes, tandis qu'elles sont imparfaites ou rudimentaires chez les adultes, qui ressemblent en cela aux Ammothéidés typiques. A vrai dire, les Ammothéidés présentent parfois des palpes plus complexes que les Nymphonidés, et c'est précisément le cas des *Leiconymphon* dont les palpes présentent neuf articles, tandis que les Nymphonidés n'en ont jamais plus de sept ; mais on sait que le nombre des articles des palpes est très variable actuellement dans les deux familles, et il n'est pas irrationnel d'admettre qu'il en était de même jadis quand ces mêmes familles se différencièrent. Dans tous les cas, les Ammothéidés sont des Pycnogonides à tarse court, à propode arqué et pourvu de griffes auxiliaires; en cela ils ressemblent à certains Nymphonidés (surtout aux *Chæto-nymphon*), et c'est à des formes analogues qu'ils doivent peut-être leur origine. Quant aux EURYCYDIDÉS, qui présentent des tarses longs, un propode droit et dépourvu de griffes auxiliaires, ils me paraissent constituer un second rameau, qui se rattache aux Nymphonidés dont les pattes ont une structure analogue (*Paranymphon*, Nymphons à long tarse). Ils se distinguent d'ailleurs par les connexions de leur trompe, qui est ramenée sous le corps et articulée sur une saillie céphalique, mais c'est là, sans doute, un caractère d'adaptation secondaire. Au surplus, les Eurycydidés ont un palpe de dix articles (et non de dix-sept, comme un lapsus [1906, 10; 1910, 31] me l'a fait antérieurement dire), ce qui les différencie nettement des Nymphonidés où ce nombre est loin d'être atteint. C'est en me basant sur ce fait que je fus porté tout d'abord (1910, 31) à considérer comme provisoirement acceptable la série des Ascorhynchomorphes proposée par M. RAY LANKESTER (1904, 225); mais on peut répéter à ce sujet ce que j'ai exposé plus haut en parlant des Ammothéidés, et peut-être convient-il de considérer les deux familles comme un rameau des Nymphonomorphes.

La série des PYCNOGONOMORPHES est avant tout caractérisée par ses formes robustes, ses pattes courtes et épaisses, dont la partie coxale est à peu près aussi longue que le fémur, encore que les articles en soient réduits ou subégaux; elle se distingue en outre des précédentes par la position des orifices sexuels qui se localisent sur les pattes postérieures au lieu de se trouver sur plusieurs paires de pattes et parfois sur toutes. La série se limite à une seule famille, celle des PYCNOGONIDÉS; elle débute par les *Pentapycnon* qui sont décapodes. La série se continue par les *Pycnogonum*, qui sont des octopodes tout à fait normaux et qui semblent être, à l'heure actuelle, les formes les plus évoluées, non seulement de la série, mais du groupe tout entier des Pycnogonides. Je ne crois pas, en effet, qu'on puisse ranger dans les Pycnogonidés, avec M. LOMAN (1908), les genres *Böhmia* HOEK, *Hannonia* HOEK et *Ryncho-thorax* DOHRN, qui présentent tous des ovigères dans les deux sexes et certains même des palpes (*Böhmia*, *Rhynchothorax*) ou des chélicères (*Böhmia*, *Hannonia*); par tous leurs caractères, ces genres sont fort différents des Pycnogonidés et se rangent parmi les Nymphonomorphes.

Étant donné ce qui précède, on peut exprimer la classification et les affinités de la sous-classe des Pycnogonides par le tableau suivant :

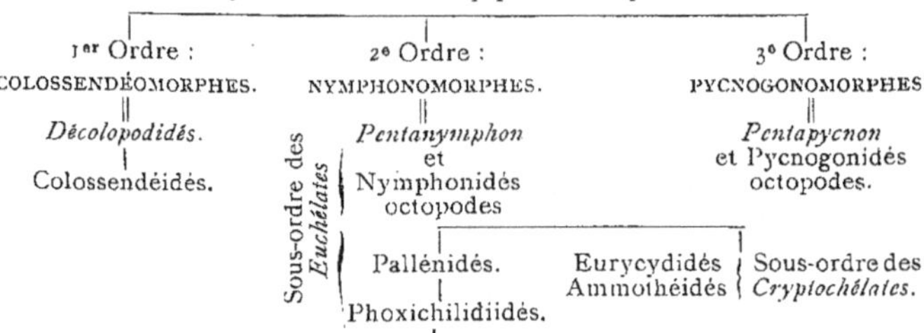

Chacune des séries constitue des ordres qui commencent par un type décapode, toutes les autres formes étant octopodes.

Cette classification emprunte au système de M. SARS (1891) les groupements fondés sur la présence, l'atrophie ou la disparition totale des appendices céphaliques; mais au lieu de considérer ces caractères comme primordiaux, elle les range au second plan pour diviser l'ordre des Nymphonomorphes en trois sous-ordres, *Euchelata, Cryptochelata, Achelata*, qui ont naturellement une étendue plus faible que ceux de même nom établis par l'éminent zoologiste norvégien.

L'atrophie des appendices buccaux est évidemment un phénomène secondaire, car elle s'est produite à des stades divers de l'évolution : de fort bonne heure chez les Pycnogonomorphes, puisqu'elle est réalisée déjà chez les *Pentapycnon;* un peu plus tard chez les Colossendéomorphes, où elle se manifeste chez les Colossendéidés; beaucoup plus tard chez les Nymphonomorphes, où elle caractérise les Phoxichilidés qui sont les formes terminales de la série. Abstraction faite des différences morphologiques, qui sont importantes et nombreuses, il est impossible, à cause du *Pentapycnon*, d'établir un lieu génétique entre les Phoxichilidés et les Pycnogonidés réunis par M. SARS dans son groupe des Achélates; et à cause du genre *Decolopoda,* il en est de même des Colossendéidés, que M. SARS a réunis aux Ammothéidés et

Eurycydidés dans son groupe des Cryptochélates. Mais ces observations ne pouvaient être faites à une époque où l'on ignorait absolument les Pycnogonides décapodes, et le système de M. SARS reste comme une belle œuvre dont on doit conserver le principe en donnant à celui-ci une valeur subordonnée.

On doit en dire autant du système proposé par M. RAY LANKESTER (1904) à une date plus récente, quoique antérieure à la découverte des Pycnogonides décapodes. Dans ce système comme dans le nôtre, les Pycnogonides sont divisés en trois groupes: les *Nymphonomorphes* (Nymphonidés et Pallénidés), les *Ascorhynchomorphes* (qui correspondent aux Cryptochélates de M. SARS) et les *Pycnogonomorphes* (*Pycnogonum*, *Phoxichilus* et *Hannonia*).

Nous avons conservé ces termes empruntés à M. POCOCK, mais en leur donnant une autre étendue et en remplaçant le nom d'Ascorhynchomorphes par celui de Colossendéomorphes proposé par M. COLE. Les Colossendéomorphes de M. COLE (1905) correspondent aux Ascorhynchomorphes de M. RAY LANKESTER et aux Cryptochélates de M. SARS (1), mais leur nom convient à merveille à la première de nos séries (Décolopodidés et Colossendéidés) et c'est pourquoi il est bon de le conserver.

Tandis que les systèmes précédents se rapprochent du nôtre par le groupement des Pycnogonides en séries ou ordres, ceux de DOHRN (1881), de M. HOEK (1881) et celui tout récent de M. LOMAN (1908) s'en éloignent par ce fait, qu'ils se bornent à établir des familles. DOHRN et M. LOMAN rapprochent à juste titre les Phoxichilidés et les Phoxichilidiidés, M. HOEK les *Phoxichilidium* et les *Pallene*, mais il me paraît peu rationnel de réunir en une même famille, avec M. LOMAN, les Colossendeidés, Eurycydidés et Pycnogonidés, avec M. HOEK, les *Colossendeis*, *Ammothea* et *Ascorhynchus*. DOHRN se rapprochait davantage de la vérité en établissant quatre familles dans le groupe des Pycnogonides (Ammothéidés, Phoxichilidés, Nymphonidés et Pycnogonidés).

Je reviens à la classification proposée dans cette note. Les trois séries ou ordres qui, d'après cette classification, constituent la sous-classe des Pycnogonides représentent des types morphologiques bien distincts et toujours fort accentués, au moins dans leurs formes

---

(1) La seconde série établie par M. COLE est celle des *Pycnogonomorphes*, qui comprend tous les autres Pycnogonides.

primordiales. Et comme chaque série débute par une forme décapode, c'est-à-dire ancestrale au point de vue du nombre des segments, on peut dire que *la sous-classe s'est différenciée en types morphologiques à une époque où elle se trouvait encore au stade primitif caractérisé par au moins cinq segments munis de pattes.* C'est là un exemple de différenciation très précoce dont les êtres vivants offrent à coup sûr peu d'exemples.

Cette différenciation précoce me paraît propre à expliquer le peu d'étendue et la grande homogénéité de la sous-classe des Pycnogonides.

Chaque groupe du règne animal semble posséder en propre une certaine mesure de plasticité, de puissance adaptative, qui lui permet de se modifier en même temps que les milieux; une fois cette mesure dépensée, le pouvoir adaptatif se limite à de légères modifications de formes qui respectent les caractères morphologiques essentiels. Puisque la différenciation morphologique des Pycnogonides s'est effectuée au stade où l'animal possédait un tronc de cinq segments au moins, on doit la fixer à une époque fort ancienne, à l'époque primaire sans doute, c'est-à-dire à une période géologique où les climats et les milieux présentaient une grande uniformité. Depuis lors, climats et milieux se sont diversifiés à la surface du globe sans retentir sur la morphologie du groupe dont la puissance plastique était épuisée.

L'étude comparative des trois ordres justifie complètement cette manière de voir. Dans l'ordre des Pycnogonomorphes, les modifications des appendices céphaliques (perte des chélicères et des palpes dans les deux sexes, des ovigères chez la ♀) se sont produites de très bonne heure, à un stade où l'animal avait pour le moins cinq paires de pattes, comme le montrent les *Pentapycnon;* aussi le groupe se réduit-il à la famille peu étendue et très homogène des Pycnogonidés. Dans l'ordre des Colossendéomorphes, les modifications des appendices céphaliques sont plus réduites parce qu'elles se limitent aux chélicères, elles ont aussi été plus tardives, parce qu'elles ne frappent pas encore les *Decolopoda* ; mais elles se sont produites aussitôt après chez les *Colossendeis*, ce qui a limité singulièrement l'étendue de l'ordre. Tout autre a été l'histoire évolutive des Nymphonomorphes; dans ce groupe, la puissance plastique afférente aux appendices céphaliques s'est dépensée lentement et longuement, depuis les Euchélates munis de palpes (Nymphonidés); ainsi ont pu se produire les Cryptochélates, qui sont munis de palpes avec chélicères réduites ou nulles (Ammothéidés, Eurycydidés), les Euchélates sans palpes (Pallénidés et Phoxichilididés) et finalement les Achélates qui se limitent à la famille des Phoxichilidés où les ovigères ont disparu chez le mâle, en même temps que les chélicères et les palpes dans les deux sexes. En somme, une différenciation lente et tardive explique suffisamment la grande diversité des Nymphonomorphes; au point de vue de la réduction des appendices, cet ordre a fini par où avait commencé le groupe des Pycnogonomorphes.

Je termine en appelant l'attention sur ce fait, que les trois ordres sont particulièrement bien représentés (1) dans les eaux glacées de l'océan Antarctique, et qu'en dehors de cet océan les Pycnogonides décapodes sont inconnus. Cela porte à croire que le groupe s'est différencié d'abord au sud du continent brésilo-éthiopien, dans le vaste océan Antarctique primaire, et que ses types morphologiques étaient déjà fixés et octopodes lorsque leurs descendants se répandirent au nord dans toutes les mers du globe. Cette opinion n'est évidemment qu'une simple hypothèse, mais elle prendrait une haute valeur si, dans la suite comme par le passé, on ne trouvait pas de Pycnogonides décapodes en dehors des mers antarctiques (2).

APPENDICE.

A la veille de me rendre au Congrès entomologique de Bruxelles, lorsque cette note était déjà complètement écrite, mon ami,

---

(1) Les vrais Pycnogonidés sont rares dans les mers australes, où l'on n'en connaît que deux espèces, d'ailleurs subantarctiques, le *Pycnogonon magnirostre* Möbius et le *Pycnogonon magellanicum* Hoek. Le seul *Pycnogonum* vraiment propre aux mers antarctiques a été découvert à Port-Lockroy par le « *Pourquoi pas?* »; je l'ai appelé *Pycnogonon Gaini* (1910, 30) en l'honneur de M. Gain, le zélé biologiste qui a récolté les Pycnogonides au cours de l'expédition.

(2) Il faut rejeter cette hypothèse, car j'ai fait connaître recemment (1911), sous le nom de *Pentapycnon Geayi*, un Pycnogonide décapode qui habite le littoral de la Guyane française. (Note ajoutée au moment de la correction des épreuves.)

M. W.-E. CALMAN, a eu l'obligeance de me communiquer un intéressant article (1) relatif au *Pentapycnon* et aux Pycnogonides décapodes. Dans cet article, M. CALMAN reprend une opinion déjà soutenue par M. CARPENTER et diamétralement opposée à la mienne : il regarde, en effet, les Pycnogonides décapodes comme des formes dérivées issues, à une date récente, des Pycnogonides octopodes.

Nous voici dès lors en présence de deux thèses contraires, et je veux répondre à un désir exprimé par M. CALMAN (2) en relevant ici certaines raisons qui paraissent en faveur de la mienne.

On doit, semble-t-il, considérer comme des caractères primitifs et ancestraux la présence des appendices céphaliques, c'est-à-dire des chélicères, des palpes et des ovigères. S'il en est ainsi, les Pycnogonides décapodes présentent des caractères primitifs relativement aux Pycnogonides octopodes; dans la série des Colossendéomorphes, nous voyons en effet les *Decolopoda* présenter, avec un plein développement, les trois paires d'appendices céphaliques, alors que les *Colossendeis* n'ont plus de chélicères; dans la série des Nymphonomorphes, où l'évolution fut plus lente, les *Pentanymphon* et les Nymphonides possèdent tous les appendices céphaliques, tandis que les palpes disparaissent dans les autres familles, et les chélicères elles-mêmes dans les Phoxichilidés, où d'ailleurs les ovigères n'existent plus chez les femelles. Dans la série des

(1) L'une des raisons invoquées par M. CALMAN en faveur de son hypothèse, c'est la localisation des Pycnogonides décapodes dans les mers antarctiques. Mais depuis que j'ai fait connaître (1911) le *Pentapycnon Geayi* de la Guyane, cet argument ne présente plus de valeur, ainsi que M. CALMAN l'observe lui-même dans les lettres qu'il m'a écrites. (Note ajoutée au moment de la correction des épreuves.)

(2) L'article de M. CALMAN est intitulé « Antarctic Pycnogonons » et se trouve dans le numéro de « Nature », du 28 juillet 1910, p. 104. « La plupart des autorités, observe M. CALMAN, par exemple le Profr BOUVIER et le Profr D'ARCY THOMPSON (dans « Cambridge Natural History »), adoptent l'idée que l'état décapode est primitif et a été conservé par les membres les plus primitifs dans les deux branches divergentes du groupe. L'autre explication, d'abord suggérée par le Profr C.-H. CARPENTER et soutenue par le présent auteur, c'est que l'état décapode résulte d'une évolution récente qui s'est produite indépendamment dans les deux cas. »

Pycnogonomorphes, qui ressemble aux Phoxichilidés par la même disparition des appendices céphaliques, certains *Pycnogonum* (*P. Gaini*) présentent encore les traces de l'avant-dernier segment qui persiste bien développé chez les *Pentapycnon*. Ainsi, la règle est absolue, les Pycnogonides décapodes présentent tous des caractères ancestraux qui manquent plus ou moins aux Pycnogonides octopodes; ils ne peuvent donc dériver de ces derniers.

INDEX BIBLIOGRAPHIQUE.

1906. E.-L. BOUVIER, Pycnogonidés du « *Français* ». (« Exp. ant. franç. 1906. »)

1910 ID., Les Pycnogonides à cinq paires de pattes recueillis par la Mission antarctique Jean Charcot à bord du « *Pourquoi pas?* ». (« Comptes rendus Acad. des sciences », t. CLI, pp. 26-32, 1910.)

1905. L. J. COLE, Ten legged Pycnogonides, with remarks on the classification of the Pycnogonida. (« Ann. Nat. Hist. », [7], vol. XV, pp. 405-415, 1905.)

1881. A. DOHRN, Die Pantopoden des Golfes von Neapel und der angrenzenden Meeresabschnitte. (« Fauna und Flora des Golfes von Neapel ». III, 1881.)

1834. J. EIGHTS, Description of a new Animal belonging to the Arachnides of Latreille, discovered in the sea along the shores of the New Shetland Islands. (« Boston Journ. Nat. Hist. », vol. I, pp. 203-206, pl. VII, 1834.)

1907. E. V. HODGSON, *Pycnogonida*. (« National antarctic Exped., Nat. Hist. », vol. III, 1907.)

1908. ID., The Pycnogonida of the Scottish national antarctic Expedition. (« Trans. Roy. Soc. Edinburgh », vol. XLVI, part. I, pp. 159-188, 1908.)

1881 P. P. C. HOEK, Report on the Pycnogonidea. (« Challenger Zoology ». vol. III, 1881.)

1908. J.-C.-C. LOMAN, Die Pantopoden der Siboga-Expedition mit Berücksichtigung der Arten Australien's und des tropischen Indik. (« Siboga-Exped. », vol. XL, 1908.)

1904. E. RAY LANKESTER, The Structure and Classification of the Arachnida. (« Quat. Journ. Micr. Sc. », vol. XLVIII, part II, pp. 165-269, 1904.)

1891. G. O. SARS, Pycnogonidea. (« Norw. North Atlantic Exped. 1876-1878 », Zool., 1891.)

A note on methods of preserving Insects in tropical climates,

by F. M. Howlett, B. A., F. E. S., Christ's College, Cambridge,
Second Imperial Entomologist, Pusa Research Intitute, India.

There is little doubt that the ideal place for storing specimens in the tropical and sub-tropical zone would be a deep cellar or subterranean building at such a depth as to enjoy an equable temperature and avoid the extremes met with above ground. The difficulties would lie in providing for illumination and arranging for a constant supply of dry air at the same temperature as the interior of the building.

In ordinary laboratories, working-collections in a country such as India require constant inspection and attention, and while it may sometimes be difficult to arrange for adequate supervision, the risk of irretrievable damage from any negligence in this direction is very much greater than in temperate climates. In any case deterioration takes place, and valuable type-specimens are best preserved elsewhere than in the tropics.

Here the need is for methods such as will reduce to a minimum the amount of attention necessary to keep specimens from deterioration under tropical conditions, and the object of this short paper is merely to draw attention to some few points where experience has shown the ordinary methods to be particularly unsuitable.

Tropical conditions are generally characterised by :

1° Extremes of temperature ;
2° Extremes of humidity and dryness;

3° Abundance and rapid growth of moulds in wet seasons ;
4° Abundance of Ants, Beetles (*Anthrenus*), Mites (*Psocidæ*), and Fish-insects (*Lepisma*), etc.

We will touch as briefly as possible on the more obvious effects of these conditions.

With extreme humidity, not only do moulds flourish, but set specimens become limp and relaxed. Arrangements for very thorough drying during setting must be made, and specimens already set should as far as possible be kept fairly dry. Two methods may be used. The specimens may be enclosed in large wooden or tin cases in which hot air is kept constantly supplied from a lamp burning outside, or the air within the case may be dried by pans of calcium chloride or sulphuric acid. Carbolic acid may be used as an additional prevention against mould.

One of the great difficulties at Pusa has been the fact that no store-box has yet been found which will stand the great seasonal and daily variations in temperature and humidity, and at the same time will not be of too thick and heavy a make to permit of being handled with comfort. This is one of our pressing wants, and it is to be hoped that manufacturers may produce a cheap light box made of non-conducting material, unaffected by changes of temperature, and impermeable to moisture.

Changes of temperature induce deposition of moisture on conducting surfaces, and this alone renders the use of metal boxes impossible. Similarly moisture may condense on pins, and if these are of the ordinary kind, verdigris is practically certain to appear in time and often ruins a specimen. Nickel or silver pins should be *invariably* employed if possible. Absorption of moisture is also probably the cause of the corrosion so often seen in the part of a pin which is imbedded in the lining of the box, whether this be cork, peat, or pith. Certainly a large amount of moisture can be and is absorbed by any one of these three substances, while they also afford convenient lodging for Mites, especially in boxes which have been long in use. The Mites and mould can be kept in cheek by periodically washing the floor of the box with a mixture of equal parts of white beech creosote and a saturated solution of napthalene in chloroform. A nearer approach to the ideal method of lining boxes is secured by using paraffin wax (m-pt. 15°-20° F. above the maximum shade temperature) melted up with 25 % of naphthalene and run into the box to a depth sufficient to give a firm hold for the pin (about 15-20 mm.).

Boxes thus lined have now succesfully stood a two years trial at Pusa, and the method seems worthy of general recommendation. There is no corrosion of the pins, no necessity for treating with poisonous washes, and the boxes are completely free of Mites, which appear unable to live in them. Naphthalene paraffin wax is also used to dip labels as an effective protection against the attacks of Fish-insects; this is in place of washing them over with poisons such as corrosive sublimate, which is otherwise necessary.

Ants and Termites are usually excluded by methods of isolation, such as standing the legs of shelves or cabinets in saucers by a « moat » or channel filled with water. At Pusa a concreted water-channel about eighteen inches wide surrounds the whole of the laboratory buildings and is kept permanently filled.

Constant evaporation of alcohol and the changes of volume brought about by alterations and temperature make the preservation of spirit specimens somewhat troublesome, and there seems to be no way of evading the demand for regular inspection of all tubes and bottles which are not hermetically sealed. An effective device for closing tubes containing alcohol, which would permit of their being at any time opened and reclosed, would be most valuable.

« Viscose » bottle-caps for permanent, and plugs of plasticine for temporary closure seem useful and convenient, but the latter substance is reliable only with formalin, not with alcohol, for which paraffined corks are usually employed.

Since entomological collections in the tropics are often formed in connection with medical research, it may be well to say a word on the mounting of small pinned specimens in general, and particularly of Diptera and Mosquitoes. These should always be pinned to supports with silver or nickel pins; double-pointed pins can be bought, and are most useful, as the specimen can be removed, examined from all sides, and then replaced.

Practical experience under foreign conditions has led us to abandon altogether the use of paper discs as supports for Mosquitoes, though this method has been widely adopted. For permanent collections a much more rigid and solid support is needed. We have used papered cork or cork-carpet; « suberit » (a cork composition) covered with white enamel, which makes a good support;

polyporus, and pith; the last named is excellent except for its power of absorbing moisture and corroding the pin. We are trying pith supports impregnated with paraffin wax in the hope of overcoming this defect.

For Anopheline Mosquitoes collected for the sake of locality-records and similar objects, where detailed examination of specimens may be unnecessary, a method shown me by Dr. C. A. BENTLEY would appear useful. This consists in laying the Mosquito on a glass coverslip (or sheet of mica) and covering it with a 1 % solution of celloidin; when dry this is painted over with thicker celloidin, and the whole is backed with white enamel, making an absolutely permanent mount.

The Conservation of Types,

by Dr. W. J. HOLLAND (Pittsburgh).

It is scarcely necessary that I should undertake in this presence to attempt to define what constitutes a *type*. You all are familiar with the subject in all its phases. It may, however, be proper to remark in passing that all material, which has been made the subject of accurate investigation by an author and to which he has definitely attached nomenclatorial designations, becomes by that very fact typical, and its preservation for the use of students is important.

In the first place, the preservation of typical material is important because language, even when employed by a master of the descriptive art, often fails to give an adequate conception of the objects themselves. The personal equation enters largely into all descriptive work. Observers are not all equally competent and careful. The most competent and careful observer may fail by means of nouns and adjectives to exactly convey to the mind of a reader that which he has seen. Word-painting, even when executed by a SHAKESPEARE or a TENNYSON, does little more than suggest to the mind that which the eye teaches. Even when verbal descriptions are accompanied by well executed illustrations, there is often something left to be desired. The pictorial art has been brought into the service of science during the past one hundred years in such a manner as at times to make it appear almost superfluous to preserve original material; but even in cases where verbal description has been most splendidly supplemented by the work of the photographer and the delineator, it is often found by the accurate and painstaking student that much is left to be desired.

But not all descriptive and illustrative work is of the high order to which I have referred. Those of us who are familiar with the literature of description know well that much of it falls short of the highest ideals. A description may be so brief as to be unsatisfactory; or it may be so verbose as to be equally unsatisfactory. Brevity was characteristic of the work of the earliest authors. A conspicuous illustration of brevity so great as at times to be unsatisfactory is found in the writings of the man, whom we delight to honor as the father of the modern biological sciences, LINNÆUS. You all know how very brief and how utterly inadequate in many cases are his descriptions of living things. Only referring now to his labors in the field of entomology, it is showing no lack of respect to his memory to say that, were it not for the preservation of a knowledge of the species he named in the illustrated writings of his pupil CLERCK, and the conservation of many of the types of the species he described, and which have since been made the subject of study by such careful writers as AURIVILLIUS and his associates, it would be almost impossible to determine to what species the names he gave should be applied. It is only within the past week that I have had the opportunity to examine the entomological collections of LINNÆUS preserved at Upsala, and were it not for the fact that these collections have fortunately survived the lapse of years, we should to-day be in a measure of doubt as to the whole body of Linnean nomenclature so far as it relates to our own favorite branch of the natural sciences. The same remark holds good as to the work of « the immortal Swede » in the domain of botany. The herbarium of LINNÆUS is fortunately preserved in Great Britain, and has there become the last court of appeal whenever questions have arisen as to the species to which he applied generic and specific names.

While brevity of description may be carried to an extreme,

extreme verbosity may equally well interfere with a correct understanding of the sense of an author. I shall not attempt to cite illustrations of this, though I have some instances in my mind. They occur principally in the writings of recent authors, who are not deterred from much speaking by the cost of printer's ink.

Again, the advances made in biological research during the past fifty years under the stimulating influence of advanced thought as to the origin and development of species have resulted in sharper definitions, and the modern student is not satisfied with the broader and more generalized classification, which was satisfactory to students a century ago. We recognize that vital processes are active in the production of local races, insular varieties, and multitudes of aberrant forms. To many of these subspecific and varietal names have been given. It is often the case that the distinctions existing between the original type and the subspecies are minute, though constant. To accurately define in words these subtle differences, which are patent to the eye, is at times almost impossible. In such cases the preservation of types is of the utmost importance as a guide to the student.

In the first place then it is important that the types of species, subspecies, and varieties should be preserved because of the necessary inadequacy of merely verbal description to convey an accurate and satisfactory idea of the meaning of an author. In cases of dispute as to the intention of an author the type becomes the last court of appeal.

But there is a second reason for the preservation of types. We all know that in the realm of life things are not ever the same. There is constant ebb and flow in the stream of living things. You are all familiar with the splendid work of BATES, the eminent English naturalist, who collected upon the Amazons and especially in the vicinity of Ega. The types of the new species he there discovered are preserved in England. Nearly half a century has elapsed since he made his discoveries, and the species which he found were named. Present collections made in the region where he labored show that in the lapse of even so short a time as fifty years marked changes have taken place in the form of living things in that locality. Two hundred and fifty or more generations of

Insects have come and gone since BATES first wrought at Ega. The comparison of the Insects taken by BATES with their descendants of to-day is both interesting and instructive. Such a comparison could not be made were it not for the preservation of the types collected by BATES himself. Such material serves to guide us in our understanding of evolutionary processes.

To sum up what I have said, it is highly important to preserve typical material in the first place because without the types as material for reference and comparison it is in many cases almost absolutely impossible to be certain what was the intention of an author in naming a species; and secondly, because such material is valuable as definitely marking in the case of living species the stage of evolution reached at the time when the material was made the subject of observation.

I wish now to advance for consideration a few propositions as to the location of typical material. It is in my opinion highly desirable that so far as possible all such material should be conserved in institutions which are sufficiently well endowed and manned as to make it certain that the collections entrusted to them will be kept in good condition so long as it is possible to do so. I deprecate the retention of the types of species in private collections or in institutions inadequately endowed and carelessly administered. It is inevitable that private individuals engaged in the study of collections, which they have made, should create types. Many of the most valuable contributions made to our own science are due to private initiative and effort. Collections acquired by specialists, who have written upon them, are often exceedingly rich in types, but I maintain that it is in every way desirable that such collections should find their final resting place in the greater and well endowed museums of the world. I deprecate the placing of such collections in the custody of small and poorly endowed and equipped institutions. You will pardon me for venturing to give you at this point a concrete illustration of the disadvantages, which flow from a disregard of this matter. Some forty or more years ago one of the most eminent paleontologists of Great Britain gave his large and valuable collection of the types of the fossils he had obtained in the Siwalik Hills in India to the Museum of a small Scottish College. Years passed by. Quite recently an important question arose as to the dentition of an extinct order of Mammals the remains of which occur both in the old world and the new. I wrote to the President of the Scottish institution referred to requesting him to have a cast of the skull figured in the writings of the great paleographer made for me. Though a carefully executed and beautifully lithographed figure of the specimen exists, it is impossible from this figure to be quite sure whether the Animal possessed upper incisors or not. I received no answer. I then wrote to Dr. SMITH WOODWARD, the head of the Geological Department of the British Museum, requesting him to use his good offices in securing for me that which I wished to see. At last after long writing he received an answer from the institution stating that a tradition existed that at one time such a collection had been given to the college, but that the janitor of the building, the only one now living who seemed to have any recollection of the matter, stated that a number of years ago these « old bones » had been by order of one of the professors cleaned out of the Museum and put into the courtyard, where they were afterwards buried under the flower-beds. Dr. WOODWARD wrote me it was his intention to get permission to dig up this courtyard in order if possible to recover the types of Falconer's Fossil Fauna of the Siwalik Hills. I narrate this little incident because it is instructive, and shows in a striking manner the dangers to which invaluable and most important collections are exposed when entrusted to the custody of institutions in which there are not men who comprehend the value of such things and where there do not exist the means which are necessary for their preservation. The private student has not discharged the whole of his duty in merely naming and describing the species he has acquired. He should attend to it either in person or by testamentary provision that the types of his species be placed where they will be sacredly guarded and where they may be consulted by those who come after him.

The museums of small colleges and provincial towns are not, as a rule, proper resting-places for scientific types. They should be placed in the greater and more amply endowed institutions, where they are reasonably certain of being properly esteemed and sedulously cared for and preserved. Even in these greater institutions it is not always certain that the best of precautions will be taken. In the smaller institutions it is almost certain that sooner or later misfortune will overtake such collections. The reason for this is not hard to find. In smaller institutions in general the scientific staff consists of at best but a few persons. The Professor of Natural History is to-day an entomologist; to-morrow he is succeeded by a geologist. The geologist does not know about the collections carefully acquired by his predecessor. The boxes full of « bugs » are left to be the prey of pests and mould, while he is happy in collecting fossils. In turn he is succeeded it may be by a botanist. His time is limited, his duties are many. He is interested in his herbarium. The herbarium grows apace while the « bugs » are devoured by Mites and Anthrenus, and the « fossil bones » are dumped into the courtyard.

In Great Britain the Natural History Museum in London is becoming more and more recognized as the proper place for the deposition of types. In France the National Museum is the proper resting-place for such collections. The Royal Museum in Berlin and the Imperial Museums in Vienna and St. Petersburg are well equipped to receive and care for such collections. In the United States the National Museum in Washington, the Zoological Museum in Cambridge, the American Museum in New York, the Academy of Natural Sciences in Philadelphia, the Carnegie Institute in Pittsburgh, and the Field Museum in Chicago are in a position to carefully preserve for all time the treasures committed to their keeping. But outside of the institutions I have named there are not at the present time in America any institutions to the custody of which in my judgment such material should be consigned. College and university museums the world over are as a rule the very worst places into which to put types or typical material. There may be a few honorable exceptions, but they are very few.

There is another matter which suggests itself in this connection, and that relates to the transmission of types and typical material to experts for study. In view of the importance of this material I think it is in the main well for institutions entrusted with its custody to retain it rigidly in their care and not to send such material away from their fireproof and well munitioned rooms, exposing it to the risks of breakage and fire. This rule should not, however, be too rigidly enforced to the detriment of science. There arise now and then cases in which in the interest of science it is desirable that types should be sent from one museum to another for purposes of comparison and investigation, or to eminent specialists who require this material in the preparation of monographic papers. Each case of this sort, however, should be made the subject of definite inquiry, and the utmost precaution should be taken to insure against the loss of specimens. As a rule I think that cotypes should be selected for transmission in such cases, when they exist, and that types themselves should not be exposed to the dangers of travel and of fire on sea and land.

In this connection it remains for me to say that whenever great museums such as I have indicated entrust their collections to scientific specialists for study and determination, it should be the invariable rule that the types of the species named and described by such specialists should be returned to the museum from which they came. The claim was made a few years ago by one or two persons, who were, I am sorry to say, entomologists, that it is the unwritten but universal law that an author is always entitled to reserve for his own collections the types of species which he may name. I took occasion at the time to memorialize the leading scientists of the day as to the existence of such « unwritten and invariable law », and discovered, as I knew I would, that there existed so far as I could ascertain only three or four persons in the world who claimed that such a law obtains. The law existed only in the disordered imaginations of my correspondents, and the mere suggestion of its existence was ridiculed by the most eminent systematists of the world.

For my own part, I desire to say that I make it a rule to require of all specialists the return of the types to the Carnegie Museum, but am always ready, should there be sufficient material to justify me in so doing, to accord to students for their own private use such duplicate material as I may have at my command, and with only one or two exceptions have discovered any disposition on the part of specialists to dissent from such an arrangement.

Thanking you for your kind attention to these somewhat desultory remarks, I now invite you to the fullest and freest discussion of the subject before us.

Mimicry,

by Dr. F. A. Dixey, F. R. S. (Oxford).

At the outset of my discourse I should wish to say that I feel highly honoured by the invitation to address so notable a body of scientific men as are gathered together in the First International Congress of Entomology under the guidance of our distinguished President; and to express the hope that this may be the beginning of a long series of Congresses, which in advancing the study of Entomology will confer benefits upon the race, and contribute towards the great object of international amity.

The subject to which I propose to devote this lecture is that of Mimicry; a subject which has formed perhaps the largest part of my studies in Insect Bionomics.

It may, I think, be assumed that all naturalists are acquainted with the main features of what is known as « *Mimicry* ». But it may be doubted whether all naturalists realise how numerous are the facts which can be ranged under this head, or how complicated are the phenomena with which a full consideration of the subject brings us into contact.

We should, I venture to think, be false to all the best traditions of scientific method, if, with this great array of remarkable facts before us, we made no attempt to interpret them. It is hardly necessary for me to point out that while in the region of fact we may reasonably hope to attain a great measure of certainty, our interpretations must be to a large extent provisional. It is true that the day may come when we shall be able to speak positively, and with general agreement, as to the causes and full bionomic significance of these noteworthy resemblances; but the time is not yet,

24

and we must be content, for the present, to examine and to test, by every means in our power, those explanations that have from time to time been offered. The fuller our knowledge of the facts to be accounted for, the more nearly true is our interpretation likely to be; and this is the justification for reviewing in some detail any kind of evidence that may have a bearing on the question before us. And I will ask my audience to be good enough to observe that, when I use the term « mimicry », I do so at present, as the lawyers say, « without prejudice ».

Many cases of mimicry between Insects of different orders have long been known. The very remarkable resemblance borne by certain Moths to some of the stinging Hymenoptera long ago attracted the attention of observant naturalists. BOISDUVAL drew attention to the fact that three Butterflies belonging to three different families, namely *Limnas chrysippus*, the female of *Hypolimnas misippus*, and the *trophonius*-form of the female of *Papilio dardanus*, show a close resemblance to each other in outward aspect. Of late years very numerous instances of a similar kind have come to light. Sometimes the observed resemblance occurs between Insects of the same order but of different families, as between the Papilios and Pierines among the Butterflies; e. g., *Papilio nephalion* and *Euterpe rosacea*; sometimes between Insects of different orders. We find, for example, Ants mimicked by *Hemiptera*, *Homoptera* and *Orthoptera*, while other *Hymenoptera* are closely copied by two-winged Flies. It is needless to multiply instances of this sort, for numbers of them must be familiar to all working naturalists. And when the extraordinary prevalence of this phenomenon is once realised, it becomes impossible to dismiss the question as being merely a matter of coincidence. If we had only a few such instances to consider, we might be justified in calling them accidental. But, apart from other reasons, their very number raises the improbability of such an interpretation to so high a pitch as practically to forbid its acceptance.

Let us look at the facts a little more closely. We have seen that some of the nearest resemblances occur between Insects of different orders. We may therefore dismiss at once the idea that the likeness is merely due to affinity. At the same time there is no doubt that the element of affinity does to some extent enter into the question. We shall return to this point later.

A short examination of cases will show us that the mimicry is often confined to the female sex. It is well known that in the instances mentioned just now it is only the *female* of *Hypolimnas misippus* that resembles *Limnas chrysippus;* and the same statement applies to *Papilio dardanus*. This latter Butterfly, as is also well known, supplies us with another feature in the case. The female is polymorphic, and each form of the female is a copy of a different Danaine model. I show here a representation of a brood of this *Papilio*, all the specimens being the offspring of a single female. The *trophonius*-form, as we have seen, mimics *Limnas chrysippus,* the *hippocoon*-form resembles *Amauris dominicanus,* and the *cenea*-form is in mimetic relation with *Amauris echeria* and *Amauris albimaculata*. This curious phenomena is by no means an isolated case, as is shown by the next illustration. We have here the male of the African Pierine *Leuceronia argia,* in which sex the species is practically invariable. But the female exists in many different forms, each of which shows a resemblance to a Butterfly of no very close affinity to *Leuceronia*. The mimicked Insects belong to the genera *Belenois, Mylothris, Phrissura* and *Pinacopteryx*.

There are cases on record in which both male and female of a sexually dimorphic Butterfly are mimetic, but the respective models of the two sexes are different. I do not at the present moment recall any instance of a species where the male is a mimic and the female not.

We have then reached this point : that the female sex is more susceptible to the mimetic influence, whatever it may be, than the male. This is shown by the numerous cases of sexual dimorphism in which the female alone mimics, and also by those examples of polymorphism, confined to the female, in which each separate form assimilates itself to a different model.

We may now pass on to another consideration. In all the instances that I have shown, the forms that so resemble one another are found in the same, or nearly the same, regions and localities. In many cases they are observed to have similar habits. It has often happened that a group of Insects, diverse in affinity but closely allied in aspect, has been taken, not only on the same day and within a limited area, but actually on the same plant. The illustration I now exhibit depicts a wonderful assemblage of Insects, all characterised by the same arrangement of colours, comprising Wasps, Braconids, Moths, a Bug, a two-winged Fly, and Beetles of different families; many members of which assemblage I have myself seen settled on or flying about the same tree at East London in South Africa. And what is perhaps even more remarkable, we find that when geographical races, or representative species, inhabit different areas of the same continent, the members of these mimetic groups all change their aspect together, and in the same direction. By the kindness of Professor POULTON, I am enabled to illustrate this statement by a very beautiful series of Butterflies from Central and South America. The assemblage in question contains species of very diverse affinities, including Ithomiines, Heliconiines, Danaines, Nymphalines and the females of certain Pierines, all characterised by a peculiar arrangement of the colours red, yellow and black. While these figures are being shown on the screen, I quote from a former description of my own : « The members of this assemblage as it occurs in the northern part of Central America — Guatemala to Nicaragua — present in common a remarkable streakiness of pattern, a feature that makes them easily recognisable among the corresponding forms from other regions of the same continent. Passing on to Venezuela, we find among the geographical races, or, if we like to call them so, the representative species, that there replace the Central American forms, a tendency to the breaking-up of the streaks, and a slight encroachment of the red ground-colour upon the yellow of the apex. In Trinidad there occurs a general paling of the ground-colour, due to an increase of yellow pigmentation, and running, as before, through the entire group. Next, taking the corresponding Guiana forms, we find a further breaking-up of the streaks into spots, and also a general darkening, especially of the hind-wings, which gives a most characteristic aspect to the whole assemblage. In East Brazil we have a modification which somewhat recalls the Trinidad facies, though here the yellow streak on the hindwing is better defined, and a pale spot makes its appearance on the apex, the dark area of which is less broken up. At Egá, on the Upper Amazon, a curious dark chestnut tinge pervades the group, while in Peru a characteristic spottiness takes the place of the streaky pattern we saw elsewhere, and the apex becomes more uniformly dark. Finally, in Ecuador the streaks have all but disappeared, and even the spots have become almost blocked out by a dark infusion which now occupies, not only the apex, but also a large part of the base of the forewing, and the whole, or nearly so, of the hindwing. After a little study of some of the typical members of each of these geographical groups, it becomes easy to pronounce, with a considerable degree of confidence, upon the local habitation of a species that we may never have met with before. »

There are two genera of African Pierines, *Mylothris* and *Phris-*

sura, not very nearly allied to one another, but exhibiting in many of their species, or geographical races, a curious parallelism. Nearly every form of *Mylothris* has its own copy among the forms of *Phrissura*; and exactly as in the instance of the South American assemblage we have just been examining, the changes observed in passing from one portion of the African continent to another are alike in the corresponding forms of the two genera. Thus, as is shown by these lantern illustrations, *Mylothris narcissus* is associated in East Africa with *Phrissura lasti*, both being Butterflies with lemon-yellow hindwings and black marginal spots. A form of *Mylothris* from Uganda, white with a dark apex to the forewing, a row of dark marginal spots on the hindwing, and a basal patch of bright orange on the forewing, is accompanied by a form of *Phrissura* (*P. sylvia*) showing the same characters of colour and pattern. In the Congo region we find a form of *Mylothris* (*M. asphodelus*) similar to that just mentioned, except that in the basal patch the orange is replaced by lemon-yellow; and from the same region comes *Phrissura perlucens*, in which exactly the same change has taken place. Tropical West Africa has a form of *Mylothris* (*M. bernice*) in which the patch of basal orange takes on a darker tinge and is somewhat modified in shape. In both these respects the *Mylothris* is followed by a form of *Phrissura* found in the same locality. Lastly, there are parallel pairs of the same genera, inhabiting respectively the same localities, which show a curious barring or striping of the marginal area, accompanied in one instance by a brown coloration of the forewing, affecting the representatives of both genera.

These instances — and it would be easy to multiply them — derive their principal interest from the special resemblances, often, to our view, minute, which obtain between pairs or assemblages of different species, and which change in an identical manner when we pass from one locality to another.

Taking a more general view, we cannot avoid noticing that certain distinct systems of colouring are broadly characteristic, though with modifications, of certain definite large areas of the earth's surface. Anyone, for example, seeing a Butterfly with a uniformly dark coloration, the forewing being crossed diagonally by a crimson band (as in the representations here shown of a *Heliconius*, *H. guaricus*, and two Pierines, *Pereute leucodrosime* and *P. charops*) would in nearly every case be right if be pronounced them to be natives of the Neotropical Province, that is to say, of Central or South America. So too, the general aspect of *Mylothris* is shared not only (as we have seen) by *Phrissura*, but also by members of several other distinct genera; but all these are African. Another very recognisable type of pattern is common to several species of *Danainæ* and the females of many species of *Nepheronia*; this type is found in the Oriental and Australian Provinces, but nowhere else on the globe. It is quite true that we come upon occasional instances of the occurrence of types resembling some of these local developments of pattern in far-removed regions of the earth's surface; but such case are very rare, and in most instances may in all probability be fairly put down as accidental. There is, for example, a curious little South American Nymphaline, *Cybdelis mnasylus*, which looks very much like a miniature version of the Indian *Hypolimnas bolina*. But the hardiest framer of theories would scarcely venture to suggest any special bionomic significance in a phenomenon of this sort. It may legitimately be set down as a coincidence. The case, however, is widely different when we contrast with sporadic occurrences such as this the enormous number of instances in which the forms that so closely resemble each other inhabit the same localities, the extensive « homœochromatic » combinations all changing together as one passes from one part to another of the same continent; and also when we consider the wide prevalence, throughout a given region, of a characteristic pattern like the dark ground-colour with a crimson band of Central and South America. The facts are undeniable; their interpretation may be in doubt, but to deny that there can be any underlying principle to regulate such phenomena as these would argue a scepticism so extreme as to pass the proper limits of scientific method.

Now let us turn to a fresh series of considerations. We have already noticed the fact of sexual dimorphism in its relation to the phenomena of mimicry. There is another kind of dimorphism, examples of which are not unknown among the Butterflies of temperate regions, though its full development must be sought in the tropics. I refer to the changes which are observed in successive generations of the same Insect in correspondence with the change of season, from hot to cold or from dry to wet. An instance of this seasonal dimorphism probably well known to all is furnished by the European *Araschnia prorsa-levana*, the spring and summer emergences of which Butterfly differ so completely in aspect that it seems at first sight impossible to believe that they can be conspecific. Equally strange instances abound in the tropics, and their number has within recent years been increased by the researches in especial of Mr. G. A. K. MARSHALL, who has proved by breeding that some of the forms of the genus *Precis*, most distinct from one another in aspect, are nevertheless related to one another as offspring to parent. One of the most remarkable of these instances, *Precis octavia-natalensis* and *P. sesamus*, is here shown on the screen. It has also been proved by the same indisputable evidence that, in many cases, forms of African *Pierinæ*, notably in the genus *Teracolus*, which had previously been described and named as distinct, were merely seasonal phases of the same species. In very many, probably most, of these examples of seasonal dimorphism as exhibited by tropical Butterflies, the dry season phase is far more closely assimilated in aspect to its inanimate surroundings than is the wet; in a few instances, while the dry season form is well concealed when reposing among dead leaves or on the ground, the wet season form of the same species is comparatively conspicuous, and bears more or less resemblance to another Butterfly of remote affinity. On the other hand there is a case where the model (*Mylothris agathina*) is sexually, not seasonally dimorphic. One of its mimics (*Belenois thysa*) is both seasonally and sexually dimorphic. The male of *Belenois thysa* copies the same sex of the *Mylothris* in both seasons, but much better in the dry season than in the wet; while the female *Belenois* is a close mimic of the female *Mylothris* in the dry season, but frequently departs altogether from its model in the wet.

We have learned then that in seasonal, as in sexual dimorphism, it may happen that one phase of the species may be mimetic and the other not.

Let us now turn to the consideration of the actual nature of the resemblances themselves. The outstanding feature which must strike everyone who gives them his attention is, that they are purely superficial. Take the case of *Limnas chrysippus*, the female of *Hypolimnas misippus*, and the *trophonius*-form of the female of *Papilio dardanus*, three Butterflies which we have already noted as presenting a remarkable and even deceptive likeness in general aspect. One of these Butterflies is a Danaine, another is a Nymphaline, and the third a Papilio. I need hardly remind any of my present audience that each of these groups is characterised by certain features, which are called « structural », belonging especially to the segments and appendages of the legs, and to the number and arrangement of the veins in the wings. But do we find any mutual approach in these structural particulars corresponding to the very striking assimilation in obvious aspect? We do not; and the same remark will apply to every one of the cases that we have had under observation. Not only in the instances of resemblance between Insects of different Orders, as between Hymenoptera and Diptera, but also where the affinity is much closer and the divergence in structure is comparatively slight, we never encounter the smallest indication that the process of assimilation involves anything but superficial and easily recognisable features. Less obvious external characters and all the details of internal organisation remain unaffected, except in so far as they may assist the superficial resemblance. If there is any significance at all in the phenomena under discussion, we seem led to the conclusion that they must stand in some relation or other to the faculty of vision.

Akin to the foregoing point is the fact that in the establishment of a mimetic resemblance, the same broad and visible effect is often produced by different means. It has been established, for example, that although certain South American Pierines, as we have seen, are excellent copies of the red, yellow and black Ithomiines and Heliconiines of the same region, the red and yellow pigments of the Pierines are chemically distinct from those of their models. A still more striking illustration of the same principle is due to an interesting investigation by Prof. POULTON. There is a large number of cases in which the resemblance is in great measure dependent on an acquired, or rather secondary, transparence of an originally opaque wing. It might have been expected that this quality of transparence had been in all cases brought about in the same manner, the visual effect being practically identical. But POULTON has shown that « whereas in the Ithomiines the transparence is due to an alteration in *shape* and diminution in *size* of the minute scales which normally clothe the wing, in the Pierines the same effect is produced by a mere diminution in size, the shape remaining unaltered. The Danaines [which enter into this combination] owe their transparence to a reduction in the *number* of the scales, not to any alteration in shape or in size; while in the associated Moths the effect results, not from any change in size, shape or number of the scales, but from the fact that the individual scales themselves become transparent, and are sometimes set up vertically, so as to let the light pass between them » (the Author, in « Nature » for October 31st, 1907, p. 675). Here then we have another proof that the assimilation does not extend further than to easily obvious features.

A further point that soon impresses itself upon the observer of the phenomena of mimicry is this: that the resemblances which present themselves to his notice differ widely among themselves in respect of completeness. In some instances the superficial likeness between two Insects is marvellously close, extending to the most minute particulars. This may happen even when the affinity between the two is remote. BATES was so much impressed with the excellence of the resemblance in some cases, that he speaks of « a minute and palpably intentional likeness which is perfectly staggering ». This phrase, especially the use of the word « intentional », is no doubt open to criticism ; but most students, for example, of the neotropical lepidopterous fauna will admit its virtual accuracy. In other cases the resemblance, though sufficiently arresting, is less exact. In a further series of instances a resemblance, while certainly present, may be of so remote a kind that opinions may legitimately differ as to whether it possesses any bionomic significance at all. Between the two extremes every degree of transition is found to exist. I show on the screen specimens of *Heliconius aranea* (underside), *H. leuce*, *H. alithea* (underside) and *H. galanthus* (underside), together with *Perrhybris lorena* ♀ (underside), *Pieris noctipennis* ♀ and *Pieris locusta* ♂ (underside). Here we see examples of resemblance between *Heliconius* and Pierine as to the significance of which I am quite prepared to find that different views might be taken, though I am myself for various reasons inclined to the opinion that the likeness is what BATES would have called « intentional ». Some, again, may be disposed to doubt whether the Danaine here exhibited (*Melinda formosa*) bears more than an accidental resemblance to this *Papilio* (*P. rex*). The individuals before us are, however, both males, and their respective females, though easily recognisable as each belonging to its own male, show a mutual resemblance which is really close. I may mention that both sexes of each species, with other most interesting forms, have been well figured in « Trans. Ent. Soc. Lond., 1906 », pl. XI, and by Mr. ELTRINGHAM in his fine work on African mimicry just published.

In many cases there exists a resemblance, not to any other Insect in particular, but to a group or assemblage in general. In all these instances, it is perhaps superfluous to mention, there is no necessary dependence on affinity. But that, as before suggested, the influence of affinity cannot be entirely ignored, we see from such an example as that of the African *Acræas*, many species of which are superficially so much like one another that it requires a skilled observer to distinguish between them. The same may be said of many of the Eastern *Euplœas*. Contrast this with a group of the European *Vanessas*. These are probably as nearly allied to one another as are the *Acræas* and *Euplœas*, but though presenting in common the characteristic *Vanessa* facies, they are distinguishable from one another at a glance. There is therefore in all probability some other factor at work in bringing about the resemblance between the members of these tropical groups besides that of mere affinity.

Certain other points remain to be noticed before we can be quite sure that we are in possession of all the data needful for an explanation. It is no doubt natural to enquire as to the comparative numbers of the various forms concerned. The answer here is perfectly definite; sometimes one of a pair, or several of an assemblage showing a common aspect, is much rarer than the rest; also it often happens that some one form of the combination is much more abundant than any other constituent of the association. But on the other hand there are plenty of cases in which most, if not all, of the mutually resembling forms are common. This fact was a great puzzle to BATES, as it plainly did not fit in very comfortably with his theory. On this point I shall have more to say before concluding.

Once more; we find that these mimetic assemblages or combinations, so to call them, are not sharply marked off from one another, but show frequent passages from one to another by almost imperceptible gradations. Take for instance such a series at that now shown on the screen, which might be considerably extended. The *Papilio* at the top (*P. iphidamas*) and the *Heliconius* at the bottom (*H. venusta*) are each of them members of a large mimetic association. The yellow patch on the forewing is common to both, though its shape and position on the wing show differences; in other respects the patterns exhibit much divergence. But the three intermediate Butterflies (*Euterpe approximata*, *E. bellona*, and *E. nigrina* [underside]), which are all Pierines, show an array of connecting links which enables us to pass by an easy gradation from one extreme of the series to the other. This is only a single example of a state of things, which is constantly to be met with in the lepidopterous fauna of tropical regions.

What then have we learned in the course of this brief survey? The points may be summed up as follows :

1. The cases of resemblance between distinct kinds of Insects are very numerous — too numerous to be accidental.

2. These resemblances are to a very great extent independent of

affinity. Some of the most striking are those between Insects of different orders.

3. They are peculiarly liable to occur in Insects of the female sex.

4. They are, speaking generally, found only between the inhabitants of the same region.

5. They may affect one phase of a seasonally dimorphic Insect differently from the other.

6. No structure or detail of organisation is involved in these resemblances except in so far as a modification therein may assist in producing a superficial likeness in aspect or behaviour.

7. In the production of these resemblances the same effect is often brought about by different means.

8. Every transition exists between a likeness, which is so remote as to be fairly disputable, and a resemblance, which may even deceive a skilled observer.

9. In some cases there may be great disparity in point of numbers between the forms linked together by community of aspect. In other cases the numbers may be nearly equal.

10. The combinations of two or more forms resembling one another are in many cases not isolated, but are often connected with other combinations by a more or less complete series of gradations.

So far we have been concerned with facts. What is to be said about their explanation? We have already seen that these cases of resemblance are too numerous to be reasonably considered accidental; moreover their evident relation with conditions of sex, locality and visibility seems of itself to forbid such an interpretation.

When we consider the fact of the limitation of a given system of pattern and coloration to a particular area of the earth's surface, and especially when we examine the changes that affect a mimetic assemblage in common as we pass from one portion to another of such an area, as in the series just now exhibited of successive modifications undergone by the same general type of coloration in the passage from Guatemala to Peru, we are tempted to conjecture that geographical conditions may have some bearing on the matter. We may remember that many arctic Animals are white, and that both Birds and Mammals inhabiting desert regions are frequently assimilated in colour to their sandy surroundings. But if we attempt to find in these circumstances an analogy with the phenomena under present discussion, we are at once confronted with difficulties that may well appear insuperable. The prevailing coloration of Animals that live amid snow and sand respectively is with high probability attributed, not to the direct influence of their surrounding conditions, but to the advantage they gain from concealment whether from enemies or from prey within their respective environments, their community of coloration being, to use Prof. POULTON's term, syncryptic. But though Mr. ABBOTT H. THAYER, who surveys the subject from the point of view of an artist, maintains that the variegated patterns of the Butterflies in question similarly aid concealment, I do not think that naturalists in general will find his arguments on this head convincing. At any rate his contention does not accord with my own experience in the tropics. But even if his theory be sound with regard to Butterflies, it will not account for the resemblance of a Moth to a Hornet, or of a two-winged Fly to a Carpenter-bee. It will scarcely be denied that both Hornet and Carpenter-bee are even aggressively conspicuous. And what are we to say to the case of a Locustid, which is, so to speak, painted to look like an Ant, or to that of a Membracid, which screens itself beneath a sculptured representation of a similar model? These are not cases of syncryptic modification; nor is it conceivable that the direct influence of external conditions, even if they are similar (which may be doubted), can impose a deceptive picture or piece of sculpture upon the body of an otherwise unaltered Insect. Take again the case of a Butterfly like *Papilio dardanus*, the subject of female polymorphism. Community of external conditions can scarcely be appealed to in order to explain the likeness between each form of the female and a distinct species of Danaine, when the individuals of the same brood of the *Papilio*, all presumably exposed to the influence of identical conditions, have diverged along these three or four different channels. Taking all the facts into consideration, we must, I think, conclude that the influence of a common geographical environment, whether its influence be directly or indirectly exercised, fails to explain the phenomena of mimicry.

A view that has often been put forward, and maintained with great ability, attributes these resemblances to internal causes, which compel various species to pass through similar phases of development. These phases, it is held, must from time to time coincide, and so we may get between distinct forms a correspondence in aspect, which will present the appearance of mimicry. As a rough illustration of this view we may suppose a series of kaleidoscopes, each furnished with a similar set of fragments of glass, and all undergoing rotation together. From time to time it may no doubt happen that the patterns shown by two or more of the instruments will practically coincide. But the application of any such principle to the phenomena of Insect mimicry is attended with serious difficulty. The cases to be explained are not scanty in number, but abundant. Then again, many of the resemblances occur, as we have seen, between Insects widely separated in point of affinity. Take the Lycoid assemblage that we have previously mentioned. Is it probable that Beetles, Braconids, Wasps, Bugs, Moths and two-winged Flies should all have been impelled by internal causes to reach the same stage of colour-development at the same epoch of their phylogenetic history? And if this be considered not impossible, why should the various members of this assemblage be all found together in the same place, many of them actually on the same tree? We have already given attention to the fact that the forms resembling each other are as a general rule inhabitants of the same localities. It is not by any means clear how this is to be explained on the theory of internal causes of similar development. It is true that, as we have seen, there are sporadic cases of resemblances that have to all appearance developed independently of one another. But why should they be so few in comparison with the enormous number of instances, which occur under the conditions of a common habitat? There is no apparent reason, under the theory of internal causes, why there should be any connection between the likeness and the locality.

Those of my audience who happen to be acquainted with my writings on this subject, will have anticipated the solution of the problem which I should myself favour. I should find myself in agreement with Mr. THAYER to the extent of believing, with him, that these resemblances are of service to the forms exhibiting them, and that their establishment and survival have taken place under the control of natural selection. But I cannot follow him in the opinion that all the patterns which we have been considering, and which are so widely adopted by Insects of such different affinities, are calculated to render their possessors invisible against their background. On the contrary, and I think the experience of most observers will here bear me out, it appears to me that the Butterflies, which exhibit these brilliant and variegated colours, are for the most part conspicuous on the wing. Moreover, many of them adopt a slow, deliberate mode of flight, which seems to court observation. This is certainly the case with members of the genera *Mylothris* and *Amauris,* and with several of the *Acræas*. We have now a good deal of evidence that some of these forms are unpalatable to certain Birds, and are at any rate not taken by preference.

Probably no form is absolutely immune; it should always be recognised that these matters are relative. But it seems to be clearly established by observation and experiment that some Birds at all events avoid some of these conspicuously marked Butterflies, and that there are various degrees of preference. Certain observers, it is true, have denied that Butterflies are fed upon by Birds at all, but there exists now a considerable body of evidence to the contrary. This being so, we are led to the conclusion that the brilliant colours of these Insects are, to use Prof. POULTON's term, « aposematic », that is to say that they are warning marks, which signify to insectivorous enemies such as Birds the presence of some quality whether of taste, or of odour, or of toughness, which makes their possessors unsuitable for food. If this conclusion is well grounded, we can find in the theories of BATES and of FRITZ MÜLLER a sufficient explanation of the significance of mimicry. BATES pointed out, just upon fifty years ago, that a palatable Insect might escape attack by sailing under the false colours of an inedible species, and he was followed about twenty years later by FRITZ MÜLLER, who called attention to the fact that if Birds had to pass through an education in order to learn by trial what Insects to capture and what to avoid, the combination of unpalatable Insects into mimetic associations would protect each constituent of such an assemblage from a certain amount of experimental tasting. It has been shown, chiefly by Prof. LLOYD MORGAN, that this education of young Birds in what to eat and what to avoid is a reality and no mere assumption; and the theory of FRITZ MÜLLER may thus be said to rest on a substantial foundation. The first of these theories, that of BATES, is the theory of what may be called true mimicry. That of FRITZ MÜLLER, as has been pointed out by Prof. POULTON, is more correctly designated as *synaposematism,* or the adoption by two or more forms of a common warning pattern.

Opinions may legitimately differ as to the relative importance of these two theories; and until more data are at our disposal, it will be possible to doubt as to which of them is applicable to this or that given instance. But the theories are complementary to one another, and not mutually exclusive. And it is to be observed that both of them, equally with that of THAYER, imply the preservation, by natural selection, of appropriate variations. Let us now see how far these theories are in accordance with the facts with which we started.

1. In the first place, it is obvious that the abundance of cases

constitutes no objection. If it be granted that the possession of a common pattern is advantageous, there is no reason why its adoption should not be of frequent occurrence.

2. Nor is the fact that the resemblances are largely independent of affinity adverse to the theories of BATES and MÜLLER. Natural selection will work upon any material that comes to hand, quite irrespective of its taxonomic relations.

3. That the female sex should be more liable to enter these associations is also to be expected. It is a matter of common observation that the female of many Birds and other Animals is better protected from attack by coloration and habits than the male; no doubt, as was pointed out by WALLACE, because the life of the female, as guardian of the future brood, is especially valuable to the species. Prof. POULTON has also drawn attention to the fact that the female, being in Butterflies often more subject to individual variation than the male, gives greater scope to the operation of natural selection.

4. The fact that the forms resembling each other are usually found together, finds a ready explanation; inasmuch as it implies that they have been exposed to the attacks of the same enemies. Otherwise, the adoption of a common aspect would carry no benefit.

5. With regard to seasonal dimorphism, it is generally found that the dry-season phases, which occur when Insect-life is scarce and competition among Insect-eaters is keen, are better protected than the wet-season phases of the corresponding species. Hence, we need not be surprised to find that in some cases the wet-season phase is mimetic, while the dry-season phase adopts for its protection what is probably the more efficient method of cryptic coloration. Nor, again, is it surprising that a Butterfly like *Belenois thysa*, which is mimetic in both seasons, becomes much more strongly so in the dry.

6. The fact that the changes from the normal are all in the direction of a resemblance that is merely superficial, is strongly in favour of the theories. For these superficial modifications are plainly an appeal to vision, and it is not easy to conjecture what alien vision can be of importance to these Insects, except the vision of their actual or potential enemies.

7. That the same apparent effect is often brought about by different means is quite characteristic of natural selection; which, as we have already seen, proceeds by adopting any means that offer, irrespective of affinity, homology, or any similar consideration.

8. The existence of every transition between resemblance which is practically complete and resemblance which is so slight as to be even disputable, is exactly comparable with what may be observed in other modes of protection; as for instance in cryptic assimilation to the ground, leaves, twigs, bark or other indifferent objects. These matters, as has been so often stated, are relative. Probably no means of protection gives absolute security, but different grades exist; as indeed we should expect on any theory of evolution. And it is often observable that where one kind of protection is feeble, it is compensated for by excellence in another method.

9. The fact that forms resembling each other may be severally common, is to some extent an objection to the application to such cases of the theory of BATES, which is usually considered to postulate the comparative scarcity of the mimic. It is, however, no obstacle in the way of synaposematism; for each accession of inedible individuals only tends to increase the common safety.

10. So too, the fact, that the associations are connected with one another by intermediates, is consonant with the theory of natural selection; for these gradational forms may be looked upon in effect as sign-posts showing the course which the evolutionary process has taken. Their survival is quite explicable on the Müllerian theory; for if themselves distasteful, each transitional form would be capable of sharing protection with the nearly resembling forms on each side of it, and thus would be established a chain of mutually protective links, reaching from one inedible assemblage to another.

I am not sanguine enough to suppose that everyone in my audience will agree with the interpretation of these phenomena, which I have ventured to advocate. I must be content with having tried to put the case of those theories, which seem to me to account for the facts better than any others that have yet been developed. And I would urge in conclusion, as I did at the outset, that the data, about which there should be no dispute, are interesting and curious in the highest degree. Any rival explanation, which neither neglects nor distorts the actual facts of the case, will deserve and, I am sure, will receive the closest attention of all scientific naturalists.

The Systematics of some Lepidoptera which resemble each other, and their bearing on general questions of Evolution,

by Dr. KARL JORDAN, Tring (Herts).

The subject matter of this address being given as including the systematics of certain Lepidoptera, I am afraid that some of the members of this Congress may have been frightened away.

Classification, as we know, has the reputation of being as dry as are our cabinet-specimens — if not mouldy — and of having an interest only for those who work at the special group of Animals classified. There are even biologists of fame who, in their misguided wisdom, scoff at systematics and look down upon this kind of work as more or less fruitless. As the Entomological Congresses should be and, I trust, will be supported to a very great extent by those who are occupied in describing Insects, unravelling their synonymy and life-history, and trying to bring the immense multitude of diverse forms into natural order, I take the opportunity which this First Congress of Entomology offers of stating emphatically that sound systematics are the only safe basis upon which can be built up sound theories as to the evolution of the diversified world of live beings.

In medical and economic entomology it is considered a matter of course that research must primarily be directed to studying the distinctions between the various possible forms that may be the cause of some certain disease in Animals or plants, so as to enable us to lay our hands on the actual culprit. Haziness in the discrimination of the more or less similar species would only lead to a useless slaughter of the innocents. The point is obvious. And what is true in the case of the practical branches of our science, is no less true also in regard to the philosophical treatment of biology in general. Although we admit that it may be utterly indifferent at the time to know whether some certain form of Insect is specifically distinct, it would be equally rash for the scientist to point to this or that group of Animals as a negligible quantity, or to this and that line of research as being futile. Science must not prejudge anything and must not be negligent anywhere. We can never know beforehand to what prominence in practical life or in philosophical theories any creature may attain.

The aim of the systematist as such is the establishment, on reasoned evidence, of the degree of relationship between the forms with which he is concerned, the evidence being furnished by the specimens and the bionomics of the species and varieties to which the specimens belong. He has to weigh similarities which may mean relationship or may not, and differences which may be only recent modifications or may point to the Insect having branched off at an early period. If a classifier tries to accomplish this difficult task, if he marches along this road so studded with obstacles, he is nothing less than a speculative philosopher who constructs a cosmogony *in parvo*.

The philosophic aspect of systematics is unfortunately much obscured, and it cannot be denied — no thoughtful systematist, at any rate, will do so I think — that this is due in a large measure to an unduly great importance being attributed to the mere giving of names. Nomenclature is the servant of science, but has in many houses the position of master. However, that is a subject which another Section of the present or a future Congress will have to consider.

My object in taking up your time is to demonstrate by a few selected instances that there is more in classification than lies at the surface, by proving to you, with the help of slides, that there exists a very intimate connection between, on the one hand, the vexed question of specific distinctness, and, on the other, the elucidation of the factors of evolution to which is due the great diversity of the animal world. The very title of the book which is the basis of modern biology, the *Origin of Species*, renders it evident even to those who do not claim a special knowledge in any branch of biological science, that the discrimination between species and non-species (= varieties) must play an important part in all research which aims at the explication of the origin of species. It is unnecessary to expand on that any further.

In our researches on the systematics of Lepidoptera we have come across some facts which, in being followed up, proved to be a striking illustration of the point to which I have just referred. There is a remarkable incongruity in the systematics of many not nearly related genera of Butterflies which resemble each other, and which, without prejudice, we shall call mimetic, as is customary. From what we know of the systematics of Heterocera, which resemble members of other orders of Insects, e. g. Hymenoptera, the principle involved holds good to a great extent also there. I shall abstain from entering at any length into the classification of the genera represented on the slides. The details may be relegated to another place. The tilling of the soil is a very important proceeding, but it would hardly be interesting for you to assist in the ploughing of the fields, and I propose therefore to take you right away to the harvesting of the modest crop.

At the outset let us be clear about two points :

1. Similarities are a reality; they exist. Systematists have over and over again been misled by resemblances. The study of resemblance is as much a necessity for the classifier, — who must always be on the alert that he is not taken in by what looks alike but is different, — as the knowledge of systematics is a necessity for the theorist. The similarity remains a reality also in those species which resemble each other much more when set out in a cabinet than the live individuals at large.

2. The agreements in aspect and in the details of the structure and colour between not closely related species require as much an explanation as the differences between nearly allied species.

Now, from a purely morphological point of view and quite apart from the question whether the resemblance serves any object or does not, — such as for instance protection, — the similarities may be divided into two groups : *a*) fictitious similarity, and *b*) genuine similarity.

We will deal with the sham similarity first, as it is an agreement with which the systematist as such is not concerned. As examples of this kind of resemblance we mention the following few cases : One of the females of the Philippine *Papilio rumanzovia* ESCHSCH. (1821) bears along the abdominal margin of the hindwing a bright red stripe which recalls the red abdomen of *Papilio semperi* FELD. (1861). The same phenomenon is observed in two American Butterflies, *Gnathotriche epione* GODM. and SALV. (1878) with a red abdominal wing-margin and *Actinote neleus* LATR. (1811) with a red abdomen. *Papilio laglaizei* DÉPUIS. (1877) has an orange patch on the underside of the hindwing, which corresponds to a conspicuous orange patch situated on the underside of the abdomen of *Alcidis agathyrsus* KIRSCH (1877), which closely resembles *Papilio laglaizei* on the upperside. The long antennæ of certain wasp-like Beetles and Orthoptera are black proximally and pale distally, the pale tip giving them the appearance of being short, like the antennæ of the Wasps. Certain stick-insects, when disturbed, stretch the forelegs straight forward with the head and antennæ lying between them, poise the body on the mid- and hindlegs and open the claspers, the Insect in this attitude looking uncommonly like a Lizard. And so on. Theorists, with the exception of those who hold to natural selection as a force in evolution, reject these similarities as mere accidents or analogies which as such do not require an explanation. The cases of false similarity are, however, far too numerous to be considered in the same spirit as the apple-peel which the maiden throws over her left shoulder.

The species which exhibit genuine (or homologous) agreement with other species may be grouped for our purpose in two sections :

1. A species resembles one other species : *Homotrope resemblance*;
2. A species resembles several diverse species : *Heterotrope resemblance*.

Although Dr. DIXEY has shown us so many beautiful slides illustrative of his address on Mimicry, our first slide depicting four specimens of *Planema* and four of *Pseudacræa* may not have lost its interest, it being among Butterflies one of the most startling cases of resemblance of which I know. The eight specimens represent four species, one species of each genus having the sexes practically alike, while the other species has them unlike (see pl. XXI, fig. 1-4 and 1*a*-4*a*). The four Insects occur together at the same time of the year in Uganda. We have only lately become aware of the existence of this association, the male of *Planema macarista* being generally confounded with that of *Planema poggei nelsoni* and the female of *macarista* with some other *Planema*. All four males and the two females *nelsoni* (fig. 4) and *hypoxantha* (fig. 4*a*) have a broad orange band on the forewing and a white band on the hindwing. The other two females, *macarista* ♀ (fig. 2) and *hobleyi* ♀ (fig. 2*a*), differ from them remarkably in having a white band on the forewing. Our reasoning mind can hardly be satisfied by attributing the coexistence of these four Insects in the same limited locality to coincidence. We require some better explanation. Evolutionists account for such resemblances in different ways, which are more or less antagonistic. Firstly, it is said that inherent or constitutional forces of development modify a species in a certain definite direction. These directions of development in the various species of a family or order are few in number. The species of the same group or family or order being descendants of a common ancestor, it must necessarily happen that the line of development followed by a certain organ, f. i. the wing-pattern, is the same in a number of species (*Parallel development*), or that this organ has become but very little modified. Examples : Neuration has on the whole remained of great uniformity among the Lepidoptera and exercises a decided influence on the wing-pattern. Some authorities consider the wing-pattern of certain mimetic females, for instance in *Papilio* and *Hypolimnas*, as a relic. Several Indian Papilios have preserved tails, whereas their representatives from the Moluccas or New Guinea have lost them. The same phenomenon is met with in some Papilios from South-Eastern Brazil as compared with the forms from the Amazons. Many African Butterflies have preserved a more ancestral pattern in East Africa than at the West Coast. On the island of Celebes the majority of Papilios as well as many other Butterflies have acquired a long forewing with a strongly arched costal margin. Among island Coleoptera derived from winged forms there are many which have lost the power of flight. The loss of power of flight is frequently accompanied in Moths by shagginess of the body.

Secondly, it is further maintained that development by constitutional forces may also lead to similarity in two originally different species, if the lines of development are convergent. The lines of development must either be parallel, divergent or convergent. Convergency of the lines must especially often obtain in those cases where evolution tends towards simplification. Examples : A longitudinally striped species and a transversely striped one are very different, but the pattern becomes similar in both when the stripes break up into spots. Originally more complicated patterns are frequently so much simplified that the Insect is almost uniform in colour, for instance all white or all black.

Now, if it were true that the line of development followed by a species only depended on the constitution of the Insect, a species could not split up into several divergently developing species. But apart from that, the hypothesis does not take account of many very striking facts of resemblance. It does not explain, for instance, why the American « mimetic » Papilios of the *lysithous*-group (distantly related to *Papilio podalirius*) are similar to American Aristolochia-Papilios, while the corresponding Old World swallowtails resemble Danaids; and why generally the « mimicking » species are similar to « models » which occur in the same faunistic district. This difficulty is overcome by the theory that, thirdly, evolution stands under the direct influence of the physical environment. Development, it is said, can be accelerated, retarded, stopped, reversed, and thrown out of its direction by the surroundings acting directly on the constitution of the species, e. g. climate, food, and physical and chemical properties of the soil. « Like cause like effect » explains the similarity in non-related species. Some of the above examples might be considered as evidence for the correctness of this theory. We mention further the numerous modifications which appear under the influence of the seasons, the frequently minute variation according to locality, the pale colour of cave Insects, the results of temperature experiments, and the often great similarity in the structure of parasites (e. g., *Pulicidæ*, *Nycteribia*, *Polyctenes*, *Platypsylla*).

Lastly, an explanation of the similarities we have under consideration is offered by the theory of *Mimicry*. Although the surroundings are also in this case the agents which bring about the resemblance, the theory stands nevertheless in sharp contrast to the one just mentioned. The similarity between the mimic and the model is the result of natural selection (by enemies, such as Birds and Lizards) having weeded out the less similar specimens, which tended to render the species more and more similar to its model

In our case the Planemas are the models and the Pseudacræas the mimics. If it is true that the Planemas are protected in some measure by inherent qualities, it can also not be denied that the Pseudacræas would profit to some extent by the resemblance they bear to them. All turns on the question whether the models are protected, a point which can only be solved by observation of live specimens and their natural enemies.

However, a pattern more or less similar to that of our two pairs of species of *Pseudacræa* and *Planema* occurs also elsewhere, e. g. among American Butterflies in *Adelpha* and *Apatura*.The oblique white band of the forewing found in the females of two of the species we are dealing with are of frequent occurrence in Africa, India and America, but a combination of this band with a white central band on the hindwing is rare. If such similar markings appear also elsewhere, however rarely, it becomes plausible that the two Pseudacræas may have acquired their similar pattern independently of the Planemas. It is therefore clear that the great similarity between the two Planemas and the two Pseudacræas is not conclusive evidence as to which theory offers the correct explanation of the resemblance.

Now let us consider homotrope resemblance between two polymorphic species. I have selected *Planema epæa* and *Elymnias phegea* from the Cameroons-Congo district. The specimens represent neither various geographical nor seasonal varieties. Several authors consider the tawny *Elymnias* (pl. XXI, fig. 5*a*) and the white one (pl. XXI, fig. 9*a*) to be two distinct species. However, a careful examination of a long series of specimens, including many different grades of intermediates, leaves no doubt whatever that all these differently coloured *Elymnias* are one single species. The agreement between the *Planema* and *Elymnias* is less close than between the Planemas and Pseudacræas figured (pl. XXI, fig. 1-4, 1*a*-4*a*); especially the intermediates are no very good counterparts. Now, both species have this in common that the first and the fourth figures (pl. XXI, fig. 5, 5*a* and 8, 8*a*) of each species represent in the Gaboon the usual forms. But there is this great difference between the two Insects : In *Planema epæa*, the « model », the male is always tawny in the Gaboon district, only the female appearing in several forms. None of the females are like the male, although the first and second female (pl. XXI, fig. 6 and 7) approach the male inasmuch as the light colour of the hindwing has the same extent as in that sex. The rarity of the ♀-form represented by figure 6 may be judged from the fact that the broken specimen figured is the only one which the Tring Museum has received. Each form of the *Planema* is unisexual. On the other hand, every form of the *Elymnias* appears in *both* sexes; the five specimens figured are males, each of which could be matched by a corresponding female.

If the similarity of the two species is to be explained by the assumption that there is an inherent tendency in them to develop along definite lines, which happen to be parallel or convergent, there is still the fact to deal with that the result in one Insect is polymorphism in *both* sexes, the male and female forms being identical, whereas in the other species polymorphism is confined to the female sex. It may be argued that the *Planema* with its monomorphic male is some steps in advance of the *Elymnias*, having lost all the male forms which were similar to the females. That is a quite plausible proposition in the case of the *Planema*, however much we may differ as to the causes which have brought about the result. But if the argument is applied to the *Elymnias*, it breaks down before the array of systematic facts. The proportional numbers of specimens of the various individual forms of *E. phegea* are opposed to the assumption that there is in this Butterfly a tendency of development which will result in the male becoming monomorphic and the female remaining polymorphic as in *Planema epæa*. The frequency, in the Gaboon, of the orange and the buffish forms in both sexes and the comparative scarcity of the others can only be interpreted as meaning that there is a tendency in *E. phegea* towards breaking up into two forms, of which each would still consist of *both* sexes. In *Planema epæa*, on the other hand, the disappearance of forms 6, 7 and 9 would result in the tawny colour being confined to the male and the buffish colour to the female, which means that the lines of development in the *Planema* and *Elymnias* are different. Hence the above assumption that the two Insects follow the same line of development, *Planema epæa* being some steps in advance of *Elymnias phegea*, cannot stand scrutiny. That this is not mere speculation is proved by the systematics of the two species, if the whole range of these Butterflies is taken into account. The result of evolution in *Planema epæa* just alluded to hypothetically is an actuality in the northern faunistic district of tropical West Africa, the coast region from Sierra Leone to the Niger. Here the female is always white-banded (like pl. XXI, fig. 9), the intermediates between this form and the tawny male being absent, or at any rate not known to me. In Abyssinia, on the contrary, the female is tawny like the male. On the island of Fernando Po, in the Bay of Benin, the male has a whitish subapical band like our female figure 6, and its female is like the male except for this band being pure white. It is evident that the environment plays an important rôle in this geographical variation, and if in various districts of West Africa we further find the changes in *Planema epæa* more or less faithfully repeated in *Elymnias phegea*, we might be inclined to draw the conclusion that the similarity between the two Insects is the effect of the direct action of the environment.

However, if this agreement is due to the physico-chemical action of the surroundings, it follows that the two Insects must be similar to each other wherever they exist side by side. Is that the case?

The systematics of *Planema epæa* tell us that this species is represented in Uganda by the geographical race *paragea* (pl. XXIV, fig. 30). In this *paragea* the sexes are similar to each other, the bands being whitish in both and so much reduced that the Insect is very different in aspect from the other geographical forms of *P. epæa*. If we now compare the Uganda form of *Elymnias phegea*, we see at once that the above conclusion was hasty. The bands in the *Elymnias* are not reduced as in *paragea*, but on the contrary, at least on the forewing, much wider than in the West African specimens of *E. phegea*. There is hardly any resemblance between the Uganda *Elymnias* and *Planema epæa paragea*. This remarkable discrepancy is readily explained by natural selection. In Uganda occur several broad-banded *Planemæ* which are much more frequent than *P. epæa paragea*, which is apparently a rare Insect, and the *Elymnias* has assumed the garb of these other *Planemæ*. But if natural selection easily surmounts this difficulty, it is faced by another. According to the systematist the bands of *Planema epæa* are tawny in both sexes in Abyssinia, reduced, diffuse and whitish in Uganda, tawny in the male and white in the female in tropical North-West Africa, etc. Can selection by enemies be a possible cause of these geographical changes in a *Planema* which is protected to a great extent by its relative unpalatability? Why were these changes at all necessary in a « model »?

If the systematist were wrong, if all these forms of *Planema epæa* and *Elymnias phegea* were distinct species, it would be much easier for the philosopher to explain their characters by his special theory. Let me reiterate at this point that I am not defending or attacking any one theory, but only endeavouring to demonstrate the great importance of correct systematics for all theoretical considerations.

We come now to the second kind of genuine resemblance, which we have termed *heterotrope*. Amongst the most frequently quoted cases of similarity are certain Papilios, e. g. the African *Papilio dardanus* and the Oriental *Papilio polytes*, whose polymorphic females resemble several other Butterflies. These Insects have given rise to quite a bulky literature in the pro and contra of natural selection. As Dr. DIXEY has already shown some of them, it is not necessary to bring them again before you; nor shall I enter at any length upon the arguments advanced to explain the phenomenon of polymorphism. But in passing I should like to mention three points :

It is occasionally stated that mimetic resemblance is as a rule confined to the female sex. That is a misconception which has arisen from the facts 1) that among the Lepidoptera more females are mimetic than males, and 2) that the male is very rarely mimetic if the female is non-mimetic. The cases in which the resemblance is confined to the female sex are less numerous than those in which both sexes are mimetic.

It is further maintained in this connection that for constitutional reasons (being another sex) the female occupies a lower phyletic stage than the male and hence has preserved a more ancestral coloration, which is said to explain many instances of resemblance. The weak point in this rule is that there are so many exceptions. The female is very frequently either in advance of the male or has a coloration which the male never possessed in the phylogeny of the species, especially females which are « mimetic ». The white females of *Papilio ægeus ormenus* GUÉR. (1829), for instance, represent a direction of development of their own and certainly not an ancestral stage of the species. The black female of *Papilio glaucus* L. (1758) shows at a glance that it is a derivation from that female which resembles the male, and therefore is younger than the male as regards coloration. The females of *Papilio dardanus* BROWN (1776) exhibit specializations in several directions which the male never possessed. The black female of *Papilio hectorides* Esp. (1794), the dark females of *Papilio macareus* Godart (1819), the females of *Papilio androgeos* Cram. (1775) and *P. lycophron* Hübn. (1818?), etc., may also be mentioned as examples.

However, it remains nevertheless true that in many instances the male of mimetic females is much more advanced than these on the road towards monochromatism, and that it has followed its own road unarrested. Considering that mimetic resemblance can never be a disadvantage, natural selection by means of predaceous enemies cannot give a satisfactory answer to the question, why and how the coloration of males like those of *Papilio polytes* and *ægeus*, which are non-mimetic and represent a late phyletic stage, has been evolved. The theory of evolution by constitutional forces appears to step naturally into the breach : the male, as the less important sex for the maintenance of the species, being left alone by natural selection, followed its line of evolution, whose course depended on internal factors aided and guided by the physical environment. On the other hand, the systematics of the polymorphic females of these same species offer unsurmountable obstacles for a similar explanation of their coloration. As the same point comes up again, we propose to deal with it later, and now proceed to consider some illustrations of heterotrope resemblance which obtains in *both* sexes.

The Oriental *Papilio clytia* L. (1756) is a well-known instance of that kind of similarity. The species is strongly dimorphic, one form being spotted with white (f. *clytia*, pl. XXIII, fig. 11*a*) and the other streaked with white (f. *dissimilis*, pl. XXIII, fig. 10*a*). The sexes are practically alike in each form.

The two forms are very sharply separated in markings in most districts, although each form varies considerably in many places. The specimens figured are from South India. In North India, however, specimens of the *clytia*-form occur which bear grey streaks extending down to the base of the wings, and are clearly a connecting link between f. *clytia* and f. *dissimilis*. The sharply cut dimorphism breaks down. The fact is instructive. In order to understand a species and its bearing on general questions, it is necessary for the philosopher as well as the systematist to study the Insect from all districts of its range. Our species illustrates admirably how different the variation may be in the geographical races of one Insect or, if you prefer to call them so, in the closely allied species which replace each other in different faunistic provinces. *Papilio clytia* is represented on the Andaman Islands by a monomorphic form resembling *dissimilis*, on the lesser Sunda Islands by another form also resembling *dissimilis*, and on the Philippines by a likewise monomorphic form which on the contrary corresponds to *clytia*, while in Ceylon, India, China, Siam, etc., both the *dissimilis*- and the *clytia*-form are found together. The two forms have long been considered by systematists as distinct species. But all the evidence derived from the cabinet specimens and obtained in the field point to these different looking Butterflies being the same Insect. The streaked *dissimilis* resembles a number of Danaids, while the dark *clytia*-form is similar to Euplœids.

On Borneo, Sumatra and Java, and extending northward to Palawan and Assam there occur two Papilios which were originally the Malayan representative forms of *P. clytia*, but are now specifically distinct from it. These are *Papilio paradoxus* ZINK. (1832) and *caunus* WESTW. (1848). The former resembles *Euplœa radamanthus*, and the latter, which varies to a great extent, mimics quite a number of different species of *Euplœa*. The two Papilios are in my opinion the same Insect. They are, however, so different in aspect that it will take a long time, I think, before systematists become convinced of the correctness of my interpretation.

The kind of similarity instanced by *Papilio clytia* has not received the attention it deserves, which may be due in a large measure to the circumstance that so few cases of this kind were known. The scarcity of instances, however, is more apparent than real. Faulty systematics must bear the blame. In descriptive entomology — and in other branches of biology as well — it has often been overlooked that the mere quantity of difference between two forms is not a safe guide in the discrimination of species. Polymorphism is much more frequent than we know as yet, especially in species which bear a likeness to members of other groups. Systematists at any rate learn this much from the theory of mimicry that they must be cautious in their judgment as to specific distinctness when dealing with mimetic forms. Erycinids, certain genera of Lycænids, Castniids, certain American Hypsids which resemble Danaids, etc., should be studied from this point of view.

The most striking heterotrope resemblance among all the Papilios is found in the Brazilian *P. lysithous*. The species consists of six well defined forms which occur regularly, and a seventh of which only two specimens are known and which may be regarded as an accidental colour monstrosity. Individuals which stand intermediate between any two forms are rare, but do occur. Systematists, until recently, have considered these Butterflies as being eight distinct species. The extremes are a variety with a broad white band : f. *platydesma* R. and J. (1906), and a variety without any band : f. *pomponius* HOPFF. (1866). The four other principal forms have a white band or patch on the forewing, the hindwing also bearing a white central area or being without it. The form which probably most nearly resembles the ancestral one is f. *lysithous* (fig. 31*a*). It has the widest distribution, and the pattern of all the other forms can easily be derived from that of f. *lysithous*. The six forms, however, cannot possibly be regarded as six phyletic stages of one straight line of evolution; they represent several lines of development.

All these varieties bear a remarkable likeness to Aristolochia-Papilios of the same country. In fact, the whole group of Papilios to which *lysithous* belongs was always classified with the Aristolochia-Papilios, until an ardent believer in mimicry, E. HAASE, discovered that the resemblance was merely superficial. We have here in *P. lysithous* a similarity between a single polymorphic species on the one hand and a series of distinct species as « models » on the other. The variability of *P. lysithous* is yet more complicated than it appears from these bare statements. There is a geographical element in the resemblance. The six main forms do not all occur together; only three of them may be expected to be met with in the same district. The forms with broad band and large submarginal spots are confined to the northern provinces of South-Eastern Brazil (from Bahia to Rio), and the black f. *pomponius* (pl. XXIV, fig. 33*a*) as well as f. *rurik* (pl. XXIV, fig. 32*a*) are found further south; but the exact limits of distribution are not yet known. Now, it is significant that the distribution of at least two of the « models » has somewhat the same limits as that of the « mimics ». *P. lysithous* f. *platydesma* flies in the province of Rio, which is also the habitat of its « model » *P. ascanius*, and *P. lysithous* f. *pomponius* occurs in Paraná and further south, where is also found its model *P. perrhebus* (cf. pl. XXIV, fig. 31-33).

The majority of species to which *P. lysithous* belongs exhibit a more or less close resemblance to Aristolochia-Papilios, many of the species being sexually dimorphic, or di- and polymorphic in both sexes, e. g. *Papilio ariarathes* Esp. (1788), *P. harmodius* Doubl. (1846), *P. thymbræus* Boisd. (1836), etc.

Among the American *Nymphalinæ*, there are some genera related to our Vanessas, whose species have a very close similarity to other Lepidoptera in many and most diverse directions. Among the species of *Phyciodes*, *Gnathotriche* and allies we find many which bear a confusingly near likeness in coloration to Pierids, or Neotropids, or Acræids. There are several instances of polymorphism and sexual dimorphism among them. The facies of some of the *Phyciodes* is so much *Acræa*-like that a few species have actually been described as *Actinote* (or *Acræa*) by a very able entomologist. The resemblance of the dimorphic species is heterotrope, one form resembling one species of *Actinote* and the other form another *Actinote*. Also in *Phyciodes* and *Gnathotriche* intermediate individuals are rare, as in the Papilios we have dealt with, while there are, of course, no intermediates between the distinct species of *Actinote* which act as « models ».

Parallel cases to those of *Phyciodes* and allies are comparatively rare among the Oriental *Nymphalinæ*, but Africa possesses in *Pseudacræa* a genus which exhibits heterotrope resemblance in an exceptionally high degree. Unlike *Phyciodes*, however, its similarity is directed only towards *Acræinæ*, especially the genus *Planema*. As *Pseudacræa* contains monomorphic and polymorphic species, and as these species are either non-mimetic, or resemble one model, or bear the facies of several specifically distinct models, the genus is as fascinating for the systematist as it is interesting for the exponent of Mimicry and his adversary. We have figured (pl. XXI, fig. 1*a*-4*a*) two *Pseudacræ*, which are each similar to one species of *Planema*. We will now deal with one dimorphic and one polymorphic species and the various distinct *Acræinæ* which they resemble.

Pseudacræa clarki Butl. (1892) occurs in two forms, one bearing the pattern of *Acræa orina* Hew. (1874) and the other, described as *P. landbecki* Druce (1909), being an almost exact counterpart of *Acræa egina* Cram. (1775). The two forms of *Pseudacræa* are connected by intergradations, while the two *Acrææ* belong to different sections of the genus. Nobody who has examined a series of specimens can seriously doubt that *P. clarki* and *landbecki* are individual forms of one Insect. However, systematists will be less willing, I presume, to agree with me in the following case of polymorphism, which is indeed startling enough to make anybody hesitate to believe in the correctness of the systematics as I present them here. We figure thirteen males, all from West Africa (pl. XXII). These specimens are considered to belong at least to seven distinct species. The result of my investigation is quite different; I must regard these Insects as forms of only one single species, *Pseudacræa eurytus* L. (1758). The specimens figured do not represent geographical races, although there may be this much geographical in them that only a certain number of the forms occur in the same limited locality. Along with these *Pseudacrææ* is placed a series of thirteen males of *Planema*, each *Pseudacræa* resembling more or less closely a *Planema*. The differences in colour between the specimens of *Planema* being no greater or hardly so great as between the various forms of *Pseudacræa*, it is startling to find that the *Planema* belong to no less than twelve distinct species.

The female sex of *Pseudacræa eurytus* is as variable as the male. Some forms agree with a ♂-form, while others have a white band instead of the orange one of the corresponding male. The females of *eurytus* resemble the females of the different Planemas. The white-banded females of *Planema* are so very similar to each other that systematists have repeatedly confounded them. Even AURIVILLIUS, in his great work on African Butterflies, was unable to distinguish the females of several species, owing no doubt to the insufficient material he had at his disposal. We figure five white-banded female specimens of *Pseudacræa eurytus* together with the females of five species of *Planema* (pl. XXIII).

The phenomenon also holds good in other districts where *P. eurytus* is found. In Uganda, for instance, this Butterfly is represented by a number of forms more or less different from the West African ones. For the sake of brevity we will refer only to two, which were described as distinct species, *terra* NEAVE (1904) (pl. XXIV, fig. 28*a*) and *obscura* NEAVE (1904) (pl. XXIV, fig. 30*a*). The one resembles *Planema epæa paragea* GROSE-SMITH (1900) (pl. XXIV, fig. 30) and the other *Planema tellus platyxantha* JORD. (1910) (pl. XXIV, fig. 28). We figure an intermediate specimen (pl. XXIV, fig. 29*a*) connecting the two *Pseudacrææ*; needless to say, there is no connecting link between the two *Planemæ*.

The sweeping statements in systematics here brought forward must necessarily arouse a certain amount of scepticism. My conclusions as to the systematics of the Insects dealt with are after all based only on the study of dry cabinet-specimens. It is therefore very natural to postulate for corroborative evidence from direct observation of the live Insects. The Butterflies must be bred before we have actual proof of the correctness of my contentions, and it is to be hoped that, now attention has been drawn to the great importance of the matter for classification as well as theory, residents in tropical countries will devote some energy to the life-history of « mimetic » species. Fortunately, we are not entirely without such evidence.

We have above mentioned the polymorphism of the Brazilian *Papilio lysithous*. Three of the forms of that species have been proved by breeding to belong together (pl. XXIV, fig. 31*a*, 32*a* and 33*a*). It is not surprising that they are individual forms of one species, since intermediates are known, although we have not seen every gradation from f. *lysithous* to f. *rurik* (pl. XXIV, fig. 31*a* and 32*a*). The three « models » are Aristolochia-Papilios of the same district, and are distinct species.

The second case, an African Nymphalid, is equally instructive. The four specimens represented by the figures are two species of *Amauris* (pl. XXIII, fig. 34 and 35) and the two individual forms of *Hypolimnas anthedon wahlbergi*, namely f. *wahlbergi* (pl. XXIII, fig. 35*a*) and f. *mima* (pl. XXIII, fig. 34*a*) all from Natal. The similarity between the *Amauris* (the models) and the *Hypolimnas* (the mimics) is very close, and the difference between the two *Hypolimnas* no less astonishing. As in *P. lysithous* each form appears in both sexes; the difference being neither sexual, nor seasonal, nor geographical. No intergradations are known between f. *wahlbergi* and f. *mima*, and the Insects have always been considered distinct species until both forms were bred from the same female. However, if we study *Hypolimnas anthedon* throughout its range, we find that the two West African forms which correspond to f. *mima* and f. *wahlbergi* are connected by a very rare *Hypolimnas*, which Staudinger described as *H. dæmona* and which Aurivillius treated with « ? » as a hybrid. This supposed hybrid of the supposed two distinct species is the intermediate, which is missing in the Natal forms.

The intermediate *dæmona* appears to be exceedingly rare, which may be interpreted as meaning : 1) that it is on the verge of disappearing altogether, and 2) that in Natal the corresponding intermediate formerly existed, but has now disappeared. The significance of the occurrence of intermediates as well as of their frequent rarity is evident. The absence of intermediates between the forms of di- or polymorphic species is now and again brought forward as an argument in favour of the origin of these individual forms *per saltum*. In a survey of all the polymorphic Butterflies, however, it will be found that the absence of intergradations is the exception rather than the rule.It must also be borne in mind that it is generally due to chance that scarce intermediates fall into the hands of the collector in the tropics and that they may be absent from the locality visited and not particularly scarce in another. The variability of a species, as every field-collector knows, varies in different stations, often considerably, and it depends much on accident whether one comes across the locality where a species exhibits the widest range of variation. The absence of intermediates from our collections is therefore no proof that such individuals do not exist. Moreover, the range of variability exhibited by the comparatively few specimens which come under the eyes of the systematist, is in many cases much inferior to that shown by the same species under abnormal conditions of life, as is proved by the results of temperature experiments.

As the intermediates are opposed to the assumption of the origin of the forms *per saltum*, it appears to me evident in view of the above facts that « mutations » are merely forms which appear separated by a gap because the intermediate characters have disappeared or are suppressed. Anyhow, the theory of discontinuous mutations does not apply to the phenomena we have under consideration, on account of the intermediates to which I have constantly drawn your attention in the course of this address. I may mention in passing that very instructive series of connecting links between the individual forms are known in the American *Papilio glaucus* and *Papilio hectorides*, the African *Papilio dardanus*, and the Oriental *Papilio rumanzovia*, and that there also occur intermediates between some of the female-forms of the Oriental *Papilio polytes* and *Papilio ægeus ormenus*. And it is perhaps also not superfluous to reiterate in this connection that the geographical races, too, are nearly always found to be connected by intergradations, if a sufficiently large number of specimens is examined.

Since the theory of abrupt mutations does not fare well here at the hands of the systematics of Lepidoptera, we have to fall back for an explanation of the phenomenon of heterotrope resemblance on the theories mentioned on pages 389 and 390. Let us shortly recapitulate what is required of the theories. *Several* distinct *species* as « models » occur side by side with a number of individual *forms* of a *single* species as « mimics ». The result of evolution is therefore very different in the « models » and « mimics », but there is nevertheless a close similarity in aspect between them. The theories have to answer the question, how has this resemblance originated?

Accident may play a rôle, but it is certainly a very insignificant rôle, since otherwise the instances of resemblance between Insects from *different* countries would not be so rare as they are.

Although nobody can possibly deny that the course which evolution takes in a species depends to a great extent on the constitution of the species itself, the theory of parallel or convergent development guided only by constitutional forces and only in one direction in a species (*Orthogenesis*) breaks down before the facts presented by systematics in the case of heterotrope resemblance. For the theory assumes : 1) that the individual form of the polymorphic species is as regards coloration a phyletic stage homologous to that occupied by the « model », which is erroneous, and 2) that the various individual varieties of a polymorphic species are stages of one line of development, which they are mostly not. Moreover, the theory does not explain the fact that these similarities, with few exceptions, occur in species which fly in the same district.

The chemico-physical influence of the environment may be accepted as sufficient to account for the prevalence of a certain tone of colour in the species of a faunistic district, which is a phenomenon of frequent occurrence. The continued action of the same environment may also tend to render a variable species constant or a constant species variable. But it is inconceivable that the same chemico-physical influence on a species can break the species up into definite dissimilar *forms* which resemble each a *distinct species* living under the same influence. The theory maintains that the result attained in the species which resemble each other is the same; in this the theory is wrong. The individual forms of one species and a series of distinct species represent vastly different results of evolution. Systematics teach us that individual forms like fig. 31*a*-35*a* have an utterly different standing in evolution from the species 31-35, the difference not being one of degree only but of kind.

While these theories cannot account for the great discrepancy in the systematics of the « models » and « mimics » except by falling back on accident, there is here no difficulty at all for the theory of natural selection. The theory offers a very simple explanation by assuming that the variable ancestor of the mimicking species has been gradually modified into a di- or polymorphic species by the weeding out of those intermediate forms which did not resemble protected species already existing in the country. The frequent rarity of intermediates is direct evidence for this theory.

However, if we look closer at this explanation, we observe that the theory of natural selection does not exclude the forces of evolution on which the other theories solely depend. The ancester of the mimicking species varied in certain *directions*, which depended on the *constitution* of the species. One or the other individual variety was *accidentally* somewhat similar to a protected species. The range of distribution of the incipient mimic may have been extended, and its range of variation become larger under the *influence* of the new *environment* or the individual varieties become intensified, so that natural selection had a better field on which to operate. If one looks at evolution from this point of view, the antagonism between rival theories loses much of its acuteness.

The systematics of the Butterflies in question teach us yet more. The figures of the di- and trimorphic species placed alongside the similar figures of distinct species seem to invite directly a certain line of thought. The figures appear to say this to us. The forms *mima* and *wahlbergi* of *Hypolimnas* (pl. XXIII, fig. 34*a*, 35*a*) — to restrict our remarks to one Insect for the sake of brevity — are still one and the same species, but if evolution proceeds further, they will attain to the same degree of diversity as the two *Amauris* which they resemble already in coloration (pl. XXIII, fig. 34, 35), i. e. the FORM *mima* will become SPECIES *mima*, and the FORM *wahlbergi* SPECIES *wahlbergi*. The occurrence of intermediate specimens in West Africa (and Madagascar as well) between the forms corresponding to *mima* and *wahlbergi* and the absence of such intermediates in Natal are evidence of progressive evolution in the Insect, and the evidence might be considered as warranting the conclusion that the progressive development will ultimately result in *mima* and *wahlbergi* being modified into distinct species. The conclusion is apparently sound. In fact, exponents of the various theories of evolution have arrived at it over and over again. And yet, the systematics of di- and polymorphic species entirely refute the conclusion. The various forms of which a species is composed within the same locality, however different they may appear, are not incipient species and will never become independent of each other as long as they exist side by side in the same district. There are two criteria of specific distinctness which admit of demonstration. Firstly, if two distinct species are crossed, they produce intermediates; individuals so sharply separated as *mima* and *wahlbergi* do not produce intermediates when crossed.

Secondly, distinct species as a rule are morphologically distinguished by certain differences in the reproductive organs. In the species of *Amauris*, *Planema*, etc., which we have figured such differences are found, whereas nothing of the kind is observed in the individual forms of *Hypolimnas*, *Pseudacræa*, etc. In fact, systematists have not come across a single case of individual forms of di- or polymorphic species exhibiting even rudiments of such distinctions (1).

If we wish to find rudimentary species, systematics tell us to turn our attention to those varieties which exist under different environments. The geographical races of Lepidoptera conform to the above criteria of specific distinctness in a lesser degree than species. They are the incipient species, and their systematics present us with all degrees of distinctness, from the mere slight shifting of the average to geographical forms which have all the morphological characteristics of species.

---

(1) There exists a slight difference in the ♂ genitalia of the seasonal forms of *Papilio xuthus*.

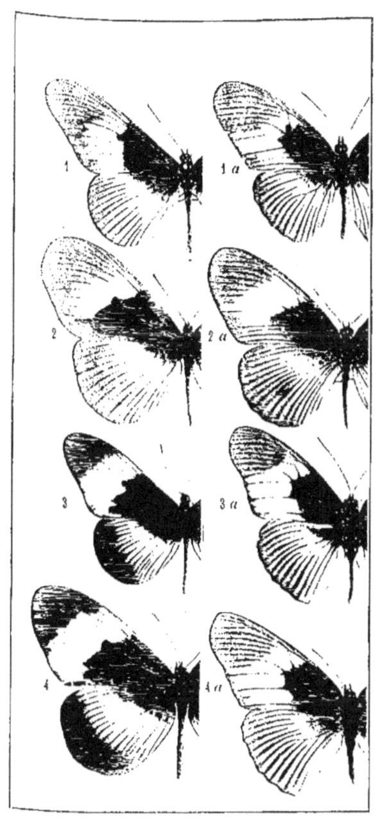

EXPLANATION OF FIGURES 1-4 AND 1a-4a; FROM UGANDA.

1. *Planema macarista* E. M. SHARPE (1906) ♂.
2 — — ♀.
3. — *poggei nelsoni* GROSE-SMITH (1892) ♂.
4. — — ♀.
1a. *Pseudacræa hobleyi* NEAVE (1904) ♂
2a. — — ♀.
3a. — *kuenowi hypoxantha* JORD. (1911) ♂.
4a. — — ♀.

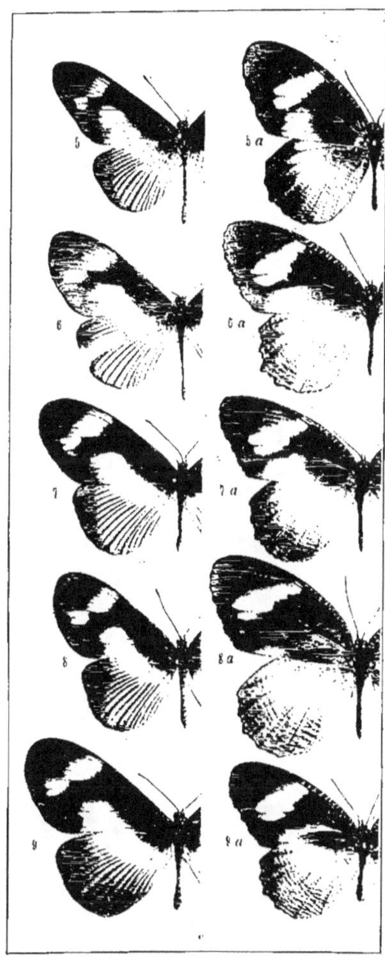

EXPLANATION OF FIGURES 5-9 AND 5a-9a.

5. *Planema epæa* CRAM. (1779) ♂ Cameroons.
6. — — ♀ Ituri forest.
7. — — ♀ Cameroons.
8. — — ♀ —
9. — — ♀ —
5a. *Elymnias phegea* FABR. (1793) ♂ Cameroons.
6a. — — ♂ Gaboon.
7a. — — ♂ —
8a. — — ♂ Congo.
9a. — — ♂ Cameroons.

Explanation of figures 12-22 and 12a-22a.

12. *Planema vestalis* Felder (1865). Gaboon.
13. — *umbra* Drury (1782). Gold Coast.
14. — *alcinoë* Felder (1865). Gold Coast.
15. — — — Sierra Leone.
16. — *elongata* Butl. (1874). Ogowe R.
17. — *consanguinea* Auriv. (1893). Congo.
18. — *excisa* Butl. (1874). Angola.
19. — *macaria* F. (1793). Sierra Leone.
20. — *pseudeuryta* G. and S. (1890). Mannyema.
21. — *tellus* Auriv. (1893). Cameroons.
22. — *epæa* Cram. (1779). Niger.

12a. Form of *Pseudacræa eurytus* L. (1758). Gold Coast.
13a. — — — Isubu.
14a. — — — Cameroons.
15a. — — — Isubu.
16a. — — — Gaboon.
17a. — — — Lambarene.
18a. — — — Gaboon.
19a. — — — Gold Coast.
20a. — — — Gaboon.
21a. — — — Cameroons
22a. — — — Niger.

Dr. K. JORDAN. — SYSTEMATICS OF SOME LEPIDOPTERA. — II.

EXPLANATION OF FIGURES 10-11 AND 10a-11a.

10. *Danaus aglea* CRAM. (1782).
11. *Euplœa coreoides* MOORE (1877)
10a. *Papilio clytia* f. *dissimilis* L. (1758).
11a. — f. *clytia* L. (1758).

EXPLANATION OF FIGURES 34, 35, 34a, 35a; FROM NATAL.

34. *Amauris albimaculata* BUTL. (1875)
35. — *niavius dominicanus* TRIM (1879).
34a. *Hypolimnas anth. wahlb.* f. *mima* TRIM. (1869).
35a. — — f. *wahlbergi* WALLENGR (1859).

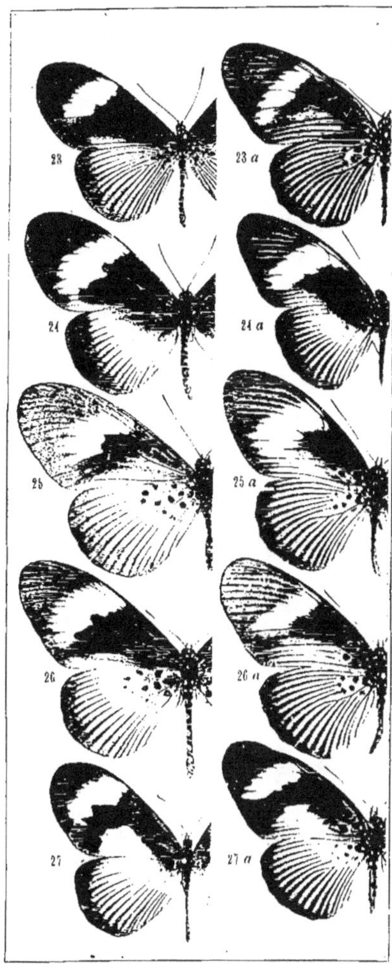

EXPLANATION OF FIGURES 23-27 AND 23a-27a.

23. *Planema epiprotea* BUTL. (1874). Cameroons.
24. *macarista* E. M. SHARPE (1906) Cameroons.
25. — *macarioides* AURIV. (1893). Cameroons.
26. — *excisa* BUTL. (1894). Cameroons.
27. — *tellus* AURIV. (1893). Cameroons.
23a. Form of *Pseudacræa eurytus* L. (1758). Gaboon.
24a. — — — Gold Coast.
25a. — — — Lambarene.
26a. — — — Gaboon.
27a. — — — Niger.

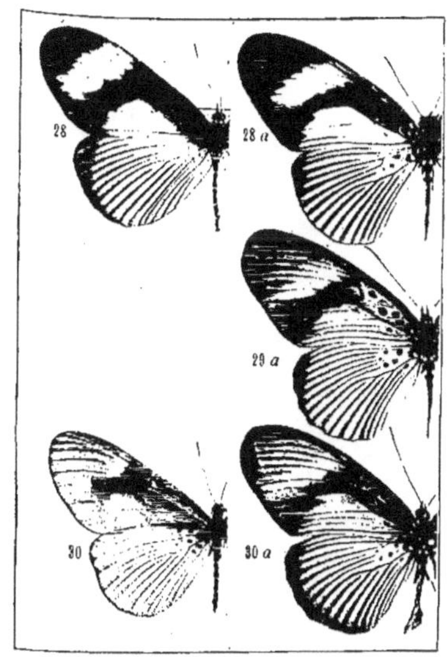

EXPLANATION OF FIGURES 28-30 AND 28*a*-30*a*; FROM UGANDA.

28. *Planema tellus platyxantha* JORD. (1910).
29.
30. — *epæa paragea* GROSE-SMITH (1900).
28*a*. *Pseudacræa eurytus* L. f. *terra* NEAVE (1904).
29*a*. — intermediate.
30*a*. — f. *obscura* NEAVE (1904).

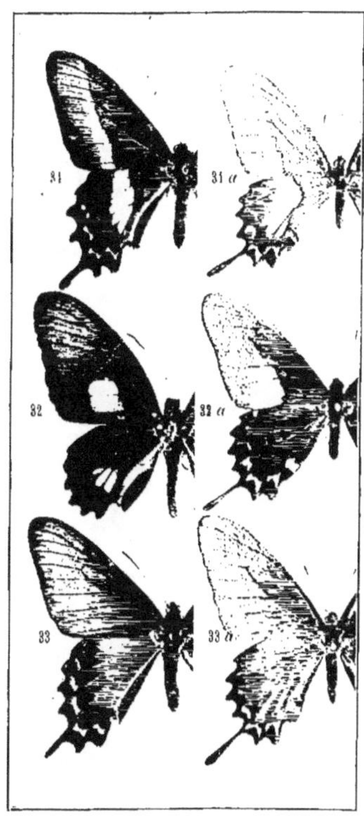

EXPLANATION OF FIGURES 31-33 AND 31*a*-33*a*; FROM RIO GRANDE DO SUL.

31. *Papilio nephalion* GODT. (1819).
32. — *chamissonia* ESCHSCH. (1821).
33. — *perrhebus* BOISD. (1836).
31*a*. *Papilio lysithous* f. *lysithous* HÜBN. (1822).
32*a*. — f. *rurik* ESCHSCH. (1821).
33*a*. — f. *pomponius* HOPFF. (1866).

Systematische Uebersicht der Monomachiden,

von W. A. SCHULZ (Villefranche-sur-Saône).

Der erste Versuch einer kritischen systematischen Bearbeitung der mit Ausnahme einer (australischen) Art im tropischen Amerika beheimateten kleinen Schlupfwespengruppe der Monomachiden wurde von SCHLETTERER unternommen. Seine Schrift darüber erschien, wie hier beiläufig bemerkt sei, Mitte März 1890 und nicht schon 1889, wie stets zitirt wird; es wäre denn, dass die Sonderabdrücke davon bereits umso viel früher verausgabt wurden. SCHLETTERER entgingen die wichtigen Unterscheidungsmerkmale, die in der Bildung der Oberkiefer und des Kopfschildvorderrandes liegen, fast ganz; andererseits hat er aus Unkenntnis ihrer Typen die vor ihm von PERTY, WESTWOOD und BRULLÉ errichteten, nur durch dürftige Beschreibungen belegten Arten meist unrichtig aufgefasst. Somit ist seine Arbeit jetzt, nach einundzwanzig Jahren, kaum mehr brauchbar. Indessen liegt es mir fern, dem verstorbenen österreichischen Auktor dafür einen Vorwurf machen und seinem Andenken nahetreten zu wollen, denn die erwähnten Mängel könnten in letzter Linie gar wol auf Ungenügendheit des ihm zur Verfügung gestandenen Materials bezw. auf die Seltenheit der Monomachiden in den Sammlungen zurückzuführen sein. Wie es in diesem letzten Punkte bestellt ist, lässt sich ersehen, wenn hier erwähnt wird, dass in den nachfolgenden naturhistorischen oder zoologischen Museen, die ich daraufhin untersucht habe, von Monomachiden nur die daneben verzeichnete Zahl von Exemplaren vorhanden ist :

London (British Museum)	4
Oxford	4
Paris	1
Brüssel	7
Copenhagen	7
Berlin	9
Stettin	2
München	16
Genf	1
Wien	8
Budapest	2
Turin	1
Genua	1.

Sonstige Museen oder Privatsammlungen, in denen sich Vertreter der genannten Wespen vorfinden, sind mir nicht bekannt geworden. Die ältesten entomologischen Schriftsteller scheinen noch keine Monomachiden gekannt zu haben ; von LINNÉ bin ich dessen nicht gewiss, da mir von seinen Hymenopterentypen bis jetzt lediglich die wenigen, im Universitätsmuseum zu Upsala aufbewahrten Stücke zu gesicht gekommen sind, wol aber von GEER, THUNBERG und J. C. FABRICIUS, deren typische Immenexemplare ich vor zwei Jahren eingehend untersuchen konnte.

Es war klar, dass, wer sich daran begeben wollte, die von SCHLETTERER offengelassenen Lücken auszufüllen und so die Monomachiden-Systematik weiterzuführen, vor allen Dingen trachten musste, die Typen und Originalexemplare sämtlicher in dieser Gruppe errichteter Formen nachzuprüfen, um sich über deren Verwandtschaft und gegenseitige Abgrenzung Rechenschaft geben zu können. Nachdem es mir nun in den letzten acht Jahren vergönnt gewesen ist, nach und nach in den verschiedenen Museen alle typischen und sonstwie authentischen Stücke aus der genannten Familie einzusehen und mir sorgfältige Notizen darüber zu machen — mit Ausnahme von *Monomachus ruficeps* CAM. (nec BRULLÉ), von dessen Originalexemplar ich nicht habe erfahren können, wo es aufbewahrt wird —, biete ich im folgenden einen neuen Gattungen-, Arten- und Unterartenschlüssel und knüpfe daran kritische systematische und biogeographische Bemerkungen. Es kann sein, dass ich dabei in der Ausdeutung mancher Formen vielleicht ein wenig kühner vorgegangen bin, als es das kleine, zurzeit auftreibbare Studienmaterial zuzulassen scheint. Wenn sich also später etwa der eine oder andere Fehler in meiner Arbeit ergeben sollte, so bitte ich, eben in anbetracht der geringen, dazu verfügbar gewesenen Individuenzahl, um Nachsicht. Immerhin ist jetzt ein Anfang gemacht, auf dem sich weiter bauen lässt.

Die Stellung der Monomachiden im Immensysteme hat seitens der Schriftsteller manchfache Deutungen erfahren. Lange gab es nur eine Gattung : *Monomachus* KLUG; die erste daraus bekannt gewordene Art (*fuscator*) hatte ihr Auktor PERTY selbst in das Genus *Pelecinus* LATR. verbracht. WESTWOOD hob 1844 (« Trans. entom. Soc. London », III, p. 252 [nicht 1843]) gleichfalls die Aehnlichkeit mit *Pelecinus*, aber auch zugleich wieder die grosse Verschiedenheit davon hervor. Von BRULLÉ wurde 1846 (« Hist. nat. Ins. Hymén. », IV, p. 533-535) ebensowenig eine bestimmte Meinung über die Verwandtschaft von *Monomachus* geäussert und nur im allgemeinen auf die Aehnlichkeit in gewissen Punkten mit *Aulacus* JUR. und *Pelecinus* hingewiesen. WESTWOOD teilte 1851 (« Trans. entom. Soc. London », [2] I, p. 217) mit, ERICHSON hätte *Monomachus* (und *Pelecinus*) als zu den Serphoiden (« Proctotrypoiden ») gehörend betrachtet. Von P. CAMERON ward dann *Monomachus* 1887 in der « Biologia Centrali-Americana » zu den Stephaniden verbracht. SCHLETTERER erklärte sich 1890 (« Berlin. entom. Zeitschr. », XXXIII, p. 197) gegen eine etwaige Vereinigung mit den Evaniiden, weil der Hinterleibstiel von *Monomachus* nicht, wie bei letztgenannter Familie, oben am Mittelsegmente, nächst dem Hinterrücken, sondern hinten am Mediansegmente, beim Ursprunge der Hinterhüften, angesetzt sei, und glaubte vorläufig einer Einreihung jener Gattung bei den Peleciniden das Wort reden zu sollen. 1893 gab ASHMEAD (« Bull. U. S. Nat. Mus. », N° 45 [« A Monograph of the North American Proctotrypidæ »], p. 345) *Monomachus* als wahrscheinlich zu den Belytiden, innerhalb des grossen Familienkomplexes der Serphoiden (« Proctotrypoiden »), gehörig aus mit dem Hinzufügen, dass, da er dieses Genus nur aus der Beschreibung kenne, ihm ein endgültiges Urteil darüber nicht zustehe. Neun Jahre später finden wir es dann in seiner Arbeit : « Classification of the pointed-tailed Wasps, or the superfamily Proctotrypidæ » (« Journal of the New York Entomological Society », vol. X, p. 243, Dec., 1902) wirklich unter die Serphoiden versetzt, aber nicht in die Familie der Belytiden, sondern in die der Heloriden, wo es zusammen mit der Gattung *Ropronia* (nicht *Roptronia*) PROVANCHER die Unterfamilie I. *Monomachinæ* bildet, die in Gegensatz zur Unterfamilie II. *Helorinæ* gebracht wird. Hatte ASHMEAD inzwischen Exemplare von *Monomachus* eingesehen, und war er dadurch zu dieser Meinungsänderung gelangt? Man weiss es nicht.

DALLA TORRE vereinigte in volumen III, pars II, p. 1088 (1902) seines « Catalogus Hymenopterorum » die soeben erwähnte Gattung mit *Pelecinus* LATR. als Familie *Pelecinidæ* HALID. Einer anderen Auffassung folgt ein Jahr später SZÉPLIGETI; er erklärt (« Annales Musei Nationalis Hungarici », I, p. 364) die Verwandtschaft von *Monomachus* mit den Evaniiden, wenn man die Hauptmerkmale (die Form des Mittelsegments [« Metanotums »], die Insertionsweise des Hinterleibes und die ausgebildete Subcostalzelle) inbetracht ziehe, für viel näher als die mit den übrigen Hymenopterenfamilien und ordnet unter Errichtung einer zweiten Gattung : *Tetraconus* (ebenda p. 389), die *Monomachinæ* als Subfamilie III den Evaniiden unter, die nach ihm die weiteren Subfamilien : I. *Gastrhyptiinæ* (verbessert aus « *Gasteruptioninæ* »), II. *Evaniinæ* und IV. *Aulacinæ* umfassen. 1904 äusserte KIEFFER in ANDRÉS « Species des Hyménoptères d'Europe & d'Algérie », tome VIIbis, p. 471, seine Ansicht dahin, dass *Monomachus* näher mit den Peleciniden verwandt wäre. Der gleiche Standpunkt findet sich in meiner Abhandlung : « Beiträge zur näheren Kenntnis der Schlupfwespen-Familie *Pelecinidæ* HAL. » (« Sitzungsberichte der mathematisch-physikalischen Klasse der königl. Bayerischen Akademie der Wissenschaften », Band XXXIII, Heft III, 1903, S. 435-450, mit Figur 8 auf Tafel I) vertreten. Jene Gattung erscheint dort mit *Pelecinus* unter der Familienbezeichnung *Pelecinidæ* zusammengefasst.

Heute nun, wo über meiner intensiven Beschäftigung mit den mehrfach angezogenen verschiedenen kleinen Schlupfwespengruppen Jahre verstrichen sind, glaube ich nicht mehr an eine wirkliche nahe Verwandtschaft von *Monomachus* (und *Tetraconus*) mit den nunmehr nur durch *Pelecinus* vertretenen Peleciniden (oder mit den Evaniiden). Die Aehnlichkeit mit diesen zwei Familien inbetreff einzelner Körperteile ist vielmehr eine rein äusserliche, hingegen bieten sich bei näherer Untersuchung zu viele und zu grosse Verschiedenheiten dar, als dass eine Einverleibung in eine von beiden zu rechtfertigen wäre. Das abweichende Vorhandensein von Adern im Hinterflügel bei *Monomachus* mag, weil dies bloss einen ursprünglicheren Zustand verrät, für die Beurteilung nicht ins Gewicht fallen, hingegen ist der Verlauf des Vorderflügelgeäders von dem bei den Peleciniden und Evaniiden sehr verschieden, wenngleich es sich schliesslich in unserer Gattung durch das Verschmolzensein der 1. Cubital- und 1. Discoidalzelle erklären lässt. Vorder- und auch Hinterleib sind bei den Peleciniden anders gebaut; die darauf bezüglichen Unterschiede zwischen ihnen und *Monomachus* habe ich bereits 1903 auseinandergesetzt. Was die Evaniiden anbelangt, so ist bei diesen noch namentlich der Hinterleib ganz anders geformt.

Die Stephaniden entfernen sich morphologisch von der ihnen in dem einen oder anderen Punkte gewiss äusserlich ähnlichen *Monomachus*-Gruppe zu sehr, als dass es nötig erschiene, in eine Erörterung der einzelnen Differenzen einzutreten. Bleiben die Heloriden und Belytiden übrig. Diese zwei Familien besitzen nun allerdings, was zum Vergleiche in erster Linie mit Anlass gegeben haben mag, wie die Monomachusse einen kürzer oder länger gestielten Hinterleib, aber der Stiel ist immer anders geformt als in der zuletzt genannten Gruppe, und vollends ist dies der Rest des Abdomens. Ferner sieht das Flügelgeäder der verglichenen beiden Familien auf den ersten Blick dem von *Monomachus* sehr ähnlich, bei näherer Untersuchung ergeben sich aber Verschiedenheiten. Weder Heloriden noch Belytiden haben keulenförmige Hinterschienen, und auch sonst sind der Abweichungen von der uns hier beschäftigenden Wespengruppe so viele, dass es nicht lohnt sie aufzuzählen. *Ropronia* PROVANCHER, auf eine nordamerikanische Art gegründet, kenne ich in Natur nicht, jedoch was in den davon vorhandenen Beschreibungen gesagt wird, schliesst die Unterbringung dieses Genus bei *Monomachus* aus. Da es endlich keine andere Hymenopterenfamilie gibt, in die *Monomachus* ungezwungen eingereiht werden könnte, so erübrigt nur, auf ihm und dem damit eng verwandten *Tetraconus* SZÉPL. eine besondere Familie : *Monomachidæ* m. zu errichten.

Ueber einige Namen aus der typischen Gattung sei an dieser Stelle folgendes bemerkt : *Monomachus apicalis* (KLUG) WESTW., *M. falcator* (KLUG) WESTW. und *M. gladiator* (KLUG) WESTW. sind blosse nomina nuda, die WESTWOOD 1844 anführte. Sie verdienen umso weniger Beachtung, als ihre Typen weder im Berliner Museum noch sonstwo mehr vorhanden sind. *M. gladiator* KLUG (nec BRULLÉ) könnte wegen der gleichen Herkunft aus Bahia vielleicht später die Type von *M. glaberrimus* SCHLETT. geworden sein.

Zitate werden hierunter nur bei Formen gegeben, die im DALLA TORREschen Hymenopterenkataloge noch fehlen.

Schlüssel für die Gattungen :

Kopfschild und Schläfen unbewehrt. (Australien und Tropisch-Mittel- und Südamerika. Mehrere Arten und beide Geschlechter bekannt) . . . . *Monomachus* KLUG (1841)

Kopfschild mit zwei seitlichen Hörnern. Schläfen an der Unterseite mit einem grossen, zapfenförmigen Höcker. (Zurzeit nur eine Species aus dem mittleren Küstenbrasilien und bloss im ♀ bekannt) . . . . *Tetraconus* SZÉPL. (1903)

MONOMACHUS

Arten- und Unterartenschlüssel für die bis jetzt bekannten ♂♂ :

1. Bewohner Australiens. (Kopfschildvorderrand etwas vorgezogen, aber mitten schmal abgestutzt. Flügel allenthalben glashell) . . . . *M. antipodalis* WESTW. (1874)

In Neotropien heimisch 2

2. Vorderflügel an der Spitze mit einem deutlichen rauchbraunen Fleck. (Kopfschild am Vorderrande mitten in einem nasenartigen Zahn endigend. Querspalte am Grunde der Oberkiefer schmal; die übrige Aussenfläche der Oberkiefer so ziemlich flach. Mittelsegment auf der Scheibe runzlig punktirt)
M. segmentator segmentator WESTW. (1841)

Vorderflügel durchweg glashell oder ausnahmsweise (*M. glaberrimus* SCHLETT.) an der Spitze undeutlich angeraucht . . . 3

3. Kopfschildvorderrand in der Mitte mit einem leichten, nasenartigen Vorsprung. Vorderflügelspitze undeutlich angeräuchert.

Mittelsegmentscheibe glänzend glatt. (Grundspalte der Mandibeln schmal, deren übrige Aussenfläche fast eben)

M. glaberrimus SCHLETT. (1890)

Kopfschildvorderrand mitten fein dreieckig ausgerandet, nur ausnahmsweise (*M. fuscator viridis* STADELM.) gerade abgestutzt. Vorderflügel allenthalben glashell. Mittelsegmentscheibe runzlig punktirt 4

4. Oberkiefergrundspalte schmal und ziemlich flach

M. lateralis WESTW. (1841)

Mandibelspalte breit und tief (*M. fuscator* [PTY., 1833]) . . . 5

5. Kopfschild vorn gerade abgestutzt. Körper hellgrün (lauchgrün), mit rotbraunen Fühlern und vom 2. Ringe an ebenso gefärbtem Hinterleibe. Vorkommen : Rio de Janeiro

M. fuscator viridis STADELM. (1892)

Kopfschild vorn in der Mitte mit zwei Zähnchen, beiderseits davon eingebuchtet. Körper unten und seitwärts schmutzig weissgelb, auf der Oberseite schwarz- bis hellbraun. Verbreitung : Callanga (1500 m hoch) in der Provinz Cuzco, Peru; Rio Beni in Bolivia; Chile. . . . . . *M. fuscator andinus* subspec. nov.

Arten- und Unterartenschlüssel für die ♀♀ :

1. Australien. Hinterleibstiel in der Anlage gerade, nur schwach gekrümmt. (Kopfschild vorn in der Mitte weit dreieckig vorgezogen und in einen nasenartigen Zahn auslaufend, jederseits daneben ausgebuchtet. Flügel gänzlich glashell)

M. antipodalis WESTW. (1874)

Neotropien. Hinterleibstiel stets deutlich nach unten gebogen 2

2. Vorderflügel an der Spitze mit einem rundlichen, scharf abgegrenzten rauchbraunen Fleck 3

Vorderflügel gänzlich glashell oder an der Spitze bloss schwach beraucht 7

3. Aussenseite der Oberkiefer neben der Grundspalte entweder zapfenförmig aufgehoben oder blasenförmig angeschwollen . . 4

Aussenseite der Mandibeln neben deren Grundspalte nahezu eben 5

4. Oberkiefer an der Aussenseite flach, neben der schmalen Grundspalte etwas zapfenförmig erhoben. Bogenförmige Ausrandung am Kopfschildvorderrande verhältnismässig schmal und flach. Gesichtkiel weniger deutlich. Körper gleichmässig rotbraun gefärbt . . . . *M. Klugi* WESTW. (1841)

Oberkiefer aussen neben der Basalspalte blasenförmig angeschwollen. Kopfschild vorn in der Mitte breiter und tiefer bogenförmig ausgerandet. Mittellängskiel des Gesichts deutlich entwickelt. Vorderkörper obenauf mit der Neigung schwärzlich zu verdunkeln, unten sich gelb aufzuhellen
M. megacephalus SCHLETT. (1890)

5. Mandibelspalte verhältnismässig etwas breiter. Kopfschildmittelteil vorn zwar auch ein wenig vorgezogen, aber dort mitten ausgebuchtet . . . . *M. bicolor* SZÉPL. (1903)

Mandibelspalte schmal. Kopfschild in der Mitte des Vorderrandes in einen starken Zahn vorgezogen (*M. segmentator* WESTW. [1841]) 6

6. Körperoberseite allenthalben dunkel. « Brasilien », Bogotá in Colombien . . . *M. segmentator segmentator* WESTW. (1841)

Kopf und Bruststück rotgelb (gelbbraun). Insel Cayenne, Obidos am Nordufer des unteren Amazonenstroms
M. segmentator gladiator BRULLÉ (1846)

Kopfoberseite nicht verdunkelt. Panama
M. segmentator Cameroni SCHLZ. (1907)

7. Oberkiefer im Basaldrittel stark hornartig aufgetrieben, davor (nach der Basis hin) der Länge nach dreieckig eingedrückt, worauf

erst die (schmale) Grundspalte folgt. Kopfschild in der Mitte des Vorderrandes mit leichtem, nasenartigen Vorsprung. Vorderflügel an der Spitze sehr schwach rauchig getrübt. Mittelsegmentscheibe polirt glatt . . . . *M. glaberrimus* SCHLETT. (1890)

Oberkiefer aussen neben der Grundspalte nahezu eben. Kopfschild in der Mitte des Vorderrandes mit zwei Zähnen bewehrt. Vorderflügel durchweg glashell. Mittelsegmentscheibe mehr oder weniger dicht runzlig punktirt 8

8. Mandibelspalte ziemlich schmal. Kopfschildzähne kräftig. Erstes Tarsenglied der Hinterbeine ein wenig länger als die Summe der drei folgenden Fussglieder . *M. variegatus* SCHLETT. (1890)

Oberkieferspalte breiter. Zähne am Kopfschildvorderrande winzig. Erstes Tarsenglied der Hinterbeine ungefähr so lang wie die drei folgenden Fussglieder zusammen (*M. fuscator* [PTY., 1833]) . 9

9. Kopf wie der übrige Vorderkörper dunkelbraun. Rio de Janeiro . . . . *M. fuscator fuscator* (PTY., 1833)

Kopf rot gefärbt. São Paulo in Südbrasilien
M. fuscator ruficeps BRULLÉ (1846)

Kopf wie auch der Thorax, einschliesslich der Beine, und das Mittelsegment, in grösserer oder geringerer Ausdehnung hellgrün (lauchgrün) gezeichnet. Rio de Janeiro
M. fuscator viridis STADELM. (1892)

Kopf und Thorax unten und an den Seiten blassgelb, oben dunkelbraun bis pechschwarz. Callanga in Peru; Rio Beni in Bolivia; Chile . . . . . *M. fuscator andinus* subspec. nov.

TETRACONUS

Einzige nur im ♀ beschriebene Art : *Tetraconus Mocsáryi* Szépl. (« Ann. Mus. Nat. Hungar. », I, p. 389, 1903).

Bemerkungen zu den einzelnen Arten :

1. Ueber die aus Melbourne stammende Type von *Monomachus antipodalis* Westw., ein ♀, im zoologischen Universitätsmuseum zu Oxford berichtete ich in meiner Abhandlung : « Alte Hymenopteren » (« Berlin. entom. Zeitschr. », LI, p. 309-310, Textfiguren *1a*, *1b*, *1c*, *1d*, 1907 [« 1906 »]). Seither konnte ich von derselben Species im kgl. zoologischen Museum in Berlin noch 1 ♂ und 1 ♀ von Westaustralien (Preiss.) und 1 ♀ aus Port Philipp in Victoria (Coulon) untersuchen. Schletterer verzeichnete 1890 als ihre sonstigen Wohngebiete « Ost-Australien » und « Süd-Australien ». Sie scheint demnach über den grössten Teil des australischen Festlandes verbreitet zu sein, wenngleich sie lokal auftreten mag. — Das bisher unbeschriebene ♂ ähnelt dem ♀, abgesehen natürlich von den durch das Geschlecht bedingten Unterschieden, und ist am Kopfschildvorderrande ebenfalls etwas vorgezogen, jedoch dort mitten schmal abgestutzt.

2. *Monomachus segmentator* Westw. Ueber die spezifische Identität der Type (♂) dieser Art im Britischen Museum mit *M. gladiator* Brullé (1846) habe ich mich bereits in der zuletzt angezogenen Arbeit p. 310-312 (Fig. 2) verbreitet und dort auch (p. 311) den von mir schon früher (« Sitzungsber. Akad. München » XXXIII, p. 447, 1903) als Unterart davon erkannten *ruficeps* Cam. (1887, nec Brullé, 1846) in *Monomachus segmentator Cameroni* umbenannt. Der gleichzeitig erwähnte *Monomachus apicalis* (Klug) Westw. wird besser unterdrückt, weil ja das betreffende ♀ im Londoner Museum kein authentisches, sondern nur ein von unbekannter Hand als fraglich gleich dieser i. l.-Form Klug-Westwoods bezeichnetes ist; die wirkliche Type von *apicalis*, die nach Westwood (1844) das Berliner Museum bergen sollte, ist, wie wir weiter oben sahen, bereits verschwunden. — In der Zwischenzeit hat sich mir Gelegenheit geboten, das typische ♀

von *Monomachus gladiator* BRULLÉ im Turiner zoologischen Museum, speziell in der dort gesondert aufgehobenen Hymenopterensammlung weiland SPINOLAS (aus der ehemaligen coll. SERVILLE), nachzuprüfen. Es ist in den plastischen Merkmalen wirklich gleich *segmentator*, in Hinsicht auf die Zeichnung weicht es aber hiervon ab, indem sein Körper gelbbraun gefärbt und nur am Hinterleibe vom 2. Segmente ab stellenweis schwarzbraun überwaschen ist. Danach kommt es denn mit meiner Auffassung (1903) von BRULLÉS Form nach 3 ♀♀ im Museum München überein und bildet eine besondere Subspecies : *M. segmentator gladiator* BRULLÉ ; hingegen ist *M. gladiator* SCHLETT. (1890), wie schon 1903 von mir hervorgehoben wurde, davon verschieden und gehört als ♀ zur typischen Unterart : *M. segmentator segmentator* WESTW. Ich sah ein Originalexemplar von « *gladiator* » SCHLETT. im Wiener Hofmuseum : es ist ein ♀ von « Brasil. » (WINTHEM), das entgegen meiner Angabe (1903) scheinbar zwei Zähne am Kopfschildvorderrande besitzt. — Zu der typischen Subspecies von *segmentator* ist nun als Synonym auch noch nach dem Befunde an der Type im Berliner Museum *Monomachus pallescens* SCHLETT. zu schlagen. Diese Type, ein ♂ von Bogotá in Colombien (LINDIG Sammler), besitzt die charakteristische schmale Querspalte am Grunde der Oberkiefer und einen vorn dreieckig oder nasenartig vorgezogenen Kopfschild. Der *M. pallescens* hingegen, den ich in beiden Geschlechtern 1903 in « Sitzungsber. Akad. München », XXXIII, p. 444-446 gekennzeichnet habe, stellt etwas davon gänzlich Verschiedenes, nämlich eine Unterart in dem Formenkreise des *M. fuscator* (PTY.) vor, worüber weiter unten näher berichtet werden wird. — Fassen wir die soeben geschilderten Verhältnisse kurz zusammen, so gewinnen wir für die Subspecies von *Monomachus segmentator* WESTW. folgendes Synonymenbild :

M. segmentator segmentator WESTW. = *M. gladiator* SCHLETT. (nec BRULLÉ) = *M. pallescens* SCHLETT. (nec SCHLZ.)

M. segmentator gladiator BRULLÉ = *M. gladiator gladiator* SCHLZ. (nec SCHLETT.)

M. segmentator Cameroni SCHLZ. = *M. ruficeps* CAM. (nec BRULLÉ).

Von *M. segmentator gladiator* BRULLÉ sah ich neuerlich im

Provinzialmuseum zu Stettin 1 ♀ aus Obidos am Nordufer des unteren Amazonenstroms, IV.-V. 1906 (HOFFMANNS leg.).

3. *Monomachus glaberrimus* SCHLETT., über den sein Auktor nur die vage Heimatangabe « Brasilien » machte, weicht von den übrigen *Monomachus*-Arten mit ungefleckter bezw. unbedeutend verdunkelter Vorderflügelspitze durch glatte Mittelsegmentoberfläche, ebensolchen Scheitel, Mittelrücken und Schildchen ab. Nach der Type im Berliner Museum, einem ♀ von Bahia (GOMEZ) sei die Urbeschreibung in folgenden Stücken ergänzt bezw. berichtigt : Oberkiefer im Basaldrittel stark hornartig aufgetrieben, davor (nach der Basis hin) der Länge nach dreieckig eingedrückt, worauf erst die (schmale) Grundspalte folgt. Mittelsegment im Gegensatz zu SCHLETTERERS Angabe an den Seiten dicht und fein gerunzelt. Aeusserlich ähnelt dieses Tier dem später aufzustellenden *M. fuscator andinus*. — Ein besonders kleines ♂ von nur 7 mm Länge im Museum Oxford, das ebenfalls aus Bahia (REED, HIGGINS, 1867) stammt, bin ich geneigt, zu *M. glaberrimus* zu ziehen. Bei ihm ist die Anräucherung der Vorderflügelspitze wenig angedeutet und der Hinterleibstiel mässig gekrümmt. Sonst stimmt es mit dem ♀ überein bis auf den wesentlichen Unterschied, dass die Mandibeln des ♂ am Grunde zwar auch nur ein schmales, spaltförmiges Loch aufweisen, dass sie auf der übrigen Aussenseite aber fast eben sind. Nervulus im selben Geschlechte schwach postfurcal.

4. Von *Monomachus lateralis* WESTW., der bisher viel Kopfzerbrechen verursacht hat, ist im Berliner Museum glücklicherweise noch die Type vorhanden. Es handelt sich um ein ♂ von « Brasil. (C.) », SELLOW Sammler : Kopfschild vorn in der Mitte fein dreieckig ausgerandet. Oberkiefergrundspalte schmal und ziemlich flach. Stirn, Scheitel und Mittelsegment dicht und tief runzlig punktirt ; Gesicht und Mittelrücken mit feinerer und etwas weitläufigerer Punktirung. Lichte (weissliche) Zeichnung auf die Oberkiefer, zum Teil Wangen und die Hüften und Schenkelringe der Vorder- und Mittelbeine beschränkt. Beine sonst ins Braune ziehend. Vorderflügel durchweg glashell. — Ob « Brasil. (C.) » etwa Zentralbrasilien bedeuten soll, ist nicht ersichtlich gewesen ; indessen weiss man, dass SELLOW vor 80-90 Jahren das südliche und mittlere Küstenbrasilien bereist hat. — *Lateralis* hat bis auf weiteres als eigene Art zu gelten, da er sich auf keine

andere der bekannten *Monomachus*-Formen beziehen lässt. Von *M. segmentator segmentator* WESTW. weicht er durch gänzlich verschieden gebildeten Clipeusvorderrand und völlig glashelle Flügel, von *M. fuscator* (PTY.) in dessen verschiedenen Formen durch schmale Oberkiefergrundspalte ab. Meine 1903 geäusserte Mutmassung, dass er mit *M. fuscator andinus* (« *M. pallescens* ») zusammenfalle, verliert also jeden Halt.

5. *Monomachus fuscator* (PTY.) habe ich in « Sitzungsberichte Akademie München », XXXIII, p. 443, 1903 nach Untersuchung von PERTYS Type im Münchener Museum für artgleich mit *M. ruficeps* BRULLÉ (1846) erklärt. Nach Ansicht des typischen ♀ von *ruficeps* BRULLÉ im Museum Paris muss ich dieser durch den Besitz eines schön rot gefärbten Kopfes ausgezeichneten Wespenform immerhin den Rang einer eigenen Subspecies zuerkennen, deren Wohngebiet der Staat Sâo Paulo in Südbrasilien ist. Im Gegensatze zu *M. fuscator ruficeps* konnte für *M. fuscator fuscator* in « Berliner entomologische Zeitschrift », LI, p. 312, 1907 erstmalig ein scharfer Fundort in Rio de Janeiro fixirt werden. Soweit war die Kenntnis von diesem Formenkreise vorgeschritten, als mir im Berliner Museum die Type von *Monomachus viridis* STADELM. (1892) zuhänden kam. Es ist dies ein ♂ aus « Brasil. SELLOW, letzte Sendung », an dem der Hinterleib vom 2. Segmente an und die Fühlerspitzen bereits abgebrochen sind. Hellgrün, mit bräunelnden Fühlern, Schienen und Tarsen. Kopfschild vorn gerade abgestutzt, mitten nicht längseingesenkt. Mandibeln eben, mit breiter Spalte an der Basis. Hinterleibstiel gerade. Nachdem so die Frage nach grünen Monomachussen einmal aufgeworfen war, kamen deren in den Sammlungen noch mehr zum Vorscheine. Da sind zunächst drei ♀♀ von *fuscator* im Wiener Museum, wovon zwei von SCHLETTERER als *ruficeps* BRULLÉ benannte, eins von « Brasil. » (WINTHEM), das andere ebenfalls von « Brasilien » (NATTERER leg.). Dieses letzte Exemplar hat aber grüne Färbung am Kopfe und an den Beinen und beiläufig einen grossen, dreieckigen Zahnvorsprung in der Mitte des Kopfschildvorderrandes und dürfte das ♀ zu *viridis* STADELM. vorstellen. Weiterhin traf ich im zoologischen Universitatsmuseum zu Copenhagen eine schöne Reihe von *fuscator* an : 1 ♀ von « Brasilia » (Mus. DREWSEN), 2 ♀♀ von Rio de Janeiro (REINHARDT leg.) und 2 ♂♂ 2 ♀♀ von Rio de Janeiro (Mus. DREWSEN). Die ♀♀ haben grüne Vorder- und Mittelbeine und verraten überdies

an den Hinterbeinen und am Kopfe sowie am Grunde des Bruststücks die Neigung zu grüner Färbung. Der Kopf ist bei ihnen hinter den Augen mässig dicht punktirt und der Hinterleibstiel stark nach unten gebogen. Die ♂♂ sind ganz grün bis auf die rotbraunen Fühler und den vom 2. Ringe an ebenso gefärbten Hinterleib; sonst stimmen sie in den plastischen Merkmalen so ziemlich mit den ♀♀ überein, namentlich auch in der Länge des 1. Hintertarsengliedes, doch ist ihre Punktirung am Hinterkopfe weitläufiger und der Hinterleibstiel länger und gerader. Fühler 14-gliedrig (der Schaft als ein Glied gerechnet). — Ein weiteres *fuscator* - ♀ aus Rio de Janeiro (CAMILLE VAN VOLXEM leg.) im naturhistorischen Museum zu Brüssel weist gleichfalls etliche grüne Färbung am Kopfe, Thorax und an den Beinen neben dem schwarzbraunen Grundtone auf, und bei noch einem ♀ derselben Art im gleichen Museum (coll. BALLION, e coll. PULS), das keine Vaterlandbezeichnung führt, überwiegt sogar das Grün am Vorderkörper bis zur Basis des Hinterleibstiels hin. Alle diese mehr oder weniger grünen Stücke beiderlei Geschlechts in den zuletzt genannten drei Sammlungen gehören nun zu STADELMANNS Form *viridis*, die nach meiner Auffassung lediglich eine Subspecies in dem Lebenskreise des *fuscator* bildet. Das Vorkommen von *M. fuscator fuscator* und *M. fuscator viridis* am selben Orte (Rio) kann hiergegen nicht geltend gemacht werden, denn bekanntlich sind die Verbreitungsbezirke von Tierunterarten nicht immer scharf gesondert, sondern greifen öfter übereinander.

Schon weiter oben wurde erwähnt, dass der von mir in «Sitzungsberichte Akademie München», XXXIII, p. 444-446, 1903 als *pallescens* SCHLETT. in beiden Geschlechtern beschriebene *Monomachus* mit dieser SCHLETTERERschen Form nichts zu tun hat und in Wirklichkeit als Subspecies zu *M. fuscator* (PTY.) gehört. Indertat sind an den betreffenden Exemplaren alle für *fuscator* charakteristischen plastischen Merkmale wahrzunehmen. Verschieden, und zwar erheblich verschieden, ist nur der Färbungs- und Zeichnungsstil, wegen dessen und wegen der so gut gesonderten geographischen Verbreitung längs der südamerikanischen Andenkette, die Errichtung einer Unterart vollauf gerechtfertigt erscheint. Diese mag von jetzt ab den veränderten Namen :

Monomachus fuscator andinus subspec. nov.

führen. Ausser den typischen Exemplaren im Münchener Museum (von Callanga in Peru) lernte ich nachträglich davon ein ♀ vom Rio Beni, La Paz-Reyes, Bolivia (BALZAN leg. 1891) im städtischen naturhistorischen Museum in Genua kennen. Ferner steckt im Museum Oxford unter dem i. l.-Namen « *valdivianus?* » ein *Monomachus* ♀ aus Chile (REED) von 15.5 mm Körperlänge, mit deutlich postfurcalem Nervulus, glashellen Flügeln und schwach gebogenem Hinterleibstiel, das ich gleichfalls damit identifizire, ebenso wie zwei ♀♀ im Britischen Museum aus dem nämlichen Lande, von je 15 mm Länge, bei denen jedoch das Gesicht und die Oberseite von Kopf und Bruststück abweichend grossenteils weisslichgelb geblieben sind. Alle drei chilenischen ♀♀ unterscheiden sich überdies von den typischen peruanischen ♀♀ dadurch, dass bei ihnen die Fühler nicht einfarbig pechschwarz, sondern gegen das Ende rotgelb sind. Es muss einstweilen dahingestellt bleiben, ob dies nicht etwa eine besondere chilenische Teilform ist.

6. *Monomachus Klugi* WESTW. repräsentirt eine durch grosse, flache Mandibeln ausgezeichnete gute Art, deren Unterschiede von dem nahestehenden *M. megacephalus* SCHLETT. die voraufgegangene Tabelle zeigt. Ausser WESTWOODS Type im Berliner Museum, einem ♀ von « Brasil. » (GERMAR), und einem zweiten ♀ ebendort, ohne Herkunftbezeichnung, sah ich ein mit diesen Exemplaren übereinstimmendes ♀ mit leider der gleichen vagen Heimatangabe « Brasilien » im Museum Stettin ein. Ferner zwei ♀♀ im königlichen Museum Brüssel (coll. BALLION, e coll. PULS), wovon eins auch keinen präziseren Fundort, das andere überhaupt keinen solchen trägt. — *Klugi* ist eine der grössten Arten dieser Gattung — die Brüsseler ♀♀ messen 18 und 19 mm Länge —, mit eher noch breiterem und mächtigerem Kopf als *megacephalus* SCHLETT. und durchweg rotbrauner Körperfärbung. Ihre Thorax- und Mittelsegmentseiten zeigen einen feinen Belag dichter und anliegender, glänzend weissgrauer Flaumhaare. Der Verbreitungsbezirk von *Klugi* ist noch festzustellen.

7. Meine Beschreibung von *Monomachus megacephalus* SCHLETT. in « Sitzungsberichte Akademie München », XXXIII, p. 446-447, 1903 deckt sich mit dem nachträglichen Befunde an SCHLETTERERS typischem ♀ im Genfer naturhistorischen Museum (aus « Brésil ») ausser etwa in dem unbedeutenden Punkte, dass der Nervulus an dieser Type nicht interstitiell, sondern leicht postfurcal steht. Die damalige Beschreibung wäre höchstens dahin zu ergänzen, dass der Ausschnitt am Vorderrande des Kopf-

schildes verhältnismässig breit und tief ist, breiter und tiefer als bei *M. Klugi* WESTW. ♀. — 1903 konnte ich Rio de Janeiro als ersten sicheren Fundort von *M. megacephalus* SCHLETT. bestimmen. Seither sind mir noch im Museum Brüssel 2 ♀♀ dieser selben Species vorgekommen, die der früh verstorbene belgische Entomolog CAMILLE VAN VOLXEM ebenfalls bei Rio gefangen hatte. Dass gerade von Rio de Janeiro die meisten süd-amerikanischen *Monomachus*-Arten bekannt geworden sind, erklärt sich wol dadurch, das solche dort bisher am meisten gesammelt wurden.

8. Mit der Type von *Monomachus bicolor* SZÉPL. (« Ann. Mus. Nat. Hungar. », I, p. 388, 1903) im Budapester Museum, einem ♀ aus Espirito Santo, einem Küstenstaat des mittleren Brasilien (O. STAUDINGER & A. BANG-HAAS Verkäufer, 1898), gelangt man nach meiner Tabelle vom Jahre 1903 auf *M. segmentator* (*«gladiator»*) *gladiator* BRULLÉ, aber SZÉPLIGETIS Art ist hiervon verschieden durch die etwas breitere Spalte am Grunde der Mandibeln und durch die Form des Kopfschildes, dessen Mittelpartie am Vorderrande zwar auch ein wenig vorgezogen, aber mitten ausgebuchtet, also nicht in einen Zahn verlängert ist. Ferner ist bei ihr der Hinterkopf nicht polirt glatt, sondern fein runzlig punktirt, und ihr im übrigen glänzend glatter Mittelrücken zeigt einige grobe Punkte. — Aussenfläche der Oberkiefer von *bicolor* so ziemlich eben, mit kräftigen, aber mässig dicht stehenden Punkten besetzt. Dasselbe gilt von der Punktirung des Gesichts, das seinerseits den üblichen Mittellängskiel nicht entbehrt. Stirn knitterig-quergerunzelt. Schläfen glatt, mit wenigen Pünktchen. Längseindruck am Hinterkopfe verwischt. Abstand der hinteren Nebenaugen von den Netzaugen etwas grösser als voneinander. Vorderrücken und Schildchen polirt glatt. Mittelbrustseiten fein und engstehend punktirt. An den Hinterbrustseiten ist die Punktirung kräftiger und gedrängter, und sie fliesst hier runzlig zusammen. 1. Tarsenglied der Hinterbeine gleichlang der Summe der drei folgenden Fussglieder. Nervulus im Vorderflügel nahezu interstitiell. Mediansegment dicht und stark knitterig gerunzelt. Hinterleibstiel deutlich, aber nicht übermässig stark gebogen und länger als der folgende Ring. — Körper rotgelb (gelbbraun) gefärbt, nur Hinterleib und Fühler schwarzbraun. Vorderflügel mit grossem, rauchbraunen Spitzenfleck.

9. *Monomachus variegatus* SCHLETT. : die Type im Wiener Hofmuseum, ein schon vielfach zerbrochenes und von Raubinsekten angezehrtes ♀ von « Brasil. » (SCHOTT leg.) hat in Wirklichkeit am Hinterkopfe deutliche und tief reingestochene Punkte, nahezu glatten Vorderrücken und dicht und grob runzlig punktirtes Mittelsegment. Die beiden Vorderrandzähne des Kopfschildes sehr deutlich ausgeprägt, also nicht « kaum bemerkbar », wie es in SCHLETTERERS Urbeschreibung heisst. Drei weitere ♀♀ derselben Art im gleichen Museum, aus Rio Grande do Sul im äussersten Süden Brasiliens (STIEGLMAYR leg.), sind frischer, und bei ihnen stehen die Punkte des Hinterkopfes dichter, und sie sind auch enger aneinandergerückt; ausserdem zeigt bei einem der Riograndenser die hintere Hälfte des Vorderrückens oben am Seitenrande Längsrunzlung. Durch letztes Merkmal und durch die dichte Runzlung des Mittelsegments dieser drei STIEGLMAYRscher Stücke ist *Monomachus eurycephalus* SCHLETT. gegeben, der demnach lediglich ein Synonym von *M. variegatus* SCHLETT. darstellt. — Die Mandibelspalte ist bei dieser Species tief und mässig breit, die übrige Aussenfläche der Oberkiefer eben, mit groben Punkten sparsam besetzt. — Unmittelbar nach Aufnahme obiger Notizen in Wien kam ich nach Berlin und besah dort die im zoologischen Museum verwahrte Type von *M. eurycephalus* SCHLETT., ein mit « Brasil., SOMMER, No. 11551 » etikettirtes ♀, das sich denn auch tatsächlich als identisch mit *M. variegatus* SCHLETT. ♀ erwies. Die vom Auktor namhaft gemachten Unterschiede sind winzig und wertlos, insbesondere ist an der Type von *eurycephalus* das Mittelsegment in Wirklichkeit dicht und grob querrunzlig punktirt. *Variegatus* gilt, weil dieser Name sieben Seiten früher veröffentlicht wurde. — Ein ♀ von *M. variegatus* aus « Brasilien » (coll. BALLION, e coll. PULS) besitzt auch das Museum Brüssel. *M. segmentator segmentator* WESTW. und *M. fuscator andinus* SCHLZ. gleichen dieser Art zum Verwechseln, die Aehnlichkeit betrifft aber nur die Färbung und Zeichnung, während die strukturellen Merkmale von ihr unberührt bleiben.

10. *Tetraconus Mocsáryi* SZÉPL. ist die grösste bekannte Monomachide : das typische ♀ im Museum Budapest, aus Espirito Santo (O. STAUDINGER & A. BANG-HAAS Verkäufer, 1898) ist ein kräftiges, dunkelbraunes Tier von gut 22 mm Länge, mit mächtigem Kopf und grossem, rauchbraunen Spitzenfleck der Vor-

derflügel. Es ähnelt am meisten *Monomachus Klugi* WESTW. ♀, unterscheidet sich aber hiervon scharf durch den Besitz je eines grossen, zapfenförmigen, geraden, nach hinten gerichteten Höckers an den Schläfen, unweit des Mandibelgrundes, an der Grenze gegen die Kopfunterseite, und eines beiderseits stumpf zapfenförmig nach aussen vortretenden und mitten der Länge nach tief eingedrückten Kopfschildes. Ob diese Differenzen indessen generischen Unterscheidungswert bedingen, wie SZÉPLIGETI wollte, oder bloss einen spezifischen, wird erst noch die Zukunft lehren müssen.

Spalte am Grunde der Oberkiefer von *T. Mocsáryi* mässig breit, deren übrige Aussenfläche in ganzer Länge eben oder vielmehr gleichmässig schwach gewölbt, polirt glatt, mit wenigen, starken Punkten. Ausschnitt in der Vorderrandmitte des Kopfschildes derselben Art tief, dreieckig. Hinterkopf dicht runzlig punktirt, ebenso das Gesicht, die Stirn und der Scheitel. Gesicht mit Mittellängskiel, der in einer schwachen Vertiefung liegt. Schläfen längs der Netzaugen polirt glatt, mit wenigen groben Punkten bestanden, nach der Kopfunterseite hin etwas reichlicher punktirt. Die von den Nebenaugen bis zum Kopfhinterrande ziehende eingedrückte Längslinie deutlich ausgeprägt. Hintere Nebenaugen von den Netzaugen doppelt so weit wie voneinander entfernt. — Vorderrücken oben in der abgetrennten hinteren Hälfte und in den herabgebogenen Seitenzipfeln fein punktirt. Mittel- und Hinterbrustseiten dicht runzlig punktirt; auf jenen ist die Punktirung fein, auf diesen einen Grad kräftiger. 1. Tarsenglied der Hinterbeine etwas länger als die drei folgenden Fussglieder zusammen. Nervulus der Vorderflügel schwach postfurcal. — Mittelsegment grob und dicht knitterig gerunzelt. Hinterleibstiel kräftig nach unten gebogen und länger als das folgende Segment.

Variation in the use of certain scientific terms and changes in the spelling of scientific names,

by HENRY H. LYMAN, M. A., F. E. S., F. E. S. A.

In a short paper entitled « Type and Typical », which I contributed to « The Canadian Entomologist » and which was published in volume XL, page 141, I wrote :

« These terms are used in such different senses by different authors, that confusion is sometimes caused, and it is much to be desired that some authoritative body of naturalists should accurately define their proper use, and then that all other naturalists should accept the decision and conform to it, even if it does not agree with their own individual opinions. »

Mr. CHARLES OWEN WATERHOUSE proposed selecting one specimen only as type and calling all other specimens that the author had before him, when describing a species, co-types, and has many followers.

Mr. OLDFIELD THOMAS and others use the word co-types for all the specimens before an author when no single type is selected, saying « the type in such cases equalling the sum of the co-types ». Mr. THOMAS added : « No species would have both type and co-types, but either the former or two or more of the latter ». Mr. THOMAS proposed the word para-types to be used for the specimens which Mr. WATERHOUSE called co-types. Others again use the word type or types for all the specimens an author retains in his own collection or specially deposits in some national museum, and call those distributed to other museums or private collection co-types.

Other terms which have been proposed are meta-type, homotype and topo-type, some of which seem quite unnecessary.

Another point on which I think a pronouncement should be made is the practice of certain authors of altering the spelling of specific names in order to make them conform more closely to their ideas of strict Latinity, as for instance : Walkeri changed to valkeri, Williamsi to villiamsi, Blakei to blacei, thus entirely obscuring the meaning and intention of these names. I humbly contend that fixity of nomenclature is of far more importance than a strict adherence to a pseudo-latinity. Villiam is not the Latin equivalent of William, and I can see no use in pretending that it is.

On the British coinage the late King's name appears as Edwardus, and if that is good enough Latin for the Royal Mint, I think that Walkeri should be good enough Latin for entomological literature and collections.

I do not suppose that this First International Congress of Entomology will be prepared to take individual action on these points as they concern the whole body of naturalists, but I invite your attention to them in the hope that some representation may be made to the general body of zoologists.

One Hundred Years of Entomology in the United States,

by Prof. Dr. HENRY SKINNER (Philadelphia, U. S. A.).

So far as ascertained the earliest articles on American Insects by a native of the country were written by the celebrated Pennsylvania botanist JOHN BARTRAM, who published on the nests of Wasps, the great black Wasp of Pennsylvania, the Dragon-Fly of Penna., the yellow Wasp and observations on *Cicada septendecim*. The former observations were in the « Philosophical Transactions (1745-1763) ».

His son WILLIAM also published on Insects in a work entitled « Travels through North and South Carolina, Georgia, East and West Florida, etc. (1791) ».

MOSES BARTRAM wrote articles on the native Silk Worms of America and on *Cicada septendecim*. The first mentioned paper was read before the American Philosophical Society in 1768.

The first two volumes of the transactions of this Society contain a few articles on Insects. In volume one is an article by Colonel LANDON CARTER of Sabine Hall, Virginia, communicated by Colonel LEE of Virginia, entitled « Observations concerning the Fly-Weevil, that destroys the Wheat, with some useful discoveries and conclusions concerning the propagation and progress of that pernicious Insect, and the methods to be used for preventing the destruction of grain by it ».

There are a few notes on injurious Insects in the work published in 1799 by Dr. BENJAMIN SMITH BARTON, which he called « Fragments of the Natural History of Pennsylvania ». The same author also wrote on the Honeybee. These articles appeared in the « Transactions of the American Philosophical Society » in 1793 and 1805.

In 1806, FRED. VAL. MELSHEIMER, a minister of the Gospel, living in Hanover, York county, Pennsylvania, wrote « A Catalogue of the Insects of Pennsylvania ». This work is called part first and is a list of the Coleoptera and contains the names of 1363 species and 111 genera. The difficulties of the study at this time must have been very great as there were many nondescript genera and species.

European books were high priced and difficult to obtain and congenial associates almost unknown. MELSHEIMER says the requests from European entomologists were urgent and ardent for American material, and that this was an inducement for American entomologists to make themselves more intimately acquainted with the productions of their country. MELSHEIMER, also says he, was much indebted to Prof. KNOCH in Brunswick, in Germany, who aided him in the classification and arrangement and that he corresponded with him for many years.

THOMAS SAY, who was born in Philadelphia in 1787, has been called the father of American entomology and, as far as this study is concerned, he was a very remarkable man and in his short life accomplished much. He was evidently a born naturalist, as when a boy his greatest delight was in collecting Butterflies and those coleopterous Insects whose variegated or splendid colors seldom fail to arrest the attention of the most careless observer. His work on American entomology, bearing the date 1824, was the first of its kind to appear, and he said of it : « It is an enterprise that may be compared to that of the pioneer or early settler in a strange land, whose office it is to become acquainted with the various productions exhibited to his view, in order to select such as may be beautiful either as regards his physical gratification, or his moral improvement and in order to counteract the effects of others that may have a tendency to limit his prosperity ». SAY, like some of his successors, thought it necessary to offer an apology for his studying entomology and he quotes from HARRIS as follows : « As there is no part of nature too mean for the divine Presence, so there is no kind of subject having its foundation in nature that is below the dignity of a philosophical inquiry ».

He certainly did pioneer work, for when he became a member of the Academy of Natural Sciences of Philadelphia; in 1812, the collection consisted of « some half a dozen common Insects, a few corals and shells, a dried toad Fish and a stuffed Monkey ».

It is said of him that he followed the maxim of SENECA, that « We have a sufficiency when we have what nature requires » and that the expenditure for his daily food for a considerable time amounted to no more than six cents a day.

In addition to being a writer and teacher, he was also an explorer and in 1818 visited the Sea Islands and adjacent coast of Georgia, and penetrated into East Florida. This journey was shortened on account of hostile Indians. He was also zoologist to the expeditions sent out by the Government of the United States under major LONG. He died in 1834 at the early age of 47. It is difficult at the present time to appreciate the greatness of a man who has to blaze the trail under every difficulty, and when we read the hundred years backward and see what has been accomplished, we can but feel grateful to and honor SAY and the other pioneers. Of those who cultivated the study in the time of SAY we can only think of TITIAN R. PEALE, of Philadelphia, and JOHN F. MELSHEIMER with whom SAY corresponded.

In 1833 PEALE published his « Lepidoptera Americana », an illustrated work that was very creditable for that early day. SAY's work was recognized abroad, and he had the honor to be a member of the Linnæan and Zoological Societies of London and the Société Philomatique of Paris. In his own city he was curator of the Academy of Natural Sciences of Philadelphia and the American Philosophical Society, professor of natural history in the University of Pennsylvania and of zoology in the Philadelphia Museum.

Prof. S. S. HALDEMAN, whose articles appear in the publications of Academy of Natural Sciences from about 1842 to 1859, was another Pennsylvanian of note who did excellent work. He spoke of the character of the times and lamented the lack of general works adapted to the country and also said the student was retarded by the extent of the subject and the want of instruction in educational institutions. The good seed sown in Pennsylvania was destined to bear additional fruit and in 1859 the American Entomological Society was founded.

It lacked funds to carry on the work, but what it needed in this respect, it made up in energy and the devotion of its members. The Proceedings were put into type and printed by the members, and it produced or aided some of the greatest of the world's entomologists.

Drs. LE CONTE and HORN did splendid work in the Coleoptera and E. T. CRESSON was the pionneer systematist in the Hymenoptera. He was virtually the founder of the Society and still takes an active interest in its work and welfare. This Society and its publications have made a great impression on the study of entomology in America.

The limits of this paper will not permit any further detail as to what has been done in Pennsylvania, which may be called the cradle of entomology in America. There were developed other centres in various parts of the country and it is necessary to refer briefly to these.

In 1841 Dr. THADDEUS WILLIAM HARRIS, of Cambridge, Mass., published his work entitled. « A Treatise on some of the Insects Injurious to Vegetation ». It formed one of the scientific reports published by the State. His directions from the Governor of the Commonwealth were as follows... « It is presumed to have been a leading object of the Legislature, in authorizing this Survey, to promote the agricultural benefit of the Commonwealth, and you will keep carefully in view the economical relations of every subject of inquiry. By this, however, it is not intended that scientific order, method, or comprehension should be departed from. At the same time, that which is practically useful will receive a proportionately greater share of attention than that which is merely curious; the promotion of comfort and happiness being the great human end of all science ».

This valuable work was extensively used and played an important part in the education of more than one of our present day workers in entomology and it passed through a number of editions.

An important center was developed in the far west, when Dr. HERMAN BEHR left Germany for San Francisco and began his papers in the « Proceedings of the California Academy of Sciences » and in the « Proceedings of the American Entomological Society ». At the present time his successors are doing good work, and the names of FALL, FUCHS, BLAISDELL, VAN DYKE and WRIGHT are well known. The Pacific Coast Entomological Society is active and efficient and doubtless has a great future. Dr. BEHR's first paper in this country was published in 1855, and in the next ten or twenty years a number of men commenced to contribute to the science, and while they were by no means a large army, the study was being pursued in a number of places.

In 1861 Prof. P. R. UHLER had commenced his valuable studies on the Hemiptera, which has been his life work so far as entomology is concerned. In 1869 Dr. A. S. PACKARD gave to American entomologists his « Guide to the Study of Insects », and no serious student of the subject was without this work. He was a prolific writer, and we can only mention his « Text-book of Entomology » and his « Studies on the Bombycine Moths ». In 1867 a man commenced to write on the Noctuidæ. I refer to A. R. GROTE of Buffalo, New York. He was an untiring systematist and described a considerable portion of our species in the family.

In the late fifties that indefatigable author SAMUEL H. SCUDDER, whose many contributions to entomology mark him as one of our greatest students, commenced his work. His work on the Orthoptera, his « Butterflies of the Eastern U. S. and Canada » and his works on fossil Insects will make his name immortal in entomology.

Another Pennsylvanian who acquired a great reputation was Dr. HERMAN STRECKER of Reading. He was probably known to every collector of Lepidoptera the world over and amassed an enormous collection, which now adorns the Field Museum of Natural History in Chicago.

In 1878 he published his « Butterflies and Moths of North America » and in 1872 commenced his illustrated work entitled « Lepidoptera, Rhopaloceres and Heteroceres ». The last parts did not appear until 1900.

In the early sixties, WM. H. EDWARDS, one of the world's greatest lepidopterists, commenced writing and describing species and life-histories and did splendid work for more than thirty years. His publication on American Butterflies has never been surpassed, either as to the value of the contents or for beauty and accuracy of the illustrations. When he began there was very little known of the stages of our American Butterflies.

The first catalogue of Lepidoptera was the work of Dr. JOHN G. MORRIS, and appeared in 1860 and was published by the Smithsonian Institution in Washington, D. C. Dr. MORRIS was a delightful old gentleman when the writer became acquainted with him, and lived in Baltimore, Maryland.

All honor is due to that great German entomologist who was born in Kœnigsberg and finally made America his home. Dr. HERMAN AUGUST HAGEN became professor of Entomology at Harvard College in Cambridge, Mass. His principal contribution to the entomology of this country was the « Synopsis of the Neuroptera of North America, with a List of South American Species ». The « Bibliotheca Entomologica », by his pen, is consulted every day in the year. Another distinguished European entomologist who did a large amount of work in America was C. R. OSTEN SACKEN. Every entomologist should read that most interesting book entitled « Record of my Life Work in Entomology ».

For twenty-one years Baron OSTEN SACKEN was Secretary of the Russian Legation in Washington and resided in the United States. He served as Secretary of the Russian Legation in Washington and as Consul general of Russia in New York. These twenty-one years were, as far as regards entomology, principally devoted, in collaboration with Dr. H. LOEW, to the task of working up the Diptera of North America north of the Isthmus of Panama. The large and valuable collection of Diptera made by OSTEN SACKEN he most generously presented to the Museum of Comparative Zoology at Cambridge, Mass.

It would necessitate a number of volumes to enumerate the works that have appeared and mention can only be made of some of the men who have also done acceptable work. In the Hymenoptera may be mentioned Mc COOK, NORTON, BLAKE, FOX, ASHMEAD, BASSETT. In the early days Dr. BRACKENRIDGE CLEMENS, V. T. CHAMBERS and C. H. FERNALD furnished much information on the Micro-Lepidoptera.

The museums have had a phenomenal growth and at the present time are palatial buildings, costing in some cases millions of dollars.

The principal entomological collections are in the National Museum at Washington, D. C.; the Academy of Natural Sciences of Philadelphia; the American Museum of Natural History in New York; the Museum of Comparative Zoology in Cambridge,

Mass.; the Carnegie Museum in Pittsburg, Penna.; the Field Museum of Natural History in Chicago, Illinois; the Museum of the Brooklyn Institute of Arts and Sciences; the Boston Society of Natural History. Many colleges also have fine museum collections and there are many private collections of note.

To enumerate the important works published in recent years would take entirely too much time and space.

A number of important journals have appeared but have been discontinued, but during the time of their life were important factors in entomological work. Among those that have ceased to exist may be mentioned: « The Practical Entomologist »; « The American Entomologist »; « The Bulletin of the Brooklyn Entomological Society »; « Papilio »; « Entomologica Americana »; « Insect Life ». At the present time the journals of America devoted to Insects compare favorably with those published in any part of the world. « The Transactions of the American Entomological Society »; « Journal of the New York Entomological Society »; « Annals of the Entomological Society of America »; « Journal of Economic Entomology »; « Proceedings of the Entomological Society of Washington »; « Psyche » and « Entomological News » are all doing excellent work for the cause.

The most hopeful and interesting outcome has been the public recognition of the value and benefits to be derived from the study.

This appreciation by the people has been largely brought about by economic entomology in general and particularly by that branch known as the transmission of disease by Insects. The demand for trained entomologists is greater than the supply, and we are therefore getting a better class of men to take up the study. The educated college man, the trained student, is taking the place of those who took up entomology for the love of it and gradually developed into the professional entomologist. Many of our colleges have courses in entomology, and the number of students taking the course are increasing every year. When one State finds it necessary to expend a million dollars a year and also receives aid from the National Government, which appropriates three hundred thousand dollars toward the work, we can readily understand how immense some of the problems are and why entomologists are in demand.

Economic entomology has had a prodigious growth in America. The principal workers in this branch in the early days have been HARRIS in Massachusetts, ASA FITCH and J. A. LINTNER in New York, B. D. WALSH in Illinois and C. V. RILEY in Missouri. RILEY'S « Missouri Reports » and his activity as United States Government Entomologist were of an epoch making character, and he has been considered the greatest of the world's economic entomologists.

The amount of valuable work done by the present Chief Entomologist of the Government and his able corps of assistants is very large and deserves our admiration and commendation. When we think of the enormous loss from Insects in the United States annually, estimated at ten hundred millions of dollars, we can readily understand the large sums appropriated by the Government and many of the States for the purpose of preventing this great loss. The damage done by the San Jose Scale, the Gypsy Moth, the Cotton Boll Weevill, has been very great. They have however served to call the attention of the people to value of all entomological work. In the next ten years the study of entomology will be on a plane never dreamt of by the pioneers in the study. This international Congress is in itself evidence of the great advance made, and we can confidently look forward to seeing our study respected and honored by all humanity.

Experimental Entomology. Factors in Seasonal Dimorphism,

by F. MERRIFIELD, F. E. S. (London).

I felt considerable diffidence when invited to take a part in one of the sectional meetings of this important international gathering of entomologists, but was informed that it was desired that experimental entomology should be represented, and as this is one of the few subjects connected with entomology on which I have had some experience, though little as compared with that of others such as STANDFUSS, FISCHER and BACHMETJEW, who are prevented from being present, I have ventured to do so. It is not necessary for me in this assembly to dwell on the importance of the experimental side of observation in reference to the many problems that present themselves to investigators into the reasons why organisms come to be what they are. That their structure and their life habits are owing mainly to the external conditions to which they or their ancestors have been subjected may be taken as a fact. But the external natural conditions and the internal adaptabilities of the organisms are so numerous, so complicated and delicate, so varied and so mingled together, that, in order duly to appreciate cause and effect, it is absolutely necessary to resort to the means which experimental work affords of isolating different factors, and thus learning by which of them the results we see have been brought about. Isolation is the necessary means for unravelling the tangles presented by nature, many of them very beautiful, but for which disintegration is necessary if the path is to be cleared Those who have resorted to the process of experiment must often have been astonished at the completeness with which a very simple one has sometimes disposed of a theory having much to recommend it.

Seasonal Dimorphism.

SUMMER AND WINTER PHASES. ASPECT AND LIFE-HABIT.

I think I can best justify the occupation by me of the time and thoughts of members of this International Congress, if, instead of any further remarks on the general subject of experimental entomology, I proceed to the application which I have been able to make of experiment for the purpose of throwing light on the fascinating subject of the factors in seasonal dimorphism. For many years, though with much interruption owing to the pressure of other demands, I have been engaged in experiment, with a view to ascertain some of these factors and their relative importance. The facts they present are sufficiently remarkable. Between two successive generations the differences are so great that, but for knowledge of their parentage, the succeeding generation would often be taken for a different species altogether. I have found it expedient for the purpose of this exposition to divide these differences into two classes. One class of differences, all which I include in the general term « aspect », comprises general colouring, pattern or markings, shape, and size (this latter I would rather consider as « mass », since bulk is in a sense more significant than expanse, and is also much more easily ascertained by weighing); the other, and as I shall venture to submit the more fundamental class of differences, may be summed up in the term « life-habit ». The species on which most of my experiments have been tried are the well-known *Araschnia levana* and the common *Selenia bilunaria*, the one generation of each species appearing as an imago in the spring or very early summer, the other in the height of summer.

Let me now mention the names by which I propose in this paper to distinguish these two generations. They are commonly spoken of as the *first* and *second broods*. I am going to adopt for them the terms *winter phase* and *summer phase*. The form commonly called the first « brood » is really the second brood of the previous year, passing in that year its larval and great part of its pupal life, while what is commonly called the second brood passes its whole life in the spring or first part of the summer of one and the same year. As to their life-habits, I sum them up as follows: *winter phase* — egg and larva in summer and early autumn, pupa in winter, imago in spring, period of life eight to nine months; *summer phase* — whole life (egg, larva, pupa and imago) in spring and first part of summer, period of life three to four months.

Let me here shortly describe the difference in aspect of the two phases. In *A. levana* the winter phase (*levana* proper) is orange brown chequered with blackish brown spots, the summer phase (*prorsa*) is uniform black with a conspicuous white oblique fascia extending across both pairs of wings, usually with a few small orange-brown spots, or traces of them; this phase is generally larger, sometimes considerably so, than the winter phase; I have not found that it is usually greater in weight. In *S. bilunaria* the imago of the winter phase is dull brown with a strongly marked pattern of darker hue, that of the summer phase is bright chestnut, verging, especially on the under side, on orange, the pattern somewhat different from that of the other phase, and not so strongly marked. But the great difference in what I have called aspect, is in size or mass; from a large number of weighings made of several broods, it may be taken as about the difference between 13 or 14 grammes and 18 grammes, i. e. the winter phase is something over 30 % heavier than the summer phase.

Temperature the cause of the different phases.

We come now to the question of the causation of these two different phases. Without by any means doubting that other influences than temperature, such as moisture, are, under some circumstances, more particularly in tropical and semitropical countries (1), the determining cause, I do assert that in the species operated on by me, and, as I am convinced, in many other species, the appropriate temperature, and nothing more, applied in the right stage, is sufficient in itself to cause the Insect to belong

(1) In Southern Africa, for example, as shown by the experiments of Mr. Guy Marshall.

to either phase, in all respects; that is to say, both as regards aspect, including size and therefore weight or mass, and life-habit, and even to divert the Insect, after being launched in the direction of one phase, into the other.

To avoid constant reference to the thermometer with its two scales (Fahrenheit and Centigrade), I have classified the different temperatures used in my experiments as follows :

1. « Spring » temperature, a mean between 50° F. (10° C.) and 56° F. (13° C.), usually about 54° F. (12° C.) to 56° F. (13° C.);
2. « Cool summer » temperature 60°-62° F. usually about 60° F. (15°-16° C.);
3. « Full summer » temperature, 65° F. (18°-19° C.);
4. « Forcing » temperature, 75°-80° F. (24°-27° C.).

My temperatures are mean temperatures, obtained by shifting the objects from time to time, between a refrigerator usually about 46° F. (8° C.) and a room at a temperature of about 66° F. (19° C.), varying the time of exposure — usually once in 24 hours, but sometimes once in 2 or 3 days or even more — to suit the mean temperature desired. This is all that my opportunities permitted; the temperatures so arrived at can only be considered approximate; and it does not follow that exposure for 6 hours at 40° F. and 18 hours at 60° would have the same operative effect as 12 hours at 50° and 12 at 60°, though the mean would be the same. But it is more nearly in accordance with nature than a fixed temperature would have been, because, as we know, there is almost always a great difference between the day and night, often amounting to 20 degrees F. (11 degrees C.) or more, especially in the summer months.

Not « Alternation of generations ».

It has sometimes been suggested that the cause of difference between the two phases is a congenital one, that under ordinary circumstances, where seasonal dimorphism occurs, it is a case of alternation of generations. This is certainly incorrect as regards the two species on which most of my experiments have been tried, *A. levana* and *S. bilunaria*. Of the latter I have had four successive generations within about six months, all of the pure summer phase, and I believe there is not in either of the species even a tendency to alternate generation.

Stages in which temperature operates.

The next question that arises is : What is the right stage at which the determining temperature should be applied? It must of course be before the time when the imago is formed and perfected. I have tested this in *S. bilunaria,* — using temperatures which are adequate to operate effectively in the stages of egg onwards, — as far back in the ontogeny as the lifetime of the parents, before the union between them, and therefore from an early stage of the germ plasm.

Egg stage.

As to the first of these stages, that of the egg, I have seen no reason to believe that the phase to be assumed depends on what takes place in that stage (or in the earlier ontogeny). In 1894, at the suggestion of my friend Dr. Dixey (now the President of the Entomological Society of London), I tried a long series of experiments with eggs of *Selenia tetralunaria*, exposing them for from seven to forty days, to various temperatures ranging from 80° F. (27° C.) to 33° F. (1° C.), but this differential treatment, though, as might have been expected, greatly accelerating or retarding the hatching of the egg, made no difference whatever in the habit or aspect of the perfect Insect. Since then I have tried similar experiments with *S. bilunaria*, and with similar want of effect. The stages, in which this external influence can operate, may therefore be considered to be those of the larva and pupa; the former being the stage of growth, never I think less than many hundredfold, sometimes many thousandfold, comprising also considerable changes in structure and sometimes in habits; the latter a stage of external fixity and quiescence, with vast internal changes of structure, but loss instead of gain in mass.

Pupal stage.

The very conspicuous results on the facies of Lepidoptera produced by exposing the pupæ to abnormal temperatures have directed much attention to the pupal stage, and I will first address myself to that stage. Some of the most striking of these effects have been caused by the application of extreme temperatures, such as the organism would rarely or never be exposed to in nature, and, though the results have been exceedingly valuable in, it may be said, helping to establish the mutual relations of diverse species, and in affording a basis for theories as to their phylogeny and their future evolution, I submit that this class of experiment can throw but an insufficient light on the processes by which, with normal temperatures proper to the seasons of the year, seasonal dimorphism has actually been caused.

PUPAL STAGE IN *A. levana.*

Great changes of aspect, as is well known, can be caused by subjecting the pupa of the summer phase of *A. levana* to a low temperature; by this treatment the imago may be completely converted in general aspect from the *prorsa* to the *levana* form, or to the intermediate form called *porima.* But as regards life-habit, — that is the conversion of it by such means, if applied only while in the pupal stage, into the winter phase *with its characteristic of the normal long pupal period,* — this has not that I know of been found practicable; it is certainly extremely difficult, and has always failed with me. The actual pupal period has indeed in such cases been greatly lengthened by the low temperature, but it has never in my experience produced that state of resolute resistance when afterwards exposed to a warmer temperature, — a resistance often to the death (1), — which the normal winter phase shows. And as to the pupa of the winter phase, the conversion of that into even the summer phase *aspect* or facies, that of *prorsa,* presents great difficulty.

PUPAL STAGE IN *S. bilunaria.*

I think I may say the same of *S. bilunaria.* Cooling the summer phase protracts the pupal period and, if considerable in degree

(1) See my presidential address, January 1906. (« Proc. Ent. Soc. Lond. », 1905, XCII, XCIV-XCVI, CX, CXI.)

and long persisted in, alters its pattern so as to produce a close resemblance to that of the winter phase, which emerges in spring, and gives it a general darkness of colour resembling that phase, but of course cannot give it the much greater mass of the winter phase; and warming at the proper time the winter phase pupa gives it the characteristic colouring, and in a measure the pattern, of the ordinary summer form, but does not alter its habit of long pupal period or its great difference in mass.

LARVAL STAGE.

With these remarks on the insufficiency of temperature, brought to bear during the pupal stage alone, to alter the life-habit of the Insect, I pass on to the effect of temperature brought to bear during the larval stage.

A. levana; SUMMER PHASE LARVÆ.

And first as to the summer phase of *A. levana*, those individuals which under normal circumstances would have emerged in the summer as *prorsa*. It will be seen that in all but very advanced stages of larval growth these can readily be converted by a low temperature into the complete winter phase.

On June 26th, 1908, I received from North Germany several hundred larvæ which I judged from their size to be in their 3rd, 4th and 5th instars. Between 150 and 200 were placed at a low spring temperature, 50° F. (10° C.) as a mean. Half of them, which I will call A, were left there until they pupated, and were then moved to a warm room (about 67° F. 19-20° C.); the other half, which I will call B, were left there also as pupæ.

As to the A larvæ, 74 pupated, a first lot of 20, which I judged to be within 3 or 4 days of being full fed and to be in their 5th skins, pupating in from 12 to 18 days; 15 emerged in 13 to 15 days, all in aspect *porima* or *prorsa*, except one of *levana* aspect and one nearly so, one died and 4 went over to the winter. Of a second lot of 20, which I judged to have been 6 or 7 days short of being full fed, probably in their 4th skins, and pupating in 20 to 26 days, 4 only emerged, in 11 to 12 days, all of *porima* aspect, 7 died and 8 went over to the winter. Of the remaining or 3rd taking, 34 in number, all younger than the others, and pupating in 27 to 50 days, 11 died and 23 went over to the winter.

The B larvæ, 67 in number, were treated in the same way as A, except that their pupæ were left at the low temperature for 41 to 52 days, when some appeared to be on the point of emerging, and they were then all removed to the room temperature, where 7 of them emerged in from one to 7 days more, all being of *levana* aspect or neary so; of these 7, 6 were from the first taking of 20, consisting of those larvæ judged to be within 3 or 4 days of being full-fed, and one from the second taking of 20, consisting of larvæ judged to be 6 or 7 days of being full-fed. The rest of the first and second takings, and all the third taking, numbering 27, died or went over to the winter. Of the whole taking of 67, 26 died and 36 went over to the winter (1).

Some of the younger larvæ were transferred for the last week or two of their lives to the warm room temperature, pupating there in from 3 to 9 days, but none were thus reconverted to the summer phase. It may be taken therefore that all the June summer phase larvæ which were more than 6 or 7 days short of being full-fed, were converted by a temperature of 50° F. (10° C.) to the full winter phase.

On June 13th, 1910, several hundred larvæ of *A. levana* were received from North Germany, the most advanced, being apparently in their 5th skins, pupating at the room temperature from day to day in from one to 10 days, and such of these as were left as pupæ at that temperature emerged there in about a fortnight after pupation (2). A number of those which appeared to be in

---

(1) This experiment indicates that, when the larvæ had not been subjected to a low spring temperature until in an advanced period of larval life, the prolongation of the low temperature into the pupal period caused a larger proportion to assume the full winter phase, i. e. including the life-habit, than the exposure to the low temperature during the larval stage only had done, so that under these special circumstances the life-habit in this species may be affected in the pupal stage.

(2) September 25th. Some of these pupæ were on pupation placed at a spring temperature of about 55° F. (say 13° C.) for about 6 weeks and then at the room temperature; none had emerged when my paper was read on August 4th, but having since been transferred to the room temperature, all except a few that died emerged in from 1 to 2 weeks time, showing that they remain of the summer phase as regards life-habit. In aspect most are *porima*, but a few are *levana* and one *prorsa*.

their 4th skins were placed at a low spring temperature (about 50° F., say 10° C.), pupating there in from 14 to about 30 days and being afterwards transferred to the room temperature; not one has emerged, as they would certainly have done had the larvæ remained at summer temperature. I expect the whole will « go over » to the spring (1).

A. levana, WINTER PHASE LARVÆ.

Next as to winter phase larvæ. On August 18th and 25th, 1908, I received some hundreds, mostly about 1/3 of an inch (say 1-2 centimetres) long. 25 were forced at 80° F. (27° C.); of these only 5 emerged as summer phase, all of course of *prorsa* aspect. All the others of those received « went over », though placed at a spring temperature for various periods and then transferred to higher temperatures.

A. levana, CONVERSION OF ONE PHASE INTO THE OTHER.

These experiments, and many other published experiments both with the summer and the winter phase (2), show how difficult it is to convert the winter phase of *A. levana* into the summer phase, even if it is by no means in a very advanced larval stage. But this last conversion can be readily effected if the transfer to the higher temperature is made in a very early larval stage; for some larvæ from eggs laid by the summer phase in the latter part of July 1909 and which would therefore under normal circumstances have pupated late in August or in September as winterphase *levana*, forced as larvæ at 80° F. (27° C.) from the time of hatching, and pupating in 16 to 20 days, all emerged in 2 to 3 weeks, at the outdoor temperature, as pure *prorsa*.

The particulars I have given afford, I submit, sufficient proof that either phase of *A. levana* can be made to assume the other

---

(1) September 25th. All are still in pupa.

(2) WEISMANN, Studies in the Theory of Descent (English translation by Prof. MELDOLA, vol. I, appendix 1, pp. 117-125); « New Experiments [in 1883-1884 and 1886] on the Seasonal Dimorphism of Lepidoptera » (English translation by NICHOLSON) reprinted from « The Entomologist », January-August 1896.

phase, by temperature applied at a suitable part of the larval stage, which is the stage on which I lay special stress (1).

S. bilunaria, SAME TEMPERATURE DURING WHOLE LARVAL LIFE.

That the same takes place with *S. bilunaria*, I have abundant evidence. In experimenting on this species it had occurred to me that the whole normal difference of mean temperatures in larval life between the two phases being not more than 4° or 5° F. (say 2° to 3° C.), it was not probable that *mere* difference of temperature of so small amount would be enough to account for the two phases, and I tried some experiments with the special view of ascertaining this, by subjecting larvæ to the respective temperatures during their whole life. A large brood of *S. bilunaria from the same parents* was divided from the time of hatching, one part being placed during the whole of larval life at the mean temperature of 54° F. (12° C.), the other at the mean temperature of about 60° F. (15°-16° C.) till pupation, both being then brought into outdoor temperature. Those kept at the lower temperature emerged in an average time of 118 days, their range being 67 to 190 days, those at the higher temperature in 184 days, their range being 79 to 208 days (2). Similar experiments with other broods had similar differential results and establish that « spring » temperature applied to the larva during its whole life tends to shorten the pupal period as compared with a « cool summer » temperature.

---

(1) I think the facts recorded by Prof. WEISMANN as to his experiments on the July and August larvæ of *A. levana* accord with my views so far as to show that it was only when the Insects were placed at the high temperature in their *larval* stage that there was any considerable conversion from the winter (*levana*) phase to the summer (*prorsa*) phase. To establish an exact parallelism with nature, or one as nearly approaching to it as the inherent difficulties will permit, experiments applying high temperatures, but lower than 80° F. (27° C.), on young July and August larvæ appear desirable.

(2) As to the difference in weight or mass see later.

But if a « forcing » temperature is so applied to the larva, the pupal period is greatly shortened (1).

STAGES IN LARVAL LIFE WHEN TEMPERATURE EFFECTIVE.

Experiments in large numbers have been made by me with the view of ascertaining at what stage in larval growth the low and the high temperatures respectively should be applied to be effective, larvæ being transferred from one temperature to the other in different stages, in the 1st, 2nd, 3rd, 4th, and 5th instars, both with *A. levana* and *S. bilunaria*. Without troubling you with details, I may state that experiments made upon many hundreds show conclusively that, when the larvæ are subjected to a « spring » temperature during the early part of their lives, the proportion of those, whose pupal period is absolutely the short one (that is the one proper to the pure summer phase), is much larger than in those whose young larvæ were at the « cool summer » or « full summer » temperature; and when the usual duration of that short summer phase period has been exceeded by both, its duration has in general been considerably shorter in the former case, where the first part of larval life was at « spring » temperature, than where the young larvæ were at the « cool summer » or « full summer » temperature.

I have not been able to say, however, that there has been any transfer of *summer phase S. bilunaria* larvæ from the lower to the higher temperature in the *fifth* instar, the reason being that I am nearly sure that under these circumstances there was *no fifth instar*. My time for observation, owing to other engagements, was usually limited to an hour or two in the early morning or late afternoon, and I found it impossible to note when the larval skins were changed, but certainly what I had thought, during the growth at the lower temperature, to be their fourth instar proved

(1) This may explain why in countries like Scotland, where the summer temperature is low, our *S. bilunaria* is « single brooded » i. e. has but one, the winter phase, and why in other countries, in very hot summers, an additional summer phase brood is often interpolated in seasonally dimorphic species, making three generations within the year.

to be their last one. This, with other reasons, led me to test the *mass* of the pupæ by weighing.

Test of mass, or weight.

Over 350 pupæ have been weighed in 25 different lots, 12 of these lots having been during the early part or the whole of larval life, at the higher temperature (60° in 1906, 65°, as afterwards mentioned, in 1910), and 13 at the lower or 54°. The former averaged about 30 % more in weight than the latter. Nearly all show the same gulf of separation in weight between those at the higher, and those at the lower temperature; not one of the latter lots showed so great a weight in its average individual as any one of the former lots. Six of the 25 had been at the respective temperatures during the whole larval life, 3 of them at the lower and 3 at the higher, the remaining 19 lots during varying portions of the earlier life only; in their case, taken as a whole, the difference in weight was slightly less than with the 6. Two of the 6 were the case before mentioned where the eggs laid by a single parent were so numerous that they were sufficient to divide between the two temperatures of 54° and 60°, the respective weights here being 134 and 181 grammes, an excess of about 35 %.

Experiments of 1910, higher temperature raised.

Most of my experiments, especially with *S. bilunaria*, were conducted with the object of imitating as nearly as might be the temperatures to which the two phases are in nature subjected during their larval stages. For these purposes in 1909 I had placed those intended to represent the summer phase larvæ at the supposed mean of about 53° or 54° F. (12° C.), afterwards raised to about 61° or 62° F. (16°-17° C.); those intended to represent the winter phase larvæ at about 61°-62° F. (16°-17° C.), afterwards lowered from this to about 58° F. (14°-15° C.). I became satisfied afterwards that certainly the higher one of these temperatures was too low to represent nature. Moreover, a closer examination of my apparatus showed me that what I thought 61°-62 F. was in many cases really about 60° or 61° F. Therefore, when subsequent experiments in 1910 were tried, the higher of the temperatures was raised, 65° F. (18°-19° C.) being the mean then aimed at, perhaps rather too high for my purpose. This made little difference in the *general* results, but had the effect of causing considerably larger proportions of the different lots to emerge within the shorter pupal periods, and above all, to increase their mass. In this way I have obtained some of enormous size; some at 65° F. (18°-19° C.) till their fourth skins, and then at the lower temperature, averaged in weight 27 grammes, instead of the usual average of 18, and several reached 30 and even more, with a wing expanse only just under 2 inches (fully 50 millimetres).

[The experiments of 1910 were far from completion when this paper was read, so that, though the pupal weight was known and included in my statements, the pupal period could not be. I am now able to subjoin a tabular statement, which will sufficiently show the difference in pupal period caused by the substitution of 65° F. (18°-19° C.) for 60° F. (15°-16° C.) as the higher temperature, also the difference in result caused by the order of precedence in which the different temperatures were applied. It will also show, by comparison with the information previously given as to weight, another fact which is not to be disregarded, viz., that the effect on pupal period is much less regular than the effect on weight or mass.

Tabular statement.

I. — *Whole larval life at same mean temperature.*

YEARS.	NUMBER OF		TEMPERATURES.	NUMBER emerging in short pupal period.
	lots.	individuals.		
1909-10. . . .	3	38	54° F. (12°-13° C.)	1
"	2	15	60° F. (15°-16° C.)	0
Total 1909-10. .	5	53	—	1
1910	1	13	65° F. (18°-19° C.)	3
Totals 1909 and 1910 . .	6	66	—	4

II. — *Larvæ transferred from one to another temperature.*

YEARS.	NUMBER OF		TEMPERATURES.		NUMBER emerging in short pupal period.
	lots.	individuals.	Early life.	Later life.	
1909-10 . .	4	84	54°F.(12°-13°C.)	60°F.(15°-16°C.)	14 = 1/6
»	5	83	60°F.(15°-16°C.)	54°F.(12°-13°C.)	1 = about 1/80
Total 1909-10.	9	167	—	—	15
1910 . .	6	113	54°F.(12°-13°C.)	65°F.(18°-19°C.)	55 = nearly 1/2
»	4	34	65°F.(18°-19°C.)	54°F.(12°-13°C.)	7 = about 1/5
Total 1910 .	10	147		—	62

By « short pupal period » is meant that which I consider represented, in those experimented on by me, the period proper to the summer phase, i. e. 16 to about 40 days according to the temperature, the average of the 81 individuals being 23 days, the average of the rest, 299 in number, was over 150 days, and not more than about 20 of these were under 70 days. Many of those of 1910 are still (25 Sep.) in pupa.]

Conclusions.

I will now state certain conclusions to which my experiments have led me as to factors in seasonal dimorphism in climates where the main difference between the seasons is that of temperature.

1. The life-habit — the long or short pupal period — constitutes the broad distinction between the winter and the summer phase. It is in the *Selenias* — and doubtless many other *Heterocera* —

usually associated with great difference in mass (1), and probably in the number of larval instars.

2. It is in the larval stage that the life-habits and mass of the two phases are, as a rule, determined, but in the pupal stage the facies may be very materially affected, and even transformed.

3. The larvæ, when first hatched, can develop into either the winter or the summer phase, according to the conditions to which they are subjected, temperature being the governing if not the only one.

4. The effect of temperature is different according to the different periods of life in which the larva is subjected to it.

5. The operation of temperatures on larvæ is by no means in direct proportion to their degree. It depends in great measure on the period of larval life in which they are applied.

6. In *S. bilunaria* a forcing temperature 75°-80° F. (24°-27° C.) or one that is considerably higher than is usual in nature, applied during the whole of larval life causes large size, and a short pupal period.

7. A mean « spring » temperature, applied during the whole of larval life, shortens in this species the pupal period more than does a mean temperature several degrees higher.

8. Applied during the earlier part of larval life only, a « spring » temperature in *S. bilunaria* and some other Heterocera promotes smallness of mass; and, in these species, also in *A. levana*, short pupal period; and probably in *S. bilunaria* it reduces the number of instars from 5 to 4. A summer temperature so applied promotes in *S. bilunaria* large size and also, unless several degrees above 60° F. (15°-16° C.), in both long pupal period.

---

(1) This difference in mass has been observed on by Prof. POULTON (« Trans. Ent. Soc. Lond. », 1902, p. 415) and Dr. DIXEY (« Ib. ». 1903, p. 159) and by others as indicating that in some cases of seasonal dimorphism conditions antecedent to the pupal stage must have been the cause.

9. Applied during the latter part of larval life only, a « spring » temperature promotes in *S. bilunaria* large size and long pupal period.

10. Applied during pupal life, a high temperature tends to give the facies proper to the summer phase, a low temperature the facies proper to the winter phase; but the life-habit of the summer phase, under normal circumstances, usually carries with it the subjecting of the pupæ to a high temperature, that of the winter phase to a low temperature.

11. There is great individual variation in nature as to the periods during which individuals, even those belonging to one brood, may remain in the egg, larval and pupal stages. Thus, in general June larvæ of *A. levana* belong to the summer phase, and August larvæ to the winter phase, some individuals, however, often going over the border line in each case; and these variations have a tendency to form into groups so as apparently to split a whole brood into two or more groups. I think the application of unusual changes of temperature increases the tendency to splitting.

I well realise that seasonal dimorphism is a very large subject, and that mine are but small contributions to it, but I offer them in the hope that they may be of use to others able and willing to follow up the investigation with greater precision than has been within my reach. They may be regarded as building stones, out of which, aided by materials supplied from many other climates and by many other minds, a fairly complete and symmetrical fabric may hereafter arise.

A new Aphis-gall on « Styrax japonicus SIEB. et ZUCK. »,

by Prof. C. SASAKI (Rigakuhakushi).

The trees (*Styrax japonicus* SIEB. et ZUCK.) called « Yegonoki » by the vernacular are widely distributed in Japan. In May or June, they bear very often a large number of galls on their branches or branchlets.

The galls are vulgarly called Nokoashi (Cat's foot) from their close resemblance to a cat's paw.

These galls are closely allied to those found on the *Styrax Benzoin* DRYANDER collected by Mr. A. TSCHIRCH in the island of Java. They are produced by an Aphis named *Astegopteryx styracophila* as suggested by Dr. KARSH in Berlin.

According to Mr. A. TSCHIRCH (1), the galls are formed on flowers or leaf-buds of *Styrax Benjoin*, as may be seen from the following extract from his work : « Blüthengallen (pl. XXV, fig. 1-4) entwickeln sich zu einer Zeit, wo die Höcker der Kelch-,

---

(1) A. TSCHIRCH, Ueber durch *Astegopteryx*, eine neue Aphidiengattung, erzeugte Zoocecidien auf *Styrax Benzoin* DRYAND. (« Bericht d. Deutsch. bot. Gesellschaft », 1890, VIII, pp. 48-52.)

Blüthen- und Staubblätter eben erst angelegt, die Fruchtblätter kaum angedeutet sind und das die Blüthe tragende Zweiglein noch ganz kurz ist. Dieses Zweiglein bleibt auch in der Folge kurz (pl. XXV, fig. 5), stellt also sein Wachsthum bald nach dem Eintreffen des Thieres ein, eine bei von Blattläusen befallenen Trieben häufige Erscheinung. Etwa auf halber Höhe jedoch sind die Blattränder nicht miteinander verwachsen, sondern nur fest aneinander gedrückt (pl. XXV, fig. 6 und 10)... Dieser Theil ist auch äusserlich sofort kenntlich, denn hier liegt eine breite, seichte Furche, in deren Mitte der Spalt sich findet (pl. XXV, fig. 6). Gegen die Spitze dieser Furche hin ist in der Mitte einer flachen Erhebung ein rundes Loch zu bemerken (pl. XXV, fig. 6, L), welches den in der Taschengalle befindlichen Läusen als Schlupfloch dient. »

The Insects which produce the galls on *Styrax japonicus* are evidently a species belonging to the genus *Astegopteryx*, but the coloration of the body, number of ring-markings on the antennal segments, the presence of cornicles, the formation and shapes of the galls, etc., differ from *A. styracophila*. These characters appear sufficiently distinct to justify me in giving the present Insect a new specific name *Nekoashi*.

Formation of galls.

If we examine the trees of *Styrax japonicus* towards the middle of May, many of the terminal buds of their branches or branchlets may be found to have grown larger than the normal ones, and be composed of some number of smaller deformed leaves when looked at from the outside. These are really the beginning of the formation of galls (pl. XXV, fig. 1). In each of these galls is generally imprisoned a single larva of diminutive size (pl. XXV, fig. 2). This is the larva which makes its way into the bud and initiates the formation of galls.

As the buds grow larger and more swollen, the modified leaves increase in number and size, and assume the shape of either horn-like processes or partly expanded leaves (pl. XXV, fig. 3).

Within the gall there is now found a deep excavation containing a larva of the 2nd stage (pl. XXV, fig. 4) together with a single exuvia.

Further developed galls (pl. XXV, fig. 5) have each a widened disc-like summit and a central hollow. The periphery of the disc is provided with two rows of more or less modified leaves or of finger-like processes formed of the same. The leaves forming the inner row are about eleven in number and finger-like in shape, while those of the outer row retain the shape of leaves more or less. The finger-like leaves of the inner row lie on the surface of the disc with their free ends directed inwards, so that they all meet at the centre of the disc.

If we cut open the galls at the level of the insertion of the finger-like leaves, there may be found a hollow disc, whose periphery is marked with several depressions, each occupying the base of the corresponding finger-like leaves. In the cavity of the disc there is always found a single wingless viviparous female (pl. XXV, fig. 6) with two exuviæ, and a number of larvæ laid by it. Each of these larvæ (pl. XXV, fig. 7) usually rests in the depression of a finger-like leaf, from which they derive nourishment.

By the beginning of June, the galls have become more developed, having assumed a depressed oval or round form, and their stalk is strongly compressed and marked with some longitudinal streaks. The finger-like leaves or processes growing on the periphery of the central disc are now more swollen than before, and have their ends directed towards the center of the disc, where they all meet together just as in the buds of Chysanthemum flowers (pl. XXVI, fig. 16). Outside the circle of these processes, there lies another circle of slender horn-like processes, as well as the deformed leaves. The central cavity of the disc becomes wider and deeper than in the previous stage.

In each of the finger-like processes, the larva (pl. XXV, fig. 7) develops into that of the 2nd stage (pl. XXV, fig. 8), and the latter becomes later a wingless viviparous female (pl. XXV, fig. 9). This viviparous female reproduces parthenogenetically the young within the finger-like processes (pl. XXV, fig. 10 and pl. XXVI, fig. 11).

In the last part of June, the finger-like processes grow in size, and form a nearly spindle-shaped sac. The central hollow disc becomes shallow, and the sacs are directed outwards, looking like a fully opened flower. In this case, a single gall is about 2 cm. in diameter, and of a light greenish color (pl. XXVI, fig. 17) although some (pl. XXVI, fig. 18) are much larger with a diameter of about 4 cm. Each of the sacs lying around the central disc contains usually a single wingless viviparous female (pl. XXV. fig. 9) together with the numerous larvæ laid by it (pl. XXV, fig. 10 and pl. XXVI, fig. 11).

In the middle of July, the galls attain their full growth, the diameter reaching about 6 cm. (pl. XXVI, fig. 19). Each spindle-shaped sac now assumes an elongated oval form and is of a light greenish yellow color. The larvæ contained in a single sac (pl. XXV, fig. 10 and pl. XXVI, fig. 11) develop into pupæ (pl. XXVI, fig. 12) and when the winged Insects are ready to emerge, the free rounded end of the sac opens, and the Insects fly out in succession. These are all females and produce the young parthenogenetically (pl. XXVI, fig. 14) either in- or outside the galls. These larvæ after moulting once pass into the 2nd stage (pl. XXVI, fig. 15).

Metamorphosis of « Astegopteryx Nekoashi ».

Larva of 1st stage in a newly formed gall : Length 0.492 mm., breadth 0.276 mm. Body elongated oval, amber yellow with dark brownish markings, and covered with a few hairs. Head large, occupying nearly one third of the body, eyes greyish, 3 on each side of the head, lying on a nearly triangular blackish marking.

Antennæ 4-jointed, the 1st and 2nd nearly equal in size, the 3rd longer than the 2nd, and the 4th twice as long as the 3rd, bearing 2-3 stiff hairs at its end.

Rostrum stout and long, 3-jointed, the 3rd or terminal segment much elongated and pointed and extending beyond the base of the 3rd pair of legs. Tarsus with 2 claws and 3 pairs of digitules of variable lengths (pl. XXV, fig. 2).

Larva of 2nd stage : Length 0.756 mm., breadth 0.422 mm. Body nearly spindle-shaped and swollen, dull greenish yellow with scattered greyish green markings. Three simple eyes on each side of the head, dull brown, and placed in a triangular blackish marking.

Antennæ similar in shape and number of segments to those of the 1st stage. Rostrum short, stout and extending beyond the base of the 2nd pair of legs. Tarsus with 2 claws, but no digitules.

WINGLESS VIVIPAROUS FEMALE. — Length 1.14 mm., breadth 0.72 mm. Body oval, dorsally swollen, light greenish yellow, sparsely covered with hairs and with snowy white mealy secretions.

Head nearly hemispherical, eyes simple, light red, 3 on each side. They are placed on a nearly circular blackish marking. Antennæ 5-jointed, 1st and 2nd joints short and stout, 3rd longest, the 4th and 5th nearly equal in length and each nearly one half as long as the 3rd. The 5th or terminal joint bears a slight notch at the middle of one side and has a blunt end with a small bristle. Rostrum 3-jointed, reaching beyond the insertion of the 2nd pair of legs. Legs rather stout. The hind legs are larger than the other two pairs. Tarsus with 2 claws (pl. XXV, fig. 6).

Larva of 1st stage laid by the female (pl. XXV. fig. 6) : Length 0.67 mm., breadth 0.30 mm. Body oval, depressed, its anterior half light yellow, posterior half light orange yellow, head very large and broadening towards the posterior edge. Antennæ 4-jointed, the terminal segment much elongated, with a blunt end bearing a few bristles. Ocelli colorless, 3 on each side of the head, at the edge of a brownish red irregular marking. Rostrum 3-jointed, pointed and extending as far as the base of the 3rd pair of legs. Legs all nearly equal in size, with two claws and two digitules (pl. XXV, fig. 7).

Larva of 2nd stage : Length 1.116 mm., breadth 0.666 mm. Body light yellow, with scattered orange yellow markings, and covered with white mealy secretions. Head rather small. Antennæ 4-jointed with a few hairs, the 2 basal segments short and thick, the remaining 2 longer, and the 4th or terminal one bluntly pointed and beset with short stiff hairs. Three ocelli on each side of the head, at the periphery of a crimson marking. Rostrum 2-jointed and extending as far as the base of the 2nd pair of legs. Legs rather short and stout, with 2 claws, bearing 2 long hairs at their base (pl. XXV, fig. 8).

WINGLESS VIVIPAROUS FEMALE. — Length 1.35 mm., breadth 0.845 mm. Body light orange yellow. The other characters are similar to those of the larva (pl. XXV, fig. 8), but dorsally the sides of the body are covered with white mealy secretions forming a sort of band. This mother Insect (pl. XXV, fig. 6) lays viviparously a certain number of larvæ.

Larva of 1st stage laid by the female (pl. XXV, fig. 9) : Length nearly 0.636 mm., breadth 0.30 mm. Body oval, flattened, light greenish yellow, and covered with a few hairs. Head rather large, nearly hemispherical. Antennæ thick, stout, and composed of 4 segments, the two basal ones thick and short, 3rd and 4th each nearly twice as long as the basal one. The 4th or terminal one is abruptly narrowed near the end, which is blunt and bears 4-5 short stiff hairs. Ocelli : three on each side of the head, each three embracing between them a deep crimson area.

Rostrum 3-jointed and extending as far as the insertion of the 2nd pair of legs. Legs rather stout; each bears two claws and two long digitules (pl. XXV, fig. 10).

Larva of 2nd stage : Length 0.876 mm. breadth 0.42 mm. In advanced stage length 1.08 mm. and breadth 0.648 mm. Body light dull yellow, the sides of the abdomen loosely covered with white filamentous secretions, but when advanced the white secretions thickly cover the body. Head hemispherical, comparatively large in size. Antennæ composed of 5 segments, 1st to 3rd short and thick and nearly equal in size. The 5th or terminal on twice as long as one of the remaining segments, with an abruptly narrowed end. Rostrum with a blackish apex, reaching to the base of the 2nd pair of legs. Eyes simple, 3 on each side of the head as usual, and colored deep crimson. Between the three eyes a light crimson marking. Legs all nearly equal in size, and colored light yellow; tarsus with 2 claws and 2 long digitules (pl. XXVI, fig. 11).

Larva of 3rd stage (pupa stage) : Length 1.62 mm., breadth 0.81 mm. Body long oval, orange yellow, excepting the head and wing-cases, which are light yellow. Antennæ 5-jointed, basal two joints short and stout, the 3rd nearly twice as long as the 1st or 2nd segment. Rostrum short, 3-jointed, with a blackish apex, and extending just beyond the base of the 1st pair of legs. Ocelli crimson, 3 on each side of the head. Each three embrace between them a triangular blackish marking. Below each group of ocelli there appear for the first time in this stage a large number of reddish wedge-shaped markings arranged with the pointed ends towards the group of ocelli. These markings are the first trace of the compound eyes of the future winged Insect. Legs all nearly equal in size, and provided with 2 claws and 2 digitules. Wing-cases present at the sides of the thoracical region (pl. XXVI, fig. 12).

IMAGO. — All females and no male. Length 2.34 mm., breadth 3.79 mm. Head blackish brown, thorax dark brown, abdomen purplish red. Eyes black, with a supplementary eye : ocelli 3, light ochre brown (fig. 13*a*). Antennæ blackish, 5-jointed, the two basal segments short and stout, the 3rd twice as long as the 4th and provided with 29 raised ring-markings. The 4th and 5th nearly equal in length, the former with 14 and the latter with 13 ring-markings. The apex of the 5th segment is smooth and beset with a few stiff hairs. Rostrum dark brown, rather short, and extending as far as the insertion of the first pair of legs. Wings transparent, veins yellow. Fore wing long and wide, and twice as long and over twice as broad as the hind wing. The 1st and 2nd oblique veins reach the subcosta at one point, the 3rd bifurcated near its end with inconspicuous base. Stigma long, greyish green. Hooklets on the hind wing 2. Wings horizontal at rest. Legs of moderate size, blackish brown, claws 2, digitules 4 (pl. XXV, fig. 13).

Larva of 1st stage (laid by the winged female) : Length 0.432 mm., breadth 0.252 mm. Body long oval, light orange yellow, dorsally with dark orange markings. Head large, nearly hemispherical, as wide as the rest of the body. Antennæ rather long and stout, and composed of 5 segments, the two basal ones short, the 3rd and 4th nearly equal in size, and each twice as long as one of the basal two. The 5th half as long as the 4th, with a few hairs at its end. Beyond the distal end of the 4th segment lies a single sensory disc (olfactory organ ?).

Eyes simple, deep crimson, 3 on each side of the head close to the triangular blackish space. At the frontal edge of the head lie two large horn-like processes with their basal half marked with transverse wrinkles. Rostrum light orange yellow with a blackish apex, reaching to the base of the 3rd pair of legs. Legs comparatively long and stout, with 2 claws and 2 digitules at their end (pl. XXVI, fig. 14).

Larva of 2nd stage : Length 0.73 mm., breadth 0.35. Body long oval, more or less depressed, light orange red, and covered with white short filamentous secretions. Head large, nearly as broad as the thorax. Antennæ rather stout and composed of 5 segments, the basal one shorter and thicker than the other segments, the 2nd,

3rd and 4th are nearly equal in size, the terminal or 5th segment twice as long as the 4th, but its distal half much reduced in size so as to form a finger-like process. At the junction of the basal and distal halves of the 5th segment lies a sensory disc.

Eyes deep crimson red, 3 on each side of the head, each surrounded with a broad blackish ring. The horn-like processes situated at the frontal edge of the head of that of the previous stage have entirely disappeared. Rostrum 3-jointed, light orange yellow, with a light blackish apex. On the 5th abdominal segment there lies a pair of short but broad cornicles having a disc-like expanded end (pl. XXVI, fig. 15).

The further stages of this larva are still unknown, but it is most probable that it produces the larvæ which initiate the formation of the galls the following spring and repeat the cycle described above.

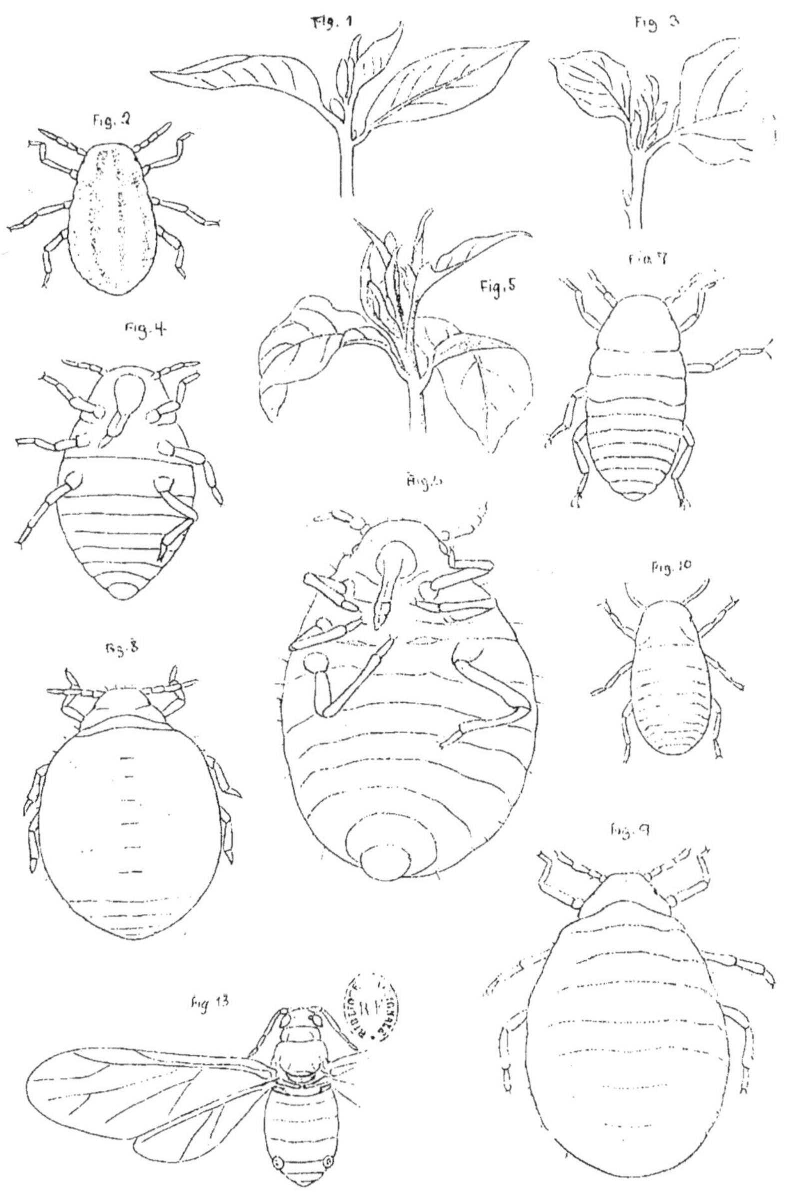

C. SASAKI. — A NEW APHIS-GALL. — I.

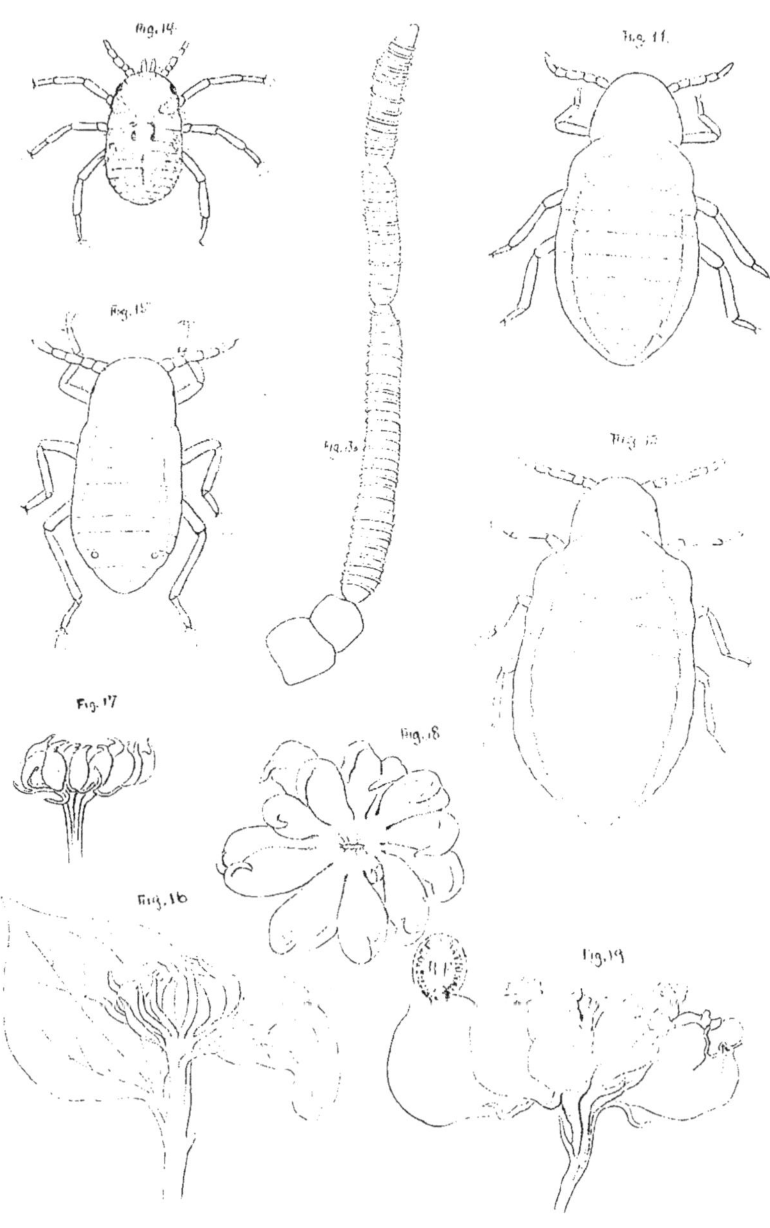

C. SASAKI. — A NEW APHIS-GALL. — II.

Sobre la nidificacion, la biologia y los parasitos de algunos Esfégidos,

por RICARDO GARCIA MERCET (del Museo de Historia Natural de Madrid).

Voy á tener el honor de comunicar al Ier Congreso de Entomología algunas observaciones sobre la nidificación, la biología y los parásitos de ciertos Esfégidos. Estas observaciones, confirman, amplian ó rectifican lo que hasta ahora se ha escrito sobre la vivienda y los primeros estados de las especies en que me he de ocupar.

Stigmus Solskyi.

No conozco ningún trabajo sobre la nidificación y biologia de este pequeño Esfégido. Sobre la de su congénere el *S. pendulus* han expuesto opiniones contradictorias, de un lado DAHLBOM y GIRAUD; de otro DUFOUR, PERRIS y RUDOW. Los primeros le suponen un Insecto afidívoro; los segundos le califican de parásito, afirmando haberlo obtenido de tallos secos donde habian previamente nidificado especies de los géneros *Trypoxilon*, *Psen* y

Mimesa. Tambien le consideran parásito de las *Osmia* y de algunas arañas.

Por nuestra parte, y con relación al *Stigmus Solskyi*, podemos asegurar que construye sus nidos en los tallos secos de la zarzamora (*Rubus amœnus*) formando en ellos galerias con perqueños ensanchamientos ó cámaras en los que deposita las provisiones (Afididos) con que se ha de alimentar su descendencia. Estas celdas son ovaladas ó en forma de tubo alargado y sinuoso, algunas miden de 5 á 6 m.m. de longitud por 1'5 - 2 m.m. de anchura, y aparecen recubiertas de una borra ó pelusa blanquecina en toda su extensión. En ninguno de los tallos habitados por este Esfégido se han visto nidos de *Trypoxilon* ni de *Mimesa* ni de *Psen*; ni de los nidos de *Trypoxilon* en los tallos del *Rubus* ha salido tampoco el *Stigmus*. Esto nos permite afirmar que el *Stigmus-Solskyi* no es parásito de los *Trypoxilon* ni debe serlo de la *Mimesa* ni del *Psen* sinó que construye y aprovisiona sus nidos como es costumbre entre los Esfégidos del grupo á que pertenece.

Su larva es ápoda, vermiforme, de color amarillento y de 5'1 m.m. de longitud. Consta de cabeza y 13 segmentos corporales. La cabeza es esférica, completamente lisa, y solo se puede ver en ella el labro, bilobulado, y la extremidad de las mandíbulas, quitinosa y negra. Los anillos del cuerpo están estrangulados entre sí y los que han de corresponder al abdomen del Insecto perfecto aparecen recorridos por un surco longitudinal, á cuyos lados hay una prominencia tuberculiforme. Todo el cuerpo aparece estriado longitudinalmente, excepto en el dorso, donde las estrias son transversales.

La ninfa es de color amarillento, como la larva, ofrece el aspecto general de las ninfas de los Esfégidos y representa con bastante exactitud el Insecto que de ella ha de salir. Las alas son rudimentarias; el primer segmento del abdomen más estrecho que el siguiente; los anillos abdominales 1-3 ofrecen un marcado reborde apical; los anillos 3-5 presentan una impresión ó surco longitudinal en el centro del dorso. El vientre está recorrido en toda su extensión por una quilla, y una linea saliente se observa también á cada lado del abdomen. Longitud de la ninfa 5'2 m.m.

PARASITO. — De un nido de *Stigmus Solskyi*, se ha obtenido un ejemplar de *Ellampus parvulus*, que habia devorado la larva del Esfégido.

Seguramente otros *Ellampus* y aun otros *Crisidos* serán también parásitos de los *Stigmus*, pues cada especie de *Chrysis* no es parásita de una especie determinada de Esfégido ó de Euménido, sino que puede vivir indistintamente sobre unas ú otras. En la provincia de Madrid de los nidos de *Odynerus sociabilis* y *O. Hispanicus*, hemos visto salir las siguientes especies del género *Chrysis* : *C. austriaca*, *C. hybrida*, *C. pustulosa*, *C. Chevrieri*, *C. ignita*, *C. æstiva*.

Diodontus minutus.

Esta especie nidifica en la tierra, construyendo cámaras ó aposentos que miden 6-7 m.m. de longitud, por 3-3'5 de anchura. Estas celdas son lisas y pulimentadas y encierran restos de las provisiones que sirvieron á la larva para su alimentación. La larva es extraordinariamente parecida á la del *Stigmus Solskyi*, lo que no tiene nada de particular, puesto que tambien los adultos de una y otra especie se asemejan mucho. La longitud de esta larva es un poco mayor que la del *Stigmus*, llegando á medir 6 m.m.

Parásito del *D. minutus* es el *Ellampus auratus*.

Trypoxilon scutatum.

La nidificación de este pequeño Esfégido no difiere en nada de la de su congénere el *T. figulus*. Construye sus nidos en los tallos secos de la zarzamora (*Rubus amœnus*) y los approvisiona con los mismos materiales. Su larva y su ninfa se parecen extraordinariamente á las del *T. figulus*, diferenciándose solo por su menor tamaño.

Hemos visto un nido de *T. scutatum*, parasitizado por la *Chrysis cyanea*. El Crisido á su vez, sufría los ataques de un Calcídido, la *Eurytoma tibialis*, y esta á su vez era parasitizada por un pequeño Proctotrúpido, cuya determinación no ha podido hacerse todavia.

Pelopœus destillatorius.

La nidificación de los *Pelopœus* ha sido estudiada y descrita por varios náturalistás, entre ellos REAUMUR, DURY, LUCAS, SAUNDERS, EVERSMAN, MAINDRON, ANDRÉ, FERTON, &. y de los parásitos de estos Esfégidos han hablado ABEILLE y MOCSARY. El trabajo de MAINDRON sobre los nidos y las costumbres de los *Pelopœus* es de lo más completo y exacto que se ha escrito, y los dibujos que le acompañan reproducen con fidelidád los primeros estados de estos animales y la habitación que construyen para pasarlos. Poco se puede añadir á lo dicho por tan concienzudos observadores; sin embargo por lo que se refiere á una de las especies que viven en mi pais, el *P. destillatorius* ILLIG., creo que sobre sus larvas y ninfas y sobre sus parásitos puedo comunicar á este sabio Congreso algunas noticias que completen las ya dadas á conocer por algunos de los naturalistas que he nombrado.

El *P. destillatorius* es una especie muy abundante en la región central y en el E. de España, que se puede recoger en los meses de Junio y Julio sobre las flores de la *Thapsia villosa*, del *Peucedanum stenocarpum*, de la *Ruta montana*, del *Eryngium campestre*, del *E. maritimun*, del *Daucus carota*, del *Foeniculum vulgare*, &.

Forma sus nidos, como las demás especies del género, con tierra humedecida, á la que comunica mayor cohesión alguna sustancia segregada por el animal. Los nidos se construyen ó adosan sobre las grandes piedras ó sobre las tapias pedregosas bien expuéstas al sol. Estos nidos, de volumen variable, ofrecen exteriormente aspecto irregular y encierran un número de celdas ó cavidades que no baja de tres ni pasa de siete. Las celdas ofrecen forma cilíndrica ó fusiforme, son perfectamente lisas, aparecen como barnizadas ó recubiertas por una capa lustrosa; miden de 27 á 30 m. m. de longitud por 5-6 m. m. de anchura, y en ellas deposita la ♀ que las construye, primero las provisiones que han de servir de alimento á su prole y despues el germen del individuo que las ha de consumir. Las provisiones consisten en arácnidos de la familia de la Epéiridos, que la obrera inmoviliza y anestesia por medio del aguijón, y el número de arañas depositadas en cada nido varia de 5 á 6. Con este repuesto la celdilla queda completamente llena y en disposición de recibir el huevo que ha de albergar. Depuesto este sobre una de las arañas, el *Pelopæus* cierra la celda con el barro que construyó el nido y se ocupa en aprovisionar otro de los alojamientos que preparó.

La larva que nace de estos huevecillos es de color blanco amarillento, gruesa, ápoda, ancha en el centro y estrechada hacia sus extremos, de unos 20 m. m. de longitud por 5 m. m. en su mayor anchura, que corresponde al 6º y 7º anillos del cuerpo, y consta de cabeza y 13 segmentos corporales.

La cabeza es pequeña, redondeada, lisa y lustrosa sobre la frente y el vértice y sin otros órganos ni aparatos visibles que el de la masticación. El clipeo es transverso, más ancho que largo; el labro, transverso también, ofrece dos lóbulos en su extremo inferior. Las mandibulas son fuertes, corneas, bidentadas; las máxilas, redondeadas, presentan en el ápice dos tubérculos cónicos, de los cuales el exterior es más corto que el interno. El anillo protorácico ofrece cuatro mamelones, dos sobre el dorso y otros dos laterales. El resto del cuerpo presenta estrangulaciones interanulares. Sobre cada segmento se observa un surco transversal, y á cada lado una protuberancia ó mamelón. Todo el cuerpo aparece estriado longitudinalmente. El último anillo, más estrecho que los anteriores, pero algo mayor que la cabeza, es liso, casi semiesférico y está desprovisto de protuberancias.

La ninfa es de color amarillento y mide unos 20 m. m. de longitud. Las alas, cortas todavia, apenas pasan de la cara posterior del segmento medio; el peciolo, todo él mucho más grueso y fuerte que en el Insecto adulto, ofrece tres quillas carnosas, una dorsal y dos laterales. El abdomen, más ancho que en la fase perfecta, presenta á cada lado de los anillos 2-7 un mamelón carnoso muy pronunciado. El 8° anillo (la ninfa observada es de ♂) lleva las armaduras genitales. Todos los anillos, excepto el último, presentan sobre el borde posterior de la cara dorsal una doble fila de papilas mameloniformes.

El capullo ninfal, de consistencia como papirácea, es de color de caramelo y está cerrado en uno de sus polos por un grueso tapón parduzco, de aspecto y color escrementicios. Mide de 21 á 24 m. m. de longitud.

De los nidos del *P. destillatorius* hemos obtenido dos parásitos : la *Chrysis Tachzanowsckyi* y el *Stilbum siculum*.

En las celdas parasitizadas se observan muchos restos de las provisiones que acumuló la ♀ propietaria del nido. El capullo del parásito, mucho menor que el del Esfégido, ha reemplazado al de este. Los capullos del Crisidido son oviformes y se abren por uno de sus polos, que se separa de un modo regular, constituyendo como un opérculo ó tapadera.

En el tomo III de « Species des Hyménoptères d'Europe et d'Algérie », por EDMUNDO ANDRÉ, al hablar de la nidificación de los *Pelopœus*, expone este autor la sospecha de que tal vez las especies del subgénero *Chalybion* construyan sus nidos en otra forma y con otros materiales que los verdaderos *Pelopœus*. Funda su hipótesis el Sr. ANDRÉ en el hecho de haber visto un nido de *Chalybion violaceus* construido en un trozo de madera agujereada.

Yo no sé ciertamente si la opinión sustentada por el fundador de « *Species des Hyménoptères* » ha sido ya rebatida por algunos naturalistas, pues no he tenido tiempo de revisar todas las notas sobre biología entomológica que desde entonces se han publicado, pero por si acaso nadie lo hubiese hecho, daré cuenta aquí de algunas observaciones verificadas por mí mismo en Manila sobre la nidificación del *Chalybion bengalensis*. Esta especie, muy común en las Islas Filipinas, construye sus nidos, como los verdaderos *Pelopœus*, con tierra humedecida y á la que da consistencia una baba obscura que segrega por la boca el Animal. El *Ch. bengalensis* construye sus nidos dentro de las habitaciones, adosándolos á los marcos de las puertas, á las cornisas de los armarios y de las camas, y estos nidos tienen la misma forma, el mismo aspecto y la misma disposición interior que los que construyen los *Pelopœus*, Yo he tenido ocasión de ver, varias veces en Manila, en mi misma casa, fabricar sus nidos á la ♀ del *Ch. bengalensis*. La primera materia para la construcción, la tierra humedecida, la tomaba de un jardín immediato, transportándola á pequeños fragmentos, sostenidos entre la boca y las patas anteriores. Por medio de las mismas patas, y de las mandíbulas, fijaba el mortero sobre la cornisa de madera elegida para sustentáculo de la futura habitación, y colocando fragmento sobre fragmento llegaba á construir el nido. Las arañas para aprovisionarlo las encontraba también en el jardín próximo, y era de ver el celo desplegado por esta infatigable obrera en la busca, captura y acarreo de los comestibles para su futura prole. La presencia del hombre no parecía importarle poco ni mucho, y si un nido le era destruido á medio construir, ella emprendía de nuevo la fabricación del albergue sobre el mismo punto de donde la habían querido arrojar.

Estas observaciones demuestran que los *Chalybión* construyen y aprovisionan sus nidos como los verdaderos *Pelopœus* y que el nido de madera agujereada de que hablaba M. Edmundo André, debe constituir un hecho excepcional.

La tendencia á construir sus nidos dentro de la vivienda humana, que se observa en algunos *Pelopœus* y *Chalybión* exóticos, debe constituir para estos Animales un medio de esquivar el ataque de sus enemigos y parásitos ó de hacer más difícil la persecución á estos.

En efecto : Mas de 20 nidos de *Chalybión bengalensis*, fueron recogidos por mí, en Manila, dentro de habitaciones, y de ninguno de ellos se vió salir otra cosa que individuos de la especie propietaria. Esta habia escapado por completo al ataque de sus parásitos. En cambio, en España, de tres nidos de *Pelopœus destillatorius*, encontrados sobre piedras en el campo, dos estaban parasitizados, uno por la *Chrysis Tachzanowsckyi* y otro por el *Stilbum siculum*. Se vé, por lo tanto, que la vivienda humana ofrece á los *Pelopœus* unas condiciones de seguridad que no encuentra el Insecto en la naturaleza, y se comprende que una vez habituada una especie á la convivencia con el hombre busque en el albergue de este la protección necesaria para su descendencia y seguro desarrollo ulterior.

Progress of Economic Entomology in India,

by H. MAXWELL-LEFROY M. A., F. E. S.,
and F. M. HOWLETT, B. A., F. E. S.,
Imperial Entomologists, India.

We propose in this papers to give an account of the measures that are being taken to check crop pests in India and of the efforts that are being made to bring to the attention of the Agricultural Community in India the necessity for and value of preventive and remedial treatment for injurious Insects. In order to be clear we must first deal with the circumstances of the case, since both Indian agriculture and the conditions of the Indian agriculturist are very different from those of European countries.

Agriculture in India is, to an extraordinary extent, the staple industry of the country. The proportion of persons engaged in it is 62 %, the cultivated area is 382,975 square miles and, compared with it, manufacturing arts and industries are very small. The great proportion of the people live in villages, are an agricultural people pure and simple, and lead extremely simple lives.

For our purposes, there are three classes of people concerned, the small holder, tenant or cultivator, the zemindar or large land-owner, and the European planter. The last is so small an item as to be negligible; there is an English community producing tea in Assam, North Bengal, etc., another producing tea, coffee, rubber, etc., in South India, another producing Indigo and staple crops in Behar. The last community is, in essence, a landowner community similar to the zemindars, not so much directly engaged in cultivation as having rights over land, administering it and taking a proportion of the produce of the cultivators on it. The European planters, however, are in a position to understand and profit by scientific improvements, and, though a conservative and slow body of men, are able to exert an influence in the districts and are an important medium for introducing new methods and improved practices. The cultivator is the dominant class, a body of men of immense number, who live on the land and seldom move about, who have an extremely limited education, who have inherited a singularly perfect system of agriculture in most cases, and who are an extremely difficult class to reach or to move in any practical manner. With them is the far smaller body of zemindars or landowners, who really farm on a larger scale or administer large areas, and who are a wealthy and well-educated class. Even to them, the improvements of science appeal only slowly, and then only when placed before them in an extremely practical and simple manner. Were there a body of men from their own community, trained in scientific agriculture, able to adapt it to the needs of the country and to the peculiar conditions of their agriculture, the benefits of scientific knowledge and its practical application might be put to them in a way they would appreciate and understand. They would in turn influence the immense cultivating class below them, the vast class of men cultivating only from a half to ten acres of land each, or belonging to the villages which each cultivate communally a few hundred acres and whose experience is bounded by the limits of that one village and the dozens like it that surround it. It is for such men that an endeavour has to be made to devise improvements, and it is to such men that the results of these endeavours will have to be brought home, in the hope that the prosperity of the country may increase and the yield of the land be enhanced.

Naturally the resources of such a community are not very large and no appliance but the simplest is available. To the practical entomologist from Europe, accustomed to spraying machinery, insecticides and mechanical appliances, the resources of these villages appear very small. Nearly all their possessions are made of wood, bamboo, or pottery. Brassware is locally made, and simple iron appliances. But nothing else is available; machine-woven cloth and kerosene oil have penetrated to all the villages; iron cane-mills and improved ploughs to some; but to think of insecticides and of spraying in connection with these villages is ridiculous. For a certain class of pests, light-traps have been found suitable; but such were not really possible because an appliance giving a good bright light, that will keep alight in a high wind, does not exist in the villages, and only near large towns have proper glass-chimneyed lanterns been seen or are available. In the big towns anything may be bought; in the thousands of villages, isolated from the towns, nothing is available but what can be made of bamboo, wood, cloth and rope, and the only insecticidal substance is kerosene; even for making kerosene emulsion, soap is not generally obtainable, and one is driven to reject all such appliances and to turn to something far simpler.

In India, as in other tropical countries, the losses from Insects are considerable; no statistics are available because the recording staff competent to attribute a failure of crops to the proper source is not existent. The value of cotton lost annually, that might be saved, is perhaps two million pounds; in the epidemic of Bollworm in the Punjab and Sind in 1906-1907, the loss of the crop over a large area was from 40 to 100 %, and the value of the crop lost was somewhere about £ 2,000,000. In the Locust campaign of 1903-1904, £ 14,000 was spent in rewards for destroying Locusts in one province, which was regarded as a form of State relief to those who lost crops eaten by Locusts. In parts of Eastern Bengal and Bengal, the rice crop is sometimes a failure over a considerable area or the loss is a very heavy proportion; so too with other staple crops and it is no exaggeration to say that in some districts certain crops are not grown because the experience has been handed down that crop is a failure, originally owing to Insect-pests which were not checked, the reason being now forgotten. Now that the Agricultural Department, with no knowledge of this inherited art, are introducing new crops to new districts and testing crops on experimental farms, this is being found out, and it is certainly true that the immunity of a crop from Insect-pests has probably been a determining factor in many cases whether that crop was grown in that district or not. In the past another crop was substituted for a crop liable to attack, and though the reason is now forgotten, it is probable this was in the past fully taken into account. At the present day, staple crops suffer severely from epidemic crop pests at intervals of several years, they also suffer from minor pests regularly, but not to such an extent as to prevent their cultivation. Both of these forms of loss can, we believe, be prevented wholly or in part.

The cultivating classes have inherited a system of agriculture that is in most things extremely efficient and the result of ages of inherited experience. But in questions of Insect-pests and the like, this experience is dominated by what are, to us, totally erroneous ideas about life and what is also, to us, an incomprehensible aversion to directly taking life. An intelligent cultivator, growing sugar-cane under irrigation on an extremely sound system with good manure, believes that the Cane-borer comes with the well water used for irrigation; he has no conception that it has hatched from an egg, that the Caterpillar turns to a chrysalis and so to a Moth, that the Moths are male and female, couple and lay eggs. He regards the whole thing as a mystery, not comparable with the life of any other animal; he will, as likely as not, call in a priest to check it; the priest will perhaps write four texts from holy writings, place them one at each corner of the field to confine the evil influence and then remove one to let out the influence which the texts have incommoded. Or he will pay a man of certain caste to plough a line across the field at night, the man having to be starck naked. In some parts Locusts are believed to be the incarnation or the manifestation of a particular deity and for each one killed, a hundred will come; it is quite likely that this has occurred, a small swarm of which a few were killed being followed later by a much larger swarm, but where we see no connection, he sees a definite sequence of events. A case came up where a man freed his rice field of a pest by a simple mechanical method; his crop benefitted but soon after his cow died, and to that village the one was a consequence of the other. It will take much to convince that village that crop pests can or may be checked. This attitude towards crop pests, bound up with the inherited instincts and religious beliefs of the people, gives way when the cultivator can be shown the results of checking pests; but to do this in every village of India, to wait till a definite case comes and then to prove that it was the treatment that checked the pest is an impossible task. With a well educated community, much can be done by personal influence, by the former giving it a trial and being able to see for himself the rationale of the treatment; even to begin to do this with the Agricultural Community of India cannot be thought of yet. We cannot here dwell on this point, though it is the crucial one in India, the very little we have said cannot convey an impression of the attitude which is ingrained in nearly all classes in India and it can be only realised by actual contact with the people themselves; we cannot and do not really know what their attitude is, we can only get glimpses of it, but the attitude is a real factor that is all important in practice. Where money is really concerned, where one can absolutely show that money can be saved, the scruples disappear and the necessity of obtaining daily bread induces the cultivator to adopt a remedy; but it must reach that point, and the very idea of constantly and always checking crop pests or trying to understand them is absolutely foreign to the minds of the agricultural classes in India.

Government. — In India, Government exerts an influence and takes care of minute things in a way scarcely known elsewhere, and seeing that it is as a section of Government that an Agricultural Department comes in to benefit agriculture, we must be clear as to the means which Government has at its disposal. The system in India is an extraordinary one, in great measure taken over by the British from previous native rulers; roughly speaking, the Government is primarily a revenue-collecting body; in each village is one or more officials, one concerned with the general management of village affairs, another with the revenue payable on each little bit of land, and so on; over a group of villages is another official, who in turn reports to a higher official controlling groups of these groups, and he reports to an officer in charge of a subdivision; two, three or more subdivisions constitute a district in charge of a single official (the collector), usually an Englishman, who is responsible to the Government for everything concerning the welfare of his district. In one village let us say the crop of rice is suffering from a pest; the village official reports it, as it is an all important thing to them to get the revenue remitted on that crop, and the report goes up through the other officials to the collector, who calls upon the Agricultural Department to advise, or to check the pest. If the attack is not limited to one village, but is wide spread, then notice comes quickly. In ordinary cases nothing more follows; the information is recorded for future guidance; perhaps a man is sent to investigate. If it is a bigger attack, then measures are taken, if possible; the case is thoroughly investigated; precautions, if possible, are suggested and the head of the district sets in motion the machinery and from official to official goes down the order to do so and so, until it reaches and is discussed in every little village. In the Bollworm campaign in the Punjab, instructions were sent down to cut down and burn all cotton-plants during the winter so as to leave none to harbour the pest, this was done in thousands of villages, better done in districts where the English official showed he meant it to be done, but done on a very large scale. There were sent out 80,000 copies of a circular in Punjabi and Hindi, explaining why this and other measures were to be carried out, and thus what was to be done, and the reason why was brought to the knowledge of the people in all the villages that had lost their crop of cotton.

This gives a rough picture of the Government machinery and it is in an emergency an extraordinarily good machine. When the expert side of agriculture has been more developed, this machinery will spread the work of the expert throughout the country. Government in India takes note of very small things, and in every district a weekly report is sent in on the condition of crops, mentioning among other things if there are losses from Insect-pests. We are now beginning to know what the Insect-pests are, and are slowly beginning to see how they are to be met.

The Agricultural Department has been founded to study and improve the methods of Indian agriculture, to test new methods and introduce new crops, to adopt and teach the best methods of western science and, in short, to try to make two blades grow where one grew before. It will do this by direct teaching and practice, and through the medium of the district officer and the system of Government outlined above. It is at present a new thing having been founded in recent years and having only recently been expanded. For each province there is one or several English officers, trained in agriculture, who deal with general agricultural problems, there is also a chemist, and a botanist, both highly trained Englishmen, and a native staff to assist them. There is also a native staff to deal with entomology and mycology, under the direction of the botanist or some other officer. In addition, there is a Department under the Government of India, which deals less with local problems than with bigger questions best dealt with at one centre; and in this Department is the Entomological Section, which largely directs and controls the work all over India In this Section is done all the scientific work, the study of life-histories, the testing of remedies and methods; in the Provincial Departments is done the work of investigating outbreaks, giving advice and help on the spot and accumulating local information.

For us in the beginning the first thing was to know what the pests were; we had the published records in Indian Museum Notes; we have also had a system of reporting the occurence of Insect-pests; in all districts, the responsible officers were asked to report pests and send specimens; this is done but in a peculiar manner; in a group of villages a pest will occur say on paddy; this is reported, with a view to revenue remission; probably a subordinate native official remembering the orders sends a man out with a bottle of spirit to get some of it; if he is a careful man he goes himself, but in many cases he sends in whatever his ignorant constable catches; this goes up with a report and eventually reaches the Department. In addition to this, which experience has taught us to accept with caution, we have now a number of trained men actually investigating what the pests are, and we have secured the interest and help of many English and well-educated native officials, of missionaries, planters, business men and others. We get a stream of reports and specimens, and we are ourselves constantly investigating. We now know a great deal, we think we know all the major pests and many minor ones, and though we occasionally get new ones, the list of pests attached is probably fairly accurate (corrected departmental list is attached).

The next thing is the life-history. This is worked out in the insectary at Pusa, then checked wherever it occurs and worked out for the different seasons. With this comes the question of hibernation, parasites, influence of climate, seasonal occurrence, etc. We are seeking to find what conditions produce outbreaks, but it must be years before we can make much progress for so diverse a country as India. Nevertheless we are making steady progress all the time and for many species we have accurate information for the whole year, for the whole of India. At the same time we consider the question of remedies; this is so much a matter of local conditions that it is largely the work of the men stationed in the different parts of India who alone can thoroughly know and study local conditions. If it were a question for instance of spraying, we could at Pusa work out the best method and we always test and work out methods at Pusa. But when it comes to checking a pest by burning stubble or preventing hibernation or using trap crops, it must be worked out on the sport with regard to local agricultural practices. In India, the conditions of land tenure make it a matter of necessity that cooperation must be secured in the simplest protective action against crop pests; it is not a question of farms or of landowners. We have to deal with people growing crops on little holdings of an acre or two, and where the unit elsewhere might be a farm, with us it must be a village; so too the system of Government is such that we can work with such units, and it is the reason why, where among farmers you might have compulsion by law, you have in India persuasion by Government officers backed by a refusal to remit the revenue if the crop is deficient and if no measures have been taken for the common good. In case of loss of crop from any cause the proper officers must be convinced in order that the full assessment of land revenue or rent may not be taken and to secure remission : if the proper officers are convinced that the measures to be enforced are really workable and efficient, they have means of exerting powerful influence, and it is usually possible to get action taken on a very large scale, provided the remedy is shown to be really worth while. There is no legislation on this subject as in some other countries, and it is not required. The following are some cases that have occurred of the successful application of remedial measures on a large scale.

In 1903-1904, the *Bombay Locust* occupied 12,000 square miles of Bombay and then spread out over 140,000 miles to lay eggs. In this case, all the districts involved took the matter up, with advice and instructions from the entomological staff. In this case the whole staff of Government subordinates is used if necessary, and the European officials in charge of districts were hard at work. The main things done were to destroy the Locusts at night when they were torpid, to collect coupling Locusts, to mark down egglaying places and collect eggs, and to destroy the Hoppers. For practically all these Government offered rewards fixed from week to week; for instance, for young Hoppers, as much as 8 annas (pence) per pound was paid, but as the Hoppers got larger, this was reduced gradually to 2 annas or 1 anna. The bag-method, then invented, was and has since been most valuable, and probably by far the most useful discovery in applied entomology as it affects India, as it is used for a great variety of pests. Very largely the action of the officials was to let all know about the rewards, to stimulate the people to collect, and to pay the rewards. In all £ 14,000 was paid out in the form of rewards. A full account was published (« Mem. Agric. Dept. Entom. », vol. I, n° 1), and the reader should especially note Appendix A, which gives a very good idea of the extraordinary efficiency of the Government machinery in India.

HAIRY CATERPILLARS. — In particular tracts, a noted pest is the swarms of Arctiid Caterpillars that appear regularly soon after the first rain. What actually occurs is that when the first shower falls, the Moths come out from the ground and lay eggs. These Moths can be captured by means of light traps, and if these are used exactly at the right time for each tract, very much good can be done. This method is now being practised in two areas; the first year or two years the Department have carried it out; after that it is left to the people to carry on (see « A. J. I. », vol. III, p. 152, and vol. V, n° 3.)

THE POTATO-MOTH. — For some years, seed potato has been imported from Europe owing mainly to the difficulty of preserving seed through the moist hot weather and with these have been imported the notorious Potato-Moth (*Lita solanella*), which attacks both the plant and the tuber. It was found that this pest rendered the storage of seed still more difficult and the matter was taken up. A long series of storage trials was done at Pusa from April to October; those tubers were then planted and the yields noted. A small number out of the methods tried were selected for their efficiency both as regards preventing the access of the Insect and as not encouraging the tubers to rot. Of these the best for local use was selected in one area in the Central Provinces and tried in several villages, the Department bearing the cost and doing the work for the growers. The results were so satisfactory that in the next year the same method was voluntarily adopted by the growers, a fieldman from the Department being sent simply to see that they did the treatment properly. This method or modifications of it are taking on in other localities and will probably spread over the whole potato-growing areas (see « A. J. I. », vol. V, p. 19, article by H. M.-L. and G. EVANS).

RICE GRASSHOPPER. — This pest does serious damage to the rice crop, very much reducing the yield. At the present time it is being successfully checked by a very simple method. A bag is made of very coarse open gunny cloth (jute); one side is kept open by two bamboos and the bag is held by these and drawn through the rice. As the crop is grown in water, the work is slow, but in practice there are four men; two draw the bag so that the lower edge is just below the surface, two more go in front and a little outside, with a switch to brush the paddy; they frighten the Grasshoppers in towards the bag and, as they are sluggish, they are mopped up in the bag and then destroyed. This method has been thoroughly worked out, thoroughly done for one season at one place, and is now done by the people themselves, who bear the whole expense and who subscribe the necessary funds. Wherever the Rice Grasshopper is a pest, it can be met by this method, and it is now solely a question of demonstrating the process year by year, as it is rendered necessary by destructive outbreaks of the pest, till all affected areas know of it and do it.

BOLLWORM IN COTTON. — The cotton-plant in India is attacked by three species of Bollworm, two *Earias* and a *Gelechia*. In 1905, the cotton crop over 700,000 acres of the Punjab was a failure. Investigation showed that Bollworm *Earias* was enormously abundant, so much so that in many areas every flower bud was eaten before it opened, and the crop was reduced to one-eighth or a quarter of what it should have been. The loss to the growers was about £ 2,000,000. The measures resolved upon and carried out by the Punjab Agricultural Department were : 1) to destroy all the plants still standing in January, to prevent hibernation of the Insect in the resting larva or pupa stage; 2) to plant bhindi *Hibiscus esculentus* as a trap crop and destroy it before the cotton flowered; 3) to reestablish the parasite, believed to have been killed in the preceding winter by the abnormal cold, whose absence we believe led to the outbreak. The last is the most important, and we looked to it mainly. The first measure was done by putting pressure on in every district and by the free use of leaflets, 80,000 of which were sent out. It was probably fairly well done over the greater part of the area. The second was a failure, because bhindi is a garden crop, requires frequent watering when young, and the bulk of the people did not understand it. In a few places where there were market gardening castes it was grown and was probably useful. The third was done by sending in boxes of bhindi pods containing Bollworms from places outside the Punjab, some coming over 1,000 miles. The box was so arranged that parasites could get out through wire-gauze of such a mesh to keep in the Moths. Whether from this cause or not, the parasite was reestablished in the next season and the Bollworm reduced to its normal level. The strongest reason for believing the parasite was the real cause, was shown by the neighbouring area of Sind, to which no parasites were sent and in which Bollworm raged unchecked the next year; in the following year, the parasite was sent to Sind, and the Bollworm in that season became normal. An account of the Punjab campaign will be found in « Bulletin n° 1 of 1907 » (Dept. of Agric. Punjab, 1907). We may draw attention to the International importance of Bollworm; India might produce very large quantities of cotton, especially the long-stapled cottons now required. India does not, very largely because Bollworm in India is a serious bar to the growth of any but the existing short stapled cottons which mature rapidly and offer little scope for Bollworm attack.

MANGO MEALYBUG. — In India, two very important trees flower and fruit between March and July. These are the mango and the jackfruit. Both are attacked by a Giant Mealybug (*Monophebus octocaudata*), which has a very peculiar life-history; it is one-brooded; the eggs, laid in April, do not hatch till November and the Bugs take till April to become mature. They require very large quantities of food at a time when the mango and the jackfruit trees are putting out new shoots and flowering; settling on the shoots they suck out the sap rising to the fruits and the fruits drop off while quite small, the crop of mangoes and jackfruit being then very small. Till now the matter has been looked on as hopeless, and even an entomologist would express doubts over a Mealybug in the warm climate of India having only one brood a year; but now we have worked it out and there are two simple remedies; the first is to band the trees and prevent the Bug getting up to develop or *down* to lay eggs; the second is to destroy the eggs laid in the debris at the foot of the trees, these eggs lying there from April to November (see « Mem. Agric. Dept. », vol. II, n° 6). These methods are now in use in North India where this pest occurs.

These are instances of the work that has to be done and which is being done in India now. We are not doing this violently all over India, but where we can; where a pest turns up and really does harm enough to make the people want to do something, we try to organise a campaign. The work is at its very beginning, and against many pests we cannot even suggest the lines of work, as the preliminary study has never been done, but this is being done and as the class of work is being better understood, we are getting nearer to being able to tackle our pests by direct active work of this kind.

It must be remembered that in India only a very little can be done by publications of any kind, by the ordinary methods useful elsewhere. We must go to the cultivator, show him, put pressure to bear on him and, as a rule, actually do the work for him at the start. If therefore we appear to have made little success, this must be borne in mind, and we cannot hope to make any but the very slowest progress. An interesting example of this in seen in the account of the work against Til Stem Borer by C. U. PATEL (« A. J. I. », II, p. 153).

INSECTICIDES. — When the Section commenced work, insecticides were very little known or used at all in India. Paris green and kerosene emulsion were used to a small extent. We introduced Crude oil emulsion, first having it made in England, and then, as the sales increased, getting it made locally. As a standard contact poison it is very effective and useful; it finds a large sale at 6/4 (8 s. 4 d.) per 5-gallon drum, making between 150 and 300 gallons of wash. It is also used for domestic Animals and against Fleas, etc. For a stomach poison, we introduced lead arseniate, which is locally made and sold at 1 Re. (1 s. 4 d.) a pound. It has a small sale only and we are now using lead chromate, an insecticide found after an exhaustive inquiry with the aid of a chemist (R. S. FINLOW, Esq.).

For special work we have found rosin compound a valuable insecticide, and on tea soaps are used. During the past seven years we have tested many patent and proprietary insecticides sent to us by the agents for firms in England and Germany; we have not yet found one that can be sold in India at such a price as to take the place of the above, and none have any use in India. Among other insecticides, we test particularly those offered for use against white Ants, and among these we have recommended one which is really effective and whose use we stimulate. The following are some that we have tested :

McDOUGAL's, V. 1, V. 2, Carbolineum, Cardo, Glifts, Apterite, BLOMER's solution, Vermisapon, Atlas compound, Solignum, Fichtenin, ROBINSON's tetramulsion.

It is open to any firm or individual to send insecticides for trial, and as we receive a large number of enquiries from users and are ourselves large users, we exert a very considerable influence upon the trade and use of insecticides, endeavouring to keep out useless rubbish made only to sell and to maintain the business at a high level. So also with sprayers; we maintain a register of makes stocked in India, we test them all, and we recommend, both to the trade and to the user, what we consider good makes. The largest sale is for knapsack sprayers, which are being more widely used, but spraying is restricted.

FUMIGATION is not practised in India for either buildings or plants, but there is a limited amount done for grain, using carbon bisulphide. It is never likely to be widely used inland on account of the cost, the damage of the material and the immense cost of carriage by rail. The treatment of grain and seeds stored in bulk against Weevil, etc., will be on other lines, not using fumigants at all probably, and we have sought to replace carbon bisulphide by some other fumigant in industrial use such as for factories, etc. Some such liquid as carbon tetrachloride will probably be used. Fumigation of plants and buildings with hydrocyanic acid and of grain with carbon bisulphide is practised at Pusa, and if there is scope for it, will be adopted more widely.

A distinct branch is concerned with Insects that affect the health of man and his domestic Animals; the Insects parasitic on Animals, the Insects that are known to convey disease, the Insects which suck blood and which are possible disease-transmitters, are being studied and worked at; this work is in its infancy; but it has reached the stage where we know what the more common blood-suckers and parasitic Insects are, and for a number of them have worked out life-histories and know their habits. The difficulties of the Mosquito problem in India are well known, and in view of the impossibility of using ordinary insecticidal remedies in any but a very small minority of cases, many of the difficulties with Cattle-pests are much increased. It is probable that progress with this branch of entomology will have to be made on different lines from those on which the treatment of crop pests has gone, and that when the initial work has been completed, it will be found that only by further investigation on more purely physiological lines will it be possible to make more than a slow advance in certain directions.

Few countries suffer a greater loss of potential energy from Insect attacks than does India from the almost universal prevalence of Mosquito-borne malaria, and in this one direction alone, in spite of the good work that has been done by many medical officers, there is ample scope for detailed bionomic research of a kind which might have far-reaching results.

The work at Pusa is carried on in connection with the Medical and Veterinary Departments, and we are able to give technical assistance to others who are working at parasitic and disease-transmitting Insects.

Publications are helping us, even publications in English, as they get translated into the vernaculars and filter down into the newspapers. « Indian Insect Pests », published in 1906, mainly reached European officials, planters, etc., but also some well-educated men; much of it has been translated into vernaculars and an edition in Bengali is just out, while others in Hindi, Telugu, Gujarati, etc., are following. We have also published articles in the « Agricultural Journal », and short articles or leaflets in the vernacular « Agricultural Magazines » now conducted in some provinces. In special cases we issue special leaflets; in the Punjab Bollworm Campaign, for instance, 80,000 copies of a leaflet were issued. In connection with these publications, and separately, we publish colour plates with very short explanations; these are now being more widely used for exhibitions at farms, at shows, at museums, schools, etc., and in some cases they are published in the agricultural papers or weekly newspapers. A good colour plate is appreciated for itself, though it is doubtful whether we have yet got the right stamp of plate to appeal to people. Very many people in India, for instance, cannot see a photograph, but we are finding the type of plate that appeals, and we find that a really good picture of a plant is recognised. We shall in time meet these difficulties, and as we have not yet reached one per cent of the people, we may take our present efforts to be tentative and experimental rather than far-reaching educational influences. We are also tentatively trying how far lantern-slide shows will be understood, but it is still doubtful whether these will be more than a show, though the impassioned appeal of the lecturer may incidentaly fix one idea in the head of many of the audience while they admire the to them incomprehensible pictures. Still a change has become evident even in seven years; many have read something, many are interested, and the next generation will probably be different; it is a new thing that interests and though we have not got to forming societies or reading records of captures, we have numbers of influential people who know that a Caterpillar becomes a Moth which lays eggs, and who take all occasions to teach others. These are straws which we clutch at and heap together as showing that even with one worker to about 10 millions of people something is being done and we sometimes think that per man we do more than in civilised places where entomologists are numerous enough to form societies. You must remember that literally and absolutely in 300 millions of people there is not one man who, apart from our and similar Government Departments, takes any interest in the subject or has ever done so. The mere idea of taking an interest in nature is foreign to all but such peoples as the Khasis and Lepchas, who are born naturalists. So that any progress we make will be in the face of a good deal of difficulty. We shall make more rapid progress when the effect of the students trained by the Agricultural Colleges becomes more marked. In some cases, these men become officials and regard their training simply as an avenue to Government Service, forgetting it all as soon as the examination is over; but in other Colleges, genuine agriculturists are sending their sons, and these men, going back to their land and having local influence, will do much to advance any branch of agricultural knowledge, including ours.

In addition to the work with destructive Insects, we have taken up work with *productive Insects;* sericulture is a big industry, so is lac, and there are possibilities in Bee-keeping. We have studied sericulture less with the view of improving existing industries, which is the work of technical experts, than of developing new lines or of transferring industries to new localities. India is very large and some branches of sericulture are extremely localised. Eri silk, for instance, seems to offer possibilities in other parts of India than in Assam, where alone it exists. We have worked it up and made improvements, and are now a centre for teaching those who want to take it up and are able to advise as to the suitable localities, etc. We sell two machines (patented) for cleaning coccons and for spinning, and a home industry in Eri silk is springing up.

In Mulberry silk, we have as yet done little but experiment, but our experiments are working out to lines of improvement that may be very valuable. In all branches of sericulture we are able to advise not only the private individual but the various Governments, and until some other organisation takes up this work, we shall continue to be engaged in it.

Lac is an important article of Indian commerce produced to the value of over three million pounds. At present it is mainly a forest product, not properly grown, but collected wild in jungles. We have shown that it can be produced more cheaply by proper cultivation on trees growing on pastures and waste lands in agricultural tracts. Such trees are valuable now chiefly for firewood, and the lac crop in no way interferes with this. As in Eri silk, we now offer a short course of training at Pusa, but as soon as lac cultivation on these lines spreads, we shall probably transfer this work to the various provincial centres, as is being done with Eri silk.

Bee-keeping is an Insect industry that has been very little developed, but it is probably suited only to the hill tracts of India and so comes less into our sphere than into the sphere of the Forest Department. Progress is being made with it in one province, but not in India generally.

A subject in which this Conference should be interested is that of legislation for preventing the entry of pests from other countries. One of us has dealt with that subject in the West Indies (« West Indian Bulletin », vol. III); in India, the climatic conditions are so adverse, the agricultural conditions so different that we do not fear Insect introductions quite to the same extent that some other countries do.

In conclusion, we may point out that we are only beginning and our aim is to lay the foundations surely and well and to provide for expansion as the different lines of work show their value. We cannot pretend to deal with a country so vast as India, with its enormous stretches of land and its 180 million agriculturists, by means of a staff of thirty odd men. But compared with 1903, when there was one man with an artist and a setter, we have made progress, and it is from that point of view we must take it. We have tried to describe also what may interest this Congress, the development of an entomological Section that had so to speak a clean sheet, a free field, no competition, no varied institutions doing bits here and there; imagine taking a new country with a vast and organised system of Government supervision of a very paternal kind, in what way would one work in entomology as an applied subject, what lines would one take up, what lines would one omit? To have one organisation, even a small one, with which to cover a country in which nothing was done at all, in which no entomology but systematic work had been done, and with that to make entomology an applied subject, makes one dream dreams, but we have, we think, followed natural lines, and have had little time to dream. First the pest, what he is, his life-history, his hibernation and his checks; then to try remedies not in one place but wherever opportunity arises; then to be ready for an outbreak, to adapt the campaign to local practices and to organise measures against him; then to spread a knowledge of this and kindred things over a province. It is inevitable that one then gets drawn into education, and, for the vast influence it brings, into such productive things as silk, for they appeal far more to our people and bring us into touch with them very much more quickly than anything else. Above all, to answer every enquiry courteously, to miss no opportunity of showing the value of the work and to be able to give sensible advice whether it be on the subject of Ants eating the sugar or an outbreak of army-worm over 100,000 acres.

In figures our work comes out to this; our list shows 104 important pests, of which we know the life-history now of 74, of 75 we have prepared plates such as are shown, of 65 the plates are published and available for every-day use, for 40 we have simple easy methods of prevention or destruction, and against another 21 are now working out in the field measures which will probably be effective with modification to suit local circumstances and which we expect to be able to apply over large areas of country.

We have said nothing here as to our technical difficulties, but they are not of the less importance, and it has been a very big item in our work merely to get our species identified and to keep up the technical side of the subject. We owe a great debt to those workers in Europe and America who have examined our collections, identified the species and enabled us to have a solid basis of accurate classification to work on. We have a quarrel with systematic entomologists as such owing to their pedantic changes of nomenclature; but we freely admit our debt to them, and one of our activities is concerned solely with making collections and sending them to workers in Europe for the furtherance of purely systematic work. How important this is an example will show. *Schœnobius bipunctifer* is a familiar species of Pyralid, a yellow Moth with a black spot on each forewing. If you are working on it as the Rice Stem Borer, you get it in abundance and can breed it, but you get also an Insect very closely resembling *Chilo simplex*, which is the Stem Borer of maize, sorghum and other cereals allied to rice. The ordinary worker concludes that rice is attacked by two species, *Schœnobius bipunctifer* and *Chilo simplex*; but it has been proved that this Insect, closely resembling *Chilo*, is the male of *Schœnobius*, its venation being quite distinct. Now the whole of our practical measures would be radically different if we thought that *Chilo* usually attacked rice as well as maize and sorghum; but we know now that the pest in rice is distinct, and from a practical point of view, that for *Chilo* we need not take into account the rice plant, nor for *Schœnobius* the maize and sorghum plants. Any of you, who have done economic work, will realise the enormous practical importance of this, and it is for this reason that our work must rest on the basis provided by the systematist. This is not, of course, the sole reason, but it is the one that really counts, and it is for this that we give so much of our time to aiding systematic work. We have large reference collections and, so far as we can, we send these to workers elsewhere. A collection is a necessity for teaching, and if you are going to give a year's course of entomology, you must be well up in the Insects of the country. We have had to publish « Indian Insect Life » to meet the needs of our students and fellow-workers in India, and it gives a good idea of the difficulties we have that it should have been necessary for an economic department to do such work as that in order to lay the foundation for future economic work.

What the lines of future development will be, it is for the moment impossible to say. The time is ripe for decentralisation, as to develop the results obtained at Pusa and to extend and adapt them to every province is altogether impossible with the staff at our disposal. The most natural course would seem to be that «executive» work, collection of data, and organised campaigns should devolve upon the provincial staffs, and that they should be strengthened up to the required standard of efficiency for such work to be adequately carried out We have tried to give here a rough sketch of past development, not the possibilities of the future.

On Dr. C. A. Wiggins' Researches on Mimicry in the Forest Butterflies of Uganda (1909),

by EDWARD B. POULTON, F. R. S.,
Hope Professor of Zoology in the University of Oxford, Fellow of Jesus College, Oxford.

The object of the present paper is to investigate the numerical and other relations between the models and mimics of three interesting associations, each grouped around one or more species of the Acræine genus *Planema*. The whole of the material was captured, between May 23rd and August 31st, 1909, in patches of native forest within a few miles of Entebbe, Uganda. The examples themselves are well known, the most important having been described by Mr. S. A. NEAVE, M. A., B. Sc., in 1904 (1) and 1906 (2). The instances of mimicry unrecognised at the time of Mr. NEAVE's memoirs have been described by Mr. HARRY ELTRINGHAM in his monograph on Mimicry in the Butterflies of Africa (Oxford, « Clarendon Press », 1910), where beautiful coloured illustrations of nearly every form mentioned in the present paper will be found.

(1) « Nov. Zool. », vol. XI, pp. 323-363.
(2) « Trans. Ent. Soc. Lond. », 1906, pp. 207-224.

Mr. NEAVE's two papers owed their existence to the generosity and energy of my friend Dr. C. A. WIGGINS of Entebbe, and the present paper is equally due to his kindness. Dr. WIGGINS has, in part by his own efforts, in part by the employment of native collectors, obtained collections of Forest Butterflies on a large number of dates. These collections are ample enough to enable the enquirer to judge of the numerical relationships. Special care was taken to eliminate, as far as possible, the natural preferences of the collector for one species rather than another. Dr. WIGGINS' own captures are clearly distinguished from those of the natives in the columns of the following tables. Upon Dr. WIGGINS has fallen the severe labour of recording the data of time and place with precision, and packing the collections so that no mistake in interpretation could be made.

My assistants in the Hope Department of the Oxford University Museum, and Mr. A. CANT, who has given most efficient help in the « setting », have worked hard and successfully to render the data permanent and these great collections available for the student. I wish particularly to thank my friend Mr. H. ELTRINGHAM for very much kind help in unpacking, determining and recording the collections, and Dr. KARL JORDAN for his valuable opinion on some of the most difficult of the species.

The present paper is only an instalment; for Dr. WIGGINS has most kindly promised to continue his important work until the conclusions derived from material collected during certain months have been tested by collections made in the same period of another year. But I have not wished to withhold the whole of the results until his work is complete, and I have chosen the First International Entomological Congress as the occasion on which to bring forward this preliminary statement.

The three *Planema*-centred Combinations, which form the subject of this paper, are by no means the only mimetic associations in the forests near Entebbe. Thus the species of the Danaine genera

Amauris and *Danaida* furnish dominant models, as do species of *Acræa*, and the Geometrid Moth, *Aletis*. I have chosen the *Planema*-centred Combinations in the belief that they shed more light than any of the others on various aspects of mimicry : on the relation of mimicry to sex, on polymorphism in mimicry, on secondary mimicry or the resemblance to a mimic rather than to its primary model, on the Batesian and Müllerian hypotheses. Apart from these interesting but elaborate questions, the evidence here brought forward will perhaps satisfy those who are still doubtful, that the models and mimics tabulated in the present paper do habitually fly together on the same days and in the same forests.

I now propose to set forth in three tables the specimens of the three *Planema*-centred Combinations, taken between May 23rd and August 31st, 1909 :

TABLE I

CONTAINING MEMBERS OF COMBINATION I, CENTRED BY *Planema poggei nelsoni* AND THE MALE OF *Planema macarista*.

CAPTORS AND DATES OF CAPTURE IN 1909.	Primary models, belonging to the Acræine genus *Planema*.		Acræine mimic with its non-mimetic male.		Nymphaline mimics.			Papilionine mimic: a female.	TOTALS.
	Male of *Planema macarista*.	Male and female of *Planema poggei nelsoni*.	*Aurivillii* female of *Acræa alciope*.	Non-mimetic male of *Acræa alciope*.	Male of *Pseudacræa hobleyi*.	Male and female of *Pseudacræa kuenowi hypoxantha*.	Male and female of *Pseudacræa albostriata*. The mimetic variations are marked « M. ».	Female f. *planemoides* of *Papilio dardanus dardanus*.	
C. A. WIGGINS, May 23rd	»	»	2	»	1	»	»	»	3
— — 24th	3	»	1	2	2	»	1	»	9
— — 30th	2	»	1	»	»	1♀	»	»	4
— — 31st. One *Ps. albostriata* capt. by Native.	1	»	1	1	1 (1)	»	3	»	7
C. A. WIGGINS, June 13rd	»	»	»	2	1 (2)	»	»	»	3
— — 27th	1	1♂	1	2	3	»	4 (1 M.)	1	13
— — 29th	1	»	»	2	2	»	3 (1 M.)	»	8
— July 11th	1	1♂ 1♀	»	1	2	»	»	1	7
C. A. WIGGINS, July 18th	»	»	»	2	2 (3)	»	»	1	5
— — 24th	»	»	1	2	»	»	»	»	3
Native collector, August 4th	»	»	1	»	1	»	»	»	2
— — 6th	2	»	»	»	1	1♂	»	»	4
C. A. WIGGINS (9) — 7th. Native collector (5).	9 (native 2)	»	1 (native	2 (native)	1	»	1	»	14
Native collector, August 9th	»	»	4	2	»	»	»	»	6
— — 10th	»	1♂	1	3	»	»	»	»	5
— — 11th	2	»	3	»	»	»	1 (1 M.)	»	6
— — 12th	»	»	2	1	2	»	»	»	5
— — 13th	1	»	1	1	1	»	»	»	4
C. A. WIGGINS, — 14th	9	1♂	3	1	3	»	»	1	18
Native collector, — 16th	»	1♂	4	1	1	»	»	»	7
— — 17th	3	1♂	4	2	1	»	»	»	11
— — 18th	1	»	4	2	1	»	1	»	9
— — 19th	3	2♂	1	1	2	»	»	»	9
— — 20th	3	»	3	2	»	»	»	»	8
— — 21st	1	1♂	4	2	»	»	»	»	8
C. A. WIGGINS (9) — 22nd. Native collector (2).	2 (native 1)	»	3	»	4 (native 1)	»	2	»	11
Native collector, August 24th	2	»	1	1	»	»	»	»	4
— — 25th	7	1♂	»	»	»	»	3	»	11
— — 26th	6	»	»	»	»	»	»	»	6
— — 27th	3	»	1	»	»	»	1	»	5
— — 28th	6	»	»	»	1	»	»	»	7
C. A. WIGGINS (9) — 29th. Native collector (6).	12 (native 3)	»	»	»	2 (native)	»	1 (native)	»	15
Native collector, August 30th	»	»	1	»	1	»	»	»	2
— — 31st	»	1♂	»	»	»	»	»	»	1
TOTALS	81 (4)	12	49 (5)	35	36 (6)	2	21 (7)	4	240

(1) The anal area of both hind wings shorn off symmetrically.

(2) The bar crossing the hind wings of this specimen, of one of the two captured June 29th, and one August 4th, is fulvous and not white on the upper surface: on the under surface it is of the normal white.

(3) One of these specimens is a female with the colouring of a male; the other — evidently a mimetic male of the usual form — is represented by a right fore wing found upon the ground.

(4) The females are the principal primary models of Combination II.

(5) In addition to these *aurivillii* females, 2 examples of a modified western type of female were taken on August 14th and 1 on August 18th.

(6) The females enter as mimics into Combination II. See however note 3.

(7) It is to be noted that there are only 3 incipient mimics in these 21 specimens.

TABLE II

CONTAINING MEMBERS OF COMBINATION II, CENTRED BY THE FEMALE OF *Planema macarista* AND OF *Planema alcinoe*.

CAPTORS AND DATES OF CAPTURE IN 1909.	Primary models, females of the Acræine genus *Planema*, with the male of one, not mimicked at Entebbe.			Acræine mimics, primary and secondary, with their non-mimetic males.				Nymphaline mimic: a female.	TOTALS.
	Female of *Planema macarista*.	Female of *Planema alcinoe*.	Male of *Planema alcinoe*.	*Carmentis* f. of female of *Acræa jodutta*.	Non-mimetic male of *Acræa jodutta*.	White-marked female of *Acræa althoffi*.	Non-mimetic male of *Acræa althoffi*.	Female of *Pseudacræa hobleyi*.	
C. A. WIGGINS, May 23rd	»	1	»	»	1	»	»	»	2
— — 24th	»	»	»	1	»	»	»	»	1
— — 30th	1	»	»	»	1	»	»	2	4
C. A. WIGGINS (2), May 31st Native collector (1).	»	»	»	1 (native)	1	»	»	1	3
C. A. WIGGINS, June 27th.	»	»	»	»	»	»	»	1	1
— — 29th.	2	»	»	»	4	»	»	»	6
— July 11th.	1	»	»	1 [1]	1	»	»	1	4
— — 18th.	»	»	»	1	3	»	1	»	5
— — 24th.	»	»	1	1 [2]	3	»	1	»	6
C. A. WIGGINS (5), August 7th Native collector (8).	5 (native 2)	2 (native)	1	1	2 (native)	»	1 (native)	1 (native)	13
Native collector, August 10th.	1	»	»	3	2	»	»	1	7
— — 11th.	»	»	1	3	2	»	»	»	6
— — 12th.	»	»	»	2 [3]	4	»	»	2	8
— — 13th.	»	»	»	3	3	»	»	1	7
C. A. WIGGINS, — 14th.	5	2	5	1	»	»	1	1	15
Native collector, — *16th*	[illegible]	[illegible]	[illegible]	5	4	[illegible]	[illegible]	[illegible]	11
— — *17th*	»	»	»	3	6	»	1	»	10
— — *18th*	»	»	»	2	8	»	1	1	12
— — 19th	1	»	»	2	1	»	»	3	7
— — 20th	»	1	»	3	3	»	»	1	8
— — 21st	»	»	2	3	4	»	3	1	13
C. A. WIGGINS (8), — 22nd Native collector (8).	4 (native 1)	»	2 (native 1)	1 (native)	3 (native 2)	»	3 (native)	3	16
Native collector, August 24th	2	»	2	6	4	2 [4]	1	1	18
— — 25th	2	1	1	»	»	»	»	»	4
— — 26th	»	1	2	2 [5]	1	»	»	1	7
— — 27th	»	»	2	»	»	»	»	»	2
— — 28th	1	»	1	»	1	»	»	3	6
C. A. WIGGINS (13) — 29th Native collector (5).	9 (native 4)	2 (native 1)	7 (native 1)	1 (native)	»	»	»	2 (native 1)	21
Native collector, August 30th	2	»	2	1	»	»	1	»	6
— — 31st	2	1	1	»	»	1	»	»	5
TOTALS	39 [6]	11	30	47 [7]	62 [8]	3 [9]	14 [10]	25 [11]	234

(1) This specimen was captured in July, probably on the 11th. [Certainly July 11th, E. B. P. October 1st, 1911.]

(2) The pale area of the hind wing, extending on to the fore wing, is yellowish and not white, and thus transitional towards the colour of the male.

(3) The hind wing markings approach those of the *tellus*-like forms of female in Combination III.

(4) The pale markings of one of these females are yellow and not white, thus following the male of *A. jodutta* as a model, instead of its female.

(5) Both hind wings of one specimen shorn through symmetrically. The straight line of the cut suggests the beak of a Bird.

(6) The males are the principal primary models of Combination I.

(7) Another form of female, only half as numerous as the *carmentis* form, enters as a mimic into Combination III.

(8) These non-mimetic males, with the exception of 5 captured May 30th and June 29th, are repeated in Combination III, which contains 9 additional specimens taken August 9th and 10th.

(9) It is to be noted that 2 out of these 3 specimens are secondary mimics of the *carmentis* female of *A. jodutta*, rather than mimics of the two primary *Planema* models. The third *althoffi* female mimics the male *jodutta* : see note 4. A commoner form of female *althoffi* enters as a mimic into Combination III.

(10) These males, with the exception of that captured on August 30th, are repeated in Combination III, which contains an additional specimen taken on August 9th.

(11) The males, together with a single female with male colouring, enter as mimics into Combination I.

TABLE III

CONTAINING MEMBERS OF COMBINATION III, CENTRED BY *Planema tellus platyxantha*.

CAPTORS AND DATES OF CAPTURE IN 1909.	Primary model, *Planema* (*Acræinæ*). — Male and female of *Planema tellus platyxantha*.	Acræine mimics, primary and secondary, with their non-mimetic males.				Nymphaline mimic: male and female. — Male and female of *Pseudacræa terra*.	Papilionine mimic: a female. — Female f. *niobe* of *Papilio dardanus dardanus*.	TOTALS.
		Tellus-like f. of female of *Acræa jodutta*.	Non-mimetic male of *Acræa jodutta*.	Fulvous-marked female of *Acræa althoffi*.	Non-mimetic male of *Acræa althoffi*.			
C. A. Wiggins, May 23rd	1 ♀	»	1	»	»	»	»	2
— — 24th	»	»	»	1	»	1 ♀	»	2
— — 31st	1 ♂	1	1	»	»	1 ♂	»	4
— June 27th	2 ♂	»	»	»	»	»	»	2
— July 11th	2 ♂ 1 ♀	»	1	»	»	2 ♀ [1]	1 [2]	7
— — 18th	»	»	3	»	1	»	»	4
— — 24th	»	2	3	»	1	»	»	6
Native collector, August 6th	1 ♂	»	2	»	»	1 ♂	»	4
C. A. Wiggins (12), August 7th Native collector (11).	18 ♂ (native 7)	1	2 (native)	»	1 (native)	1 ♀ (native)	»	23
Native collector, August 9th	»	1	7	»	1	»	»	9
— — 10th	3 ♂	2	2	»	»	»	»	7
— — 11th	1 ♂	3	2	»	»	1 ♀	»	7
— — 12th	3 ♂	»	4	»	»	1 ♀	»	8
— — 13th	2 ♂	»	3	1	»	1 ♂	»	7
C. A. Wiggins, — 14th	8 ♂ 3 ♀	»	»	»	1	3 ♀ [3]	»	15
Native collector, — 16th	1 ♂ 3 ♀	2 [4]	4	»	»	»	»	10
— — 17th	»	2	6	»	1	»	1	10
— — 18th	2 ♂	3	8	»	1	»	»	14
— — 19th	5 ♂	»	1	»	»	»	»	6
— — 20th	2 ♂	»	3	»	»	»	»	5
— — 21st	3 ♂	1	4	1	3	»	»	12
C. A. Wiggins (9), August 22nd Native collector (5).	4 ♂	»	3 (native 2)	1	3 (native)	2 ♂ 1 ♀	»	14
Native collector, August 24th	2 ♂ 1 ♀	3 [5]	4	1	1	1 ♂ [6]	»	13
— — 25th	3 ♂ 2 ♀	»	»	1	»	»	»	6
— — 26th	1 ♂	»	1	»	»	»	»	2
— — 27th	1 ♂	»	»	»	»	1 ♀	»	2
— — 28th	»	»	1	1	»	1 ♂	»	3
C. A. Wiggins (6), August 29th Native collector (7).	10 ♂ 1 ♀ (nat. 5 ♂)	1 (native)	»	1 (native)	»	»	»	13
Native collector, August 31st	»	1	»	»	»	»	»	1
TOTALS	89 (75 ♂ 14 ♀)	23 [7]	66 [8]	8 [9]	14 [10]	18	2	220

[1] One of these specimens was captured in July, probably on the 11th. [Certainly July 11th, E. B. P. Oct. 1st, 1911.]

[2] Captured together with the female *Planema tellus platyxantha* in a single sweep of the net.

[3] One of these females is transitional towards the form *Ps. obscura* Neave.

[4] One of these females is a very dark form with the fulvous markings greatly reduced.

[5] On this date, and on August 11th, 14th and 17th, single examples of a *Danaida chrysippus*-like form of female *A. jodutta* were captured. These forms combine the *tellus*-like pattern with a white subapical bar to the fore wing, like that of the mimetic females in Combination II.

[6] This interesting specimen is transitional, on the under as well as on the upper surface, towards the male of the form *Ps. hobleyi*.

[7] See note 7 on p. 489.

[8] See note 8 on p. 489.

[9] These females are secondary mimics of the females of *A. jodutta* in the same Combination rather than of the primary model *Pl. tellus platyxantha*. See also note 9 on p. 489.

[10] See note 10 on p. 489.

Combination I, centred by *Planema poggei* DEWITZ, subsp. *nelsoni* GROSE-SMITH, **and the male of** *Pl. macarista* E. M. SHARPE **(Table I).**

This deeply interesting association was described by S. A. NEAVE in « Trans. Ent. Soc. Lond. », 1906, p. 218, and its members figured in the accompanying plate X. Two of the mimics figured by the author were not captured in the period now under review and are therefore excluded from Table I: *viz.* a form of *Elymnias phegea* F., and an outlying member, the female of *Precis rauana* GROSE-SMITH. The plate also contains excellent figures of *Pseudacræa kuenowi hypoxantha* JORD., the male of *Ps. hobleyi* NEAVE (1), and the *planemoides* TRIMEN, female form of *Papilio dardanus dardanus* BROWN. The distinction between *Planema poggei nelsoni* and the males of *Pl. macarista* at Entebbe was not recognised, and the former, correctly represented on plate X, figure 1, was supposed to be the only model. The *aurivillii* STAUD. female of *Acræa alciope* HEW., is represented in figure 2 as the male of *Planema aurivillii* (2). The confusion into which all authors had been led by this puzzling mimetic form was finally cleared up by Mr. ELTRINGHAM (3) in 1909 (« Proc. Ent. Soc. Lond. », LXVII-LXIX). The incipient mimicry exhibited by a small proportion of the specimens of *Pseudacræa albostriata* LATHY was also unknown at the time when Mr. NEAVE was writing his paper.

I have already alluded to Mr. ELTRINGHAM's recently published work with its beautiful coloured figures of nearly all the species in all the three tables of the present paper. With these and Mr. NEAVE's excellent uncoloured figures the members of Combination I can be studied under very favourable conditions.

---

(1) By a clerical error printed as *nolleyi* on the plate.

(2) Correctly placed as an *Acræa* in the text, p. 218, and description of plate X. p. 223. The author was unfortunately compelled to hurry his departure for Africa at the time when this paper was being brought out, and was unable to spare the time for careful revision.

(3) During the present year, 1911, Dr. G. D. H. CARPENTER has bred 8 *A. alciope* males and 5 *aurivillii* females from 13 small larvæ found on a single leaf of the food-plant on Damba Island in the Victoria Nyanza, near Entebbe.

The two primary *Planema* models of Combination I.

It is noteworthy that, in addition to the differences in form and direction of the bar crossing the fore wing in the two species, the fulvous tint is of a deeper shade in the male *macarista* than in *poggei nelsoni*. It can scarcely be doubted that there has been synaposematic approach between the two models; but their relationship over a much wider range must be studied before a confident decision can be reached. In the forests near Entebbe the male *Pl. macarista* is seen to predominate greatly over the other model, 81 specimens being shown in Table I, as against 12, including a single female, of *Pl. poggei nelsoni*. These proportions are sure to be different, and may even be reversed, in other parts of the common range of the two species. The strongest evidence will however be supplied if it can be shown that one of these models exists in a modified form outside the range of the second, and that the likeness to the other increases when it enters its area It is slightly more probable that *Planema poggei nelsoni*, with both sexes alike, has influenced rather than been influenced by the male *Pl. macarista*.

The striking and peculiar colour-scheme is borne by both sexes of *poggei nelsoni*, while the female of *macarista* possesses an appearance common to several female Planemas.

It is possible that, on its under surface, the male *Planema alcinoe* FELD. is an incipient mimic of the two primary models. A careful comparison between *alcinoe* of Entebbe and of the parts of the west coast, where *Pl. poggei* and *macarista* are unknown, will probably settle the question. In the meantime the male *alcinoe* is placed beside its female, which is a model in Table II.

Acræa alciope HEW. : the *aurivillii* STAUD. female form.

The resemblance of this female *Acræa* to the male of *Planema macarista* is equally remarkable on both surfaces, the basal area of the hind wing under side exhibiting the characteristic Planemoid brown triangle, over which are scattered black spots corresponding to those of the male *alciope*, but larger in size. Certain points in the mimetic association deserve fuller consideration.

When a particular pattern is restricted to one sex in both model and mimic, it is common for the female to resemble a female, the male a male. Thus the likeness founded on similarity of pattern is heightened by similarity in the habits and movements which are characteristic of sex. A good example is afforded by *Pseudacræa hobleyi* with its male mimicking a male *Planema* and its female the female of the same *Planema*. The female of *Acræa alciope* is a well-marked exception to this rule; for on the tropical west coast of Africa it mimics the dark male Planemas of the type of *Pl. alcinoe*, while at Entebbe its *aurivillii* female mimics the male of *Pl. macarista*. There can be no doubt that the western females are ancestral as compared with those at Entebbe; for the displacement of the older form is not complete. Thus, in addition to the 49 *alciope* females recorded in Table I, 2 specimens, captured on August 14th and 1 on August 18th, are modifications of the western type, perhaps altered by incipient mimicry of *Planema tellus platyxantha* JORD., or by secondary mimicry of the *tellus*-like females of *Acræa jodutta* F. The presence of modified western forms of the female *alciope* side by side with its other females, that are mimics of the male *Pl. macarista* and both sexes of *Pl. poggei nelsoni*, is extremely interesting.

Another interesting point is the great relative abundance of the mimic. It will be seen by a glance at Table I on pages 486-487 that 49 mimetic females and 35 non-mimetic males of *A. alciope* were taken in the period under review, together with 93 models : *viz.* 81 males of *Planema macarista* and 12 examples (11 ♂ 1 ♀) of *Pl. poggei nelsoni*. The relative numbers of the mimic, its place in a highly distasteful group of Butterflies, and the fact that its non-mimetic male itself supplies the model for a female Lycænid (*Mimacræa fulvaria* AURIV.) combine to render the *aurivillii* female of *A. alciope* a remarkably striking example of Müllerian or synaposematic resemblance.

The question indeed arises as to whether the whole of this remarkable likeness is to be accounted for by approach on the side of the *Acræa* or whether the male of *Planema macarista* has not itself undergone some diaposematic change in the direction of its mimic. I omit consideration of *Planema poggei nelsoni* to which the *aurivillii* female bears a much less perfect resemblance.

There are two features in the pattern, that is common to the upper surface of all or many *macarista* males and all or many *aurivillii* females of *alciope*, which are certainly ancestral in the latter. Whether they are also ancestral in the *Planema* is a matter for future enquiry, involving the examination of long series of local forms of *Pl. macarista* over all parts of its range and the careful comparison of this species with its near allies. With the facts as we know them at Entebbe there are good reasons for believing in a possible diaposematic relationship; but the whole subject is so complex and difficult that I should not venture to make a more definite statement until further investigations have been undertaken.

1. *The broad fulvous band which in many specimens borders the white bar of the hind wing.* — This feature is, in my experience, much commoner in the *Acræa* mimic than in the *Planema* model, but, when it *is* present in the latter, bears the strongest superficial likeness to that of an *aurivillii* female in which the marking is developed to an equal extent. Comparison with the pattern of the male *alciope,* and especially with that of western forms of the female and their modification at Entebbe, proves beyond doubt that the feature is ancestral in this latter species.

It is worthy of remark that a similar brown margin to the discal white bar of the hind wing is developed in a small proportion of the white (*carmentis*) females of *Acræa jodutta.*

2. *The serrated outer margin of the lower (inner marginal) half of the fulvous bar crossing the fore wing.* — This feature, which goes far to produce the remarkably close resemblance between the male *macarista* and the female *alciope*, also deserves consideration as a possible instance of diaposematism. Not only are the serrations relatively deeper and more numerous (4 crests and 3 valleys to 3 crests and 2 valleys in the *Planema*) in the *Acræa*, but they are shown by comparison with other forms of the female and with the male, in which however they are far less marked, to be undoubtedly ancestral. On the other hand the feature is found, although not always developed to an equal extent, in the female as well as the male *Planema*.

The serration is produced in a manner common in Butterflies, viz. by the extension of a coloured area outward along the veins, leaving a V-like indentation in each internervular region. It is not so much the serration itself, as its restriction to a corresponding part of the pattern, that constitutes evidence of mimetic approach from one side or from both.

Pseudacræa hobleyi Neave, and its forms.

This very interesting series of mimetic forms, on a superficial examination, seems to be opposed to Dr. Karl Jordan's conclusions on the polymorphism of certain mimetic Pseudacræas, published in the present volume, but, subjected to a more searching enquiry, it tends to support them. It is certainly difficult to believe that a sexually dimorphic form such as *Pseudacræa hobleyi* — its males mimicking two Planemas with a fulvous bar crossing the fore wing, its females mimicking two Planemas with a white bar in a similar position — can belong to the same species as *Ps. terra* Neave, in which both sexes mimic the fulvous and black *Planema tellus platyxantha* with a pattern entirely different from the models of *hobleyi*. These latter models, male as well as female, possess the well-known, black-spotted, brown triangle at the base of the hind wing under surface, and so do the two sexes of *Ps. hobleyi*. *Planema tellus*, on the contrary, has the black spots without the triangle, and so has *Ps. terra*.

In spite of these difficulties, the fine series of Pseudacræas provided by the generosity of Dr. C. A. Wiggins has enabled me to produce evidence in support of the remarkable and unexpected conclusion at which Dr. Jordan has arrived.

In the first place, the investigation of a long series of specimens shews that the sexual dimorphism of *Ps. hobleyi* is not complete. A female (1) example captured July 18th (Table I) possesses the colour and to a large extent the pattern of the male. It is a mimic of the male and not the female of *Planema macarista*. And this is not an isolated example, for I have since met with others. The whole of the material presented by Dr. Wiggins will enable me to estimate with a fair degree of accuracy the proportion in which such exceptions occur at Entebbe.

---

(1) I owe the recognition of the sex of this specimen to my friend Mr. Harry Eltringham, to whom it had been submitted by my assistant Mr. A. H. Hamm. The shape of the wings is characteristically female, and the form of the anterior or costal section of the V-shaped fulvous bar crossing the fore wing resembles the female pattern rather than the male.

Secondly, intermediate varieties occur between the pattern of the male *Ps. hobleyi* and that of *terra*. Thus, in three males of *hobleyi*, taken respectively on June 13th, June 29th, and August 4th (Table I), the bar of the hind wing is fulvous and not white. The specimen of June 29th is especially remarkable in this respect, the bar of the hind wing being of a tint very nearly as deep as that of the fore wing. The costal end of the hind wing bar is often fulvous in *hobleyi*, although this feature is probably not transitional towards *terra*, but mimetic of a common variety of the male *Pl. macarista*.

A far more remarkable and significant form was captured on August 24th. Although intermediate, it is much nearer to *Ps. terra* than to the male *hobleyi*, and is therefore placed in Table III. On the upper surface, the hind wing is that of *terra*, while the fore wing bears a pattern representing the fusion of the inner marginal fulvous area of *terra* with the V-shaped fulvous bar of the male *hobleyi*. Its colour is the paler tint of *terra* rather than the richer fulvous of the male *hobleyi*. The under surface reproduces the pattern of the upper except for one important feature — the appearance of a reddish tint at the base of the hind wing — clearly transitional towards the reddish brown triangle of *hobleyi*.

A very slight trace of this reddish tinge can generally be detected, especially in the costal region, at the base of the hind wing under side of *Ps. terra* in both sexes.

Furthermore, Dr. WIGGINS' series of captures between May 23rd and August 31st, 1909, not only exhibits transition between *Ps. hobleyi* and *terra*, but also between *terra* and *obscura* NEAVE, the rarest of these mimetic forms of the *hobleyi*-group at Entebbe, and a mimic of the rarest *Planema*-model, *Pl. epæa paragea* GROSE-SMITH. The specimen of *Ps. terra* taken August 14th (Table III) is beautifully transitional towards *obscura* (1). The upper surface colour and pattern of this intermediate specimen bear a close resemblance to those specimens of the *carmentis* female of

(1) About a year later, on September 8th, 1910, an intermediate form between *Ps. terra* and the female *Ps. hobleyi* was taken by Dr. WIGGINS' native collector. The specimen, which has not yet been set, and cannot therefore be thoroughly examined, apparently possesses the white fore wing bar of the female *hobleyi* combined with the hind wing pattern of *terra*.

A. jodutta, in which all the white markings except the subapical bar are replaced by yellowish. An example of this latter form of the female *jodutta* was captured on July 24th (Table II).

Dr. JORDAN's conclusions, based on a study of the male genitalia, are thus supported by a study of the Entebbe material. But inferences so important should be submitted to the ultimate test supplied by the method of breeding before they are accepted as final. However firm the proof, — and Dr. JORDAN'S is certain to be of the firmest, — the cooperation of various lines of evidence is demanded by a conclusion so far-reaching as one which unites into a single interbreeding community, continuous over nearly the whole of the forest areas of Africa, all the divergent forms of the most important mimetic group in the genus *Pseudacræa*, — from *imitator* (1) in the south-east and *rogersi* (2) in the east, through the Uganda forms, *hobleyi, terra*, and *obscura*, to the numerous mimetic modifications of *eurytus* L., on the west coast and the great tropical forest region of the continent.

Pseudacræa kuenowi DEWITZ, subsp. *hypoxantha* JORD.

Dr. JORDAN concludes that this species, which is rare at Entebbe, is entirely distinct from the forms of *Ps. hobleyi*. Its superficial resemblance in both sexes to the far more abundant males of *Ps. hobleyi* is so extraordinarily close as to suggest that there has been a secondary mimetic approach. Such an interpretation is more probable than one based on the hypothesis of arrested divergence; for it is unlikely that details in the shape of the fore wing bar in one sex of a single member of the *hobleyi* group are specially ancestral, and it is these very details that are so precisely reproduced in both sexes of *kuenowi* I may add that the bar of the male *hobleyi* is extremely variable in shape, and that it is the commonest and most characteristic form which appears in the less variable *kuenowi*.

---

(1) *Pseudacræa imitator* TRIMEN.

(2) *Ps. rogersi* TRIMEN.

Pseudacræa albostriata LATHY.
(In Dr. JORDAN's opinion *albostriata* is the Uganda subspecies of *dolomena* HEW.)

It is possible that in *albostriata* we meet with a sudden mutation sufficient to provide the foundation for the evolution of mimetic resemblance. Only 3 individuals exhibiting a very rough likeness to the pattern of the male *Planema macarista* were present in the series of 21 captures recorded in Table I. But even in these mimics, rough as they are, it is improbable that we are witnessing the results of a sudden mutation, and hard to resist the conclusion that there has been some moulding by natural selection. The position and form of the white bar crossing the hind wing, the retention of the fulvous tint at its costal extremity and the replacement of this colour by dark fuscous beyond both margins of the bar, form a combination of mimetic features in whose simultaneous and spontaneous origin it is difficult to believe.

Papilio dardanus dardanus BROWN, **female**
f. *planemoides* TRIMEN.

With the exception of *Ps. kuenowi hypoxantha*, this was the rarest mimic in Table I, only 4 specimens being captured between May 23rd and August 31st. The numerical proportion borne by *planemoides* to the other female forms of *dardanus* at Entebbe requires a much longer period of time and larger number of captures for its determination. It may be stated however that the *hippocoon* F. form, mimicking *Amauris niavius* L., is far commoner than any other, indeed probably far more so than all the others together. Although, in my experience, always in small numbers, *cenea* STOLL, and *trophonius* WESTW., occur at Entebbe, together with *niobe* AURIV., the form which enters Combination III.

The fore wing bar of *planemoides* resembles that of *Pl. poggei nelsoni* rather than of the male *macarista*. Its shape is not very like that of *poggei*, but is even less like *macarista*. The fulvous bar of an Entebbe form of *Elymnias phegea* F., which enters Combination I, resembles that of *planemoides* rather than the primary models. It is possible that some secondary mimetic approach has taken place, a question that can probably be decided by the examination of a sufficient series of this form, together with a comparison between it and the white-barred form of *phegea* at Entebbe and on the west coast.

Mimacræa poultoni NEAVE.

H. ELTRINGHAM has remarked on the great variability of this species, of which the commonest form is doubtless, as NEAVE originally suggested, a beautiful mimic of *Acræa sotikensis* E. M SHARPE. Some of the forms of this Lycænid at Entebbe perhaps gain advantage as outlying members of Combination 1; for their upper surface pattern is such that, upon the wing, there would probably be considerable resemblance to the members of this association, especially to the female of *A. alciope*. Only a single specimen of this *Mimacræa* was captured in the period under review, on June 27th, but it is an example of the form referred to above.

Combination II, centred by the females of *Planema macarista* E. M. SHARPE, **and** *Pl. alcinoe* FELDER **(Table II).**

Mr. S. A. NEAVE (loc. cit., p. 212) described one of these primary models, viz. the female of *Pl. godmani* (1) BUTL. (= *alcinoe*), and its beautiful Nymphaline mimic, *Pseudacræa tirikensis* NEAVE (= *Ps. hobleyi*, female). He also mentions on the same page the white-barred form of *Elymnias phegea* (not taken in the period with which the present paper deals), placing it among the mimics of the Danaine Butterfly *Amauris niavius*. I do not doubt that the

(1) The female *Planema* spoken of by Mr. NEAVE was in reality *Pl. macarista*, which had not at the time been separated from *Pl. alcinoe* in the Entebbe district. The same error was repeated by the present writer in « Proc. Ent. Soc. Lond. », 1909, pp. LXIII-LXIV. We owe the correct sexing and determination of the two species in this area to the recent researches of Dr. KARL JORDAN.

black and white females of Planemas and, through them, their Acræine, Nymphaline, Elymniine and Papilionine mimics have been greatly influenced by the black and white species of the African Danaine genus *Amauris*.

The black and fulvous pattern of the male Planemas is almost certainly ancestral, and the appearance in the females of black and white — a pattern unusual in the *Acrœinæ* — has probably been brought about by the influence upon selection of the dominant Danaines, as indeed the present writer suggested in 1897 (« Report Brit. Assoc. Adv. Sci., Toronto Meeting », 1897, pp. 689-691).

In this sense the models of Combination II may be considered as mimics, their mimics as secondary, and the mimics of their mimics as tertiary. But the resemblance of female Planemas to *Amauris niavius* is rough, while that of the mimics to the female Planemas is strong and precise, so that I have preferred to treat the Combination as a separate category with the Planemas for its primary models. And I have no doubt that the white-barred forms of *Elymnias phegea* are somewhat outlying members of this Combination. They even exhibit traces of the *Planema*-like brown triangle at the base of the hind wing under surface. The indication of this characteristic feature cannot be doubted, vague and indistinct as it is, in common with Elymniine mimicry in general (1).

No other mimics belonging to Combination II had been received from Dr. WIGGINS when Mr. NEAVE's paper was written. Coloured figures of nearly all the forms recorded in Table II are given in Mr. ELTRINGHAM's work.

The two primary *Planema* models of Combination II.

The female of *Pl. macarista* was far more abundant than that of *Pl. alcinoe,* the numbers being 39 to 11. The patterns of the two models are remarkably alike and, upon the wing or in the resting position at a little distance, would be indistinguishable. Without further enquiry it is impossible to decide whether either species has

(1) An excellent account of the « sketchy » mimicry, that is characteristic of the *Elymniinæ*, Oriental as well as African, is given by H. ELTRINGHAM (« Mimicry in African Butterflies », p. 50).

acted as a model for the other. The pattern which they bear in common is characteristic of the females of a group of western Planemas, and we may here be witnessing an example of arrested divergence among the descendants of a common ancestor, which arose under the influence of the black and white Danaines. The possible incipient mimicry of the models of Combination I by the male of *Pl. alcinoe* has been spoken of on page 493.

The mimetic females of *Acræa jodutta* F.

Three different female forms of this species — all mimetic — were captured at Entebbe during the period under consideration in the present memoir : 47 white-marked mimics (the *carmentis*-form) of the females of *Planema macarista* and *Pl. alcinoe* (Combination II), 23 fulvous-marked mimics of *Planema tellus platyxantha* (Combination III), and 4 fulvous-and-white-marked mimics of *Danaida chrysippus* L. These latter combine the *tellus*-like pattern with the white subapical bar of *carmentis*. Their comparative rarity is probably due to the fact that *D. chrysippus* is not nearly so common as the Planemas in the forests where these collections were made and where *A. jodutta* abounds.

Assuming, from its close resemblance to the pattern of the non-mimetic male, that the white-marked *carmentis* is the older of the two principal female forms, it is a tempting hypothesis to suppose that the *tellus*-like female has been produced from it by a sudden colour « mutation », in which the white has been replaced by fulvous. But the patterns are by no means identical, and the differences are such as to increase the likeness to their respective models. The fulvous area covers far more of the hind wing (extending further towards the base as well as towards the margin) of one form than the white does of the other, while the area due to the invasion of the fore wing by the fulvous of the hind is also larger, and of a different shape from that produced in the other form by the invasion of the white. It is unreasonable to suppose that these features, nicely adjusted as they are to the pattern of a species in another genus, sprang into existence, ready-made and complete, and at the same time as the change in colour.

I have recently been given the opportunity of comparing, on a fairly complete scale, the pattern of the white-marked *carmentis*-form of female in the Lagos district with those of Entebbe. In the absence of *Planema tellus* the fulvous form of female is relatively rare near Lagos and, when it occurs, presents a somewhat different pattern, probably following the male of *Planema epæa* CRAM. as a model. My experience of this latter form of the female *A. jodutta* on the west coast is however insufficient to justify a safe conclusion. The white-marked females of the west differ from those of Entebbe in the greater extension of the white patch caused by the invasion of the fore wing by the pattern of the hind. I have observed this feature not only in many specimens captured in the Lagos district, but also in 16 females of a family recently bred by my friend Dr. W. A. LAMBORN. These shewed no appreciable individual difference in the extent of the white patch in question. This extension of part of the pattern in the west corresponds with the same marking in the western model, the female *Planema epæa*. The two white-marked female Planemas at Entebbe are, on the other hand, without this invasion, or, at the most, exhibit (in *Pl. alcinoe*) but a faint trace of it. They are followed at a distance by the *carmentis* female of the Entebbe *jodutta* with its reduced fore wing patch. Reduced as this patch is, and rendered greyish by overspreading black scales, it still influences not only the white females of *Acræa althoffi* DEWITZ (see p. 504), but also appears to affect the variable female of *Pseudacræa hobleyi*, which in this respect commonly possesses the pattern of the *Acræa* mimic rather than that of the *Planema* model. The slight difference between the western and the Entebbe *carmentis* female of *jodutta*, and its correspondence with the difference between their respective models is also adverse to the mutation theory of the origin of mimicry recently advanced by Prof. PUNNETT, unless indeed a « mutation » may be only another name for a small variation such as the Darwinian has, for half a century, believed to supply the material for mimetic modification.

As regards the base of the hind wing under surface, both white and brown females of *jodutta* commonly possess rather larger spots than the males, thus tending in the direction of their respective models, while the white-marked female also clearly exhibits an early stage in the development of the characteristic brown triangle.

It is interesting to note that one small but important detail in the mimetic resemblance of these two females exists ready-made in the fore wing of the non-mimetic male, viz. the notch just below the centre of the inner border of the subapical bar. In this respect the brown female is a better mimic than the white, because the corresponding notch is in a similar position in *Planema tellus platyxantha*, whereas in the two white-marked female Planemas it is above rather than below the centre of the bar.

In the *chrysippus*-like forms of the female *A. jodutta* we have merely a combination of the characteristic pattern of the *tellus*-like female with the subapical bar and adjacent white striæ of the *carmentis* female. Natural selection has here been clearly restricted to the combination of existing characters. Among all possible combinations, the one that reproduces the pattern of *Danaida chrysippus* has alone established itself.

Acræa althoffi Dewitz.

Fifteen males and 11 females of this species were captured between May 23rd and August 31st, 1909. They afford strong evidence of the Müllerian character of the association between the Acræas and the Planemas in two out of the three Combinations. Two of the females are white-marked and mimic the *carmentis* female of *A. jodutta*, itself a mimic of two female Planemas, while one with yellow markings mimics the non-mimetic male of *A. jodutta*. These 3 examples are included in Table II. The 8 remaining females, included in Table III, mimic the fulvous females of *A. jodutta*, themselves mimics of *Planema tellus platyxantha*. While the females of *althoffi* prove that the *jodutta* mimics of Planemas are themselves able to act as models, the peculiarly brilliant upper surface colouring of the male *althoffi* is evidence of the unpalatability that is characteristic of the whole sub-family of *Acræinæ*, so far as it has been experimentally tested.

It is of importance to produce detailed evidence in support of the conclusion that the female forms of *althoffi* are in reality mimics of mimics rather than mimics of primary models.

The upper surface pattern of the white-marked females has been derived from that of the male by the transformation of the red fore wing markings into white, and by a similar modification, accompanied by a broadening and a loss of sharply defined margins, in the yellow bar of the hind wing. The changes described above are accompanied by a marked diminution in the depth of the black ground-colour, a diminution which renders distinct 3 black spots below the cell of the fore wing. These spots, although distinct on the under surface of the male, are, on its upper surface, barely distinguishable from the dark ground-colour. The general result of all these changes is to produce a pattern which, in its delicate translucency, in the gradual melting of the light markings into the dark ground-colour, especially in the hind wing, and in the invasion of the fore wing by the pattern of the hind, presents a far closer resemblance to the *carmentis* female of *jodutta* than to the females of *Planema macarista* and *Pl. alcinoe*. There is furthermore a close correspondence in size between the females of *althoffi* and *jodutta*. On the under surface the evidence of secondary mimicry is even stronger; for, in addition to the changes described on the upper surface, the brown triangular area at the base of the hind wing of the male is almost lost and the pronounced black spots reduced in number and size. The result is to produce a close resemblance to this element in the pattern of the *carmentis* female. The basal markings of the hind wing under side of the male *A. althoffi* would, if retained in the female, have produced a likeness to the two female Planemas, and the change in the direction of their mimic has involved a departure from the pattern of these primary models.

The single yellow-marked example (August 24th : Table II), which is an evident mimic of the male *A. jodutta*, possesses a pattern similar to that of the two white-marked females, save for the suppression of the linear marking in the cell of the fore wing, a change which renders the mimicry of *jodutta* still more complete. The strong and obvious resemblance of this yellow-marked female to the non-mimetic male of *jodutta* affords further evidence that the white females mimic the females of the same species rather than the primary models of Combination II.

The 8 fulvous females of *A. althoffi* are also better mimics of the *tellus*-like female of *A. jodutta* than of *Planema tellus* itself. The two former agree in the shade of the fulvous markings, which is deeper than that of the *Planema*. In the spotting at the base of the hind wing under surface the female *althoffi* is a good mimic of both *tellus* and *jodutta*, although somewhat better of the latter with its smaller spots, a fact which is all the more significant because of the relatively large size of the spots in the male *althoffi*, to which attention has been already directed.

The production of the fulvous-marked mimetic female of *althoffi* has involved a wider departure from the pattern of the male than that described in the paler mimetic females. The markings in the basal half of the fore wing of the former have not only been changed in tint, but they have also coalesced into a broad fulvous area interrupted only by the three black spots below the cell. A point of special interest is the development of one or two fulvous striæ in the black border at and above the anal angle, representing the less peripherally placed striæ that are very characteristic of the female *jodutta*. In the hind wing also, the departure from the pattern of the male has been greater; for the fulvous area, representing the yellow band, is far more extended than the pale areas of the other females — both white and yellow.

In concluding this account of the mimetic forms of *A. althoffi* I wish to call attention to the resemblance of the under surface of the male to that of the *aurivillii* female of *A. alciope*. There is considerable likeness, in colour as well as in pattern, between the parts exposed in the resting position. If this suggested resemblance be confirmed by observations in the field, the males of *althoffi* should be given a place in Table 1, and, being mimics of a mimic, would provide additional evidence that the likeness of the female *alciope* to the Planemas of Combination I is Müllerian. On the upper surface, too, the pattern of the male *althoffi* suggests an outlying association with Combination I, and the white patch on the hind wing of *Acræa admatha leucographa* Ribbe may have become the characteristic of the Uganda form under the influence of the same Combination. There also appears to be a likeness which may be significant between the under surface of the male of *A. althoffi* and that of *A. pelasgius* Grose-Smith.

Other members of Combination II.

The beautiful mimetic *Pseudacræa*, the female of *Ps. hobleyi*, has been already considered on page 496. It is here only necessary to speak of its relative abundance, 28 specimens having been taken, as compared with 39 and 11 of the two models respectively. Accepting Dr. Jordan's conclusions, we find the following captures of three forms of a single species of *Pseudacræa* at Entebbe between May 23rd and August 31st, 1909 :

Male *hobleyi* form . .	36	*Planema* Models . .	93
Female *hobleyi* form .	28	— — . .	50
Both sexes *terra* form .	18	— — . .	89
Totals. . .	82		232

The mimics were thus well over 1/3 of the models — a remarkably high proportion — and one far more consistent with the Müllerian than the Batesian hypothesis.

In addition to the white-barred *Elymnias phegea*, one or two examples of a Hypsid Moth, *Deilemera acræina* H. Druce, which enters this Combination, have been found in Dr. Wiggins' collections, although not in the period we are now considering.

Combination III centred by *Planema tellus* Auriv., subsp. *platyxantha* Jord. (Table III).

Little need be said about this Combination, of which the members have been incidentally dealt with in the preceding sections. The beautiful mimicry of *Pseudacræa terra* was described by Mr. S A. Neave (loc. cit., p. 218) and, together with the model and other mimics of the Combination, figured by Mr. H. Eltringham

The only other members of Combination III present in the collection studied by Mr. Neave were the forms of *Acræa encedon* L., referred to below, and the fulvous-marked females of *A. jodutta*. The mimicry of the latter is so perfect that, at the time, it was mistaken for its model by Mr. Neave as well as other naturalists, including Mr. Roland Trimen and the present writer.

The numbers of the *Planema*-model of Combination III are noteworthy, being much greater relatively to mimics than in either of the other Combinations.

A single mimic, a form of *Acræa encedon encedon* L., figured by Mr. Eltringham, has not yet been sufficiently studied to be given a place in Table III. I refer to the examples in which the subapical bar of the fore wing is fulvous instead of white, the pattern bearing the same relation to the typical *encedon* that the *niobe* female bears to the *trophonius* female of *Papilio dardanus*. Corresponding changes in the *Acræa* and the *Papilio* have converted a mimic of *D. chrysippus* into a mimic of *Planema tellus*.

Apart from the important facts as to the occurrence and relative proportions of models and mimics, the chief result of the present paper appears to me to be the strong support it affords to the Müllerian as opposed to the Batesian theory of mimicry. This is seen in the numbers of the mimetic forms, in the group to which some of them belong, — the *Acræinæ*, — and above all in the existence of secondary mimicry or the mimicry of mimics. The females of *Acræa jodutta* also clearly prove that polymorphism in mimicry does not furnish evidence in favour of a Batesian interpretation; for *jodutta* is certainly a distasteful member (1) of a distasteful group, and is itself mimicked by *Acræa althoffi*. The perfection of the resemblance attained by the Acræine mimics also refutes another conclusion which, like the last, has been held by distinguished naturalists, *viz.* that a precise and detailed likeness is inconsistent with a Müllerian interpretation.

In concluding this paper I wish again to express my warm thanks to my friend Dr. C. A. Wiggins, who has given me the opportunity of working out these most interesting collections and has generously presented them to the Hope Department of the Oxford University Museum, where they will be readily accessible to all naturalists.

---

(1) *Acræa jodutta* has recently been shown to be distasteful by some hitherto unpublished experiments carried on in the Lagos district of Southern Nigeria by my friend Dr. W. A. Lamborn.

Economic Entomology in Trinidad,

by F. W. URICH, Entomologist, Board of Agriculture, Trinidad.

Historical.

On the reorganization of the Agricultural Department and the formation of the Board of Agriculture late in 1908 provision was made for the appointment of an entomologist. Previous to that time there was no official economic entomologist, but good work was done by some members of the Field Naturalists Club, and the late Superintendent of the Botanical Department, Mr. J. H. HART, F. L. S. Occasional articles on entomology appeared in the publications of these two bodies and called attention to local Insects in their relation to agriculture or public health. In connection with the latter principally Mosquitoes were dealt with. The following passage taken from the annual address of the President (Dr. L. O. HOWARD) at the Sixth Annual Meeting of the Association of Economic Entomologists of America gives a summary of the standing of economic entomology in Trinidad in 1894 :

« No official recognition of economic entomology has yet been reached in this island, but a very active organization, known as the Trinidad Field Naturalists Club, has been established, which is well worth mention in this connection, since its President, Mr. H. CARACCIOLO, and its Secretary, Mr. F. W. URICH, have interested themselves especially in the subject of economic entomology and are labouring to interest the Government. His Excellency the Governor occasionally attends the meetings of the Club and by the institution of prizes for essays and by similar means a widespread interest in economic entomology is being roused. »

Professor P. CARMODY was appointed Director of Agriculture in 1908, and the writer was appointed entomologist under the Board of Agriculture in January 1909. It may here be mentioned that the Board of Agriculture, of which the Governor is President, consists of planters and others interested in agriculture. Besides advising the Government in matters agricultural, the Board has its own staff consisting of a mycologist, an entomologist, agricultural instructors and inspectors. It also possesses a well equipped laboratory where research work is carried on by its employees.

The entomological work that is done at present is principally devoted to agricultural pests, but work is also undertaken in connection with Insects in relation to disease, and also in co-operation with the Education Department to help in the scheme of nature study, successfully inaugurated by the Inspector of Schools, Lieutenant-Colonel J.-H. COLLENS, V. D. The details of the work done is best discussed under different headings.

Agricultural Work.

Besides reporting on specimens sent in by planters, special visits are paid to plantations where practical demonstrations are given in spraying and controlling Insect pests. Special investigations are also carried on. It is hardly in the scope of a paper like this to consider all the Insect pests affecting our cultivations, but one or two may be mentioned. The main product of the island is cacao, the fields of which harbour many Insects; in fact these estates are the entomologist's best hunting ground.

Fortunately most of the Insects found there are not injurious and on the whole, with one or two exceptions, Insect life is fairly well balanced, there being a fair proportion of natural enemies always present. The most serious pest is the Cacao Beetle (*Steirastoma depressum*), which seems to affect young trees in recently opened up districts. Cacao Beetles eat the bark of trees, and their larvæ destroy young branches of the trees. The tender young leaves of a cacao tree attract quite an army of Caterpillars and Beetles of several families, which, although they do not constitute serious pests, require watching. In the same manner the leaves attract gnawing Insects, the flowers and young pods are injuriously affected by many sucking Insects. The Leaf-hopper (*Horiola arquata*) is the most important of these. It is fairly common not only on cacao but on several other plants. It has been observed on mango where it does a certain amount of damage to the young fruit and flowers. On cacao it appears to be responsible for the withering of some of the young pods. *Heliothrips rubrocinctus* is another Insect seldom absent from cacao estates, but which only does occasional damage. It lives and breeds on pods and young leaves.

There are quite a number of species of Ants to be found on cacao plantations and they certainly rank immediately after the Cacao Beetle (*Steirastoma depressum*) in destructiveness. Inhabiting the ground are the *Atta cephalotes* LINN., which strip young trees of their leaves in a very short time ; before the use of carbon bisulphide for destroying their nests, they were very troublesome and were the subject of an Ordinance empowering proprietors to enter lands of each other to destroy nests. Nowadays they are not spoken of much, but the destruction of the nests forms part of the regular routine work of every plantation. The other Ant living in the ground, but which does a great deal of good, is the hunting Ant *Eciton foreli* and other species, that occasionally sweep through estates and destroy numbers of harmful Insects. The tree inhabiting Ants, whose habits have not yet been studied in detail, are capable of a great deal of harm in connection with their fondness of the secretions of plant lice, Scale-insects, Tree-hoppers and Caterpillars of Lycænid Butterflies.

Among the chief offenders may be mentioned *Pheidole radoszkowskii*, several species of the genera *Cremastogaster*, *Dolichoderus* and *Azteca*, the two latter being universal and numerous. Besides encouraging the sucking Insects mentioned they serve as means of spreading fungoid diseases. In connection with sugar cane, the next staple product of the island, besides the pests of old standing, such as *Diatræa saccharalis* and *Sphenophorus piceus*, two Insects deserve special mention, they are a Cercopid *Tomaspis postica* and a Moth *Castnia licus*. Both of these Insects have adapted themselves to sugar cane and have attracted considerable attention by their numbers.

Tomaspis postica, or as it is called locally « Froghopper », was so destructive in some districts of the island that the Board of Agriculture voted the sum of £ 625 for carrying on experiments for the control of these Insects, and it was decided that I should devote myself entirely to their study. I have been able to work out their life history, illustrations of which will be found on plate XXVII.

The nymphs of *T. postica* puncture the roots of the canes and cause them to die, thus depriving the plant of its means of nourishment, and the Caterpillars of *C. licus* tunnel in the root stocks and stalks of the growing canes. Trinidad cane fields suffer a great deal from the attacks of *T. postica*, and *C. licus*, although bad enough in Trinidad, appears to be worse in British Guiana. Both Insects are extremely difficult to deal with on account of their continuous generations and overlapping of broods only occasionally interrupted by extremes of weather. In connection with the control work of these Insects some good results are anticipated by the use of a parasitical fungus (1) on the *Tomaspis postica*. Spores of this fungus were scattered broadcast over a small field and the proportion of dead Insects was so encouraging that arrangements have been made to grow the fungus on a large scale. My recent discovery of the eggs and methods of oviposition will also tend to make the control of this pest more easy. As far as *C. licus* is concerned, the only method that has been found of use is catching the adults by gangs of boys. According to Mr. QUELCH 236,357 Moths were caught in six months in 1904-1905 on one estate in Demerara, to-day on the same estate the catch for seven months is 52,278. In Trinidad on the same estate 91,684 were caught in 1909 and 36,914 in 1910 for the same periods of six months.

The guard against the introduction of any Insect or fungoid pests from abroad there is an Ordinance which requires the inspection of all plants or their parts before they are allowed to be brought into the island, this Ordinance is now being amended and in addition to inspection of all plants, seeds, etc., before landing it will also provide for the inspection of all nurseries. There are other Ordinances which are made to ensure the destruction of injurious Insects on plantations, so for instance the Locust Ordinance which compels a proprietor to destroy all Locusts on his own property. The reason for this Ordinance is that sometimes, luckily up to now at remote periods, swarms of migratory Locusts invade the

---

(1) With regard to the name of this fungus Mr. J. B. RORER, the Mycologist of the Board of Agriculture, writes as follows : « Material sent by COLLENS to the U. S. Dept. of Agriculture was determined as *Oospora destructor* and *Penicillium anisotliæ*, while specimens, which were sent to Kew in 1908, were examined by MASSEE, who diagnosed the fungus as a new species to which he gave the name *Septocylindrium suspectum* ».

island from the continent of South America, and this Ordinance ensures their speedy destruction. It also provides for the controling of any outbreaks of local species.

The Ordinance in connection with *Atta cephalotes*, or « Parasol Ants » as they are called locally, has already been referred to. Another Ordinance prohibits absolutely the importation of the Mongoose, and Wild Birds are protected by law.

Insects in relation to disease.

In connection with this branch of entomology Mosquitoes were studied. They may conveniently be divided into two groups : town Mosquitoes and field Mosquitoes. The two Mosquitoes always found in towns are *Stegomyia calopus* and *Culex fatigans*. Occasionally a large specimen of *Megarhinus trinidadensis* makes its appearance in a house and occasions some consternation by its size. But these Mosquitoes do not bite or suck blood, and in their larval stage do good by devouring *Culex* and *Stegomyia* larvæ. There was a time when it was not possible to be in peace in Port of Spain on account of the attacks of *Stegomyia calopus* during the day and *Culex fatigans* at night. Through the introduction of regulations making it an offence against the law to have larvæ in any water receptacles about houses and the energy and tact of the Medical Officer of Health, Dr. J. E. Dickson, it has been possible to materially reduce the number of Mosquitoes in Port of Spain and to-day in many parts of the town *Stegomyia* is rarely seen, and *Culex fatigans* rarely felt at night. This work of destruction has also been much helped by the excellent system of drainage of the town. No *Anopheles* are found in the town proper. In country districts control work is not possible with such good results, but the country people are getting to appreciate the benefit of keeping away Mosquitoes and use nets more freely than before. They also know the value of the little fish *Girardinus guppyi* in keeping down larvæ in water barrels about houses. In country districts in places surrounded by many trees with bromeliæ growing on them Mosquitoes are plentiful and more troublesome during the day than at night. The species occuring in such localities and biting only during the day are *Hæmagogus* several species, *Sabethoides*, *Stegomyia sexlineata*, *Wyeomyia longirostris*, this last species being particularly troublesome. In places where trees and scrub have been cleared to within 100 yards of the house and the draining of the surrounding land attended to, there is hardly a Mosquito to be felt day or night.

The Anopheles generally make their appearance at night, but they appear to confine their attacks to the first part of the night. An exception to this is *Anopheles lutzii*, which lives in *Bromelia* water and bites during the day in shady places. In the woods proper the most troublesome species during the day are *Psorophora*, *Janthinosoma*, *Culex serratus*, *Culex confirmatus*, *Culex scholasticus*, and several species of *Wyeomyia*.

Educational Work.

In connection with the Annual School Shows organized by the Education Department, specimens of injurious Insects are exhibited. In the past special attention has been paid to Mosquitoes and these were exhibited alive in tanks with large labels attached. The following are exact copies of two of the series :

Wrigglers or Mosquito Larvæ.

CULEX FATIGANS.

The night-biting Mosquito of Trinidad.

Breeds in any foul water about houses and stables, is very common in cesspits filled with water in the suburbs of Port of Spain and country towns.

As this kind breeds in water, it *can* be exterminated by pouring kerosine oil on it.

TIME OF DEVELOPMENT.

Egg stage	1 to 1 ½ days.
Larval stage	3 to 4 —
Pupal stage	1 to 1 ½ —
	4 to 7 days.

The little brown masses floating on the water are the egg rafts.

These mosquitoes transmit the parasite which causes elephantiasis. — Barbados leg.

Fish which eat Mosquito Larvæ.

GIRARDINUS GUPPYI.

Belly fish, Fat belly, Millions.

Females of plain silvery colour. Abdomens distended with young.
Males gaily coloured with black, blue and red markings.

Habitat. — In all streams of the Island, in estate drains, ponds, etc., in any kind of water whether clean or dirty.

Use. — Eat Mosquito larvæ voraciously and a few kept in any collection of water about houses will prevent Mosquitoes breeding.

Besides mounted specimens of adults and larvæ in formalin, enlarged photographs of all adults and larvæ and their parts are exhibited.

EXPLANATION OF PLATE.

Life History of *Tomaspis postica.*

Eggs are inserted by means of the ovipositor into the tissue of withering cane or grass leaf sheaths. They are 0.80 mm. long × 0.24 mm. broad.

FIG. 1. — Eggs recently laid, front view. Colour light yellow. Anterior end bears a « hatching lid », which is white in this stage, but becomes jet-black with age.

FIG. 2. — Egg after nymph has escaped by pushing « lid » open.

FIG. 3. — Side view of egg just before hatching of nymph. Colour of shell white, eyes of nymph, which are red, and abdominal segments show through shell when viewed by transmitted light.

FIG. 4. — Newly hatched nymph. Colour light yellow, eyes red. An internal red organ gives abdomen a reddish tinge on the sides.

FIG. 5. — Full grown nymph before changing to perfect Insect. Colour, head and thorax light brown. Wing pads black. Abdomen light yellow, almost white. Eyes claret.

FIG. 6. — Nymph on rootlet of cane in the act of surrounding itself with spittle (enlarged).

FIG. 7. — Adult male. Colour, head and thorax dark bronze, wings dark brown, transverse bands dull yellow.

FIG. 8. — Adult female.

FIG. 9. — Side view of proboscis of adult (enlarged).

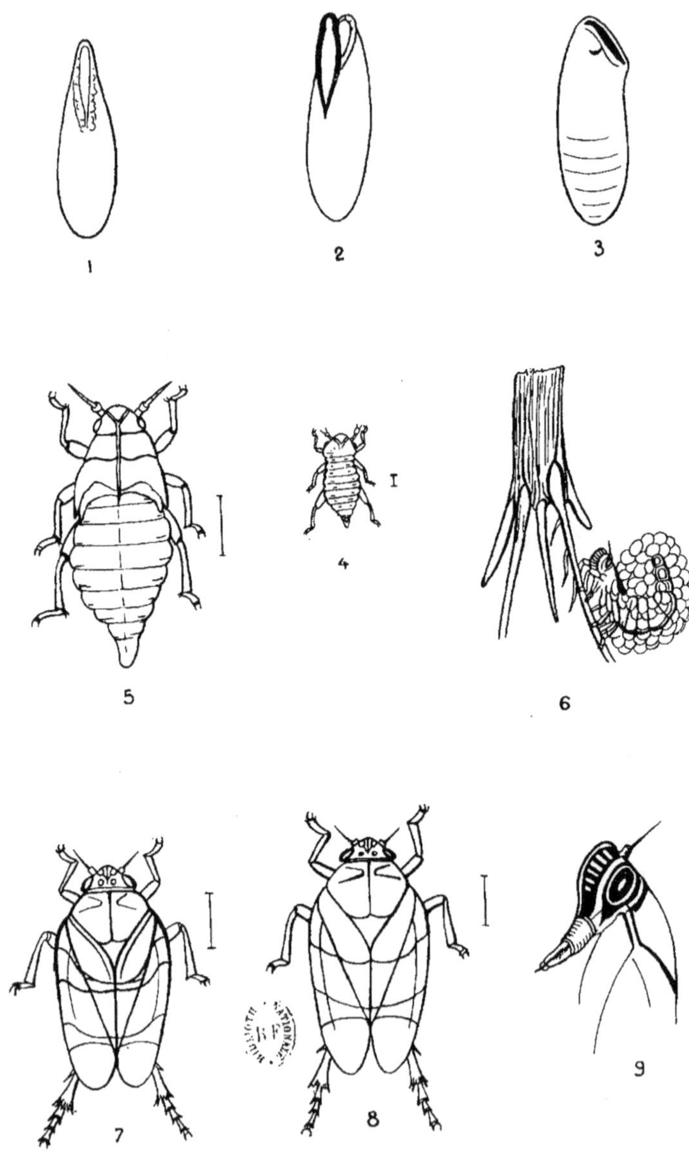

P. L. Guppy Del.

F. W. UHRICH. — ECONOMIC ENTOMOLOGY IN TRINIDAD

TABLE DES MATIÈRES

BIBLIOTHEQUE NATIONALE DE FRANCE
3 7511 00161273 1

www.ingramcontent.com/pod-product-compliance
Lightning Source LLC
LaVergne TN
LVHW010119230826
846091LV00001BA/84

* 9 7 8 2 0 1 4 4 4 0 9 5 9 *